NUTRITION IN KIDNEY DISEASE

Nutrition ◊ and ◊ Health

Adrianne Bendich, PhD, FACN, Series Editor

Handbook of Nutrition and Pregnancy, edited by **Carol J. Lammi-Keefe, Sarah C. Couch, and Elliot H. Philipson,** 2008

Nutrition and Health in Developing Countries, Second Edition, edited by **Richard D. Semba and Martin W. Bloem,** 2008

Nutrition and Rheumatic Disease, edited by **Laura A. Coleman,** 2008

Nutrition in Kidney Disease, edited by **Laura D. Byham-Gray, Jerrilynn D. Burrowes, and Glenn M. Chertow,** 2008

Handbook of Nutrition and Ophthalmology, edited by **Richard D. Semba,** 2007

Adipose Tissue and Adipokines in Health and Disease, edited by **Giamila Fantuzzi and Theodore Mazzone,** 2007

Nutritional Health: Strategies for Disease Prevention, Second Edition, edited by **Norman J. Temple, Ted Wilson, and David R. Jacobs, Jr.,** 2006

Nutrients, Stress, and Medical Disorders, edited by **Shlomo Yehuda and David I. Mostofsky,** 2006

Calcium in Human Health, edited by **Connie M. Weaver and Robert P. Heaney,** 2006

Preventive Nutrition: The Comprehensive Guide for Health Professionals, Third Edition, edited by **Adrianne Bendich and Richard J. Deckelbaum,** 2005

The Management of Eating Disorders and Obesity, Second Edition, edited by **David J. Goldstein,** 2005

Nutrition and Oral Medicine, edited by **Riva Touger-Decker, David A. Sirois, and Connie C. Mobley,** 2005

IGF and Nutrition in Health and Disease, edited by **M. Sue Houston, Jeffrey M. P. Holly, and Eva L. Feldman,** 2005

Epilepsy and the Ketogenic Diet, edited by **Carl E. Stafstrom and Jong M. Rho,** 2004

Handbook of Drug–Nutrient Interactions, edited by **Joseph I. Boullata and Vincent T. Armenti,** 2004

Nutrition and Bone Health, edited by **Michael F. Holick and Bess Dawson-Hughes,** 2004

Diet and Human Immune Function, edited by **David A. Hughes, L. Gail Darlington, and Adrianne Bendich,** 2004

Beverages in Nutrition and Health, edited by **Ted Wilson and Norman J. Temple,** 2004

Handbook of Clinical Nutrition and Aging, edited by **Connie Watkins Bales and Christine Seel Ritchie,** 2004

Fatty Acids: Physiological and Behavioral Functions, edited by **David I. Mostofsky, Shlomo Yehuda, and Norman Salem, Jr.,** 2001

Nutrition and Health in Developing Countries, edited by **Richard D. Semba and Martin W. Bloem,** 2001

Preventive Nutrition: The Comprehensive Guide for Health Professionals, Second Edition, edited by **Adrianne Bendich and Richard J. Deckelbaum,** 2001

Nutritional Health: Strategies for Disease Prevention, edited by **Ted Wilson and Norman J. Temple,** 2001

Clinical Nutrition of the Essential Trace Elements and Minerals: The Guide for Health Professionals, edited by **John D. Bogden and Leslie M. Klevey,** 2000

Primary and Secondary Preventive Nutrition, edited by **Adrianne Bendich and Richard J. Deckelbaum,** 2000

The Management of Eating Disorders and Obesity, edited by **David J. Goldstein,** 1999

Vitamin D: Physiology, Molecular Biology, and Clinical Applications, edited by **Michael F. Holick,** 1999

Preventive Nutrition: The Comprehensive Guide for Health Professionals, edited by **Adrianne Bendich and Richard J. Deckelbaum,** 1997

Nutrition in Kidney Disease

Edited by

Laura D. Byham-Gray, PhD, RD

Department of Nutritional Sciences
University of Medicine and Dentistry of New Jersey
Stratford, NJ

Jerrilynn D. Burrowes, PhD, RD, CDN

Department of Nutrition, C.W. Post Campus,
Long Island University, Brookville, NY

Glenn M. Chertow, MD, MPH

Division of Nephrology, Stanford University
School of Medicine, Stanford, CA

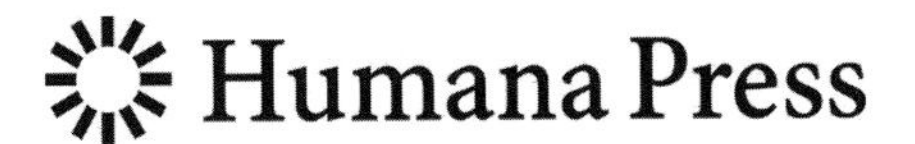

Editors
Laura D. Byham-Gray
Department of Nutritional Sciences
University of Medicine and
Dentistry of New Jersey
40 East Laurel Rd.
Stratford, NJ, USA
e-mail: laura.byham-gray@umdnj.edu

Jerrilynn D. Burrowes
Department of Nutrition
C.W. Post Campus
Long Island University
720 Northern Blvd.
Brookville, NY, USA
e-mail: jerrilynn.burrowes@liu.edu

Glenn M. Chertow
Division of Nephrology
Stanford University
School of Medicine
Stanford, CA, USA
e-mail: gchertow@stanford.edu

Series Editor
Adrianne Bendich
GlaxoSmithKline Consumer Healthcare
1500 Littleton Road
Parsippany, NJ, USA
Adrianne.4.Bendich@gsk.com

ISBN: 978-1-58829-781-5 e-ISBN: 978-1-59745-032-4

Library of Congress Control Number: 2007941657

Printed on acid-free paper

9 8 7 6 5 4 3 2 1

springer.com

Dedications

Laura dedicates this book to her husband, Steven, and her daughters, Erin and Jillian. Jerrilynn dedicates this book to the thousands of patients with kidney disease who will benefit from the information embedded in these pages. The goal is to improve the health and well-being of patients with kidney disease through optimal nutritional practices.

Series Editor's Introduction

The Nutrition and Health™ series of books have, as an overriding mission, to provide health professionals with texts that are considered essential because each includes: (1) a synthesis of the state of the science, (2) timely, in-depth reviews by the leading researchers in their respective fields, (3) extensive, up-to-date fully annotated reference lists, (4) a detailed index, (5) relevant tables and figures, (6) identification of paradigm shifts and the consequences, (7) virtually no overlap of information between chapters, but targeted, inter-chapter referrals, (8) suggestions of areas for future research, and (9) balanced, data-driven answers to patient /health professional questions that are based upon the totality of evidence rather than the findings of any single study.

The series volumes are not the outcome of a symposium. Rather, each editor has the potential to examine a chosen area with a broad perspective, both in subject matter as well as in the choice of chapter authors. The international perspective, especially with regard to public health initiatives, is emphasized where appropriate. The editors, whose trainings are both research and practice oriented, have the opportunity to develop a primary objective for their book, define the scope and focus, and then invite the leading authorities from around the world to be part of their initiative. The authors are encouraged to provide an overview of the field, discuss their own research, and relate the research findings to potential human health consequences. Because each book is developed de novo, the chapters are coordinated so that the resulting volume imparts greater knowledge than the sum of the information contained in the individual chapters.

Nutrition in Kidney Disease edited by Laura Byham-Gray, Jerrilynn Burrowes, and Glenn Chertow is a very welcome addition to the Nutrition and Health™ series and fully exemplifies the series' goals. This volume is especially timely as over 10% of the US adult population currently suffers from chronic kidney disease and the number is expected to increase as the major comorbidities, obesity and diabetes, continue to increase around the world. As only one example cited in this critically important volume, even though 7% of the adult US population has diabetes, 54% of the patients on kidney dialysis have diabetes.

This text has many unique features, such as highly relevant case studies, that will help to illustrate the complexity of treating the patient with kidney disease and/or reduced kidney function. However, the volume is also relevant for the non-practicing healthcare providers as there are in-depth discussions of the basic functioning of the kidney, demographics of the different kidney diseases, and conditions that affect the kidney. There are also clear, concise recommendations about dietary intakes and use of drugs and supplements across the stages of kidney disease. Thus, this volume provides the broad knowledge base concerning kidney anatomy, physiology, and pathology required by the practicing health professional as well as those professionals who have an interest in the latest, up-to-date information on the consequences of loss of kidney function, treatment of kidney disease, and disease implications on morbidity and mortality.

Nutrition in Kidney Disease serves a dual purpose of providing in-depth focus on the nutritional aspects of treating individuals throughout the lifespan who have lost some or all of their kidney functions as well as examining the current clinical modalities used in treating kidney disease and the consequences of the treatments on nutritional status. The book is organized as a stand-alone text that provides the historic beginnings of nutritional interventions in patients and reflects upon the necessity of these historic practices even today in developing countries where dialysis and/or kidney transplants, expensive drugs, and other disease management tools are not readily available and medical nutritional support remains the primary care available to patients with kidney disease.

The three editors of this volume, Laura Byham-Gray, Jerrilynn Burrowes, and Glenn Chertow, are internationally recognized leaders in the fields of clinical nutrition and renal disease research, treatment, and management. Each has extensive experience in academic medicine and collectively, they have over 500 peer-reviewed publications and numerous awards for their efforts to improve the care of those with kidney disease. The editors are excellent communicators and they have worked tirelessly to develop a book that is destined to be the benchmark in the field of nutrition and kidney disease because of its extensive, in-depth chapters covering the most important aspects of the complex interactions between kidney functions, diet, obesity, cardiovascular disease, autoimmune disease, and diabetes as examples, and the impact of loss of kidney function on other disease states. Additionally, the nutritional consequences of loss of kidney function in infants, children, pregnant women, and the aged are examined in depth in separate chapters that also include potential solutions to the nutritional deficits specific for those with kidney dysfunction.

Nutrition in Kidney Disease contains 25 chapters and each title provides key information to the reader about the contents of the chapter. In addition, each chapter begins with a list of concise learning objectives as well as key words. The introductory chapters provide readers with the basics so that the more clinically related chapters can be easily understood. The editors have chosen 46 well recognized and respected chapter authors who have included complete definitions of terms with the abbreviations fully defined for the reader and consistent use of terms between chapters. Key features of this comprehensive volume include the eleven case studies at the end of the relevant chapters and the inclusion of the answers to the case study questions at the end of the book. The volume includes over 100 detailed tables and informative figures, an extensive detailed index, and more than 1500 up-to-date references that provide the reader with excellent sources of worthwhile information. Moreover, the final chapter contains a comprehensive list of resources in print as well as via the internet including a complete listing of the practice guidelines that have been developed under the auspices of the National Kidney Foundation's Kidney Disease Outcomes Quality Initiative (KDOQI); protocols from the American Dietetic Association concerning nutrition therapy for the non-dialysis patient; tables of general as well as specific nutrient contents of foods for individuals with different stages of kidney disease; an extensive list of reliable internet sites as well as examples of relevant assessment tools for the health provider.

Unique to this volume is the inclusion of chapters dealing with prevention as well as treatment of kidney disease. The text is divided logically into five consistently organized parts: I. Foundations for clinical practice and overview that contains information about kidney function, historic perspective of nutritional support for kidney disease patients, demographics, and overall assessment tools; II. Chronic kidney disease (CKD) during stages 1–4 in adults that contains a first section on disease prevention and the role of hypertension, diabetes, and dyslipidemias in increasing the risk of kidney disease development and nutritional strategies that can reduce this risk. In section 2, nutritional and pharmacological treatment of individuals with stages 1–4 kidney disease is reviewed. Part III, Stage 5 chronic kidney disease, is also divided into two sections with treatment options of dialysis and transplant examined with respect to the nutritional needs and consequences. Section 2 looks at the most commonly seen management issues in patients with end stage renal disease and includes chapters on protein-energy malnutrition, nutrition support, anemia management, bone and mineral consequences, and other chronic diseases seen in CKD patients. Part IV delves into the nutritional needs of patients that have CKD and become pregnant, those that have kidney malformations or malfunction in infancy, childhood, or adolescence, or at the end of life in section 1. Section 2 includes separate chapters on acute renal failure, nephritic syndrome, and kidney stones. The final Part V: Additional nutritional considerations in kidney disease contains individual chapters on dietary supplements, compliance with dietary programs, the value and complexities of outcomes research, and the extensive chapter on resources for the practitioner. Thus, there is no question about the comprehensive coverage of the complex field of nutrition and kidney disease that is found in this important volume.

In conclusion, *Nutrition in Kidney Disease* edited by Laura Byham-Gray, Jerrilynn Burrowes, and Glenn Chertow provides health professionals in many areas of research and practice with the most up-to-date, well referenced volume on the importance of defining the nutritional status of the patient with decreased kidney function regardless of cause and the critical value of medical nutrition evaluation, treatment support, and management for patients with CKD and related chronic diseases that can affect human health. This volume will serve the reader as the benchmark in this complex area of interrelationships between diet, supplements, specific nutritional products for enhancing kidney function, and the functioning of all organ systems that are intimately affected by renal disease. Moreover, these physiological and pathological interactions are clearly delineated so that students as well as practitioners can better understand the complexities of these interactions. The editors are applauded for their efforts to develop the most authoritative resource in the field of "Nutrition in Kidney Disease" to date and this excellent text is a very welcome addition to the Nutrition and Health™ series.

Adrianne Bendich, PhD, FACN

Foreword

The interrelationship between nutrition and chronic kidney disease (CKD) is one that has enticed scientists, physicians, and nutritionists from all over the world. No other disease entity draws the same intensity of study and interdisciplinary treatment as CKD. There are so many facets to address: definition of normal function and metabolism, identification of abnormalities and their consequences, what intervention to provide and how. Each component, basic science, medicine, and nutrition, involves a spectrum of investigation and/or treatment that is all encompassing, with the goal of forestalling progression of disease and allowing continuation of life. The interdisciplinary requirement for successful treatment and management of the individual with CKD is another unique attribute of this amazing field.

The text *Nutrition in Kidney Disease* edited by Byham-Gray, Burrowes, and Chertow well exemplifies this description of the CKD specialty. This outstanding book includes information from the many areas that impact the individual with CKD. There are sections on CKD epidemiology, pathophysiology, metabolism, and diet therapy. The chapters are written by experts in the field. The authors provide insight into these areas as well as state-of-the-art information for students, professionals, and other healthcare professionals interested in nutrition and kidney disease.

This complex discipline simplifies to one element: the individual patient. Individualizing the diet and the treatment leads to the response of maximizing positive outcomes. *Nutrition in Kidney Disease* provides vital information to provide exceptional care for extraordinary people: individuals living with CKD.

D. Jordi Goldstein-Fuchs, DSc, RD
Kidney Nutrition Specialist, Sparks Dialysis, Sparks, NV
Co-Editor-In-Chief, *Journal of Renal Nutrition*

Preface

The field of kidney disease has evolved over the years to encompass a broad and sophisticated knowledge base. There has been a proliferation of scientific information and technical advances in the field. The clinician involved in the care of patients with kidney disease must have a vast knowledge of nutrition management of the disease. The purpose of this book is to provide a comprehensive reference on the practice of nutrition in kidney disease. It is our belief that this book will become a useful reference and tool for the practicing clinician in the fields of nutrition and nephrology, as well as other disciplines whose research, practice, and education includes nutrition and kidney disease. This book will also be a current resource for undergraduate and graduate level nutrition and allied health profession students, medical students and residents, nutrition and allied health clinicians, including general practitioners, nephrologists, educators, and researchers.

ORGANIZATION AND CONTENT

Nutrition in Kidney Disease is organized into five sections with a variable number of chapters based on breadth and depth of information. Part I addresses kidney function in health and disease. It provides a historical perspective of the emerging science in nutrition in kidney disease over the past several decades, and it defines and forecasts health care trends and outcomes in kidney disease. A comprehensive review of the components of the nutrition assessment is also provided. In Parts II and III, in-depth information on the prevention of common disorders associated with chronic kidney disease, current treatment options based on the latest scientific evidence, and management of comorbidities such as protein-energy malnutrition, anemia, and bone disease are covered. Part IV presents the nutrition concerns of special needs populations such as through the life cycle–pregnancy, infancy, childhood, adolescence and the elderly, and nutrition management of disorders such as acute kidney injury, nephrotic syndrome, and nephrolithiasis. Part V addresses additional nutritional concerns in kidney disease such as complementary and alternative medicine, cultural issues affecting dietary adherence, and outcomes research.

In an attempt to make this volume as practical as possible, a wide variety of tables, resources, practical tools, clinical practice guidelines, and Internet websites are compiled into one chapter.

FEATURES

The chapters in this textbook have been designed with special features to enhance learning. Each chapter begins with an abstract, key words, learning objectives, and a content outline, and ends with a summary. Up-to-date references for more in-depth review are included at the end of each chapter. This list provides the clinician and student with an extensive source of reading for continued study. In addition, several chapters end with a case study, which can be used to assess knowledge of the content area within the context of didactic curricula. These provide thought-provoking, illustrative questions that will add to the student's learning and clinical application of the material. The answers to the case studies are provided at the end of the book. The problems posed in these chapters enable the clinician and the student to apply the chapter material to "real-life" nutrition-related problems.

The chapters have been written by a collaborative group of distinguished dietitians and physicians in the specialized field of kidney disease and clinical nutrition who have devoted their careers to the care of patients with kidney disease. This collaborative effort is a testament to the interdisciplinary approach that is used to provide care to this unique patient population. It is our belief that this book will be used to guide and enhance the care of the patients we serve.

Laura D. Byham-Gray, PhD, RD
Jerrilynn D. Burrowes, PhD, RD, CDN
Glenn M. Chertow, MD, MPH

Disclaimer

Considerable attention has been taken to confirm the accuracy of the information presented and to describe generally accepted practices. However, the authors, editors, and publisher are not responsible for errors or omissions or for any consequences from application of the information in this book and make no warranty, expressed or implied, with respect to currency, completeness, or accuracy of the contents of the publication. Use of such information in a particular patient care situation remains the professional responsibility of the respective practitioner.

The authors, editors, and publisher have exerted every effort to ensure that drug selection and dosage indicated in this text are in accordance with current recommendations and practice at the time of publication. However, in view of ongoing research, changes in government regulations, and the constant flow of information relating to drug therapy and drug reactions, the reader is to check the package insert for each drug for any change in indications and dosage and for added warnings and precautions. This is particularly important when the recommended agent is a new or infrequently employed drug.

Some drugs presented in this book have Food and Drug Administration (FDA) clearance for limited use in restricted research settings. It is the responsibility of the health care provider to ascertain the FDA status of each drug planned for use in their clinical practice.

Acknowledgments

We would like to thank Adrianne Bendich for granting us this wonderful opportunity and Humana Press for making *Nutrition in Kidney Disease* a reality. We also express gratitude and appreciation to our contributors for their commitment and patience throughout this process.

Contents

Contributors

VINOD K. BANSAL, MD • *Division of Nephrology and Hypertension, Loyola University Medical Center, Chicago, Illinois*

JULIE BARBOZA, MSN, RD, APRN-BC • *Comprehensive Cardiovascular Care Program, Harvard Vanguard Medical Associates, Boston, Massachusetts*

JEFFREY S. BERNS, MD • *Renal-Electrolyte and Hypertension Division, University of Pennsylvania School of Medicine, Philadelphia, Pennsylvania*

JUDITH A. BETO, PHD, RD • *Department of Nutrition Sciences, Dominican University, River Forest, Illinois*

JERRILYNN D. BURROWES, PHD, RD, CDN • *Department of Nutrition, C.W. Post Campus, Long Island University, Brookville, New York*

LAURA D. BYHAM-GRAY, PHD, RD • *Department of Nutritional Sciences, University of Medicine and Dentistry of New Jersey, Stratford, New Jersey*

GLENN M. CHERTOW, MD, MPH • *Division of Nephrology, Stanford University School of Medicine, Stanford, California*

WM. CAMERON CHUMLEA, PHD • *Department of Community Health and Pediatrics Lifespan Health Research Center, Wright State University Boonshoft School of Medicine, Dayton, Ohio*

DAVID B. COCKRAM, PHD, RD • *Abbott Nutrition Regulatory Affairs, Abbott Laboratories, Columbus, Ohio*

PATRICIA DIBENEDETTO-BARBÁ, RD, MS, LDN, CSR • *Renal Nutrition Consultant, Honolulu, Hawaii*

WILFRED DRUML, MD • *Department of Medicine III, Division of Nephrology, Vienna General Hospital, Vienna, Austria*

JOHANNA T. DWYER, DSC, RD • *Friedman School of Nutrition Science and Policy, Frances Stern Nutrition Center, Jean Mayer USDA, Human Nutrition Research Center on Aging at Tufts University, Boston, Massachusetts*

GARABED EKNOYAN, MD • *Renal Section, Department of Medicine, Baylor College of Medicine, Houston, Texas*

D. JORDI GOLDSTEIN-FUCHS, DSC, RD • *Sparks Dialysis, Sparks, Nevada*

Haewook Han, PhD, RD, LDN, CSR • *Department of Nephrology, Harvard Vanguard Medical Associates, Boston, Massachusetts*
Kathy Schiro Harvey, MS, RD, CSR • *Puget Sound Kidney Centers, Mountlake Terrace, Washington*
Mary Kay Hensley, MS, RD, CSR • *DaVita Dialysis, Schererville, Indiana*
Bertrand L. Jaber, MD, MS • *Department of Medicine, Caritas St. Elizabeth's Medical Center, Tufts University School of Medicine, Boston, Massachusetts*
Kamyar Kalantar-Zadeh, MD, MPH, PHD • *Department of Medicine and Pediatrics, David Geffen UCLA School of Medicine, Torrance, California*
Marcia Kalista-Richards, MPH, RD, CNSD, LDN • *Department of Nutrition Services, Easton Hospital/Sodexo Company, Easton, Pennsylvania, Cedar Crest College, Allentown, Pennsylvania*
George A. Kaysen, MD, PHD, FASN • *University of California Davis, Davis, California, Department of Veterans Affairs, Northern California Health Care System, Mather, California*
Mary Pat Kelly, MS, RD, GNP • *Department of Veteran's Affairs, Palo Alto Health Care System, Gero-Psychiatric Nursing Home Service, Mento Park, California*
Pamela S. Kent, MS, RD, CSR, LD • *Genzyme Renal, Vermilion, Ohio*
Kishore Kuppasani, MS, PA-C • *Department of Medicine, University of Medicine and Dentistry of New Jersey-New Jersey Medical School, Newark, New Jersey*
Kristie J. Lancaster, PHD, RD • *Department of Nutrition, Food Studies and Public Health, New York University, New York, New York*
Orfeas Liangos, MD • *Department of Medicine, Caritas St. Elizabeth's Medical Center, Tufts University School of Medicine, Boston, Massachusetts*
Linda McCann, RD, CSR, LD • *Satellite Healthcare, Inc., Mountain View, California*
Graeme Mindel, MD • *Division of Nephrology, Barnes Jewish Dialysis Center, Washington University, St. Louis, Missouri*
Joni J. Pagenkemper, MA, MS, RD, LMNT • *Creighton University, Omaha, Nebraska*
Robert N. Pursell, MD • *Department of Nephrology, St. Lukes Hospital, Bethlehem, Pennsylvania*
Diane Rigassio Radler, PHD, RD • *Department of Nutritional Sciences, University of Medicine and Dentistry of New Jersey, Newark, New Jersey*

ALLURU S. REDDI, MD • *Department of Medicine, University of Medicine and Dentistry of New Jersey-New Jersey Medical School, Newark, New Jersey*

SHARON R. SCHATZ, MS, RD, CSR, CDE • *DaVita Dialysis, Lumberton, New Jersey*

DONNA SECKER, PHD, RD, FDC • *Department of Clinical Dietetics and Division of Nephrology, The Hospital for Sick Children, Toronto, Canada*

JEAN STOVER, RD, LDN • *DaVita Dialysis, Philadelphia, Pennsylvania*

ARTHUR TSAI, MD • *Renal-Electrolyte and Hypertension Division, University of Pennsylvania School of Medicine, Philadelphia, Pennsylvania*

KAREN WIESEN, MS, RD, LD • *Barnes Jewish Dialysis Center, Washington University, St. Louis, Missouri*

JANE Y. YEUN, MD, FACP • *Department of Veterans Affairs, Northern California Health Care System, Mather, California, University of California Davis, Sacramento, California*

I Foundations for Clinical Practice and Overview

1 Kidney Function in Health and Disease

Alluru S. Reddi and Kishore Kuppasani

LEARNING OBJECTIVES

1. Describe the gross and microscopic structure of the kidney.
2. Discuss the various functions of the normal kidney.
3. Define and discuss the various renal syndromes, such as acute kidney injury, chronic kidney disease, nephrotic and nephritic syndromes, tubulointerstitial diseases, and vascular diseases of the kidney.

Summary

The kidneys are paired organs located retroperitoneally in the lumbar region and perform three major functions: (i) maintenance of fluid and acid–base balance; (ii) removal of nitrogenous waste products; and (iii) synthesis of hormones, such as renin, erythropoietin, and active vitamin D_3 (calcitriol). The functional unit of the kidney is the nephron, which consists of a renal corpuscle, the proximal tubule, the loop of Henle, the distal tubule, and the collecting duct. The renal corpuscle consists of the glomerulus and Bowman's capsule. Plasma is filtered in the glomerulus to form protein-free ultrafiltrate. About 60% of this ultrafiltrate is reabsorbed in the proximal tubule. The loop of Henle participates in countercurrent multiplication of urine concentration. The distal tubule generates hypotonic fluid in the tubular lumen, causing hypertonic medullary interstitium. The collecting duct plays an important role in potassium (K^+) secretion, urinary acidification, and water reabsorption in the presence of antidiuretic hormone. When the structure of the kidney is disturbed by a pathologic

From: *Nutrition and Health: Nutrition in Kidney Disease*
Edited by: L. D. Byham-Gray, J. D. Burrowes, and G. M. Chertow

process, its functions are altered. These changes in the kidney result in an increase in serum creatinine levels (e.g., acute kidney injury, chronic kidney disease, or tubulointerstitial disease), proteinuria (e.g., nephrotic syndrome), and hematuria (e.g., glomerulonephritis). Also renal vasculature is affected, causing hypertension and thrombotic microangiopathies.

Key Words: Nephron; glomerulus; renal function; chronic kidney disease; nephrotic syndrome; nephritic syndrome.

1. INTRODUCTION

The kidneys perform three major functions *(1)*. As regulatory organs, the kidneys precisely control the composition and volume of the body fluids and maintain acid–base balance as well as blood pressure by varying the excretion of water and solutes. As excretory organs, the kidneys remove various nitrogenous metabolic end products in the urine. In general, the kidneys filter plasma in the glomerulus to form a protein-free ultrafiltrate. This ultrafiltrate passes through the various tubular segments where reabsorption of essential constituents and secretion of unwanted products occur. As endocrine organs, the kidneys produce important hormones, such as renin, erythropoietin, and active vitamin D_3 (calcitriol). In addition, the kidneys participate in the degradation of various endogenous and exogenous compounds. In order to understand these functions, it is essential to examine the gross and microscopic structure of the kidneys.

1.1. Anatomy of the Kidney

The kidneys are paired, bean-shaped structures located retroperitoneally in the lumbar region, one on either side of the vertebral column *(1–3)*. The lateral edge of the kidney is convex, while the medial aspect is concave with a notch called the hilum. The hilum receives the blood and lymphatic vessels, the nerves, and the ureter. The hilum contains a cavity, the renal sinus, where the ureter expands to form the renal pelvis. The normal adult kidney is about 10-12 cm long, 5-7 cm wide, and 2-3 cm thick, and it weighs 125–170 g.

Each kidney is composed of the parenchyma and the collecting system. The parenchyma consists of an outer cortex and an inner medulla. The medulla is divided into an outer (toward the cortex) and an inner medulla (toward the pelvis). The collecting system includes the calyces, renal pelvis, and the ureters. The major calyces unite to form the renal pelvis. The renal pelvis drains into the ureter, which connects the kidney to the bladder.

The basic structural and functional unit of the kidney is the nephron. There are about 600,000 (range 300,000–1,400,000) nephrons in each kidney. Each nephron contains specialized cells that filter the plasma, then selectively remove, reabsorb, and secrete a variety of substances into the urine. The nephron consists of a renal corpuscle, the proximal tubule, the loop of Henle, and the distal tubule. The collecting duct is not part of the nephron because it is embryologically derived from the ureteric bud, whereas the nephron is derived from the metanephric blastema. However, the collecting duct is commonly included in the nephron because of its related function.

1.2. Renal Corpuscle

The renal corpuscle consists of the glomerulus and Bowman's capsule. Generally the term "glomerulus" is widely used for the entire corpuscle. The glomerulus is composed of a capillary network lined by an inner thin layer of endothelial cells: a central region of mesangial cells surrounded by collagen-like mesangial matrix, and an outer layer of visceral epithelial cells. The endothelial and epithelial cells are separated by the glomerular basement membrane (GBM).

The GBM is a dense fibrillar structure, which is the only anatomical barrier between blood and urine. Biochemical studies of the GBM showed that it contains predominantly type IV collagen, proteoglycans, and laminin. Collagen provides the structural framework, whereas proteoglycans, such as heparan sulfate, confer a negative charge to the GBM. Because of this negative charge, filtration of albumin is curtailed. The Bowman's capsule, which is a double-walled cup surrounding the glomerulus, consists of an outer layer of parietal epithelial cells. Between the visceral epithelial and parietal epithelial cells is a space called Bowman's space. The glomeruli are located exclusively in the cortex, which undergo pathologic changes in several disease conditions.

The endothelial cells line the glomerular capillaries and are separated by large (70 nm) fenestrations. These fenestrations limit filtration of only cellular elements such as erythrocytes but not water or proteins. The epithelial cells, also called podocytes, represent the visceral layer of the Bowman's capsule. The podocytes have foot processes that cover the GBM. These foot processes are separated by a thin diaphragmatic structure called the slit diaphragm or slit pore.

The renal corpuscle is responsible for the ultrafiltration of the blood, which is the first step in urine formation. In this process, medium and small sized molecules are allowed to pass through into the Bowman's space, while large sized molecules, such as proteins, are left behind.

To enter the Bowman's space, the ultrafiltrate must pass through the fenestrae of the endothelial cells, the layers of the basement membrane, and the slit diaphragms of the foot processes.

1.3. Proximal Tubule

The Bowman's capsule continues as the proximal tubule, which is lined by cuboidal or columnar cells with a brush border on their luminal surface. The brush border consists of millions of microvilli, which markedly increase the surface area available for the absorption of solutes and water through cells (transcellular transport) or between cells (paracellular transport) or both. The proximal tubule reabsorbs about 60% of the ultrafiltrate. Several electrolytes (Na^+, K^+, Cl^-, HCO_3^-), minerals (Ca^{2+}, HPO_4^{3-}) amino acids, glucose, and water are reabsorbed in the proximal tubule. Also, secretion of organic acids and bases occur in the proximal tubule. The proximal tubule is susceptible to insults such as renal ischemia and nephrotoxins, resulting in altered kidney function.

1.4. Thin Limb of Henle's Loop

The proximal tubule continues into the medulla as the thin descending limb of Henle's loop. The loop then bends back and becomes the thin ascending limb of Henle's loop. The thin descending limb is more permeable to water and less permeable to NaCl. As a result, water moves into the interstitium and makes the ultrafiltrate more concentrated than in the proximal tubule. In contrast, the thin ascending limb is impermeable to water, but permeable to NaCl. Therefore, the ultrafiltrate becomes dilute and the medullary interstitium hypertonic. Thus, the thin descending and ascending limbs participate in the countercurrent multiplication of the urinary concentration process.

1.5. Distal Tubule

The distal tubule includes the thick ascending limb of Henle's loop and the distal convoluted tubule. The thick limb runs from the medulla into the cortex up to its parent glomerulus, where it forms the macula densa, a component of the juxtaglomerular apparatus that secretes renin. The thick ascending limb of Henle's loop is responsible for the reabsorption of Na^+, K^+, and Cl^- in the ratio of 1:1:2. The reabsorption of these electrolytes is dependent on the Na/K-ATPase located in the basolateral membrane. NaCl reabsorption also occurs in the distal convoluted tubule. Both segments of the distal tubule are normally impermeable to water, and thus the fluid formed in the distal tubule is hypotonic. The impermeability of the distal tubule to water, combined with active transport of Na^+ and Cl^- out of the thick ascending limb, makes the medullary interstitium hypertonic. The distal tubule is connected to the collecting duct by the connecting tubule.

1.6. Collecting Duct

Depending on its location in the kidney, the collecting duct can be divided into the cortical, outer medullary, and inner medullary portions. The epithelium of the collecting ducts contains two types of cells: principal (65%) and intercalated (35%) cells. The principal cell is the predominant type of cell lining the collecting duct system. In the cortical collecting duct, principal cells are responsible for K^+ secretion and Na^+ reabsorption. This function is only partly regulated by aldosterone, because some of the cells are capable of K^+ secretion in the absence of this hormone. Intercalated cells are involved in H^+ ion and HCO_3^- secretion. Transport of water occurs in all segments of the collecting duct in the presence of the antidiuretic hormone or vasopressin. Figure 1 summarizes the functions of various segments of the nephron.

1.7. Interstitium

The renal interstitium, a space between tubules, is comparatively sparse. It increases from the cortex to the medulla. In humans, the fractional volume of the cortical interstitium ranges from 12% in younger individual to 16% in older subjects. In the medulla, the interstitial volume increases from the outer to the inner medulla in the range of 4 to approximately 30%.

Two types of interstitial cells have been described in the cortex: type 1 cortical interstitial cell, which resembles a fibroblast, and type 2 interstitial cell with mononuclear or lymphocyte-like structure. Between the cells is a space that contains collagen and fibronectin. It is believed that the peritubular fibroblast-like interstitial cells secrete erythropoietin. Type 2 interstitial cells are believed to represent bone marrow-derived cells. Three types of interstitial cells have been described in the medulla. None of these cells is the site of erythropoietin; however, some cells (type 1 medullary interstitial cell) contain lipid droplets, which are believed to have hypotensive effects. All medullary interstitial cells synthesize proteoglycans that are present in the interstitium.

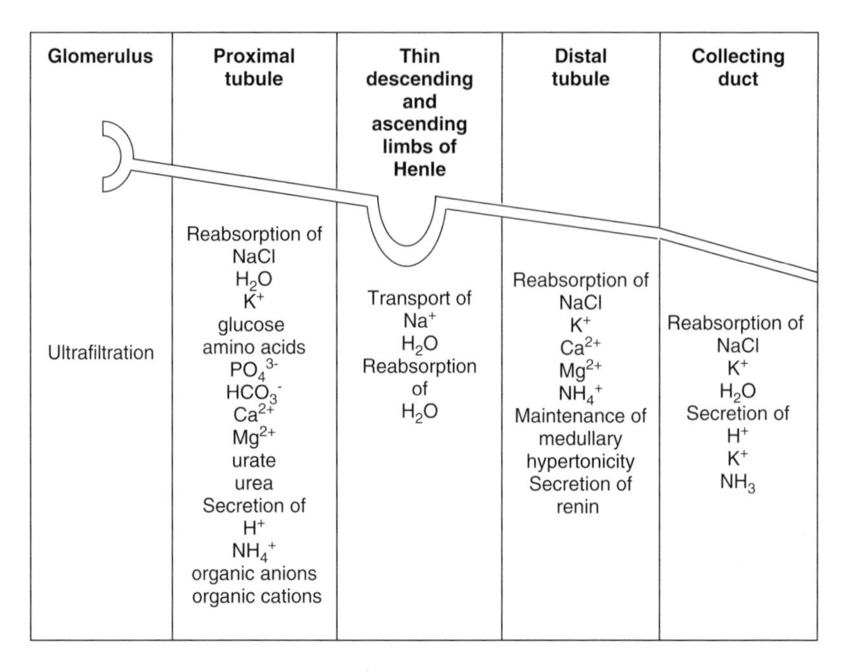

Fig. 1. Schematic representation of nephron segments showing the structural-functional relationships.

1.8. Blood Supply

Each kidney is usually supplied by one renal artery arising from the abdominal aorta. After or before entering the hilum, the renal artery divides into an anterior and a posterior branch; both of them give rise to a total of five segmental arteries. The segmental arteries are end arteries, and occlusion of a single artery results in infarction of the area supplied. These segmental branches form the interlobar arteries in the renal sinus, which follow the curvature of the kidneys to form arcuate arteries. From these arteries arise interlobular arteries that course radially through the cortex toward its surface. The interlobular arteries give rise to the afferent arterioles, which divide into five to eight lobules and form the glomerular capillaries. The loops of these capillaries reunite to form the efferent arteriole of the glomerulus. The efferent arteriole leaves the glomerulus as a short unbranched segment before it branches into capillaries. These capillaries, which supply blood to the proximal and distal tubules of the cortex, are known collectively as the peritubular capillary network. The efferent arterioles of glomeruli located in the juxtamedullary cortex and near the medullary region not only supply blood to their own tubules but also run deep into the medulla. These long, thin vessels are called arteriolae rectae, or straight arterioles. They form a loop with straight venules or venulae rectae of the medulla to form the vasa recta of the kidney. Thus, the blood supply to the medulla is entirely derived from the efferent arterioles of the juxtamedullary glomeruli. The capillaries of the outer cortex converge to form the stellate veins which drain into the interlobular veins, then into the arcuate and interlobar veins, and finally into the renal vein.

2. CLINICAL EVALUATION OF KIDNEY FUNCTION

Currently, determination of serum creatinine and blood urea nitrogen (BUN) concentrations, and estimation of glomerular filtration rate (GFR) remain the most important tests to assess the kidney function in clinical practice. GFR can be measured directly by radioisotope methods or indirectly from serum creatinine concentration as estimated GFR (eGFR), using the Modification of Diet in Renal Disease formula. Although serum creatinine concentration of 0.8–1.2 mg/dl and a BUN concentration of 10 mg/dl are considered normal, their values vary with muscle mass and protein intake as well as the functional status of the liver. Therefore, an eGFR is recommended for evaluation of kidney function. Most clinical laboratories provide both serum creatinine and eGFR to the physician for assessment of kidney function. An eGFR of 60 ml/min/1.73 m^2 or less is considered chronic kidney disease (CKD). In addition to eGFR, urinalysis provides an assessment of glomerular, tubular, and interstitial functions of the kidney. The presence of albuminuria, hematuria, and red blood cell (RBC) casts in a well-performed urinalysis indicates significant glomerular disease. Determination of urine pH and urine osmolality is helpful in assessing the kidney's ability to acidify as well as concentrate and dilute the urine. A renal biopsy is needed to assess the pathology of the kidney in disease states.

3. KIDNEY FUNCTION IN DISEASE STATES

When GFR is decreased due to functional or structural damage to the kidney, a variety of functions and also the structure of the kidney are altered. These functional and structural disturbances are briefly discussed below.

3.1. Fluid, Electrolyte, and Acid–Base Disturbances

When GFR is below normal but not low enough, the kidneys try to maintain relatively normal fluid, electrolyte, and acid–base balance. However, when GFR is severely decreased, the kidneys retain Na^+, water, K^+, Mg^{2+}, PO^{3-} and H^+ ions, resulting in edema, either hyponatremia or hypernatremia, hyperkalemia, hypermagnesemia, hyperphosphatemia, and severe metabolic acidosis. Hypocalcemia results from decreased calcitriol production by the kidney. The patients also develop hypertension due to retention of Na^+ and water. Anemia and bone disease are commonly seen in patients with low GFR.

3.2. Acute Kidney Injury (Acute Renal Failure)

Acute kidney injury (AKI) is defined as an abrupt decrease in renal function, resulting in accumulation of nitrogenous (creatinine and BUN) and non-nitrogenous waste products. It develops over a period of hours to days. Although the clinical markers for AKI are serum creatinine and BUN concentrations, the precise increase in serum creatinine that defines AKI remains elusive. Studies have shown that even a small increase in serum creatinine levels is associated with increased morbidity and mortality. For example, it was reported that an increase in serum creatinine by ≤ 0.5 mg/dl was associated with a 6.5-fold increase in hospital mortality, while an increase in serum creatinine of 0.3–0.4 mg/dl was associated with only 70% increase in mortality risk. Even the length of the hospital stay was prolonged by AKI *(4)*.

Based on the increase in serum creatinine and urine output, the Acute Dialysis Quality Initiative group recently proposed the RIFLE system, which classifies AKI into three severity categories (R = risk; I = injury; F = failure) and two clinical categories [L = loss; E = end stage renal disease (ESRD)]. However, this classification requires further validation *(5)*.

The causes of AKI are divided into three major categories: prerenal, renal, and postrenal. Prerenal azotemia is due to decreased renal perfusion, caused by hypovolemia, decreased effective circulating volume, renal artery disease, and/or altered intrarenal hemodynamics. A variety of intrinsic renal disorders due to an acute insult to the renal vasculature, glomerulus, tubules, or interstitium can cause AKI. Acute tubular necrosis remains the major form of AKI due to intrinsic renal disease, and it is caused by renal ischemia and exposure to nephrotoxins, such as drugs or contrast material. Postrenal AKI is due to obstruction to the urinary system either by intrinsic or extrinsic masses.

Treatment of AKI includes volume repletion in hypovolemic conditions and elimination of the causative agent or disease process. Some patients may require hemodialysis or other renal replacement therapies, such as continuous venovenous hemodialysis. AKI is usually a reversible process. However, some patients progress to CKD.

3.3. Chronic Kidney Disease (Chronic Renal Failure)

CKD is defined as a gradual decrease in renal function over a period of several months to years. Diabetes, hypertension, glomerulonephritis, cystic kidney diseases, and tubulointerstitial diseases (TIDs) are the major causes of CKD. Approximately 6.2 million people are estimated to have a serum creatinine level of $\geq$1.5 mg/dl. Unlike in AKI, serum creatinine level does not represent the extent of renal disease in subjects with CKD. Therefore, either actual determination of GFR by radioisotope methods or eGFR is used to assess the severity of kidney disease in a CKD patient. Based on these methods of GFR, a staging system and action plan for CKD was developed (Table 1).

There are several risk factors for the progression of CKD, including hypertension, diabetes, hyperlipidemia, excessive protein intake, smoking, anemia, and genetic predisposition to kidney disease. CKD is one of the major risk factors for cardiovascular disease. Conservative management of CKD includes (i) control of blood pressure (<130/80 for patients with no proteinuria and <120/80 for those with proteinuria >1 g/day), using dietary sodium restriction <100 mEq/l, and antihypertensive agents such as angiotensin converting enzyme-inhibitors (ACE-Is), or angiotensin receptor-blockers (ARBs) as well as a low dose diuretic; (ii) maintenance of HbA_{1c} < 7% in diabetic patients; (iii) restriction of protein intake <0.8 g/kg/day; (iv) maintenance of LDL <100 mg/dl; (v) maintenance of hemoglobin approximately 11–12 g/dl and control of bone disease; and (vi) smoking cessation. Dialysis or kidney transplantation is required if the patient progresses to ESRD.

3.4. Nephrotic Syndrome

This syndrome is characterized by proteinuria > 3.5 g/day, hypoalbuminemia, edema, hyperlipidemia, and lipiduria. The nephrotic syndrome is caused by (i) either primary (idiopathic) or secondary (known) glomerular diseases; (ii) drugs, such as nonsteroidal antiinflammatory drugs, heroin, and gold; and (iii) bacterial, viral, and parasitic infections. Among secondary glomerular diseases, diabetes is the leading cause of nephrotic syndrome in adults. The primary

Table 1
Stages of Chronic Kidney Disease and Proposed Actions *(6)*

Stage	*Description*	*GFR ($ml/min/1.73\,m^2$)*	*Action*	*Related terms*
1	Kidney damage with normal or↑ GFR	≥90	Diagnosis and treatment Slow progression of CKD Treat comorbidities Cardiovascular disease risk reduction	Albuminuria, proteinuria, hematuria
2	Kidney damage with mild ↓ GFR	60–89	Estimate progression	Albuminuria, proteinuria, hematuria
3	Moderate ↓ GFR	30–59	Treat complications	Chronic renal insufficiency, early renal insufficiency
4	Severe ↓ GFR	15–29	Prepare for renal replacement therapy	Chronic renal insufficiency, late renal insufficiency, pre-ESRD
5	Kidney failure	< 15 (or dialysis)	Renal replacement therapy	Renal failure, uremia, ESRD

CKD, chronic kidney disease; ESRD, end stage renal disease; GFR, glomerular filtration rate.

diseases that cause nephrotic syndrome are minimal change disease, focal segmental glomerulosclerosis, membranous nephropathy, and membranoproliferative glomerulonephritis.

Proteinuria is caused by losses of charge and size in the GBM. Also, a decrease in protein (nephrin, podocin, and α-actinin) or mutation in genes that encode these proteins in slit diaphragm can cause proteinuria. Hypoalbuminemia is due to renal loss of albumin and increased catabolism. Both hypoalbuminemia and increased salt and water retention lead to edema formation. Hyperlipidemia is secondary to increased hepatic synthesis and decreased degradation of lipoproteins.

The patients with nephrotic syndrome are at risk for thrombotic complications, infections, cardiovascular disease, and skeletal abnormalities. Management of nephrotic syndrome includes salt and water restriction in edematous patients, ACE-Is or ARBs for proteinuria and control of hypertension, low protein diet to improve serum albumin level, proteinuria, and renal function, vaccination to prevent infection from encapsulated organisms, prevention of thrombosis by avoiding prolonged immobilization and volume depletion, and control of hyperlipidemia. Anticoagulation is recommended in high-risk patients. Immunosuppressive therapy is reserved for patients with primary renal diseases. Elimination of the secondary cause usually improves nephrotic syndrome.

3.5. Nephritic Syndrome

The nephritic syndrome, also called glomerulonephritis, is characterized by hematuria, RBC casts, hypertension, renal insufficiency, and varying degrees of proteinuria. Based upon the etiology and pathogenic mechanisms, this syndrome can present as (i) asymptomatic hematuria or proteinuria; (ii) acute nephritis to rapidly progressive glomerulonephritis (RPGN); and finally (iii) chronic sclerosing glomerulonephritis. There are several primary (e.g., RPGN) and secondary (e.g., systemic lupus erythematosus) causes of nephritic syndrome. Treating the underlying cause by conservative (acute nephritis) or aggressive (RPGN) management and control of blood pressure remain the main stay of therapy in patients with nephritic syndrome. A substantial number of patients presents with renal insufficiency, requiring renal replacement therapy.

3.6. Tubulointerstitial Diseases

TIDs are a group of clinical disorders that affect principally the tubules and interstitium. The glomeruli and blood vessels are usually spared. It has been estimated that about 15% of all cases of AKI and

25% of all cases of CKD are attributable to TIDs *(7)*. Based upon the morphologic changes and the rate of deterioration of renal function, TIDs can be classified into acute TID or acute interstitial nephritis or chronic TID or chronic interstitial nephritis. Acute interstitial nephritis manifests as sudden onset of renal failure within days to weeks (1 day to 2 months), hematuria, mild proteinuria, white blood cell casts and at times eosinophiluria or eosinophilia. The accurate diagnosis is made by renal biopsy. Acute interstitial nephritis is caused by drugs, infections, or immune disorders. Treatment includes removing the causative agent or in some cases steroids. Hemodialysis may be necessary in some patients.

Chronic interstitial nephritis is caused by a variety of drugs, infections, vascular, metabolic, immune, and hematologic diseases, urinary tract obstruction, and heavy metals. In some cases, the cause is unknown. Clinical manifestations include hypertension, renal insufficiency, hyperkalemia, anemia (both hyperkalemia and anemia are disproportional to the degree of renal insufficiency), inability to concentrate urine, and Fanconi syndrome. Urinalysis shows mild proteinuria and absence of RBC casts. Glomeruli are affected secondarily. Pathologic findings of the kidney include progressive scarring of the interstitium, tubular atrophy, and infiltration with lymphocytes and macrophages. Removal of the offending agent or treatment of the underlying disease, and control of blood pressure with dietary sodium restriction and antihypertensive agents, and correction of anemia as well as bone disease remain the main stay of therapy in patients with chronic TIDs.

3.7. Vascular Diseases

Renal artery stenosis, hypertensive nephrosclerosis, vasculitis affecting the medium and small renal arteries, renal vein thrombosis as a complication of nephrotic syndrome, and several microangiopathic diseases such as hemolytic uremic syndrome and thrombotic thrombocytopenic purpura are some of the vascular diseases that cause altered kidney function. Appropriate management is required to prevent the progression of kidney disease.

4. CONCLUSION

This chapter has provided a brief review of gross and microscopic anatomy of the kidney and its functions in health and disease. A variety of commonly seen renal abnormalities have been discussed that require nutritional and pharmacologic intervention alone or in combination for management.

5. SUMMARY

The kidneys are paired organs located retroperitoneally in the lumbar region and maintain fluid and acid–base balance and remove nitrogenous waste products as well as synthesize hormones such as renin, erythropoietin, and active vitamin D_3. The functional unit of the kidney is the nephron, which is responsible for the formation of ultrafiltrate, reabsorption, and secretion of electrolytes and water as well as urinary acidification and concentration. When the structure of the kidney is disturbed by a pathologic process, markers of kidney function such as serum creatinine levels are elevated as in AKI, CKD, or TID, and also an abnormal amount of protein or RBCs is observed in the urine, as in nephrotic syndrome, or glomerulonephritis, respectively. Also renal vasculature is affected, causing hypertension and thrombotic microangiopathies.

REFERENCES

1. Reddi AS. Structure and function of the kidney. In: Reddi AS. Essentials of Renal Physiology. New Jersey: College Book Publishers, 1999:21–43.
2. Madsen KM, Tisher CC. Anatomy of the kidney. In: Brenner BM, ed. Brenner and Rector's The Kidney, 7th ed., vol. 1. Philadelphia: Saunders, 2004:3–72.
3. Kriz W, Elgar M. Renal anatomy. In: Johnson RJ, Feehally J, eds. Comprehensive Clinical Nephrology, 2nd ed, Edinburgh: Mosby, 2003:1–11.
4. Chertow GM, Burdick E, Honour M, Bonventre JV, Bates DW. Acute kidney injury, mortality, length of stay, and costs in hospitalized patients. J Am Soc Nephrol 2005; 16:3365–3370.
5. Bellomo R, Kellum JA, Ronco C. Defining acute renal failure: Physiologic principles. Intensive Care Med 2004; 30:33–37.
6. Levey AS, Coresh J, Balk E, Kausz T, et al. National Kidney Foundation guidelines for chronic kidney disease: Evaluation, classification, and stratification. Ann Intern Med 2003;139:137–147.
7. Kelly CJ, Neilson EG. Tubulointerstitial diseases. In: Brenner BM, ed. Brenner and Rector's The Kidney, 7th ed., vol. 2. Philadelphia: Saunders, 2004: 1483–1511.

2 Historical Perspective of Nutrition in Kidney Disease

Mary Kay Hensley

LEARNING OBJECTIVES

1. Identify the progression of dietary modifications in kidney disease in the last eight decades.
2. Describe the development of renal dietetic practice and the role of the clinical dietitian in patient care.
3. Review notable research and medical technology advances that have influenced medical nutrition therapy for individuals with kidney disease.

Summary

With the development of long-term dialysis therapies, kidney transplantation, many new medications and the advent of laboratory assay methods, the treatment for chronic kidney disease has marched forward in the past eight decades. Dietary regimes for patients with kidney disease have also changed as new technologies became available. Most of all, renal dietetics practice centered on the multiple aspects of patient care, and wellness has developed during this time.

Key Words: Kidney disease; dietetics practice; history of renal nutrition therapy; renal dietitian.

From: *Nutrition and Health: Nutrition in Kidney Disease*
Edited by: L. D. Byham-Gray, J. D. Burrowes, and G. M. Chertow

1. INTRODUCTION

This chapter reviews the dietary modifications that have been recommended for patients with chronic kidney disease (CKD) starting in the mid-1800s. It also examines dietetics practice and the development of the specialty of renal dietetics. Notable research and changes in medical technology that have influenced the dietitian's role in caring for patients with CKD is explored. The professional organizations and publications that played a role in defining renal nutrition are reviewed.

2. HISTORICAL DIETARY TREATMENT OF KIDNEY DISEASE

In the mid-1800s, Richard Bright recommended a milk diet for patients with edema and proteinuria *(1)*. Fishberg in 1930 and Addis in 1948 recommended protein restriction for uremic patients, but neither identified the biological value of the protein *(2,3)*. Many at this time believed that dietary protein restriction would decrease the workload and stress on the kidneys. In 1948, Kempner proposed a diet consisting of rice, fruit and sugar for the treatment of acute and chronic renal failure, and it became known as the Kempner Rice diet *(4)*. This diet contained about 20 g protein, 150 mg sodium and 2000 calories. The Kempner Rice Diet was also used in patients with heart disease who did not respond to salt restriction alone, but it was not recommended for people with diabetes because of the high fat and sugar content.

In 1948, Borst *(5)* reported that a protein-free, normal calorie, low salt diet improved uremia and edema in patients with advanced renal failure. As this diet did not contain usual food items and was very limited, it was not well accepted by patients and health care professionals. For example, some of the recipes and food items in the Borst diet are Borst Soup and butterballs (see Table 1).

The goal of the Kempner Rice and the Borst diets were to preserve life until the kidneys recovered; they were the alternative to dialysis in the 1950s and early 1960s. These diets emphasized the need for adequate caloric intake in severely ill patients to prevent weight loss and to increase the satiety value of the diet. The control of fluid, salt and potassium afforded by these diets was probably as important as the protein restriction. Borst soup and butterball use continued into the late 1960s and early 1970s, especially in smaller cities and rural areas where dialysis was not yet available *(6)*. Later on, renal diets were supplemented with rolls of hard candy, mints, marshmallows and jellybeans placed on hospital food trays and given to patients during dialysis treatments.

Table 1
Sample Foods in the Borst Diet

Borst Soup
- Water (1.5 l)
- Custard powder (100 g)
- Sugar (150 g)
- Butter (100 g)

Procedure: Heat slowly until hot, but do not boil. Serve in a soup bowl.

Nutrient Analysis: 1750 calories, with negligible amounts of protein and potassium. Sodium content is dependent on the salt content of the butter used.

Butterballs *(6)*
- Unsalted butter or margarine (84 g)
- Powdered sugar (100 g)
- Vanilla extract (3/4 tsp)
- Peppermint or other flavoring (four drops)

Procedure: Cream sugar and butter together. Add flavoring. Divide into 10 equal balls. Roll in sugar, if desired. Place in freezer to harden. Store in freezer until time for serving. Flavoring may be omitted and powdered soft drink mix used on the outside of the balls. Yield: 10 balls

Nutrient Analysis: 103 calories, negligible amounts of protein, sodium and potassium

By 1960, Rose and Wixon *(7)* established minimum daily requirements of essential amino acids (EAA). This discovery was made in individuals without kidney disease and contributed to the understanding of amino acid metabolism. They reported that the nitrogen from serum urea could be used in the synthesis of nonessential amino acids (NEAA) endogenously if sufficient EAA were present in the diet. Rose et al. also discussed the importance of balanced meals containing adequate carbohydrate, fat and protein to ensure overall nutrition adequacy.

In 1963, Giordano applied the concept of high biological value (HBV) protein to the renal diet *(8)*. At this time, only protein of animal origin was considered HBV. He stressed the need for a specific quality of protein as well as quantity, based on the EAA recommendations of Rose et al. Adequate caloric content with vitamin and mineral supplements were part of the diet. Giordano was able to show that uremic patients were able to maintain positive nitrogen balance and obtain relief from uremic symptoms on 2 g or 3 g of nitrogen or about 20 g of protein/day.

In 1964, Giovanetti and Maggiore achieved similar results using the same principles of a protein-restricted diet with HBV protein food sources *(9)*. A powdered synthetic EAA mixture or eggs was used. Pasta prepared from low protein wheat starch and wafers made from cornstarch were utilized to control the amount of NEAA, while supplying adequate calories. This diet, which became known as the Giovanetti or the Giordano–Giovanetti (G-G) diet, produced positive nitrogen balance, relieved uremic symptoms and maintained blood urea nitrogen levels.

In 1965, Berlyne et al. *(10)* and Shaw et al. *(11)* modified the G-G diet by utilizing an 18-g to 21-g protein diet with 12 g of protein from milk and egg sources. High caloric levels were maintained with special wheat starch products that were available commercially. This diet was supplemented with iron, multivitamins and methionine, the only EAA not naturally present in foods. Due to the restrictive nature of the diet, they recommended use of 20g protein only if blood urea nitrogen was greater than 200 mg/100 mL or if the urea clearance was less than 10 mL/min. These new diets provided a method of treatment for kidney disease when surgery or medications alone could not be used.

In 1968 and 1973, Kopple et al. *(12,13)* conducted studies comparing 20-g and 40-g protein diets of HBV in individuals with uremia. They found that both diets relieved uremic symptoms and that biochemical improvements were similar; however, maintenance of body weight was more optimal on the 40-g protein diet. Most of all, the 40-g protein diet was more acceptable to the patient because of the greater variety of food selection.

Manufacturers were able to meet the demand for high calorie, low protein food products because patient adherence to their high calorie supplements was not always consistent. Lprotein wheat starch flour (Cellu-low Protein Baking Mix, Chicago Dietetic Supply, La Grange, IL; Paygel-P Wheatstarch Flour and Dietetic Paygel Baking Mix, General Mills, Minneapolis, MN) was used to make bread, muffins, cookies, cakes and pancakes. A 40-g slice of low-protein bread supplied 0.2 g protein and 115 calories. The caloric content could be increased further with the addition of butter or margarine and jelly. However, because of the lack of gluten in the wheat starch, few patients and dietitians were able to make an acceptable bread product. Some recipes added whipped egg whites to decrease density and improve texture. Low protein pasta (Carlo Erba, Milan, Italy

Table 2
Sample Menu (40 g Protein, 20 mEq Sodium Diet) *(15)*

Breakfast	*Lunch*	*Dinner*
1/2 cup Apple Juice	One serving Vegetable Veloute[a]	1/2 cup Minted Melon Balls
2 slices French Toast	3/4 cup Strawberry Ice Cream	One serving Chicken Curry
4 t. Syrup	4 Scotch Wheat Starch Cookies	1/2 cup Rice
6 t. Unsalted Margarine	Coffee or Tea with sugar	Zesty Lettuce Salad
2 t. Milk		Soft Drink
Coffee, Sugar		

[a]Thickened white sauce.

and A-Protein distributed by General Mills) was a welcome addition to the diet as it could be flavored with cream, butter, herbs, and other seasonings. A powdered carbohydrate, protein-free supplement (Controlyte, D.M. Doyle Pharmaceutical Co, Minneapolis) was added to water, juices and soft drinks. High calorie, low electrolyte beverages (Cal-Power, General Mills; Hycal, Beecham Products, Clifton, NJ) provided 400–600 calories in less than 8 fluid ounces. Electrodialyzed whey (lactalbumin) (Wyeth Laboratories) was suggested as another source of HBV protein; it was mixed as a milk shake-like beverage with water, oil, a carbohydrate source and flavoring *(14)*.

Protein was prescribed in standard amounts that did not necessarily reflect the patient's body size or weight. Protein recommendations were based on the patient's signs and symptoms, as well as the level of kidney function. Meal patterns of 30, 40, and 50 g protein/day were published (see Table 2 for an example of a 40-g protein diet). If nausea and vomiting were present, the dietary protein prescription would be reduced to alleviate symptoms.

3. ADVANCES IN MEDICAL THERAPY AND ITS ROLE IN DIETARY THERAPY

By the end of the 1960s, the number of patients referred for dialysis increased because of advances in dialysis access devices, biomedical technology and transplantation; however, the availability of dialysis machines and the health care staff trained to manage dialysis care could not keep pace with the demand. Committees composed of hospital administrators, physicians, clergy and community representatives often reviewed individual patient records and made recommendations about who was eligible to receive dialysis. While dietitians did not participate to a great extent in the decision-making process, it was the dietitian who was often the last health care provider to see those patients who were not eligible to receive dialysis. Severely restricted diet instructions were provided to these patients upon discharge from the hospital to minimize the gastrointestinal side effects of uremia to the extent possible. The goal of these diets was to provide time for the patients to finalize their affairs.

By 1972, the United States Congress passed legislation allowing Medicare to cover 80% of the cost of dialysis treatment; two to three treatments per week became the routine. In 1974, Burton outlined specific modifications required for the various forms of treatment and stages of kidney disease *(16,17)*. These papers were published in the *Journal of the American Dietetic Association* and were readily available to all practicing dietitians. The main premise of these modifications was that the diet used for the treatment of kidney disease must be adjusted as the patient's condition or treatment changes. At this time, protein restrictions became more liberal and 50-g to 80-g protein diets became more commonplace and were met with greater acceptance by the patient and their families.

3.1. Protein

The use of EAA supplements were investigated in late 1970s as part of the daily protein requirement for acute and chronic dialysis patients *(18)*. Bergstrom and Alvestrand et al. *(19)* followed a group of patients in Sweden who were prescribed 18 g of dietary protein and an amino acid supplement that included histidine to postpone the need to start dialysis. The patients achieved positive nitrogen balance only on days when the amino acid supplement was administered, possibly identifying the role of conditionally EAAs in CKD. Hecking and Port *(20)* found that supplementation with amino acids or keto acids was indicated in only a small minority of catabolic patients, while most patients receiving maintenance dialysis did well on a mixed food diet with approximately 1 g protein/kg of body weight. Amin-Aid (Kendall McGaw, Irvine, CA) is a commercial elemental diet containing EAA plus histidine in amounts consistent with amino acid profile established by Rose and Wixon *(7)*; it became available as a total feeding or as a supplement to the dialysis diet. This diet treatment was not universally accepted probably because of the increased availability of chronic dialysis as well as reimbursement systems that made alternatives to dialysis less of a priority. More recently, in 1994, the Modification of Diet in Renal Disease (MDRD) Study looked at slowing the progression of renal failure using a very low protein diet (0.28 g/kg/day) and supplemental ketoacid analogs (0.28 g/kg/day), along with a very low phosphorus (4–9 g/kg/day) diet. The results were inconclusive *(21)*.

Research continued to support, and dietitians continued to promote, renal diets containing 1–1.4 g protein/kg of body weight during the 1980s and 1990s *(22,23)*. In 2000, the evidence-based National Kidney Foundation Kidney Disease Outcomes Quality Initiative (KDOQI) Clinical Practice Guidelines for Nutrition in Chronic Renal Failure (hereafter referred to as the KDOQI Nutrition Guidelines) were published *(24)*. These guidelines continue to be the basis for nutrition care for adults with CKD. The KDOQI Nutrition guidelines recommend 1.2 g of protein/kg/day for adult maintenance hemodialysis patients and 1.2–1.3 g/kg/day for adult chronic peritoneal dialysis patients. Please refer to Chapter 25 for the KDOQI Nutrition Guidelines.

3.2. Energy

While multiple earlier studies addressed the specific protein needs of the patient with CKD, many just described the energy and fat content as adequate or high. Sister M. Victor *(25)* analyzed diets from 20 outstanding hospitals in 1932. While the protein levels of the diets varied from 20 to 70 g and good quality protein is specified, the energy content of the diet was simply described as high. In the 1987 edition of the Clinical Guide to Nutrition Care in End-Stage Renal Disease, the section on carbohydrates and fats states adequate intake is encouraged to prevent wasting syndrome *(26)*. The section then goes on to discuss fiber and cholesterol but does not specify energy needs. In 1989, a metabolic balance study addressed specific energy levels of 25, 35, and 45 kcal/kg/day for 21 days each *(27)*. These studies determined that about 35 kcal/kg/day is needed to maintain neutral nitrogen balance and maintain body composition in patients receiving maintenance hemodialysis. This study and others were reviewed in the KDOQI Nutrition Guidelines to recommend 35 kcal/kg/day for patients receiving maintenance hemodialysis *(24)*. The same amount of energy is recommended for patients receiving chronic peritoneal dialysis except that both energy intake from the diet and that derived from the glucose absorbed from the dialysate must be considered. Adjustments for the elderly and the obese patient are discussed.

3.3. Electrolytes and Fluids

The dietary sodium prescription has been influenced by many factors in the past 50 years, but it continues to depend on the presence or absence of edema and hypertension and the kidney's ability to conserve and excrete sodium. In the 1950s and 1960s, medications to control hypertension were limited, and 250- to 1000-mg sodium diets were used to control severe edema and elevated blood pressure *(28)*. Distilled water was specified in these diets, as many sources of well water and municipal water contained excessive amounts of sodium. Vegetables like celery, beets and carrots were limited because of their natural sodium content. Aggressive diuretic therapy and frequent episodes of vomiting and diarrhea because of uremia caused the amount of dietary sodium prescribed to be adjusted frequently. Because newer anti-hypertensive agents have become available and biomedical technology has improved dialysis techniques, dietary sodium prescriptions have increased to 2000–3000 mg/day. In addition, the changing American food supply has affected dietary sodium intake significantly; this will be discussed later.

As the level of protein in the renal diet became more liberalized, potassium control also became an issue for patients receiving maintenance hemodialysis. When routine dialysis increased to two to three times per week, patients began to feel better and their appetite increased. A dietary potassium intake of 2–3 g/day was the maximum amount recommended in all meal plans for patients receiving maintenance hemodialysis *(29)*. However, dietitians realized that many patients who were receiving maintenance peritoneal dialysis did not require a potassium restriction and intake could be liberalized based on the patient's serum potassium levels.

Food lists with low and medium potassium levels were developed using standard reference materials. High potassium food lists were also developed and patients were advised not to consume these foods. Tsaltas reported that leaching or dialyzing vegetables reduced the potassium and sodium content. Patients were taught to peel, slice, and soak vegetables in water and then boil in large amounts of water to reduce potassium by approximately 50% *(30)*. Since this initial work, Burrowes and Ramer *(31)* found that the most effective method for removing potassium from tuberous root vegetables was the double cook method. This method involves placing the vegetable in a 2:1 water to vegetable ratio, bring the water to a boil, drain the water, replace with fresh water, and then bring to a boil again until the vegetable is cooked.

Fluid and mineral recommendations were reviewed in the early literature. In 1927, Norman and coworkers reviewed 165 cases in which a diet restricted in salt and water proved effective for cases of obstinate edema due to nephritis *(32)*. Yet in another article, the same authors provide contradictory statements. They recommended a 40-g protein diet for nephritis if there was retention of protein derivatives such as urea, and large amounts of fluid to wash out the poisonous substances *(33)*. A few years later in 1931, Lashmet and Enke discussed the Neutral Diet in the treatment of nephritic edema. They theorized that edema was not due to the failure of the kidneys to excrete water or chloride, but due to the alkaline ash content of the diet. It was felt that an acid ash diet would decrease edema. Lashmet and Enke recommended 45–50g of protein, 2000 calories, a slight excess of acid ash, 10–15 g of ammonium chloride in 0.5 capsules with meals and 5000 mL of fluid daily *(34,35)*.

The machines used initially for hemodialysis did not have volumetric controls, and fluid removal was not consistent. Fluid limits were very strict in the early days of dialysis, especially for patients who received dialysis infrequently or who had limited residual kidney function. Patients were instructed often to drink only what was necessary to take their medications and/or to satisfy their thirst with the fluid or moisture in food. These early diets specified fluid limits of 500–800 mL/day plus urine output *(28)*. This amount was increased in 1998 to 1000 mL/day plus output *(36)*.

3.4. Phosphorus and Calcium

Very few papers appeared in the literature prior to 1965 about dietary phosphate restriction. A renal diet study kit published by the American Dietetic Association (ADA) in 1969 does not include calcium and phosphorus modification *(29)*. Initially, physicians ordered very low phosphorus diets, which limited the amount of meat and other protein sources available to the patient, and again severely limited the variety and palatability of the diet.

Towards the end of the 1960s, the use of antacids such as aluminum hydroxide and aluminum carbonate became common, taking advantage of one of its side effects which was the absorption of phosphorus in the gut *(37)*. Dietitians welcomed this new therapy as the renal diet could be liberalized, while severe bone pain and intense itching were relieved. Binders containing a mixture of aluminum and magnesium were used in some areas. These products reduced constipation and it was thought that higher magnesium levels decreased the muscle cramps that frequently occurred during dialysis. Antacids containing calcium carbonate were regarded as ineffective during this time period *(38)*.

Recipes for cookies incorporating crushed aluminum antacid tablets were developed by dietitians. They increased adherence to the binders and improved serum phosphate control. These products became popular on hospital wards and were also made at home by patients. Dietitians and patients further developed their own cookie recipes using the antacid tables, usually replacing part of the flour with an equal amount of crushed pill powder. Each batch of cookie dough was divided equally so that one cookie contained the equivalent of two or three aluminum hydroxide pills.

Reports of aluminum toxicity were published as early as 1972, and many practitioners observed isolated incidences of dialysis-related encephalopathy and dementia *(39)*. Laboratory testing of serum aluminum began to be performed on symptomatic patients and was also used as a screening tool for the dialysis population. Municipal water supplies started to test for aluminum and discontinued use of aluminum as a purifier. Dialysis centers installed filtration systems to remove aluminum from the water used to make the dialysate. Dietitians advised patients not to use pitted aluminum cookware and other potential environmental sources of aluminum toxicity. Increased efforts were made to control dietary phosphate intake so that large doses of binders could be avoided. Reports of desferoxamine, a chelating agent, to diagnose and treat aluminum bone disease were published as early as 1981 *(40)*.

By 1985, the use of aluminum binders declined and calcium carbonate was widely used *(41)*. At this time, dietitians became more involved with individualizing the amount of binders for each patient. Titration of the binder dose to the phosphorus content of the meal or snack became common practice. Calcium citrate was used for a short period of time but was soon discontinued as citrate enhanced aluminum absorption. Calcium acetate was introduced as a binder in 1989 and was successful because of its ability to decrease calcium absorption *(42)*. The first non-calcium, non-aluminum phosphate binder was introduced in 1998, and these products continue to be used today *(43)*.

By 1978, calcitriol (1,25-dihydroxycholecalciferol), an active form of vitamin D for the treatment of osteodystrophy in CKD, was introduced. In 1986, oral forms of vitamin D available for clinical use included ergocalciferol, cholecalciferol, di-hydrotachysterol, calcifediol and calcitriol *(44)*. Dietitians monitored vitamin D therapy, but usually did not determine when to start therapy or to change the dose. An injectable form of calcitriol became available in 1987 *(45)*, and in 1998, the first vitamin D analog was introduced *(46)*. Dietitians gradually increased their role in monitoring bone disease in patients with CKD by developing protocols for calcium, phosphorus, and vitamin D dosing.

There was early concurrence that total removal of parathyroid glands in both animals and humans would result in death due to tetany. By 1950, hyper-parathyroidism and hypo-parathyroidism were defined *(47)*. The first immunoassay of human parathyroid hormone (PTH) was reported by Berson in 1963 *(48)*. By the early 1990s, levels of PTH as determined by immunoradiometric assay or immunochemiluminometric assay were routinely used for patient evaluation *(49,50)*. PTH added another tool for the management of renal bone disease.

KDOQI Clinical Practice Guidelines for Bone Metabolism and Disease in CKD were published in 2004, and they have provided direction to the clinician in managing bone disease in all stages of CKD *(51)* (refer to Chapter 25 for these guidelines).

4. CHANGING FOOD SUPPLY FOR PATIENTS WITH CKD

Just as cellophane was used to make an artificial kidney in the early days, it also changed the way food items were packaged and distributed in the United States. After World War II, the boom in packaged foods stemmed largely from the widespread introduction of new transparent wrappers that kept food fresh and allowed customers to see exactly what they were purchasing. In the 1950s and 1960s, patients with kidney disease tried these new foods, but continued to prepare most meals in their home kitchens. Canned foods were utilized and many of them were preserved by the homemaker; home-baking remained popular. When patients ate in restaurants or were a guest in another's home, food was prepared fresh, using mainly fresh ingredients. Patients with kidney disease were encouraged to ask for special preparation methods when eating away from home.

In the next two decades, a large array of new items arrived on grocery store shelves and the supermarket industry flourished. Frozen food cases at stores were now filled with a large array of products. An increasing number of meals were eaten away from home, and what we know as "fast food" restaurants today increased in popularity. The sodium content of prepared foods increased, and food labeling laws were updated to make this information available to the consumer. Renal dietitians met the challenge by publishing dining out guides and organizing classes on the topic of food selection at restaurants.

In the 1990s, the American public changed their food habits, which included more restaurant meals, in-car meals, and vending machine meals and snacks. Portion sizes increased. Foods eaten at home included a large amount of commercially prepared frozen items, restaurant carryout meals, and quick-to-fix snacks. Renal dietitians were challenged to educate their patients about the nutrient content of these new foods and to try to fit them into their patient's diet. Cooking skills and home food preparation habits declined by the end of the 1990s.

As the new millennium dawned, renal dietitians were faced with new challenges. Food manufacturers increased their reliance on phosphorus-containing food additives to enhance shelf life, to further decrease cooking or preparation time, to improve food textures, and to improve flavor acceptability *(52)*. The bioavailability of these additives is also higher than natural food sources of phosphorus. Unfortunately, there is no requirement that the phosphorus content of these products be listed on the food label. Therefore, patients with kidney disease must decipher the small print of the ingredient list to determine the suitability of a particular product. It is estimated that processed foods may contribute up to 1000 mg of phosphorus daily and this exceeds the dietary phosphorus recommendations for most patients with kidney disease. Renal dietitians continue to petition the Food and Drug Administration to make further changes in label requirements that will assist patients with CKD in making informed choices in the grocery store, especially with an increasing array of fortified and enhanced food products.

5. PROFESSIONAL ORGANIZATIONS SHAPING RENAL NUTRITION

5.1. The American Dietetic Association (ADA)

The ADA was organized at a meeting of 58 professionals in October of 1917 in Cleveland, Ohio. Present at that first meeting was Ruth Wheeler from the University of Illinois. During her term of office as president 8 years later, she founded the *Journal of the American Dietetic Association (JADA)*. Volume 2 of JADA was published in 1926 where the majority of articles were related to renal nutrition. The first article about renal nutrition was written by Dr. John Peters and discussed the use of a very low protein or no protein diet for only short period of time, and a moderate amount of salt "to spare the patient a certain amount of discomfort" *(53)*. In the spring of 1927, two dietitians, Florence H. Smith and Mary Whelan along with Dr. Keith Norman, reported on 165 cases encountered during the prior 3 years where a diet restricted in both salt and water proved effective in treating edema due to nephritis at the Mayo Clinic *(32)*. Another dietitian, Fairfax T. Proudfit reviewed a patient case with diabetes complications resulting from kidney disease *(54)*. By this time, there were approximately 1000 members of the ADA, and because this *Journal* was a membership benefit, these articles served to educate practitioners who probably did not have access to other educational materials about kidney disease. All aspects of kidney disease and its treatment have continued to be covered on a regular basis in *JADA*.

In the 1980s, the ADA Commission on Dietetic Registration began to examine areas of specialization in dietetic practice by reviewing published education materials and consistency in practice patterns *(55)*. Role Delineation Studies were conducted that showed that the care of patients with kidney disease required advanced knowledge and the performance of specific tasks unique to this patient population. These studies became the basis for Board Certification as a Specialist in Renal Nutrition. Renal dietitians were one of the first groups selected to participate in this process. This certification is granted in recognition of an applicant's documented practice experience and successful completion of a clinical problem simulation examination in the specialty area *(55)*. The first Board Certified Specialists in Renal Nutrition (CSR) were announced in early 1994.

5.2. Council on Renal Nutrition

In 1974, the National Kidney Foundation requested that a group of dietetic leaders to convene, and the Council on Renal Nutrition (CRN) was formed. The vision of this group was to develop and disseminate nutrition education materials for patients and to assist practitioners in obtaining the knowledge they needed to care for patients with kidney disease. St. Jeor reported on the first clinical meeting in 1975 *(56)*. These 3- to 5-day meetings continue to educate renal dietitians and others about the ever-changing field of renal nutrition and practice standards for CKD. In most parts of the United States, local CRN affiliates hold regular meetings for the purpose of sharing ideas and providing continuing education for its members. Many patient education tools, including renal diet books and renal cookbooks reflecting regional tastes, have been published over the years. The hallmark of each of these groups seems to be their willingness to share information, first-hand knowledge of patient care, and renal education materials with other colleagues.

The *CRN Quarterly* was the first publication of the CRN; it dealt with current research in renal nutrition and patient education. In 1990, the *CRN Quarterly* was succeeded by the *Journal of Renal Nutrition*, the premier journal which publishes renal nutrition research today. Another publication by the CRN is the *Pocket Guide to Nutrition Assessment of the Renal Patient (57)*. This concise book, now in its third edition, provides a foundation that the clinician can use to build their expertise.

5.3. Renal Dietitians Dietetic Practice Group

In 1978, the Renal Practice Group (RPG) of the ADA was formed. This organization deals with the professional concerns of the renal dietitian, including salary surveys, employment, legislation, and patient education. The RPG publishes a quarterly newsletter that features articles on renal nutrition and patient education.

The need for a uniform renal diet that could be used across the United States was identified as early as 1978, when the RPG was first formed. The project was not initiated at that time because of philosophical differences that existed among both renal dietitians and nephrologists. However, in 1987, the RPG and the CRN agreed to support jointly the development of the National Renal Diet. This long-awaited document was published in 1993 and originally consisted of six patient education booklets and a professional guide *(58)*. Another joint project of the RPG and CRN was the *Clinical Guide to Nutrition Care in End Stage in Renal Disease*, which was first published in 1987 *(59)*. The latest edition of this book was published in 2004 and reflects the evolution of nutrition management of patients with CKD *(60)*. These materials and publications have standardized renal nutrition care in the United States, and similar materials have been replicated in other countries.

6. SUMMARY

As the scope of practice for renal dietitians continues to expand, it is helpful to review how far the profession has advanced. But not surprisingly, it seems that renal dietitians have always been innovative yet practical when providing care for people with CKD.

REFERENCES

1. Osman AA. Original papers of Richard Bright on renal disease. New York: Oxford University Press, 1937.
2. Fishberg AM. Hypertension and Nephritis. Philadelphia: Lea & Fabiger, 1939.
3. Addis T. Glomerular Nephritis: Diagnosis and Treatment. New York: Macmillan Co, 1948.

4. Kempner W. Treatment of hypertensive vascular disease with rice diet. Am J Med 1948; 4: 545–577.
5. Borst JGG. Protein katabolism in uraemia: Effects of protein-free diet, infections and blood transfusions. Lancet 1948; i: 824–828.
6. Nebraska Dietetic Association. Nebraska Handbook of Diets, Normal and Therapeutic. Lincoln: Nebraska Hospital Association 1974; 142.
7. Rose WC, Wixom RL. The amino acid requirements of man. J Biol Chem 1955; 217: 997–1004.
8. Giordano C. Use of exogenous and endogenous urea for protein synthesis in normal and uremic subjects. J Lab Clin Med 1963; 62: 231–246.
9. Giovannetti S, Maggiore Q. A low-nitrogen diet with proteins of high biological value for severe chronic uraemia. Lancet 1964: 1: 1000–1003.
10. Berlyne G, Shaw AB, Nilwarangkur S. Dietary treatment of chronic renal failure: Experiences with modified Giovannetti diet. Nephron 1965: 2: 129–147.
11. Shaw AB et al. The treatment of chronic renal failure by a modified Giovanetti diet. Q J Med 1965; 34: 237–253.
12. Kopple JD et al. Controlled comparison of 20-gm and 40-gm protein diets in the treatment of chronic uremia. Am J Clin Nutr 1968; 21:553–564.
13. Kopple JD, Coburn J. Metabolic studies of low protein diets in uremia. I. Nitrogen and potassium. Medicine 1973; 52: 583–589.
14. Levin S, Winkelstein JA. Diet and infrequent peritoneal dialysis in uremia. N Engl J Med 1967; 277: 619–624.
15. Margie JD, et al. The Mayo Clinic Renal Diet Cookbook. New York: Western Publishing Co. 1974: 43.
16. Burton B. Current concepts of nutrition and diet in diseases of the kidney. I. J Am Diet Assoc 1974; 65: 623–626.
17. Burton B. Current concepts of nutrition and diet in diseases of the kidney. II. J Am Diet Assoc 1974; 65: 627–633.
18. Young G, Keogh J, Parson F. Plasma amino acids and protein levels in chronic renal failure and changes caused by oral supplements of essential amino acids. Clin Chim Acta 1975; 61: 205–213.
19. Alvestrand A, Ahlberg M, Furst P, Bergstrom J. Clinical experiences with amino acid and keto acid diets. Am J Clin Nutr 1980; 33: 1654–1659.
20. Hecking E, Port FK. Supplementation with essential amino acids or alpha-keto analogues in patients on long term dialysis. Int J Artif Organs 1980; 3: 127–129.
21. Klahr S, Levy AS, Beck GJ, Caggiula AW, Hunsicker L, Kusek JW, Stricker G, and MDRD Study Group. The effects of dietary protein restriction and blood pressure control on the progression of chronic renal disease. N Engl J Med 1994; 330: 877–884.
22. Massry S, Kopple J, Dietary Therapy in Renal Failure: Quick Reference to Clinical Nutrition. Lippincott, CT: Saunders Press 1979; 223–231.
23. Kopple J, Nutritional management. In: Massry S, Glassock R, eds. Textbook of Nephrology, Volume 2. Baltimore/London: Williams and Wilkens 1983; 8.3–8.12.
24. National Kidney Foundation. KDOQI clinical practice guidelines for nutrition in chronic renal failure. Am J Kidney Dis 2000; 35 (suppl 2): S1–S140.
25. Victor Sister M, Quality studies of therapeutic diets II. The nephritic diet. J Am Diet Assoc 1932; 8: 157–163.

26. Harum P. Nutritional management of the adult hemodialysis patient. In: Gillit D, Stover J, Spinozzi NS, eds. A Clinical Guide to Nutrition Care in End-Stage Renal Disease. Chicago: American Dietetic Association 1987; 35–36.
27. Slomowitz LA, Monteon FJ, Grosvenor M, Laidlaw SA, Kopple JD. Effect of energy intake on nutritional status in maintenance hemodialysis patients. Kidney Int 1989; 35: 704–711.
28. Henry RR. Diet therapy for hemodialysis patients. In: Bailey GL, ed. Hemodialysis Principles and Practice. New York: Academic Press 1972; 85–103.
29. Karp NR. Life in balance, nutrition in renal disease. Individual Study Kit Program. American Dietetic Association. 1969.
30. Tsaltas T. Dietetic management of uremic patients. I. Extraction of potassium from foods for uremic patients. Am J Clin Nutr 1969; 22: 490–493.
31. Burrowes JD, Ramer NJ. Removal of potassium from tuberous root vegetables by leaching. J Ren Nutr 2006; 16: 304–311.
32. Norman K, Smith FH, Whelan M. The therapeutic use of diets low in water and mineral content. J Am Diet Assoc 1927; 2: 233–245.
33. Norman K, Smith FH. The relationship of the diet to nephritis and hypertension. J Am Diet Assoc 1927; 2: 222–232.
34. Lashmet FH. Neutral diet in treatment of nephritic edema. J Am Diet Assoc 1931; 7: 224–227.
35. Enke G. The neutral diet. J Am Diet Assoc 1931; 7: 228–234.
36. McCann L. ed. Pocket Guide to Nutrition Assessment of the Renal Patient. 2nd ed. New York: National Kidney Foundation 1998; 3–5.
37. Lindholm T. Oral aluminum hydroxide in the treatment of hyperphosphatemia and acidosis in acute and chronic renal insufficiency. Acta Med Scand 1962; 7: 75–78.
38. Bailey GL, Vona JP. Pharmacodynamics in renal failure. In: Bailey GL ed. Hemodialysis, Principles and Practice. New York: Academic Press 1972; 123–125.
39. Alfey AC, et al. Syndrome of dyspraxia and multifocal seizures associated with chronic hemodialysis. Trans Am Soc Artif Intern Organs 1972; 18: 257–261.
40. Arze RS et al. Reversal of aluminum dialysis encephalopathy after desferrioxamine treatment. Lancet 1981; 2: 1116.
41. Gonella M, et al. Effects of high CaCo3 supplements on serum calcium and phosphorus in patients on regular hemodialysis treatment. Clin Nephrol 1985; 24: 147–150.
42. Mai ML, Emmett M, Sheikh MS, Schiller L, Fordtram JS. Calcium acetate, an effective phosphate binder in patients with renal failure. Kidney Int 1989; 36: 690–695.
43. Rosenbaum DP, Holmes-Farley SR, Mandeville WH, Pitruzzello M, Goldberg DI, Effect of RenaGel, a non-absorbable cross-linked, polymeric phosphate binder, on urinary phosphorus excretion in rats. Nephrol Dial Transplant 1997; 12: 961–964.
44. Harum P. Nutritional management of the adult hemodialysis patient. In: Gillet D, Stover J, Spinozzi NS, eds. A Clinical Guide to Nutrition Care in End-Stage Renal Disease. Chicago: American Dietetic Association 1987; 37–39.
45. Slatopolsky E, Weerts C, Thielan J, et al. Marked suppression of secondary hyperparathyroidism by intravenous administration of 1, 25-dihydroxy-choleccalciferol in uremic patients. J. Clin Invest, 1984; 74: 2136–2143.

46. Martin KJ, Gonzalez EA, Gellens M, Hamm LL, Abboud H, Lindberg J. 19-Nor-l-alpha-25 dihydroxyvitamin D2 (Paricalcitol) safely and effectively reduces the levels of intact parathyroid hormone in patients on hemodialysis. J. Am Soc Nephrol, 1998; 9: 1427–1432.
47. Potts JT. Parathyroid hormone: past and present. J Endocrin 2005; 187: 311–325.
48. Berson SA, Yalow RS, Aurbach GD, Potts, JT Jr. Immunoassay of bovine and human parathyroid hormone. Proc Natl Acad Sci USA 1963; 49: 613–617.
49. Gerskis A, Hutchinson AJ, Apostolou T, Freemont AJ, Billis A. Biochemical markers for non-invasive diagnosis of hyperparathyroid bone disease and adynamic bone disease on haemodialysis. Nephrol Dial Transplant 1996; 11:2430–2438.
50. Qi Q, Monier-Faugere MC, Geng Z, Malluche HH. Predictive value of serum parathyroid hormone levels for bone turnover in patients on chronic maintenance dialysis. Am J Kidney Dis 1995; 26: 622–631.
51. National Kidney Foundation. KDOQI clinical practice guidelines for bone metabolism and disease in chronic kidney disease. Am J Kidney Dis 2004; 42 (suppl 3): S1–S202.
52. Uribarri J, Calvo MS. Hidden sources of phosphorus in the typical American diet: Does it matter in nephrology? Semin Dial 2003; 16: 186–188.
53. Peters JP. The principles of diet control in nephritis with special reference to protein and salt restriction. J Am Diet Assoc 1926; 2: 137–146.
54. Proudfit, FT. A report of a case of diabetes with complications - nephrosis, nephritis hypertension. J Am Diet Assoc 1927; 2: 250–257.
55. Bogle ML, Balogun L, Cassell J, Catakis A, Holler HJ, Flynn C. Achieving excellence in dietetic practice: certification of specialists and advanced-level practicioners. J Am Diet Assoc 1993; 93: 149–150.
56. St. Jeor S. First annual meeting of the Council of Renal Nutrition. J Am Diet Assoc 1975; 67: 135–136.
57. McCann L. ed. Pocket Guide to Nutrition Assessment of the Renal Patient. New York: National Kidney Foundation 1993.
58. Renal Dietitians Dietetic Practice Group. National Renal Diet. Chicago: American Dietetic Association 1993.
59. Gillit D, Stover J, Spinozzi NS (eds). A Clinical Guide to Nutrition Care in End-Stage Renal Disease. Chicago: American Dietetic Association 1987.
60. Byham-Gray L, Wiesen K (eds). A Clinical Guide to Nutrition Care in Kidney Disease. Chicago: American Dietetic Association 2004.

3 The Changing Demographics of Chronic Kidney Disease in the US and Worldwide

Garabed Eknoyan

LEARNING OBJECTIVES

1. To discuss the changing demographics of patients diagnosed with chronic kidney disease.
2. To describe the role of evidence-based guidelines in the management of said patients.
3. To outline future directions in the pursuit of quality care in chronic kidney disease.

Summary

Since its emergence as a medical specialty in the 1960s, the focus of nephrology has changed dramatically in the past decade. What started as a discipline to provide dialysis and transplant care to patients with kidney failure in the 1970s has evolved to that of the detection and treatment of the much larger number of prevalent cases at earlier stages of the disease, in order to prevent its adverse outcomes (kidney failure, cardiovascular disease, and premature death) that occur in the course of gradual loss of kidney function. The changing demographics of chronic kidney disease (CKD) care has led to the adoption of a public health approach to the worldwide epidemic of CKD.

Key Words: Chronic kidney disease; prevalence of chronic kidney disease; dialysis; end stage renal disease; chronic renal failure; clinical practice guidelines.

From: *Nutrition and Health: Nutrition in Kidney Disease*
Edited by: L. D. Byham-Gray, J. D. Burrowes, and G. M. Chertow

1. INTRODUCTION

Whereas diseases of the kidney have tormented humans for ages and medical interest in their care and treatment can be traced to antiquity, nephrology as the science that deals with the kidneys, especially their function and diseases, is a relatively new discipline of medicine *(1,2)*. Historically, dialysis is the single most important development that focused attention on the kidneys and their failure to function *(3)*. What started in the 1960s as an exploratory effort to sustain life evolved in the 1970s into life saving renal replacement therapy (RRT) for patients whose kidney disease had progressed to kidney failure *(4)*. Since the inception of the Medicare End Stage Renal Disease (ESRD) program in 1972, the number of patients on maintenance dialysis has increased in the United States and the other parts of the world *(5)*. Currently, well over 1.3 million individuals in the world are alive on maintenance dialysis and about one-third as many with a kidney transplant (Table 1). With reported annual growth rates of about 5–8% per year, it is projected that the number of dialysis patients worldwide will double in the next decade *(6)*. What has become evident recently is the much larger number of individuals with kidney disease who are not on dialysis (~50–100 fold greater than the number of ESRD patients on dialysis), a condition that affects their quality of life (Fig. 1) and whose proper care will significantly impact the healthcare system worldwide *(6–8)*.

Table 1
Treated end stage renal failure at year end in 2004

Region	*ESRD n(pmp)*	*Dialysis n (pmp)*	*Transplant n (pmp)*
North America	492 (1505)	337 (1030)	154 (470)
Europe	473 (585)	324 (400)	149 (185)
Japan	261 (2045)	248 (1945)	13 (100)
Asia (excluding Japan)	237 (70)	196 (60)	41 (10)
Latin America	205 (380)	170 (320)	35 (65)
Africa	61 (70)	57 (65)	50 (5)
Middle East	54 (190)	39 (140)	15 (55)
Total	1783 (280)	1371 (215)	412 (65)

ESRD, end stage renal failure; *n*, number of patients in thousands; pmp, prevalence in patients per million population. (reproduced with modification from Grassman et al. *(5)*).

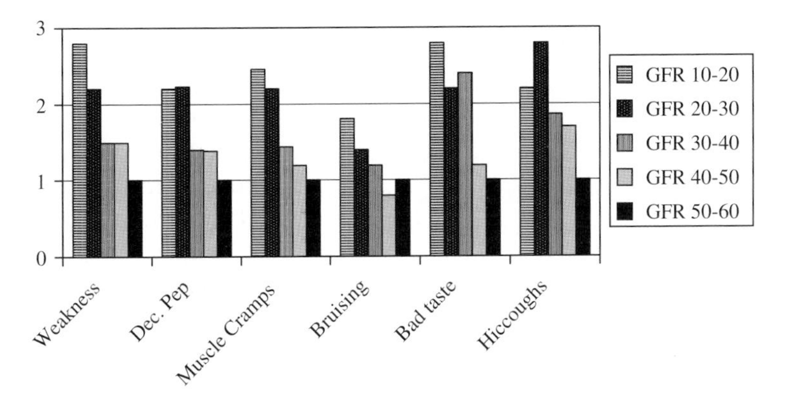

Fig. 1. Odds of having symptoms that affect quality of life and well being in 1284 patients in the MDRD study. Controlled for age, gender, race, education, income, and smoking. GFR, glomerular filtration rate in ml/min/1.73 m^2. Reproduced with permission from Rocco et al. *(7)*.

2. THE PROBLEM: A HISTORICAL PERSPECTIVE

After the widespread availability of dialysis, it became apparent in the late 1980s that the quality of life of dialyzed patients was far from adequate and their annual mortality rate (about 20% in the United States) was too high *(9,10)*. As it turned out, a principal reason for these poor outcomes was the heavy burden of comorbid conditions with which patients were being initiated on dialysis rather than the adequacy of dialysis *(11,12)*. In fact, it has been shown that the state of health in which individuals with kidney failure present to a nephrologist is a major determinant of their outcome on RRT, be it dialysis or kidney transplantation. *(13)*.

Well after maintenance dialysis became widely available and into the closing years of the past century, nephrology had categorized kidney disease by cause (glomerulonephritis, obstructive nephropathy, lupus nephritis, etc.). This approach is clearly useful when the diagnosis facilitates the specific treatment of a given disease. Unfortunately, in many cases, it is not possible to determine the exact cause of the kidney disease and often there are no specific therapies that would reverse the kidney injury due to a disease. Moreover, there are indications that the number of ESRD patients whose kidney failure is due to these traditional kidney diseases is diminishing, whereas those due to diabetes and hypertension is constantly increasing, especially in the elderly *(10,14)*. In the vast majority of these individuals, the kidney disease is asymptomatic, certainly in its early stages and well after more than half of normal kidney function is lost. Also, it has become increasingly evident that the loss of kidney function and the systemic complications that develop during the course of kidney disease (anemia, hypertension, bone disease, and cardiovascular disease) are uniform and independent of the underlying etiology of the kidney disease. Despite this realization, the situation remained confusing for quite some time as varying terminologies were used to identify and describe these patients, originating with the now abandoned historical "Bright's disease" to the still used horrible term of "pre-dialysis" patients. The cacophony of terms that came into use to describe these patients (chronic renal failure, chronic renal disease, chronic renal insufficiency, etc.) coupled with the use of the rather insensitive serum creatinine as a measure of kidney function accounts for the non-descript and confusing situation that prevailed in the first years of the current millennium *(15,16)*.

3. DEFINING THE PROBLEM

In 2002, the National Kidney Foundation's Kidney Disease Outcomes Quality Initiative (KDOQI) guidelines proposed a new definition and classification of chronic kidney disease (CKD). For the first time, these guidelines provided a uniform definition of CKD (Table 2) and a classification system based on the level of estimated glomerular filtration rate (GFR) (Table 3), rather than just that of the traditional serum creatinine level from which the estimated GFR is calculated *(17)*. Using representative databases and evidence-based analysis of the literature, these guidelines documented the increased number of complications associated with declining GFR, and the increased rates of morbidity and mortality at each of the proposed five

Table 2
Definition of Chromic Kidney Disease

Structural or functional abnormalities of the kidneys for ≥ 3 months, as manifested by
Kidney damage, with or without decreased GFR, as defined by
Pathologic abnormalities
Markers of kidney damage, including abnormalities in imaging test, the composition of the blood or urine
GFR <60 ml/min/1.73 m^2, with or without kidney damage

GFR, glomerular filtration rate.

Table 3
Stages of Chronic Kidney Disease

Stage	*Description*	*GFR (ml/min/1.73 m²)*
1	Kidney damage with normal or ⇑ GFR	≥ 90
2	Kidney damage with mild ⇓ GFR	60–89
3	Moderate ⇓ GFR	30–59
4	Severe ⇓ GFR	15–29
5	Kidney failure	<15 or dialysis

GFR, glomerular filtration rate.

stages of CKD. The entire guidelines and a user-friendly calculator of the estimated GFR are available online at http://www.kdoqi.org.

The KDOQI guidelines also proposed a clinical action plan for each stage of CKD based on accrued evidence that (i) the adverse outcomes of CKD (kidney failure, cardiovascular disease, and premature death) can be prevented or delayed; (ii) treatment of earlier stages of CKD is effective in reducing the rate of progression to kidney failure and preventing the systemic complications that develop in the course of progressive loss of kidney function; and (iii) initiation of treatment of risk factors (anemia, hypertension, and bone mineral metabolism) for cardiovascular disease at earlier stages of CKD can be effective in reducing this leading cause of mortality and morbidity of individuals with reduced kidney function *(6,17)*.

Apart from the clinical utility of the proposed action plan for the management of CKD in the KDOQI guidelines, the classification of CKD based on the estimated GFR has provided a stimulus for research on the prevalence of CKD and facilitated the performance of comparative studies on the subject. Published results since then have confirmed that (i) the increasing number of patients who develop kidney failure and require replacement therapy have several comorbidities—notably cardiovascular disease that accounts for the high mortality rates of these patients *(10,11)*; (ii) there is a substantially higher number of individuals at earlier stages of CKD in the general population *(8)*; (iii) the prevalence of CKD is high and rising because of the increasing prevalence of diabetes and hypertension, the principal underlying causes of most cases of CKD *(6,14)*; (iv) the number of patients who die at each stage of CKD is much higher than those who progress to kidney failure *(18)*; (v) the mortality rate due to cardiovascular disease is two- to four-fold higher in individuals with CKD compared to those with normal kidney function *(19)*; (vi) when compared with individuals with normal kidney function, those with CKD have increased over all non-cardiovascular mortality, as well as cause specific mortality due to pulmonary disease, infection and cancer *(20)*; (vii) for most patients with CKD, the risk of death from cardiovascular or non-cardiovascular disease far outweigh the risk of progression to kidney failure, with reported estimates of all cause mortality rates ranging from 25 to 50%, depending on the level of kidney function *(6,17)*; (viii) as clearly indicated in the CKD classification guidelines, all surviving patients with CKD do not progress through the stages of CKD, even in those with advanced stages 3–5 of CKD some 20–30% remain stable for years *(17,21,22)*; (ix) the ones who are likely, but not invariably, to progress are those with a GFR of <60 ml/min/1.73 m^2, that is, stages 3–5 of CKD *(6,23)*; and (x) the vast majority of those who do progress to kidney failure do so slowly, albeit at variable rates *(6,17,21–23)*.

This latter point accounts to some extent for the increasingly older age at which patients now present for RRT *(9)*. It also underlines the fact that the years it takes to reach kidney failure provides for unique opportunities to properly treat comorbidities, to retard the progression of kidney disease, and to detect, treat, and prevent the systemic complications of CKD at every stage of the disease that develop in the course of gradual loss of kidney function *(17)*.

4. AN EPIDEMIC OF CKD

The original data presented in the KDOQI guidelines, based on analysis of participants in the Third National Health and Nutrition Examination Survey (NHANES III), estimated that 11% (19.2 million) of the adult US population has CKD, 4.7% (8.3 million) of whom are in stages 3–5 of CKD, with a much higher prevalence in the elderly, diabetics and hypertensives *(8,17)*. Subsequent reports from other countries have confirmed these estimates. Studies in the United Kingdom and the Netherlands have shown an estimated prevalence of stages 3–5 CKD of 4.9% and 5.3% of the general population, respectively *(24,25)*, which is just about the same or quite similar to those in the United States. The reported range of CKD prevalence varies from a high in Australia of 11.2% to a low in China of 2.53% *(26,27)*. Whereas differences in study cohort, ethnicity and methods of data collection account for the variable figures reported, it is evident that the prevalence of CKD is a worldwide health problem of epidemic proportion and significant repercussions. Most disconcerting are the lost opportunities for the detection of such individuals and their treatment to prevent the risk for CVD and other comorbidities for which CKD is an identifiable and potentially remediable risk factor. Equally alarming is the considerable variability in the application of these measures in those who have been identified with CKD, resulting in further lost opportunities to improve outcomes and making lives better for patients with CKD worldwide *(6,28)*.

5. REGIONAL SOLUTIONS TO THE PROBLEM

The approach to resolve these problems has been the adoption of a public health model to CKD (Fig. 2) and the development of evidence-based clinical practice guidelines (CPG) to guide their management using algorithms. The need for guidelines stems from the current information overload, increased number of new and effective treatments and diagnostic tools, and a growing body of outcomes research. This makes it difficult for the busy practitioner desiring to practice evidence-based medicine to keep abreast of the exponential increase of a rather disparate body of information. CPGs draw on systematic reviews of the literature and make specific recommendations to assist practitioners and patients in making decisions about appropriate healthcare in specific conditions. Rigorously developed evidence-based CPGs, when implemented, can reduce variability of care, improve patient outcomes, and ameliorate the efficiency of healthcare *(29)*.

The practical specificity of guideline statements facilitates their translation into clinical practice and differentiates them from other evidence-based approaches such as meta-analyses and systematic reviews, which distill and analyze the evidence but usually do not make

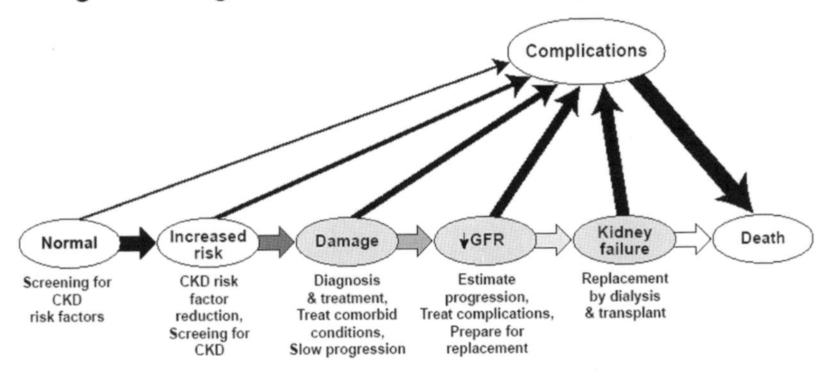

Fig. 2. Stages in progression of chronic kidney disease (CKD) and therapeutic strategies. GFR, glomerular filtration rate.

practical recommendations, and leave it to chance for the practitioner to integrate them into clinical practice. In fact, the very process of developing the evidence base of CPGs depends on meta-analysis and systematic reviews *(30)*. Essentially, it is the actionable recommendations of CPGs that make them now one of the best tools available to practice using evidence-based medicine and hence the statement that, "the implementation of rigorously developed practice guidelines can lead to even greater improvements in patient care than the introduction of new technologies" *(29)*.

Another value of guidelines is that by establishing targets of therapy and providing actionable items, they allow for the cooperative care of patients with CKD by specialists and primary care providers (Fig. 3). The care of patients with CKD by nephrologists only will overwhelm the manpower resources of any country. The algorithms provided in guidelines allow the cooperative utilization of the entire manpower of healthcare providers (specialists, primary care physicians, dietitians, nurse clinicians, social workers, and physician assistants) as well as those of informed patients in providing appropriate care, without burdening any one segment of the system.

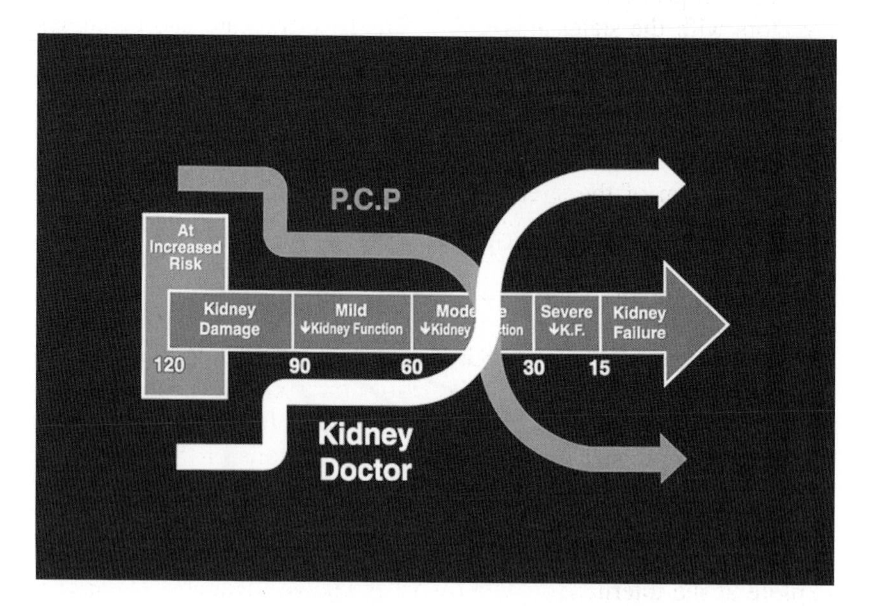

Fig. 3. The use of guideline recommendations and targets for the co-management of chronic kidney disease patients at its various stages by primary care providers (P.C.P.) and kidney doctors.

6. A GLOBAL APPROACH TO THE PROBLEM

Over the past decade, a number of national organizations have developed guidelines for the care of CKD patients. These have been instrumental in improving and facilitating the care of these patients. An important next step in the approach to CKD occurred when it became apparent to those developing nephrology guidelines that there was a need for improving international cooperation in the development, dissemination and implementation of CPGs to achieve these goals. The rationale for a global approach to the CKD epidemic is self-evident. CKD and the risks it engenders are not exclusive to countries with guidelines. The complications and problems encountered by those afflicted with kidney disease are universal. The science and evidence-based care of these complications and problems are also universal and thus independent of geographic location or national borders. With the increasing prevalence of kidney disease worldwide, strategies to improve outcomes will require a global effort directed at earlier stages of CKD. As such, there is a clear need to develop an evidence-based public health approach to the global epidemic of kidney disease *(28)*.

It was on this basis that Kidney Disease: Improving Global Outcomes (KDIGO) was established in 2003 as an independently incorporated non-profit foundation governed by an international board of directors with the stated mission to "improve the care and outcomes of kidney disease patients worldwide through promoting coordination, collaboration and integration of initiatives to develop and implement clinical practice guidelines" *(28)*. Since its establishment, KDIGO has launched several initiatives including (i) the development of a uniform grading system of the strength of the evidence and recommendation of guidelines in nephrology; (ii) the establishment of an interactive Internet database of available nephrology guidelines that allows for the comparison of recommended therapeutic targets in the posted guidelines, together with the rationale for their differences; (iii) the development of three new CPGs on Hepatitis C and CKD, bone and mineral metabolism and care of the kidney transplant recipient; and (iv) the convening of international controversies conferences on clinically relevant controversial issues in nephrology. Details of the initiatives, positions statements and guidelines of KDIGO are available online at http://www.kdigo.org.

The controversies conferences held by KDIGO are designed to explore at the international level what is known (available evidence), what can be done with what is known (recommendations and guidelines) and what do we need to know (research questions). The first of these on the definition and classification of CKD held in September

2004 recommended to (i) accept the definition and classification of CKD proposed by KDOQI; (ii) consider all kidney transplant recipients as having CKD and classified by its staging system with the added suffix of T for transplant (stage 3-T, stage 4-T, etc); (iii) identify dialysis patients by the suffix D for dialysis, independent of the stage of their CKD (stage 5-D); and (iv) report routinely the estimated GFR based on standardized serum creatinine measurement. The proceedings of this conference and its full recommendations have been published as a position statement of KDIGO *(31)*. The definition and classification of CKD is currently being updated to include a framework in which to integrate diagnosis and associations of CKD with other chronic diseases, such as cardiovascular disease, cancer, and selected infectious diseases.

The second controversies conference held in September 2005, on the definition, evaluation and classification of renal osteodystrophy, recommended that the term renal osteodystrophy be used exclusively to define the alterations in bone morphology due to CKD and that the term CKD-Mineral and Bone Disease (CKD-MBD) be used to describe the clinical syndrome that develops as a systemic disorder of mineral and bone metabolism due to CKD manifested by any one or a combination of the following: (i) abnormalities of calcium, phosphorus, parathyroid hormone or vitamin D metabolism; (ii) abnormalities in bone turnover, mineralization, volume, strength or linear growth, and (iii) vascular or other soft tissue calcification. The proceedings of this conference and its full recommendations have been published *(32)* and are posted on the web at http://www.kdigo.org.

7. SUMMARY

The reported estimates of the increasing burden of CKD present several pitfalls. The first is that associated with problems in the measurement of serum creatinine that will require standardization to avoid inaccuracies and variability *(33,34)*. The second is in the formulae used to estimate the GFR. The most commonly used, so-called Modification of Diet in Renal Disease (MDRD), formula has been validated in the United States and is most robust in the lower ranges of GFR (<60 ml/min/1.73 m^2). Its broader application will require further evaluation and the incorporation of confounding factors such as race, gender, and muscle mass *(6,16,34)*. Both of these problems are currently under active and rigorous examination and their solution should be forthcoming in the near future. In the meantime, the KDIGO recommendation of reporting GFR values below 60 ml/min/1.73 m^2 remains a reasonable alternative that would help identify those with the greatest need for special attention *(31)*.

By the same token, the adoption of a global approach to CKD based on the use of CPGs presents several challenges. One of which is the applicability of global guidelines to regions with variable resources and legislative agendas. It is felt that the development of guidelines based on the care and outcomes of patients with CKD would provide a core set of recommendations. The subsequent step-wise adoption of these recommendations tailored by local groups to available regional resources will make them accessible to everyone committed to improved outcomes.

With due diligence, focused perseverance and evidence-based solutions, all these are surmountable obstacles. That would certainly make the future care of CKD patients better than what it has been in the past.

REFERENCES

1. Eknoyan G. Rufus of Ephesus and his "Diseases of the Kidneys". Nephron 2002;91:383–390.
2. Schreiner GE. Evolution of nephrology. The cauldron of its organizations. Am J Nephrol 1999;19:295–303.
3. Peitzman SJ. Chronic dialysis and dialysis doctors in the United States: A nephrologist-historian's perspective. Semin Dial 2001;14:200–208.
4. Schreiner GE. How end-stage renal disease (ESRD)-medicare developed. Am J Kidney Dis 2000;35(Suppl 1):S37–S44.
5. Grassman A, Gioberge S, Moeller S, et al. ESRD patients in 2004: Global overview of patient numbers, treatment modalities and associated trends. Nephrol Dial Transplant 2005;20:2587–2593.
6. El-Nahas M. The global challenge of chronic kidney disease. Kidney Int 2005;68:2918–2929.
7. Rocco MV, Gassman JJ, Wang SR, Kaplan RM. Cross-sectional study of the quality of life and symptoms in chronic renal disease patients. The Modification of Diet in Renal Disease Study. Am J Kidney Dis 1997;29:888–896.
8. Coresh J, Astor BC, Greene T, et al. Prevalence of CKD and decreased kidney function in the adult US population: Third National Health and Nutrition Survey. Am J Kidney Dis 2003;41:1–12.
9. Kutner NG, Zhang R, Barnhart H, Collins AJ. Health status and quality of life reported by incident patients after 1 year on hemodialysis or peritoneal dialysis. Nephrol Dial Transplant 2005;20:2159–2167.
10. United States Renal Data Systems. Accessed March 1, 2007. Available at http://www.usrds.org/reference.htm.
11. Eknoyan G, Beck GJ, Cheung AK, et al. Effect of dialysis dose and membrane flux in maintenance hemodialysis. N Engl J Med 2002;347:2010–2019.
12. Eknoyan G. Adequacy of dialysis. Problems and challenges. Semin Nephrol 2005;25:67–69.

13. Arora P, Obrador GT, Ruthazer R, et al. Prevalence, predictors and consequences of late nephrology referral at a tertiary center. J Am Soc Nephrol 1999;6: 1281–1285.
14. Gansevoort RT, van der Heig, Stegeman CA, et al. Trends in the incidence of treated end-stage renal failure in the Netherlands: Hope for the future. Kidney Int 2004;Suppl 92:7–10.
15. Hsu CY, Chertow GM. Chronic renal confusion: Insufficiency, failure, dysfunction or disease. Am J Kidney Dis 2000;36:415–418.
16. Levin A. The advantage of a uniform terminology and staging for chronic kidney disease (CKD). Nephrol Dial Transplant 2003;18:1446–1451.
17. National Kidney Foundation. KDOQI clinical practice guidelines for chronic kidney disease: Evaluation, classification and stratification. Am J Kidney Dis 2002;39(Suppl 1):S1–S266.
18. Go AS, Chertow GM, Fan D. Chronic kidney disease and the risks of death, cardiovascular events and hospitalization. N Engl J Med 2005;351:1296–1305.
19. Levey AS, Beto JA, Coronado BE, et al. Controlling the epidemic of cardiovascular disease in chronic renal disease. Am J Kidney Dis 1998;32:853–906.
20. Fried LF, Katz R, Sarnak MJ, et al. Kidney function as a predictor of noncardiovascular mortality. J Am Soc Nephrol 2005;16:3728–3735.
21. Klahr S, Levey AS, Beck GJ, et al. The effects of dietary protein restriction and blood pressure control on the progression of chronic renal disease: Modification of Diet in Renal Disease Study Group. New Engl J Med 1994;330:877–884.
22. Coresh J, Byrd-Holt D, Astor BC, et al. Chronic kidney disease awareness, prevalence, and trends among US adults, 1999–2000. J Am Soc Nephrol 2005;16: 180–188.
23. Eriksen BO, Ingebretsen OC. The progression of chronic kidney disease: A 10-year population-based study of the effects of gender and age. Kidney Int 2006;69:375–382.
24. Anandarajah S, Tai T, deLusignan S, et al. The validity of searching routinely collected general practice data to identify patients with chronic kidney disease (CKD): A manual review of 500 medical records. Nephrol Dial Transplant 2005;20:2089–2096.
25. De Zeeuw D, Hillege HL, de Jong PE. The kidney as a cardiovascular risk marker and a new target for therapy. Kidney Int 2005;Suppl 98:25–29.
26. Chadban SJ, Briganti EM, Kerr PG, et al. Prevalence of kidney damage in Australian adults: The AusDiab kidney study. J Am Soc Nephrol 2003;14: S131–S138.
27. Chen J, Wildman RP, Gu D, et al. Prevalence of decreased kidney function in Chinese adults aged 35–74 years. Kidney Int 2005;68:2837–2845.
28. Eknoyan G, Lameire N, Barsoum R, et al. The burden of kidney disease: Improving global outcomes. Kidney Int 2004;66:1310–1314.
29. Weingarten S. Using practice guideline compendiums to provide better care. Ann Intern Med 1999;130:454–458.
30. Grimshaw JM, Russell IT. Effect of clinical guidelines on medical practice: A systematic review of rigorous evaluations. Lancet 1993;342:1317–1322.
31. Levey AS, Eckardt KU, Tsukamoto Y, et al. Definition and classification of chronic kidney disease: A position statement from Kidney Disease: Improving Global Outcomes (KDIGO). Kidney Int 2005;67:2089–2100.

32. Moe S, Drueke T, Cunningham et al. Definition, evaluation and classification of renal osteodystrophy. A position statement from KDIGO. Kidney Int 2006;69:1945–1953.
33. Myers GL, Miller WG, Coresh J, et al. Recommendations for improving serum creatinine measurement: A report from the laboratory working group of the National Kidney Disease Education Program. Clin Chem 2006;52:5–18.
34. VanBiesen W, Vanholder R, Veys N, et al. The importance of standardization of creatinine in the implementation of guidelines and recommendations for CKD: Implications for CKD management programmes. Nephrol Dial Transpl 2006;21:77–83.

4 Nutritional Assessment in Chronic Kidney Disease

Wm. Cameron Chumlea, David B. Cockram, Johanna T. Dwyer, Haewook Han, and Mary Pat Kelly

LEARNING OBJECTIVES

1. Discuss the role of biochemical parameters as a component of a comprehensive nutritional assessment.
2. Identify the biochemical parameters and the recommended frequency of measurement suggested in the Kidney Disease Outcome Quality Initiative nutrition guidelines for routine, confirmatory, and screening testing.
3. List the strengths and weaknesses of various biochemical tests in the chronic kidney disease population.
4. Discuss the major dietary and nutrient challenges that patients face at each of the five stages of chronic kidney disease that will guide the choice of dietary assessment method.
5. Use dietary intake data to counsel chronic kidney disease patients according to the Kidney Disease Outcome Quality Initiative guidelines.
6. Describe the different methods of body composition assessment that are applicable to adults with chronic kidney disease such as dual energy X-ray absorptiometry, bioelectrical impedance, and total body water.

From: *Nutrition and Health: Nutrition in Kidney Disease*
Edited by: L. D. Byham-Gray, J. D. Burrowes, and G. M. Chertow

7. State the anthropometric measurements used in assessing nutrition status in the chronic kidney disease population.
8. Understand the limitations and sources of error in evaluating and using body composition in the chronic kidney disease population.
9. Explore physiological mechanisms responsible for lesion and functional deficits.
10. Encourage a deeper nutritional thought process that considers the role of nutrient intake/disposition, drug/nutrient interaction, medical comorbidity, and nutritional cost of therapeutic interventions in the evolution of patient-specific nutrient intake plans.

Summary

Comprehensive nutrition assessments are comprised of an evaluation of the individual's body composition, biochemical tests, dietary intake and habits, and clinical profile. Integration of these methodologies is used to guide appropriate medical nutrition therapy and to monitor responses to therapeutic nutrition interventions. This chapter reviews (i) common biochemical tests used for assessment of nutritional and inflammatory status; (ii) dietary intake methodologies that provide the data needed to improve nutritional status and quality of life among patients with chronic kidney disease (CKD); (iii) body composition assessment methods useful in the CKD population which can facilitate the prescription and monitoring of appropriate clinical and nutritional therapies; and (iv) nutrition physical assessment techniques with illustrations of physical manifestations of nutrient deficiencies and excesses with a specific focus on patients with CKD.

Key Words: Nutrition assessment; biochemical tests; dietary assessment; body composition; anthropometry; nutrient-based lesions; nutrition physical examination; physical findings.

1. INTRODUCTION

Suboptimal nutritional status is common in people in the latter stages of chronic kidney disease (CKD) and is associated with increased morbidity and mortality and higher health care costs. Data from the Modification of Diet in Renal Disease (MDRD) study demonstrated that protein-energy nutritional status deteriorates as glomerular filtration rate declines. Frank protein-energy malnutrition (PEM) prior to initiation of dialysis was fairly rare, but evidence of deteriorating nutritional intake was common as kidney function declined. Among people receiving dialysis, depending on the nutritional parameters chosen and thresholds used for identifying deficits, the literature reports 18–75% of people receiving maintenance dialysis have evidence of protein deficits *(1)*. Measures of nutritional status deteriorate with time, even in the presence of excellent dialytic care and close nutritional monitoring and interventions *(2)*. Nutritional status is an important predictor of increased hospitalization rate, hospital days, and mortality. Therefore, assessing and optimizing nutritional status is important to improve patients' quality of life (QOL), improve clinical outcome, and help control cost of care.

The role of nutrition assessment is to describe a patient's nutritional status from a variety of perspectives to guide individualization of nutrition therapies. Nutrition assessments are comprised of several evaluations, including anthropometric measurements, biochemical testing, a clinical profile, and an evaluation of dietary adequacy and habits. The clinician integrates these data to determine adequacy of acute and chronic nutritional intake in relation to requirements, to determine the need for nutrition therapy and to guide and monitor responses to therapeutic nutrition interventions.

Nutrition assessments, and the optimal assessment methodologies, vary depending on the objective of the assessment. For screening purposes, simple evaluations that require minimal patient involvement and relatively little training for the person making the assessment can be used to prioritize patients for a more in-depth assessment. At the other extreme, research assessment tools typically provide very high precision at the cellular or elemental level, but tend to be more invasive and require more patient cooperation and highly trained dietetic support. Clinically useful nutrition assessment techniques balance precision, staff time, and patient effort required to provide appropriate levels of detail. The objective of nutrition assessment, and the availability of appropriate population-specific reference data, needs to be understood to select appropriate techniques.

This chapter includes discussions of biochemical, dietary intake, body composition, and clinical nutritional assessment techniques for use in screening, for routine clinical, and for research purposes. A number of common biochemical nutrition assessment parameters are abnormal in many people with CKD. Biochemical testing has the advantages of being objective, requiring minimal patient cooperation, and many measures are commonly available to clinicians. Currently, biochemical testing is the only clinically practical way to help identify the presence of malnutrition, inflammation, or both in people with CKD. Dietary intake assessment tools can be used for screening, to describe intake over a specific period of time, or to characterize usual intake. Importantly, because nutrient requirements and common nutritional challenges change over the progression of CKD, dietary records and recalls can be useful tools to help determine the effectiveness of nutritional counseling and provide specific guidance for patients. The objectives of anthropometric assessment typically are quantification of energy reserves in the forms of fat and muscle mass. Clinically useful tools, such as upper arm skinfold and circumference measurements and weight and weight change, provide practical means to monitor adequacy of nutritional intake longitudinally. Finally, the section on clinical assessment provides an in-depth review of nutrition physical examination (NPE) techniques and illustrates a number of physical manifestations of nutrient deficiencies and excesses with a specific focus on patients with CKD. NPE is a newer practice area for advanced level clinicians. However, when clinical observations are placed in the context of a patient's overall condition and with biochemical confirmation, they can be useful to identify and explain the presence of nutrition-related lesions.

2. BIOCHEMICAL ASSESSMENT OF NUTRITIONAL STATUS

Biochemical assessment offers the advantages of being readily available in most clinical settings, it is objective, and it requires only minimal patient cooperation. CKD and dialysis procedures each can influence nutritional status, limiting nutrient intake due to anorexia, dietary restrictions, socioeconomic constraints, or impaired gastrointestinal (GI) motility. In addition, CKD also exerts an indirect effect on nutritional status by increasing requirements and impairing the body's ability to down-regulate resting energy expenditure (REE) and protein turnover. Biochemical testing provides important insights into adequacy of protein and energy intake, the presence of inflammatory or oxidative stress, and nutritional adequacy over time.

The Kidney Disease Outcome Quality Initiative (KDOQI) nutrition practice guidelines recommended the use of a panel of nutritional parameters because no single index comprehensively summarizes all aspects of nutritional status *(3)*. The KDOQI nutrition guidelines recommend a battery of anthropometric, clinical, and dietary assessments in addition to the biochemical parameters listed in Table 1. The work group divided nutritional markers into tests to be done routinely, confirmatory tests to be done as needed, and screening tests requiring further confirmation. In addition to providing data for assessment of protein-energy nutritional status, other biochemical parameters frequently are of importance in the dietary and medical management

Table 1
KDOQI Biochemical Testing Regimen in Stages 2–5 CKD *(3)*

Marker type	*Marker*	*Measurement frequency*
Routine	Predialysis serum albumin	Monthly in maintenance dialysis (every 1–3 months in CKD)
	nPNA	Monthly (every 3–4 months for peritoneal dialysis; every 3–4 months in CKD)
Confirmatory	Predialysis serum prealbumin	As needed
Screening	Predialysis serum creatinine	As needed
	Predialysis serum cholesterol	As needed
	Creatinine index	As needed

CKD, Chronic kidney disease; nPNA, normalized protein equivalent of nitrogen appearance.

of patients receiving dialysis, such as glucose, glycated hemoglobin (A1C), potassium, phosphorus, and calcium, but are beyond the scope of this chapter.

2.1. Protein-Energy Nutritional Status

Assessment of protein-energy nutritional status is one of the most common applications of biochemical assessment. The most commonly used proteins for this purpose are serum albumin, prealbumin, and transferrin. Other biochemical parameters are useful as screening tools: serum creatinine, cholesterol, and bicarbonate.

Serum proteins can be broadly divided into two categories, negative and positive acute phase proteins. Circulating levels of negative acute phase proteins, such as serum albumin and prealbumin, are at their highest in well-nourished, unstressed individuals and decline in the presence of inadequate nutrition, inflammatory stress, or both. In contrast, positive acute phase proteins, such as serum C-reactive protein (CRP), normally circulate at very low levels and rise dramatically in the presence of inflammatory stress. Because both nutrition and inflammation independently influence each of these nutritional markers and respond to nutritional interventions, both must be considered when interpreting serum protein levels as nutritional or inflammatory parameters.

The expected frequency of laboratory testing should be considered in selecting nutritional markers. Where nutritional assessments are performed monthly in generally stable individuals, longer half-life nutritional parameters are logical choices. In contrast, for patients undergoing acute changes in nutritional and inflammatory state, addition of short half-life nutritional and inflammatory markers allows adjustment of nutritional interventions based on acute patient response.

2.1.1. Serum Albumin

Albumin is the most abundant protein in the blood, readily available on most biochemistry panels, and is therefore widely used as a nutritional and inflammatory marker. The half-life of serum albumin is approximately 20 days, making it a good tool for use in monthly nutritional assessments but relatively unresponsive to acute changes in nutritional or inflammatory status.

Interpretation of albumin levels is challenging in people with CKD. Both modifiable and nonmodifiable predictors of serum albumin have been identified in people with CKD. Older age, female sex, white race, presence of several chronic diseases (chronic obstructive pulmonary disease, peripheral vascular disease, diabetes mellitus, and cancer), and being the first year of dialysis are nonmodifiable factors correlated with hypoalbuminemia. Modifiable factors associated with improved albumin include smoking cessation, use of arteriovenous fistulas, or biocompatible dialysis membranes *(4)*. Longitudinal analyses of serum albumin show a decline in serum albumin in the months immediately preceding death and an improvement during the first year of dialysis *(4)*.

Serum albumin is extensively distributed in both intravascular and extravascular compartments and can be redistributed into the extravascular space, resulting in hypoalbuminemia. Serum albumin levels are also reduced in patients with metabolic acidosis. Nutrient intake, particularly protein intake, is one of several factors determining serum albumin levels *(5)*. However, serum albumin levels are maintained in otherwise healthy individuals until late in the course of protein-energy malnutrition (PEM) in the absence of underlying inflammatory stress. Serum albumin is also preserved with chronically low food intake because resting energy expenditure (REE) is simultaneously down-regulated. However, systemic inflammation inhibits this normal adaptation to protein-energy deficits. In addition, inflammation also both inhibits albumin synthesis and increases its fractional catabolic rate *(5,6)*. Thus, dietary protein and inflammation have separate and opposing influences on serum albumin. Interpretation of hypoalbuminemia must be made in the context of both nutrient intake and presence of inflammation *(6)*.

Subnormal serum albumin levels (<4.0 g/dL) have long been shown to predict both all-cause and cardiovascular mortality in people receiving maintenance dialysis *(5,7)*. Without additional information to help differentiate between nutritional deficits, inflammation, or a combination of the two, hypoalbuminemia should not be presumed to be nutrition-related. The view that hypoalbuminemia is not primarily due to nutritional deficits appears to be supported by the limited success of enteral or parenteral intervention trials to effectively correct hypoalbuminemia *(8)*. Others, however, have concluded that although both inflammation and malnutrition frequently coexist, subnormal levels of serum albumin and prealbumin are principally reflective of nutritional inadequacy *(9)*.

2.1.2. Serum Prealbumin

Serum prealbumin, like albumin, is a negative acute phase protein. Prealbumin has a half-life of about 2 days and is therefore very responsive to recent events, especially calorie and protein deficits. Thus, for patients with acute illnesses or following initiation of nutritional interventions, prealbumin can be a useful, early directional indicator of changes in nutritional and inflammatory status.

Serum prealbumin, and a decline in serum prealbumin over time, predicts all-cause mortality independently of inflammation (based on CRP) *(9)*. Predialysis serum prealbumin is directly correlated with other biochemical nutritional markers (serum albumin, creatinine, and cholesterol) as well as predialysis body weight and bioelectrical impedance (BIA)-derived reactance, body cell mass, body water, and phase angle *(10)*. Prealbumin was predictive of hospitalization over the next 12 months in a univariate analyses, though when included in a multivariate analysis, prealbumin was no longer significant *(11)*.

When the KDOQI nutrition guidelines were drafted, there was insufficient published data to conclude that prealbumin was a more sensitive or specific nutritional marker than serum albumin *(3)*. However, subsequent publications indicate that predialysis serum prealbumin is a useful addition to nutritional profiles and provides additive information to serum albumin *(10,12)*. Patients with a predialysis serum prealbumin of less than 30 mg/dL should be evaluated for nutritional adequacy *(3)*. This threshold is within the normal range

(~17–45 mg/dL) for people without kidney disease, reflecting decreased renal clearance of prealbumin in patients with CKD.

2.1.3. Serum Creatinine and Creatinine Index

Serum creatinine is a nutritional screening parameter in people receiving maintenance dialysis *(3)*. Predialysis serum creatinine concentration reflects the sum of creatinine by dietary origin (creatine and creatinine from meat) and that formed endogenously in skeletal muscle tissue less the creatinine removed by residual kidney function and dialysis. Creatinine is formed irreversibly from creatine in skeletal muscle at a fairly constant rate that is directly proportional to skeletal muscle mass. Thus, under steady state conditions of diet and dialysis, predialysis serum creatinine is roughly proportional to lean body mass. A declining predialysis serum creatinine over time in otherwise stable dialyzed patients indicates loss of skeletal muscle mass. Although not commonly used in clinical practice, creatinine index can be calculated to easily estimate fat-free body mass, especially in anuric patients *(3)*.

Serum creatinine, and a decline in serum creatinine over time, predicts all-cause mortality independently of inflammation (as measured by CRP) *(9)*. Serum creatinine levels are directly correlated with both serum albumin and serum prealbumin. The relationship between serum creatinine and mortality in maintenance dialysis patients is typically a backward "J" shape, with the lowest mortality occurring at a predialysis creatinine level of 9–11 mg/dL and rising significantly at lower levels and modestly at higher levels *(7)*. Lower levels of predialysis serum creatinine reflect low dietary intake of creatinine and creatine as well as low lean body mass. High serum creatinine typically is suggestive of inadequate dialysis. The KDOQI nutrition guidelines recommend that dietary adequacy be evaluated in patients exhibiting serum creatinine levels of less than approximately 10 mg/dL *(3)*.

2.1.4. Serum Total Cholesterol

Low serum total cholesterol is correlated with markers of protein nutritional status (serum albumin, prealbumin, and creatinine) and with mortality in most, but not all, trials. The relationship between nutrient intake and low serum total cholesterol is indirect. The presence of hypocholesterolemia, below 150–180 mg/dL, or a declining serum cholesterol concentration can be an indicator of chronically inadequate protein and energy intake *(3)*.

Serum cholesterol is primarily useful as a screening tool because its sensitivity to, and specificity for, changes in protein and energy intake is poor. Serum cholesterol also is depressed with chronic inflammation. The relationship between mortality and serum cholesterol is usually "U" shaped, with lowest mortality occurring with serum cholesterol levels of about 200–220 mg/dL in most trials and increasing for higher or lower values. A relationship between CRP and serum cholesterol has been reported, with patients at both high and low extremes of the serum cholesterol distribution having higher CRP levels *(13)*. Low levels of cholesterol and elevated CRP suggest the presence of inflammatory stress and anorexia, whereas elevated levels of both may be more reflective of cardiovascular disease (CVD).

2.1.5. Serum Transferrin

Serum transferrin is frequently used as a marker of protein-energy nutritional status in people without CKD. Compared to serum albumin, it has the advantage of a shorter half-life (about 8.5 days) and a smaller pool size, making it more responsive to nutritional deficits. However, because transferrin is also influenced by the presence of anemia, a comorbidity prevalent in people with CKD requiring maintenance dialysis, it is not recommended for nutritional assessment in patients with stage 4 or 5 CKD because its specificity for nutritional deficits is low in this population *(3)*. Transferrin is a useful nutritional parameter in patients with higher levels of kidney function. In the MDRD study, transferrin and nutrient intake gradually declined as kidney function deteriorated *(14)*.

2.2. Markers of Inflammation

Thirty to fifty percent of patients receiving maintenance dialysis have evidence of an active inflammatory response *(15,16)*. A growing body of evidence suggests that to accurately interpret serum protein status, an understanding of a patient's inflammatory state is critical. Nutritional and inflammatory stresses frequently coexist in people with CKD, and the predictive power of nutritional parameters for clinical outcome is partially or fully attenuated when nutritional parameters are adjusted for the presence of inflammation. Approximately 75% of patients with CKD have evidence of CVD. Proinflammatory cytokines and oxidative stress elicit an inflammatory response, and the presence of inflammation is closely associated with accelerated development of CVD and cardiovascular mortality. In contrast to levels of serum albumin and prealbumin, which generally rise during the first year following initiation of dialysis, circulating levels of inflammatory mediators [e.g., CRP, interleukin-6 (IL-6) and IL-10] do not improve with initiation of dialysis.

Inflammation blunts appetite and increases protein catabolism, lipolysis, and REE. The effect of inflammation on REE is subtle, increasing by approximately 10–15%, but over time, this sustained increase may result in protein-energy deficits, especially when combined with anorexia, which is a common side effect of both CKD and inflammation. Chronic inflammation in CKD patients can be caused by infections, interactions between the blood and dialyzer, contaminants in the dialysate, concomitant conditions, or a combination of these factors. In addition, dialysis itself, even when biocompatible dialysis membranes are used, results in transient inflammatory response that persists for several hours following treatment *(17)*. Thus, an understanding of inflammatory status is increasingly accepted as a key part of biochemical nutritional assessment.

2.2.1. C-Reactive Protein and Proinflammatory Cytokines

A number of positive acute phase proteins can be used to document the presence of acute or chronic inflammation. The most common and consistently used is CRP, a nonspecific marker of inflammation and proinflammatory cytokine activity. CRP has a half-life of approximately 19 hours. Its catabolic rate is not affected by inflammation, while its synthetic rate and release is markedly up-regulated during an acute phase response. Thus, levels of CRP in stressed patients can rapidly rise by several orders of magnitude. Clinically, CRP is not routinely available, but should be considered when inflammation is suspected or when response to nutritional interventions is slower than expected.

Other clinically useful markers are the proinflammatory cytokines—tumor necrosis factor (TNF-α) and IL-6. Up-regulation of TNF-α and IL-6 contributes both to wasting and the high incidence of cardiovascular morbidity and mortality in people with CKD *(18,19)*. Proinflammatory cytokine activity has numerous deleterious effects including the promotion of insulin resistance, anorexia, and oxidative stress. TNF-α stimulates lipolysis and impairs muscle synthesis, whereas IL-6 inhibits insulin growth factor-1 and plays a role in the development of sarcopenia *(19)*.

Serum albumin levels are inversely correlated with CRP *(6,20)*. However, because of its rapid turnover, CRP correlates less well with future albumin levels. In contrast, longer half-life positive acute phase proteins (e.g., ceruloplasmin or α-1 acid glycoprotein) are more predictive of future albumin levels *(6)*. CRP is predictive of hospitalization and cardiovascular morbidity and mortality, it is a marker for the presence of inflammation and CVD, and an independent predictor of all-cause and cardiovascular mortality in patients with stage 3 and 4 CKD. Serum albumin and CRP both were independent predictors of all-cause mortality, suggesting that they either act via different mechanisms (e.g., nutritional vs. inflammatory) or at different points in the inflammatory process *(21)*. CRP increases with declining kidney function prior to dialysis and continues to rise as dialyzed patients become anuric.

2.3. Nutritional Adequacy and Management

2.3.1. Protein Intake

The adequacy of protein intake can be estimated in several ways in people with CKD. Diet records or recalls can provide clinicians detailed information about food choices and can be very useful as educational tools but require a substantial effort from the patient, and thus, can be incomplete records of protein intake (see Subheading 3). Protein intake can also be estimated in clinically stable patients by determining the protein equivalent of total nitrogen appearance (PNA), also called protein catabolic rate (PCR). PNA (PCR) reflects the sum of urea generation from endogenous protein turnover and the metabolism of dietary protein minus urea removed from the body by dialysis or residual renal function and change in body urea pool size over the interdialytic interval. The urea pool accounts for changes in blood urea nitrogen (BUN) levels and changes in total body water (TBW) volume over the measurement period. Fecal and dermal nitrogen losses are disregarded because they are quantitatively fairly minor and impractical to collect and analyze under clinical conditions. Detailed procedures for computing PNA for people on hemodialysis (HD) and peritoneal dialysis (PD) are published in the KDOQI Clinical Practice Guidelines *(3)*.

Clinically, PNA (or PCR) is typically normalized so that protein intake, expressed as normalized protein nitrogen appearance (nPNA) or normalized PCR (nPCR). nPNA (nPCR) can be compared to estimated protein requirements independently of body mass. Unlike typical practice for diet records, where intake is usually normalized to actual or adjusted body weight, PNA is normalized by dividing it by V/0.58. V is the patient's urea volume which can be determined either during urea kinetics, using BIA, or anthropometric equations and provides an estimate of fat-free body mass. The 0.58 factor is the typical proportion of V as a fraction of total body weight *(3)*. The resulting value, nPNA, can be compared to dietary records obtained over the same interdialytic interval to evaluate adequacy of protein intake.

The body turns over substantially more protein on a daily basis than is reflected in nPNA, on the order of 300 g/day, most of which is resynthesized into new body proteins. Subtle increases in degradation rates or decreases in synthesis rates not offset by increased intake can result in loss of lean body mass over time. Uremia per se does not appear to significantly alter nPNA. However, the dialysis procedure itself is an acute catabolic event, resulting in decreased whole-body protein synthesis and an increase in whole-body protein breakdown that persists for several hours after dialysis sessions. In addition, there is some loss of blood and amino acids during dialysis, which increase protein requirements. The ability to increase protein synthesis to offset increased losses is limited *(22)*, which can result in significant catabolism in the presence of hypermetabolic or inflammatory states.

Care must be taken not to interpret nPNA data as a gold standard surrogate for protein intake, even though it is objectively derived. Protein intakes determined in this manner only accurately reflect dietary protein intake under steady state conditions, which typically is not the case in CKD patients undergoing renal replacement therapy. Diet records and nPNA provide independent methods for estimating dietary protein intake. As noted in Table 2, both have factors that need to be considered in their interpretation, and ideally both methods should be employed together to derive conclusions about the adequacy of intake.

2.3.2. Metabolic Acidosis

Predialysis serum bicarbonate levels should be considered when performing a nutritional assessment. Serum bicarbonate is not a nutritional marker per se, but metabolic acidosis is frequently correlated with low serum albumin *(4,20,23)*, prealbumin, and nPNA *(23)*. Acidosis reduces albumin synthesis and increases amino acid oxidation by stimulating branched chain amino acid oxidation, increasing the ATP-dependent ubiquitin-proteosome pathway, and reducing insulin-like growth factor and growth hormone receptor expression *(23)*.

The presence of acidosis must be evaluated carefully because sample collection technique and delays in analyzing the sample can cause spuriously low results. Patients with good appetites, and thus higher protein intakes, also tend to be acidotic because protein brings dietary acid along with it. Interventional studies correcting serum bicarbonate

Table 2
Interpretation of nPNA

Observation	*Interpretation*	*Consider*
nPNA exceeds DPI or is unexpectedly high	Only tentative conclusions about protein intake possible	Catabolic state • Inadequate energy intake • Presence of inflammation or inflammatory stressors (fever, infection, etc.) • Weight loss • Metabolic acidosis • Bioincompatible dialysis membrane Inaccurate diet record Low lean body mass
nPNA less than DPI or is unexpectedly low	Only tentative conclusions about protein intake possible	Anabolic state • Corticosteroid use • Recovery from infection, illness • Pregnancy or growth Inaccurate diet record Edema Excess body weight
nPNA = DPI	nPNA reflects protein intake	Conclude patient is in nitrogen balance and nPNA reflects intake if none of the above apply

DPI (g/kg/day), normalized dietary protein intake; nPNA, normalized protein equivalent of nitrogen appearance.

without changes in protein intake typically have not resulted in improved serum albumin levels, suggesting the benefit of adequate protein intake in maintaining serum albumin exceeds the potentially deleterious effect of the associated higher acid loads on reducing albumin synthesis.

2.4. Summary of Biochemical Assessment

Serum protein deficits arise from many etiologies in people with CKD *(18,24)*. Simple malnutrition is caused by insufficient nutrient intake relative to requirements and responds to correction of deficient intake. In contrast, malnutrition or wasting caused by inflammation and secondary effects of other comorbidities results in hypermetabolism and inefficient utilization of nutrients. With this form of malnutrition, simple nutritional repletion without measures to correct the underlying comorbidities and inflammation is ineffective. Because there are interactions between inflammation, anorexia, and poor nutrient intake, a multifaceted intervention strategy that optimizes dialysis delivery, energy and protein intake, corrects concomitant conditions (e.g., acidosis, anemia, uremia, medication side effects, economic concerns, dental health, etc.), and addresses inflammation and elevated proinflammatory cytokines is needed. Biochemical assessment is an instrumental part in differentially diagnosing the etiology of protein deficits in people with CKD.

3. DIETARY INTAKE ASSESSMENT OF NUTRITIONAL STATUS

3.1. Major Challenges in Dietary Intake Assessment in CKD

During the five stages of CKD, the major dietary and nutrient challenges to nutritional status vary greatly, and for this reason, the assistance of a registered dietitian (RD) is critical for improving the patient's eating-related QOL *(25–32)*. Table 3 presents likely problems with nutritional status that arise at the various stages of the disease.

Table 4 presents the nutrients of greatest concern in dietary assessment at each stage; note that in some cases, comorbid conditions such as diabetes mellitus, GI disease, and other disorders common in CKD patients may present additional problems. The choice of dietary assessment methods depends largely on the stage of kidney disease and the nutrients in need of increased attention, because it is these nutrients that are likely to be lower or higher than optimal *(33–44)*.

3.2. Dietary Assessment Tools

Table 5 presents some of the common methods to assess dietary intake and the advantages and disadvantages of each. There are three major types of dietary assessment tools that are useful in CKD. First, there are tools that help the clinician assess intakes of specific nutrients and energy. Examples include 24-h recalls and food records. Next are tools that help one to assess dietary patterns that may vary greatly from day to day (e.g. between sick and well days, or between dialysis and non-dialysis days). Examples include 2-day assisted food records and diet history. Finally, tools that assess eating-related QOL in kidney

Table 3

Major Challenges and Nutritional Status at Various Stages of Chronic Kidney Disease *(26,27,33–36)*

CKD stage[a] *GFR (ml/m in/ 1.73 m²)*	*Protein-energy malnutrition*	*Body composition (lean body mass, fat mass deficit)*	*Micronutrient deficiencies*	*Dietary intake*	*Anemia*	*Bone disease*	*Fluid retention*
1 GFR≥90							
2 GFR 60–89		?	?				
3 GFR 30–59	↑	↑	↑	↓	↑	↑	
4 GFR 15–29	↑↑	↑↑	↑↑	↓↓	↑↑	↑↑	↑
5 GFR <15 (or dialysis)	↑↑↑	↑↑↑	↑↑↑	↓↓↓	↑↑↑	↑↑↑	↑↑

[a]Adapted from ref. *(25)*. CKD, chronic kidney disease; GFR, glomerular filtration rate; ↑= increase; ↓= decrease; ? = not certain.

Table 4
Key Nutrients for Dietary Assessment in Chronic Kidney Disease *(26,80,74)*

CKD stage	*Energy*	*Protein*	*Calcium*	*Phosphorus*	*Iron*	*Vitamin D*	*Potassium*	*Other vitamins and minerals*	*Fluid*
1 GFR ≥90									
2 GFR 60–89									
3 GFR 30–59	+	+	+		+	+		+	
4 GFR 15–29	+	+	+	+	+	+	+	++	+
5 GFR <15 (or dialysis)	+	+	++	++	++	++	++	+++	++
KDOQI standards for stage 4 and 5	Age ≥ 60 years 30–35 Age < 60 years 35	GFR <25 ml/min 0.6–0.75 gm/kg ≥50% HBV	1.0–1.5 gm/day <2.0–2.5 gm/day including binder load	10–12 mg/gm protein or 10 mg/kg/day	Indivi-dualized	Indivi-dualized	Usually unrestricted unless high	DRI: B-complex	Maintain balance
Maintenance dialysis (HD, PD)	Age ≥ 60 years 30–35 Age < 60 years 35	HD ≥1.2 gm/kg PD ≥1.2–1.3 gm/kg >50% HBV	<2.0–2.5 gm/day including binder load	10–12 μmg/gm protein or <900 mg/day	Indivi-dualized	Indivi-dualized	2–3 gm/day adjusted for the serum levels	C: 60–100 mg B6: 2 mg Folate: 1 mg B12: 3 μg Other vitamins: DRI E: 15I U Zn: 15 mg	HD: Output + 1000 cc PD: Maintain Balance

DRI, Dietary Reference Intake; GFR, glomerular filtration rate; KDOQI, Kidney Disease Outcomes Quality Initiative; HBV, high biological value; HD, hemodialysis; PD, peritoneal dialysis; +, Need modification; ++, more intense modification; +++, most intense modification.

Table 5
Dietary Intake Assessment Tools for Chronic Kidney Disease (CKD) patients

Dietary Intake Methods			
Assessment method nutrient intake	*Description*	*Strengths*	*Weakness*
24-h recall	Clinician assists the patient to recall food intake of previous 24 h. Using food models or pictures can help the patient to identify portion size *(46–52,85)*	Good tool to use for large population studies and in the clinic. Useful for international comparisons of nutrient intake in both healthy and chronically ill patients. Inexpensive and easy to collect intake data for all populations, especially for illiterate patients. Random days can help to get valid estimates of usual intakes	Single recall does not represent the patient's usual intake and foods consumed infrequently may not be recorded. Elderly patients may not provide adequate information due to memory problems. Underreporting is common
Food record	Food records provide information on intakes of food and beverages (and dietary supplements) over specific periods. The most common food record includes 3 days which include 2 weekdays and 1 weekend day *(46,53,57,61,64,65,85)*	Useful to assess actual or usual food intakes. This method is widely used for dietary intake studies, especially macro- and micro-nutrient analysis	Accuracy is dependent on the number of days. Patients may change their usual diet pattern and under report intake. Patient should be literate. Underreporting is common

(Continued)

Table 5
(Continued)

Dietary Intake Methods			
Assessment method nutrient intake	*Description*	*Strengths*	*Weakness*
2-day diet diary dialysis day and non-dialysis day	Dietitian assists the food intake data specific for hemodialysis patients. Specific food models and pictures help patients to report better intake records *(30)*.	Food intake varies between non-dialysis day and dialysis day due to the dialysis schedule. Useful for comparing the dietary intake data for these days. Can also be used in CKD patients who are not in stage 5.	It often does not include weekend days during the data collection; therefore, patients can change his/her eating pattern on the weekend
Semi-quantitative food frequency questionnaire	Used to identify the food intake of specific foods over period of time (i.e., dairy products use in a day, week, month, or year). Data collection is usually self-administered and portion sizes are included in the questionnaire *(46,66–85)*	Useful in epidemiological studies and analysis includes a broad range of food intakes. Provide good comparison of specific foods, food components and nutrients with the prevalence or mortality of specific disease. Identify food patterns associated with inadequate intakes or specific nutrients. Better data collection from the study participant with faster analysis than other methods	Not very useful in CKD patients due to dietary restrictions. Making food lists and lists of dietary supplements inappropriate. Accuracy of data collection is lower

Dietary history	This consists of 24-h recalls of food intake, information of usual intake, and other information such as food allergies, aversions, and preference. The clinician can plan intervention and improve patient's intake using this technique *(46,52,57,58,85)*	Useful to describe food and nutrient intake of patients for a long period of time and estimate the prevalence of inadequate intakes. Useful tool for food policy and food fortification planning	Labor intensive and time-consuming. The results vary with the interviewers' skill of data collection

disease are available. These different tools provide the clinician with a view of what the patient experiences subjectively and how he/she feels about eating and how it is affected by the disease process from day to day. One example is the Food Enjoyment in Dialysis (FED) tool which will be discussed later *(45)*.

3.3. Collecting Dietary Intake Data with Different Tools

3.3.1. Assessing Intake on Specific Days to Assess Nutrients and Energy Intakes

3.3.1.1. 24-H Recall. The 24-h recall helps the patient to remember his/her food intake on the previous day and to quantitate it *(46–52)*. The interviewer needs to be trained to recall the patient's exact food intake. Accurate quantification is essential. Providing food models, bowls, and plate and picture charts are helpful for collecting accurate intake data. This tool assesses the actual intake of the patient. However, the 24-h recall is not sufficient to describe the patient's usual dietary pattern because it covers a short time. The recall usually takes 20–30 min to complete, but it may take longer if many different foods, ethnic foods, or mixed dishes with various ingredients are consumed. The 24-h recall is the most commonly used dietary tool in the United States, and it is used in the National Health and Nutrition Examination Survey (NHANES) *(48–49)* and the Nationwide Food Consumption Surveys of Food Intake by Individuals. The interviewer can conduct a 24-h recall on random days via telephone although accurate quantification of food intake is more difficult unless the patient is instructed on portion size in advance of the interview *(50–52)*.

The strengths of the 24-h recall include that it does not require literacy and it does not require the patient to change his or her dietary habits because the data collection is done after the patient eats. Use of dietary supplements can be easily included; these are important sources of nutrients in many CKD patients. Also, all foods can be included as choices are not confined to a food list. Serial 24-h recalls are helpful in getting an idea of patterns of intakes and the ebbs and flows of intake.

The limitations include the need for reliance on memory, need for accurate quantification of portion size, and need for a highly trained interviewer. In addition, underreporting is extremely common, and therefore, quantitative estimates need to be combined with other indices such as weight or weight change, hydration status, and so on to get a true picture of the patient's condition.

3.3.1.2. Food Records. This method provides qualitative and quantitative data on food intake. Food records are usually collected for an average of 3 days including 2 week days and 1 weekend day *(46,47,52,53)*. For dialysis patients, the record should contain dialysis and non-dialysis days and 1 weekend day because the eating pattern changes based on the dialysis treatment itself and also on appetite changes related to it. A 2-day diet diary can be used for HD patients to monitor variation in food intakes between dialysis days and non-dialysis days (Table 5) *(30)*. Instruction on time of eating, foods, beverages, portion size, snacks, methods of preparation, food ingested, and special recipes should be included to collect accurate intake data. Food models and household measuring cups and spoons are useful to estimate portion size *(46,54–56)*. If a patient participates in a research study, provision of measuring instruments or even a household scale for measuring food will help the patient record accurate data. Food records should be carefully reviewed by a dietitian immediately after completion to ensure that the detailed description of the foods and recipes is accurate as it is difficult to code the foods for later analysis if this is not done.

The major strengths of the food record are that it does not depend as much on the individual's memory, and the portion size estimates may be more accurate because the records are ideally filled at the same time of eating. Use of dietary supplements and medications containing nutrients can be included.

Common limitations include under-reporting, mistakes in conversion of weight and volume, alterations of eating habits during recording periods, and the heavy burden it puts on patients *(57–61)*. Patients must be literate to collect and record their intakes. Nutrient intakes are calculated using an up-to-date database and computerized dietary analysis system.

3.3.2. Dietary Patterns

3.3.2.1. Dietary History. The diet history is used to estimate the patient's usual food intake and meal pattern over a long period of time. The components of the dietary history include usual eating pattern, a crosscheck of frequency of consumption of specific food items, and a 3-day food record. However, the 3-day food record is usually omitted and the time periods covered by the diet history vary. A shorter time frame (less than 1 month) produces better validity, and obtaining of a dietary history for a period over 1 year is unrealistic *(46,47)*.

The major strength of the dietary history is that it provides a detailed picture of the eating pattern *(58,59,62–65)*. The weakness of the dietary history is that it requires a highly skilled interviewer. It takes 40–60 min to complete, is difficult to quantify, and is ill-suited for later data entry into a computer. Therefore, it is rarely used today for clinical and research purposes.

3.3.2.2. "Usual Intake": Semi-Quantitative Food Frequency Questionnaire. The Semi-Quantitative Food Frequency Questionnaire (SFFQ) is used to assess the frequency of consumption of food items and groups of foods during a specific reference time period, which may or may not be the period the patient actually bases his/her recalls on *(46,47,66)*. The SFFQ provides the information of usual food consumption patterns. The SFFQ evaluates usual intake of the foods rather than food intake of specific days. The SFFQ has two components, food list and frequency of consumption; seasonal variation of specific foods can be also included in it. The SFFQ is most commonly used for epidemiological studies because usual food intake patterns appear to affect the outcomes of some diseases.

The advantages of the SFFQ are that it is relatively quick to administer. It can be administered by interviewers, by telephone, via mail, or self-administered. Processing the SFFQ is inexpensive, and self-administered versions are available from some processing centers.

Limitations of the SFFQ include the fact that in CKD there often is no usual intake; intakes rise and fall from day to day as the disease waxes and wanes, so that it is impossible for the patient to identify a pattern. Also, the SFFQ portion size descriptions may be unclear, and analysis of quantities of specific nutrients can be inaccurate. The instrument is ill-suited for making estimates of total intakes; it is really designed to distinguish between groups of patients. The SFFQ should be validated and tested for reproducibility for use in minority groups *(67–73)*.

The most common SFFQs are Willett's "Harvard" SFFQ and the "Block" (National Cancer Institute) SFFQ in various forms *(66,71,72)*. These SFFQs come in different versions, and all of them are validated by concurrent comparison with another method rather than against a "gold standard" that represents actual intake. The SFFQs are not used often to assess dietary intake in patients with CKD because the multiple dietary restrictions that are involved in this population make the assumptions about commonly eaten food groups invalid *(74)*. The items on vitamin and mineral supplements are not appropriate for the types of dietary supplements common in CKD or among dialysis patients, and oral nutritional supplements are not included on usual SFFQs. The modification of these SFFQs for use in kidney disease can perhaps someday be a useful tool to assess dietary intake for

CKD patients in future studies. At present, simple food frequency questionnaires can be used which take less time. However, it is difficult to translate data from these questionnaires into quantitative estimates of patients' daily nutrient intakes.

3.4. Nutrition Quality of Life

The assessment of QOL is a novel approach to assessing dietary intake and quality. Nutrition-related QOL is the physical enjoyment of food along with its social and nurturing aspects and is a separate and distinct concept from other more general aspects of health-related QOL. It is focused on problems associated with food, eating, and nutrition and how these factors affect the patient's overall QOL. Dietary intake, nutritional status, and health-related QOL are related in CKD patients, especially in maintenance dialysis patients. The progression of kidney disease and the dialysis treatments alter patients' physical and functional status. These changes also affect patients' food preferences and intakes, presumably eventually changing their overall QOL and possibly their nutritional status *(28,29,31,75–77)*.

A new tool to measure nutrition-related QOL, the FED tool, has been developed *(45)*. The FED is a subjective and self-administered questionnaire addressing changes of appetite, taste, smell, thirst, GI symptoms, and medications that often occur in illness or other conditions. It can be used with other tools that are used for measuring QOL, such as the Medical Outcome Study Short Form 36 (MOS SF-36) *(76)* or the Kidney Disease Quality of Life (KDQOL) questionnaire *(78)*, and for measuring dietary intake, such as food records and SFFQs, to develop a fuller picture of what is going on in the patient's eating-related life. Data from the FED tool may be helpful to health care providers in crafting dietary interventions for dialysis patients that maximize QOL, as well as in monitoring changes in nutrition-related QOL. The hope is that, taken together, the use of these tools will be helpful not only in monitoring but in crafting interventions to prevent malnutrition and to improve enjoyment of food and overall QOL. However, additional research needs to be done to establish the validity and reliability of this instrument and its association with health outcomes.

3.5. Using Dietary Intake Data to Counsel Patients with CKD Using the Kidney Disease Outcomes Quality Initiative Guidelines

In our experience, dietary records supplemented by dietary recalls and some assessment of QOL are most useful *(79–81)*. The KDOQI guidelines are evidence-based clinical practice guidelines that are reasonable for counseling patients with CKD. They include nutritional components that are related to overall outcomes of CKD. It is suggested that 3-day dietary records and 24-h dietary recalls used periodically will monitor intake of protein, energy, and other nutrients in patients with CKD to prevent malnutrition. Once an appropriately detailed assessment has been obtained, the results should be compared to these guidelines and an assessment made. If intakes do not meet the guidelines, action is warranted. It is important to make sure that the intake is accurate and that low intakes are truly low and not the result of forgetfulness or an inability to complete the forms. Failure to include oral nutritional supplements and dietary supplements will make micronutrient intakes lower than what is consumed. It is important to include dietary supplement use as so many patients receive micronutrients from them, particularly in the late stages of CKD.

3.6. Newer Methods for Assessing and Planning Usual Dietary Intakes Using the Dietary Reference Intake

The state of the art for assessing and planning of usual intakes is to use the statistical techniques developed by the Institute of Medicine (IOM), National Academy of Science (NAS) in its recent publications on the Dietary Reference Intakes. These methods are usually reserved for research purposes and the assessment of large groups of subjects *(46)*. It is worthwhile to read the publication to obtain guidance for assessment of individuals as well as to become familiar with the basic techniques for use in CKD. Several publications addressing different needs are now available *(26,35)*.

3.7. Needs for Further Research on Dietary Assessment and CKD

There are several critical needs for improving dietary intake assessment in CKD *(26,35,82,83)*. They include finding simple ways for patients to record their intakes frequently and easily so that "usual" intakes can better be assessed. It is difficult for patients not to change their intake, at least for the first few days while they are keeping food records; multiple records are likely to be more representative of true intake. Additionally, in CKD, especially in the later stages of the disorder, intakes vary considerably from day to day and they are only assessed occasionally. Thus, a true picture of actual intake over the long term is very difficult to obtain. Also, true intake is more likely to be lower than recorded intakes as patients seldom record their food intakes when they are critically ill and often forget some items that they ate.

A second challenge is to find better ways to document under-reporting of intake, which occurs strikingly in obese patients, but in all patients. Over-reporting is also a problem among patients who are frequently ill because they fail to report intakes on such days.

It is also important to develop better ways to report intake of dietary supplements commonly used in CKD. The field needs more validated and better nutrition-related QOL instruments, better automated recording methods, and better dietary analysis software for both medical foods and dietary supplements commonly consumed by those with kidney disease.

3.8. Summary of Dietary Intake Assessment

Dietary assessment in CKD is challenging because dietary intakes vary greatly from day to day, dietary and oral nutritional supplements are often used, and patients may find it difficult to report intakes. The dietary components of particular concern vary somewhat depending on the stage of the disease and the nutritional status of the individual patient. In general, food records and some assessment of appetite are appropriate.

4. ASSESSMENT OF BODY COMPOSITION

Body composition can be assessed at levels ranging from the simple such as subcutaneous fat thickness to the complex such as amounts of total body potassium. The body compartments of most common interest are adipose and muscle tissues, which are available for energy needs, but among persons with CKD, these compartments can also affect chemical and water balances, the dose of dialysis and nutritional recommendations. Adults with CKD can be elderly with low body weights, low body mass index (BMI), and low levels of body fat and muscle mass *(86,87)* or African-American, overweight or obese with high BMI's and levels of body fatness *(88)*. Members of these and similar groups are likely to have CVD, diabetes, and other comorbidities associated with CKD *(89)*. Selecting an assessment method depends upon the body compartment of interest and the availability of comparative reference data. Assessing body composition in adults with CKD can facilitate the prescription and monitoring of appropriate clinical and nutritional therapies *(90)*. This chapter provides an update of body composition assessments in adults with CKD *(91)*.

4.1. Levels of Body Composition

The body's composition can be described as amounts of elements like oxygen, carbon, nitrogen, potassium, and calcium, or molecules such as water, or properties of tissues, such as density or resistance or tissue thickness, or the weight of a compartment or tissue as a percentage of body weight. Total body nitrogen and potassium are measures of protein stores, and total body calcium is a measure of total body bone mineral. Changes in amounts of nitrogen, potassium, and calcium affect persons with kidney disease and can contribute to comorbidity. Measures of isotopes of oxygen and hydrogen can be used to determine energy expenditure and the amount of TBW. At the tissue level, body composition consists of adipose, muscle, and skeletal tissues. Descriptions of these levels are based upon findings from healthy adults, but these levels and interactions among them change with age and especially with the presence of CKD and associated comorbidity. In addition, there are sex and race factors and effects on body composition that should be considered for adults diagnosed with CKD.

4.2. Measures of Body Composition

There are direct and indirect assessments of body composition. Direct assessments generally measure at the elemental, molecular, or cellular level and are accurate but frequently invasive, while indirect methods are less accurate measures at the molecular and tissue level. Neutron activation determines specific elements by exposing the body to a source of high energy neutrons *(92)*. Total body counting measures the decay of naturally occurring isotopes in the body, many of which are present in tissues at known concentrations, such as potassium 40. Computed tomography (CT) uses the attenuation of X-rays and radiographic densities of different body tissues to produce images via software. Magnetic resonance imaging (MRI) measures the energy released by the changing orientation of hydrogen protons in a strong magnetic field to produce images also from software. Both CT and MRI can quantify anatomical amounts of muscle and adipose tissues especially visceral adipose tissue *(93,94)*. MRI body composition data are available *(95)*, but reference data from this and other body composition assessments in groups of healthy adults are limited *(96)*.

A summary of assessment methods and their utility is given in Table 6. These methods are for healthy adults where body composition parameters are static. The methods applicable to adults with CKD are dual energy X-ray absorptiometry (DXA), BIA, TBW, and prediction

Table 6
Body Composition Assessment Methods

Method	*Assessment*	*Precision*	*Reliability*	*Utility*
Neutron activation	Total body nitrogen, calcium	Very high	Very high	Low
Computed tomography	Bone, adipose tissue	Very high	Very high	Low
Magnetic Resonance Spectroscopy	Adipose tissue	Very high	Very high	Moderate
Dual Energy X-ray Absorptiometry	Bone, adipose tissue, fat-free mass	Very high	Moderate	High
Total body water	Total body water, fat-free mass	High	Moderate	Moderate
Total body scanning	Total body potassium 40	High	Moderate	Moderate
Bioelectrical impedance	Total body water	Very high	Moderate	High
Anthropometry	Body size, subcutaneous adipose tissue	High to medium	Moderate	Very high

(97–100), but their use is affected by the incompatibility of the assumptions of these methods and the dynamic effects of kidney disease and dialysis on the body's composition, due in part to the unknown effects of high levels of extra-cellular fluid and solute on the quantification abilities of these methods in the predialysis state. For adults undergoing dialysis, assessments are best obtained afterwards when body fluid compartments are in a relatively normal inter-relationship.

4.2.1. Dual Energy X-Ray Absorptiometry

DXA quantifies bone mineral content (BMC) and bone mineral density of the hip and spine primarily for diagnosing osteoporosis. The differential attenuation of two low radiation energy levels through the body and subsequent computer calculations allow the quantification and total and regional analysis of adipose and soft tissues and estimates of fat mass (FM) and non-skeletal fat-free mass (FFM) *(101–103)*.

DXA is convenient for measuring body composition in healthy adults and is of potential value in dialysis patients *(104)*. DXA measures are generally independent of age-, race- and gender-sensitive assumptions, but the inherent problems and assumptions of DXA regarding levels of hydration, potassium content, or tissue density to the estimation of soft tissue values should be considered carefully for adults with CKD *(105–107)*.

DXA is an easy and convenient method for measuring body composition in most adults, and it is currently included in the ongoing NHANES *(108)*. However, it is limited by restrictions on body weight, length, thickness, and width as a function of the available table scan area with each manufacturer and type of DXA machine, that is, pencil or fan beam. Many overweight and obese adults are too wide and too thick for a whole body DXA scan with current machines *(109)*.

DXA estimates of body composition are also affected by differences among manufacturers in the technology, models and software employed, methodological problems, and intra- and inter-machine differences *(106,110,111)*. The sum of the total body weights of FFM, BMC, and FM should be within less than 2 kg of measured body weight; otherwise, concern should be raised for measurement and machine accuracy. Inter-machine and inter-method comparisons of DXA body composition estimates should be done cautiously. Regular maintenance and calibration preserve the accuracy of DXA machines.

4.2.2. Total Body Water

TBW is quantified by isotope dilution using deuterium or tritium in urine, saliva or plasma by mass spectrometry, infrared spectrometry, or nuclear magnetic resonance spectroscopy *(112–115)*. The equilibration time for isotope dilution is about 2–3 h, but this is not well documented, and variation in equilibration times with body size are poorly understood in healthy adults much less the obese and those with CKD. TBW needs to be corrected for natural abundance and isotope exchange *(111)*. Extra-cellular water (ECW) is also quantified by chemical dilution using bromide as NaBr or other chemical elements similar to chloride *(116,117)*, that estimate extra-cellular chloride space. Bromide concentration in plasma is measured by high pressure liquid chromatography.

FFM calculated from TBW assumes an average proportion of TBW in FFM of 73%, but this ranges from 67 to 80% *(118,119)*. However, as much as 30% of TBW is present in adipose tissue as ECW in normal adults, and this increases with the degree of adiposity in overweight and obese adults and those with CKD, but available reference values are out of date. These proportions are higher in women than men, higher in the obese, and can produce underestimates of FFM and overestimates of FM in healthy adults. Variation in the distribution of TBW and ECW due to obesity-associated comorbidity, such as diabetes, CVD, and kidney failure, further affects estimates of FFM and FM based on TBW.

TBW reflects urea distribution and is altered by renal insufficiency *(99,120)*. The prediction of TBW volume (V) in adults needing dialysis is used to prescribe and monitor treatment *(97)*. It is also used to calculate the dose of dialysis in the determination of Kt/V *(121–123)*, but V has frequently been predicted from equations of Watson based on non-representative White samples *(124)*. New prediction equations for TBW from anthropometry and from BIA for White, Black, and Mexican-American children and adults *(114,125,126)* have been published. The heterogeneity of comorbid conditions afflicting adults with CKD increases the errors of these equations.

4.2.3. Bioelectric Impedance Analysis

Bioelectric impedance at 50 kHz is used to estimate TBW, FFM, and Total Body Fat (TBF) from measures of the resistance of the body to a small alternating electric current at very low amperage *(127)*. Bioelectric impedance measured at multiple frequencies differentiates the proportions of intra- and extra-cellular fluid volumes. All BIA analysers employ some form of predictive modeling to obtain estimated outcomes, but large errors for adults are a limitation, especially in the presence of disease or comorbidity. BIA is used for dialysis patients in the prescription and monitoring of the adequacy of dialysis based on urea kinetic modeling *(97)*. BIA assessments of TBW in dialysis patients is an area for continued investigation *(128,129)*.

The conductor is body water, and BIA measures its impedance which is the vector relationship between resistance and reactance at a frequency. Resistance is the opposition of the conductor to the alternating current as in non-biological conductors, and reactance is from the capacitant effect of cell membranes, tissue interfaces, and non-ionic tissues *(130)*. A conductor's volume is proportional to its length squared divided by its resistance, where stature (S) is squared and divided by resistance (R) as an index of body volume, so that S^2/R, is directly related to the volume of TBW *(127)*. The impedance index is used to estimate FFM and FM, but this is based upon the fraction of 73% of TBW in FFM. Because the hydration fraction of FFM is not constant, S^2/R is combined with anthropometric data to predict body composition based upon results from other methods.

Multi-frequency BIA uses at least two different frequencies. At low frequency, the current is unable to pass through cell membranes so that low frequency currents are conducted only through ECW. At high frequency, currents penetrate cell membranes and thus are used to estimate TBW. When predicting body fluid volumes with multi-frequency BIA, the analysis may need to be adjusted for the effects of the fluid containing materials of differing conductive capacity *(131)*. The mixture effects are greatest at low frequency because the conductive volume, presumably ECW, is a smaller proportion of the total volume. This approach has been used successfully by a number of investigators to monitor changes in ECW, TBW, and FFM under a variety of conditions *(132–135)*.

BIA prediction equations for estimating body composition describe statistical associations based on biological relationships for a specific population and are useful only for adults that closely match the reference population in body size and shape. The ability of BIA to predict fatness in obese adults is limited because there is a large proportion of body mass and water in the trunk, the hydration of FFM is lower, and the ratio of ECW to TBW is increased. Prediction equations resident in commercial single-frequency impedance analysers are not recommended unless sufficient information is provided by the manufacturer regarding their predictive accuracy, errors, and the samples used to develop them. BIA can be used for segmental assessments also.

4.2.4. Anthropometry

Anthropometric methods are practical, inexpensive techniques that describe body size, identify levels of fatness and leanness in adults with CKD. The body measurements listed in Table 7 were collected in the MDRD and the Hemodialysis (HEMO) studies *(136,137)*, using standardized methods *(138)*. Standardized anthropometric techniques are necessary for comparisons among studies and for assessing nutritional status of an individual using reference data from the NHANES *(88,138,139)*. Measurements are taken on the right side, but because of a cast or an amputation, the left side can be used. As vascular access changes, circumference and skinfold measurement locations also shift. An instructive video is available *(140)*.

Stature and weight describe body size and mass. Weight is a measure of total energy stores, leanness, overweight, and obesity, and adults with high body weights tend to have high amounts of body fat and vice versa. Weight combined with stature squared or BMI is an index of overweight or obesity (and leanness). There are extensive national

Table 7
Anthropometric Measurements for Use in Assessing Chronic Kidney Disease

Measurement	*Assessment*	*Accuracy*	*Reliability*	*Utility*
Stature	Total body size	Very high	Very high	High
Weight	Total body size	Very high	Very high	High
Triceps skinfold	Adipose tissue	Moderate	Very high	Moderate
Subscapular skinfold	Adipose tissue	Moderate	Moderate	Moderate
Arm circumference	Fat-free mass and adipose tissue	High	Moderate	Moderate
Calf circumference	Fat-free mass	High	Moderate	Moderate
Knee height	Lower leg length	High	High	High
Elbow breadth	Body size	High	Moderate	Moderate

reference data available for BMI, and it has established direct relationships with levels of body fatness, and BMI values above 25 are associated with increased morbidity and mortality *(141)*. Obesity is common among adults with CKD, but sarcopenia (loss of muscle mass) can cause a normal weight adult to become "obese" because of a high percentage of body fat in a condition recognized as sarcopenic obesity *(142)*.

Skinfolds measure subcutaneous fat thickness, and abdominal circumference is an index of internal adipose tissue *(143)*. Midarm circumference and triceps skinfold are combined to calculate midarm muscle area. Midarm muscle area and calf circumference are related to levels of protein stores or used as a marker for FFM. Calf circumference is an indirect measure of muscle mass *(144)*. Elbow breadth is a limited index of frame size *(145)*, and knee height is used to estimate stature *(146,147)*. Standardized body measures should be taken immediately postdialysis.

4.3. Effects of Dialysis

Adults with CKD can be difficult to measure *(139)*, and two health technicians are frequently required to gather health information. Decreased functional status and increased comorbidity are challenges to body composition assessment methodology. Adults with CKD and those undergoing dialysis can find standing difficult and some are chair- or bed-ridden due to amputations. Recumbent anthropometric techniques can be used with those unable to stand upright *(139)*, and if stature cannot be measured, it is estimated from knee height *(147)*. For those with bilateral, below-the-knee amputations, there is presently no suitable method of estimating stature.

Anthropometric data for those receiving dialysis should include postdialysis measures of weight, stature, calf circumference, arm circumference, and triceps and subscapular skinfolds *(88)*. Adults with CKD and diabetes and especially those receiving dialysis can have large weights, BMI, arm circumferences, and skinfold thicknesses compared with those without diabetes. Adults on dialysis for 5 years or more tend to have smaller weights and skinfolds than those on dialysis less than 5 years. Adults receiving dialysis tend to be shorter, lighter and have less adipose tissue than healthy persons of the same ages based on data from NHANES II *(148)*. Limited anthropometric reference data for dialysis patients have been published *(149)*, but the recent HEMO Study anthropometric data provide a clinical reference for adults with Stage 5 CKD *(88)*.

4.4. Statistical Estimations of Body Composition

Statistical methods can be used to predict body composition for groups or individuals when other methods are not possible *(150)*. Predicting body composition requires a regression equation with FM, TBW, or FFM as the dependent variable. The selection of predictor variables depends on their biological and statistical relationships to the outcome variable, and the strength of these relationships affects the accuracy or precision of the prediction equation. Regression methods include forward selection, stepwise and backward elimination regression, and these are used if the predictor variables are not very inter-related. Stature and weight are inter-related, and if used together in the same equation, this can reduce precision and accuracy. If so, a maximum R^2 or an all-possible subsets of regression procedure are analytical choices.

Regression analysis assumes bivariate relationships between outcome and predictor variables are linear, and it is assumed that the dependent variable is normally distributed in order to allow statistical inferences about the significance of the regression parameters. A large sample results in more precise and accurate prediction equations than a small sample, but the necessary sample size is a function of the correlations between the outcome and predictor variables. The sample size required to achieve accuracy on cross-validation depends on the number of predictor variables, the bivariate relationships among the dependent variable and the predictor variables, and the variance of the dependent variable in the cross-validation sample.

Numerous prediction equations for TBW, FFM, FM and %FM from anthropometry and for TBW and FFM from BIA for non-Hispanic white, non-Hispanic black, and Mexican-American children and adults have been published *(114,125,126)*. Many of these equations are for whites only, but there are a limited number for Native-American, Hispanic-American, and non-Hispanic black American samples *(151, 152)*, and BIA prediction equations for TBW and FFM are available for non-Hispanic white, non-Hispanic black, and Mexican-American children and adults *(125)*.

4.5. Limitations and Sources of Error

It is important to account for the limitations of body composition assessments. Each method at best has an error of at least 2–3% body fat when compared with other methods *(153)*, due in part to fact that body composition in a living person is dynamic. Also, the effect of total body or regional alterations in water content on methods in adults with CKD is often not well understood. Disease states affect hydration status, and an abnormal hydration status alters the assumptions underlying body composition methods. In nephritic or CKD patients with truncal edema (ascites), BIA may underestimate body water to a substantial degree. In some with CKD, total body potassium may not correlate closely with total body nitrogen, which is an indicator of total body protein. Also, in CKD, total body potassium may not be a good estimate of FFM *(90)*.

4.6. Alterations in Body Composition with Age, Gender, and Ethnicity

In healthy adults, relationships among body constituents and compartments tend to be stable, and this allows comparisons among levels of body composition. However, these static relations become dynamic with aging and disease, and variations occur also as a function of gender and ethnicity. Changes and variations among body constituents and compartments affect estimates of body composition and increase the errors of these estimates. Changes in weight parallel energy and protein balance. Body weight varies about ±1.0 kg/day, but a consistent loss in weight of more than 0.5 kg/day over time indicates possible negative energy and/or water balance. Weight gains or losses are also associated with changes in the subcutaneous and visceral adipose tissue. In middle-aged adults and the elderly, fat redistributes to the trunk with aging.

Adipose tissue decrease on the arm and the leg with age, but subcutaneous and internal adipose tissues increase on the trunk. These changes can be associated with deterioration of limb and abdominal muscle structure or tone as well as changes in fat patterning. Post-menopausal women are reported to have more upper-body fat than pre-menopausal women so that some changes may have endocrinological significance.

There are ethnic differences in body composition also. These differences are affected by differences in socioeconomic status, diet, utilization of health care, and levels of genetic admixture. African-Americans have more dense bones and tend to have more bone mineral than non-Hispanic whites *(154)*. At the extremes of the distributions for body fatness, there are more African-American women than non-Hispanic white women.

4.7. Reference Values

The NHANES is conducted by the National Center for Health Statistics (NCHS), Centers for Disease Control and Prevention. These surveys use multiple methods of data collection that include interviews, physical examinations, physiological testing, and biochemical assessments from a representative sample of non-institutionalized civilian residents of the continental United States including Alaska *(155)*. Distribution statistics for stature, weight and selected body circumferences, breadths, and skinfold thicknesses are available from the NHANES at http://www.cdc.gov/nchs.

Estimated means from single-frequency BIA for TBW, FFM, FM, and %FM of non-Hispanic white, non-Hispanic black and Mexican-American adults from the third NHANES between 1988 and 1994 are presented in Figs 1–4 *(118)*. Current NHANES data for BMI have been published, and body composition estimates from DXA are in the current NHANES *(108)*.

NHANES III means for TBW, in Fig. 1, are estimated from BIA for non-Hispanic white, non-Hispanic black, and Mexican-American men and women. TBW means for these men and Black women are approximately 2 to as much as 9 L greater than that reported or estimated previously *(91,114,115)*. The large TBW volumes in the non-Hispanic black women and of the other groups of men and women, in Fig. 1, reflects the greater levels of obesity reported for these segments of the population and noted in Figs 3 and 4 *(118,156)*. The TBW data represent cross-sectional results of separate age cohorts, and the age trends reflect differences between these cohorts rather than slopes computed from serial measurements. There is limited serial TBW data for adults in middle age, which indicate that TBW is relatively stable in

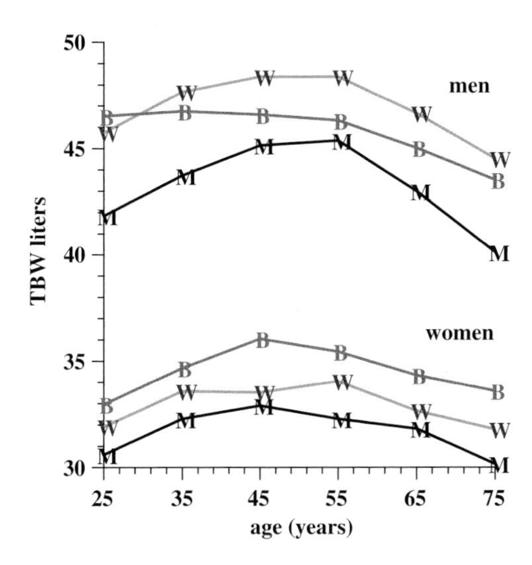

Fig. 1. Estimated means for total body water (TBW) from National Health and Nutrition Examination Survey III by 10-year age groups from 20 to 80 years for non-Hispanic white (W), non-Hispanic black (B), and Mexican-American (M) men and women.

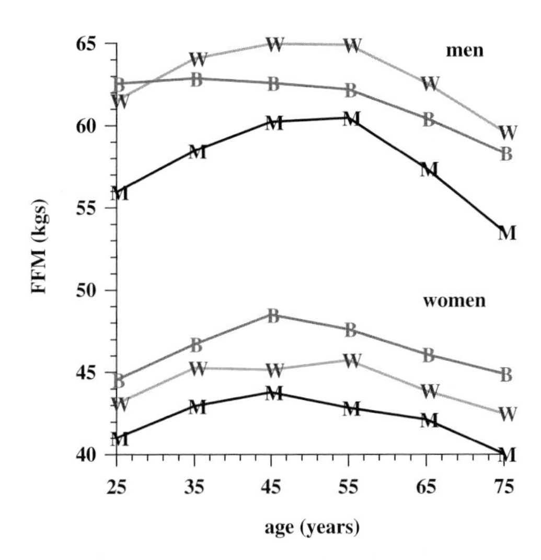

Fig. 2. Estimated means for fat-free mass (FFM) by 10-year age groups from 20 to 80 years for non-Hispanic white (W), non-Hispanic black (B), and Mexican-American (M) men and women.

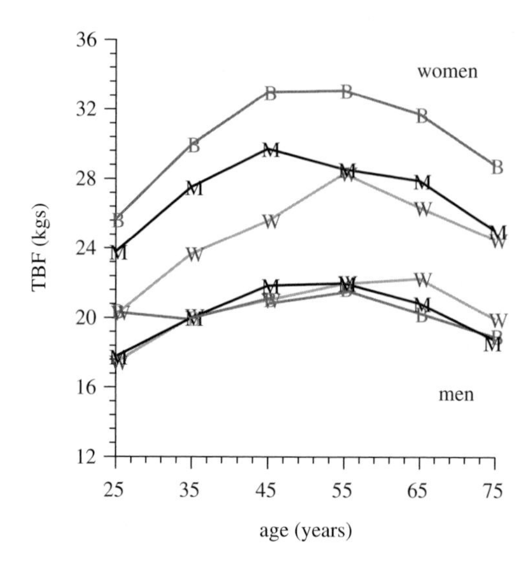

Fig. 3. Estimated means for TBF by 10-year age groups from 20 to 80 years for non-Hispanic white (W), non-Hispanic black (B), and Mexican-American (M) men and women.

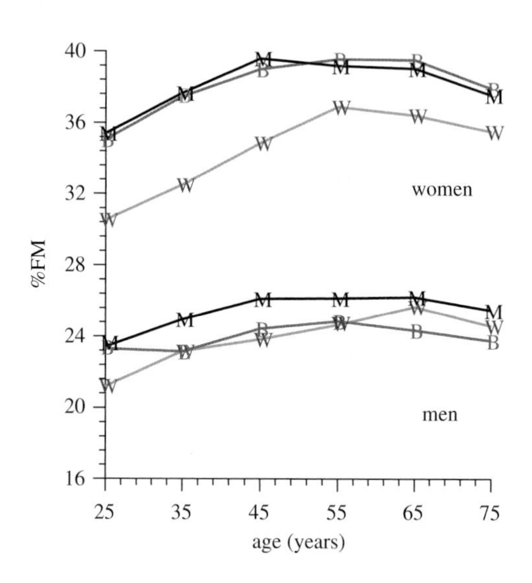

Fig. 4. Estimated means for %FM by 10-year age groups from 20 to 80 years for non-Hispanic white (W), non-Hispanic black (B), and Mexican-American (M) men and women.

healthy adults *(115)*. It is likely that TBW decreases after menopause and at older ages, but the degree of this decline is unknown.

4.8. Summary of Body Composition Assessment

There are a variety of methods available for assessing body composition, but the selection of a method depends upon what needs to be known and why. Regardless of the method selected, none are perfect, and the errors surrounding a body composition assessment should not be ignored. Errors can have clinical relevance, especially if the person is treated and observed over time, and these methods are only as good as the limits of the errors surrounding their estimates. However, the estimates are only as useful as the availability of suitable reference data from a group of persons of at least the same age, race, and gender.

What has been described has been the status quo for assessing protein-calorie nutrition, but what are our needs of improving current methods or of developing new methods? The development of accurate, inexpensive, noninvasive, portable methods of estimating body composition, body hydration, and fat patterning would facilitate greatly future investigations. Areas for particular interest for study are the complex relationships between changes in body composition and nutritional status, physiological and psychological health, resistance to acute and chronic diseases, and longevity in normal and clinical circumstances. Future work should emphasize comparisons between men and women for biological, social, and behavioral factors affecting changes with age in body composition. However, the most pressing need is for the development of reference data for anthropometry and body composition in large representative samples of African, non-Hispanic, Hispanic, and Asian Americans of all ages. Reliable and representative reference data for anthropometry and body composition are needed before reliable and valid prediction equations can be developed for epidemiologic and clinical use.

This work on Body Composition Assessment was supported by grants HD12252, HL072838, HL069995, and DK071485 from the National Institutes of Health, Bethesda, Maryland.

5. NUTRITION PHYSICAL ASSESSMENT

5.1. Physical Assessment Mandates in Chronic Kidney Disease

During the past 30 years, renal dietitians have been first encouraged, and more recently mandated by regulatory and professional organizations, to perform nutrient-focused physical assessments. The first

Medicare conditions for coverage in End Stage Renal Disease (ESRD) facilities *(157)* asked RDs "to assess the nutrient and dietetic needs of each patient." The language separating "nutrient assessment" and "dietetic needs" suggested a two-step process to identify vitamin/mineral imbalance, then translate nutritional requirements into foods, menus, and appropriate supplementation. Given the potential for both deficiency and toxicity inherent in CKD, physical assessment findings were clearly valued in evaluating micronutrient status in patients undergoing dialysis therapy.

The 1995 Comprehensive Accreditation Manual *(158)* for the Joint Commission on Accreditation of Healthcare Organizations (JCAHO) defined nutritional assessment as a "comprehensive approach to defining nutrition status which employs medical, nutrition, and medication histories; physical examination; anthropomorphic measurements; and laboratory data." In addition, "physical examination for manifestations of nutrient deficiency or excess in all high risk patients" was mandated—a population easily including end-organ failure patients. The year after the JCAHO mandates, the American Dietetic Association (ADA) identified basic physical assessment as a clinical competency for on-the-job entry-level dietetics practitioners *(159)*. Performing physical assessments remain in Standard 1 of Nutrition Assessment in the 2005 Scope of Dietetic Practice Framework of the ADA *(160)*.

The NKF KDOQI Nutrition Guidelines *(161)* validated the Subjective Global Assessment (SGA) as a useful measure of protein-energy nutrition status, concentrating on its concurrent evaluation of somatic stores and nutrient intake. Most recently, the 2005 proposed revisions to Medicare conditions for coverage in ESRD facilities *(162)* retained initial mandates, with expanded goals to "achieve and sustain an effective nutritional status." Effective nutritional status is defined as acceptable levels of "nutrients in the blood," with "clinical signs of nutrient deficiency" recognized as potential outcome measures.

With physical assessment clearly acknowledged as an essential skill in Medicare/KDOQI directives, JCAHO recommendations, and the ADA's Scope of Practice, RDs are supported by clinical and professional guidelines to pursue nutrient-focused physical examinations.

5.2. Format of the Nutrition Physical Examination

Nutrient-based lesions are observed on the external surfaces of the body—its surface anatomy *(163)*. Because nutrients play defined roles in tissue synthesis, deficiency and toxicity can modify normal structure.

Altered tissue integrity observed in the mouth, skin, hair, eyes, and nails provides clues to nutrient(s) involved, the time frame of nutritional injury, and the functional deficits which may exist. Many NPE overviews have been published *(164–167)*, but Kight's NPE has been used in patients with CKD as a template for comprehensive physical examinations *(168)*, Medical Grand Rounds *(169)*, and case study work *(170)*. The NPE begins with an examination of oral and perioral structures, skin and related structures (eyes, hair, and nails), and concludes with a systems assessment (cardiovascular, central-nervous, endocrine, GI, immunological, musculoskeletal, and renal) that allows functional assessment.

The SGA initially put forward by Detsky *(171)* and modified for the CKD population by McCann *(172)* focuses on altered fluid balance and somatic fat, muscle wasting. Changes in weight, dietary intake, GI symptoms, functional capacity, and disease processes are also considered. The tool can be scored, which provides the potential for longitudinal evaluation in response to therapy. Its application in patients with CKD was recently reviewed by Steiber et al. *(173)*.

5.3. Approach to the Nutrition Physical Assessment

Brief tutorials on physical examination techniques useful in discovering altered surface anatomy are presented. Because RDs are more familiar with a nutrient-focused approach, comprehensive nutrient assessments for vulnerable nutrients summarize nutrient disposition, dietary sources, drug/nutrient interactions, laboratory evaluation, and medical comorbidity data. Niacin, vitamin B_6 (V-B_6), and zinc have been selected for in-depth review, given their altered nutrient disposition with kidney failure (see Figs 5, 7, 9).

Historical findings in deficiency/toxicity, functional deficits, and biochemical validation of lesions documented in patients with stage 5 CKD are described for each vulnerable nutrient. Early literature is frequently cited, as deficiency in humans has long been documented and human research committees today are loathe to approve depletion studies on nutrients already known to be essential. When understood, mechanisms of deficiency and toxicity are described. Newer, more salient research in kidney failure is reviewed, chosen specifically for its future application in this disease. Nutrient-based lesions identified in niacin, V-B_6, and zinc are given in Figs 6, 8, and 10, respectively.

5.4. Physical Examination Techniques

Classic physical assessment involves inspection, palpation, percussion, and auscultation, generally performed in that order *(174)*.

Vitamin/mineral imbalance in dialysis patients affects tissue integrity of the mouth, skin, scalp, eyes, hair, hair follicles, and nails. Inspection and light palpation are techniques used most frequently. While presentation of vitamin and mineral lesions can be discrete, they are generally bilateral in the absence of injury.

5.4.1. Examination of the Mouth

The lips are smooth, a deeper color than the face, with a clear vermilion border. Vertical cracking (cheilosis) or erosion at the corners (angular stomatitis) may be observed in active lesions; scarring may be present if deficits have occurred in the past. Breath odor suggests gingivitis if dank with decayed food, oral Candida if yeasty in people with diabetes or iron deficiency *(175)*, or elevated BUN if ammoniacal *(174)*. An understanding of the patient's dental hygiene, time since the last dental visit, and bleeding with brushing will help determine the etiology of oral lesions. Cracked, fractured teeth at the gum line become sources of infection and abscess. Ill-fitting dentures following weight loss adversely affect ability to chew and may cause maceration at the corners of the mouth. The extended tongue will reveal color and texture. Usual pigment is deep pink, with fingerlike projections (papilla) scattered evenly over the surface of the tongue. Erosion of papilla at the tip and lateral aspects of the tongue may suggest early nutrient deficiency; balding with complete atrophy and lesions suggests long-term disease *(176)*. The bald tongue on nutrient replacement will develop new papilla in the central portion of the tongue first, with papilla returning at the tip and lateral aspects last.

5.4.2. Examination of the Scalp, Hair, and Skin

The scalp is observed at the natural separation in the hair and is intact with natural oil. The patient is asked to show areas of soreness or itching. Frequency of hair washing affects texture of hair and may correlate with itchiness. Hair should be evenly distributed without patchiness on the scalp. Absence of hair on the lower extremities may suggest vascular compromise. The hair shaft is straight, whether round or flat, emerging easily from a smooth hair follicle. A magnifier will help establish whether the shaft is coiled or the follicle is hyperkeratosed. Skin along the hairline, behind the ears, and between the eyebrows is in tact with adequate lubrication; in nutrient deficiency, these are primary sites for seborrheic-like dermatitis. Skin around the neck and shoulders, along the arms and down the lower extremities, is assessed for skinfold, muscle integrity.

Observing the scalp at the hair part, the skin at the midpoint of the upper arm, and above the lateral malleolus (ankle) will provide easy reference points for future comparison. Periorbital and sacral tissues may be sites for edema; pitting at the ankle or sock line upwards toward the knee may be graded.

5.4.3. Examination of the Eyes

The tissue adjacent to the eyes should be intact, without inflammation or swelling. The lateral palpebral commissure (corner of the eye) is generally darker in color, without redness or irritation. Eyes are adequately bathed, without excessive tearing. The lower palpebral conjunctiva is examined by placing the index finger beneath the lower eyelid and gently easing the facial tissue downward, as the patient looks up. Normal conjunctiva shows a rich pink capillary bed beneath a pale anterior rim; pallor is defined as little or no red color beneath the anterior rim *(177)*. The iris meets the white sclera at the limbus. A raised yellow plague (pinguecula) is a normal finding that may emerge along the horizontal plane of the eye at the limbus in aging *(174)*. However, clear glass-like crystals, observed by glancing a light off the surface of the eye at the limbus in the 3:00 and 9:00 positions of the eye, suggests calcific band keratopathy resulting from calcium/phosphorus imbalance *(178)*.

5.4.4. Examination of the Nails

Normal nail plates are clear in color, smooth in texture, and convex in both directions. Longitudinal dyschromic bands or vertical ridging in the nail plate are normal variants. Rubbing a gloved thumb over the nail plate from side to side, nail bed to tip will reveal abnormalities in shape and texture. Spooning of the nail plates (koilonychia) suggests iron deficiency or environmental exposure to water *(179)*. Horizontal ridging is abnormal, and of particular significance when found as Beau's Lines, singular arcuate erosions seen on all nail plates *(180)*. Pitting may occur in psoriasis. The nail bed is pink in color, with a half mooned lunula at the proximal nail. The nail bed blanches when light pressure is applied at the nail tip, with capillary refill in less than 3 seconds. Tissue around the nail is smooth without signs of inflammation or infection.

5.5. Physical Findings and Functional Deficits in Niacin, V-B_6, and Zinc Imbalance

5.5.1. Niacin

Pellagra was prevalent for many years with the understanding that it occurred with poverty and corn intake. Joseph Goldberger of the United States Public Health Service reproduced pellagrous symptoms in dogs, leading to the discovery that yeast could prevent and cure pellagra *(181)*. Bean *(176)* graded the symmetrical, bilateral dermatitis

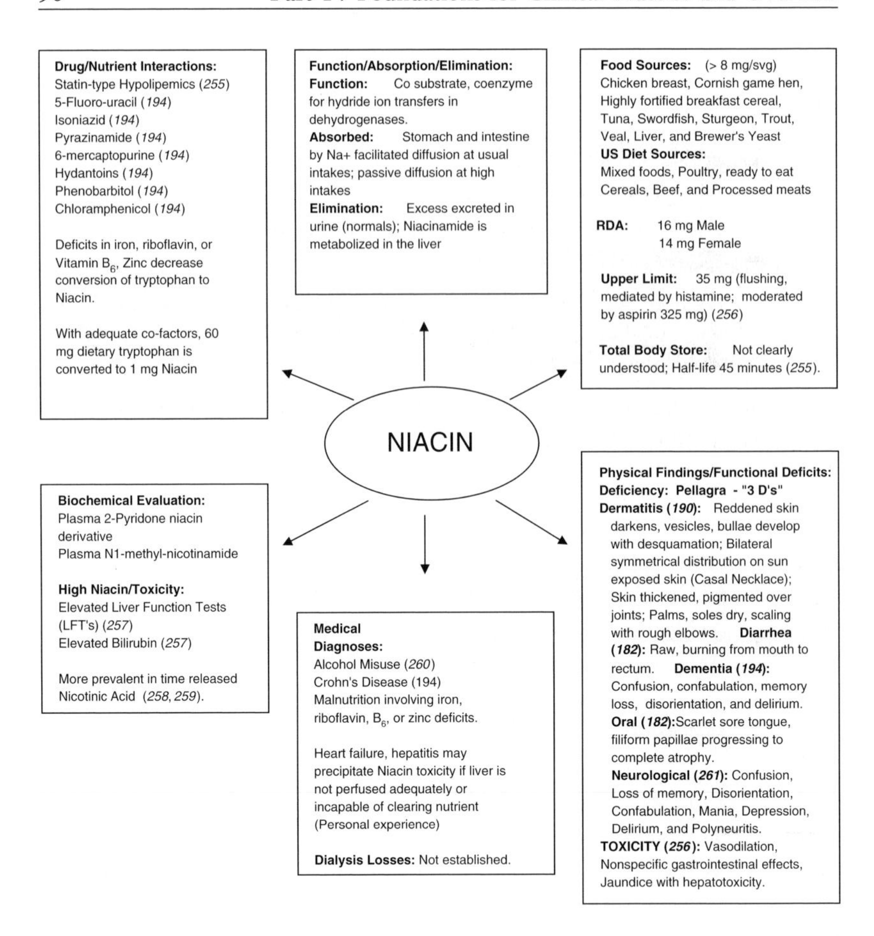

Fig. 5. Comprehensive niacin assessment in renal failure. If not annotated otherwise, data cited from ref. *(188)*.

of sun-exposed skin, and redness, inflammation and papillary atrophy of the tongue found in niacin deficiency. In 1952, Goldsmith et al. *(182)* astutely monitored niacin depletion in women eating a corn or wheat diet, supplemented with tryptophan to ensure minimal nitrogen balance *(183)* and adequate B-vitamins, without niacin. Lesions developed within 40–135 days in isolated niacin deficiency, with half the subjects experiencing some symptom by day 50. Oral lesions occurred with and without characteristic blistering, necrotic skin associated with pellagra *(182)*. Goldsmith coined the term "Pellegra sine Pellagra" from the historical literature *(184)*, documenting deficiency without skin lesions.

Pellagrins continue to elude diagnosis when they present without skin lesions. A necroscopy study diagnosed 20 cases of pellagra with

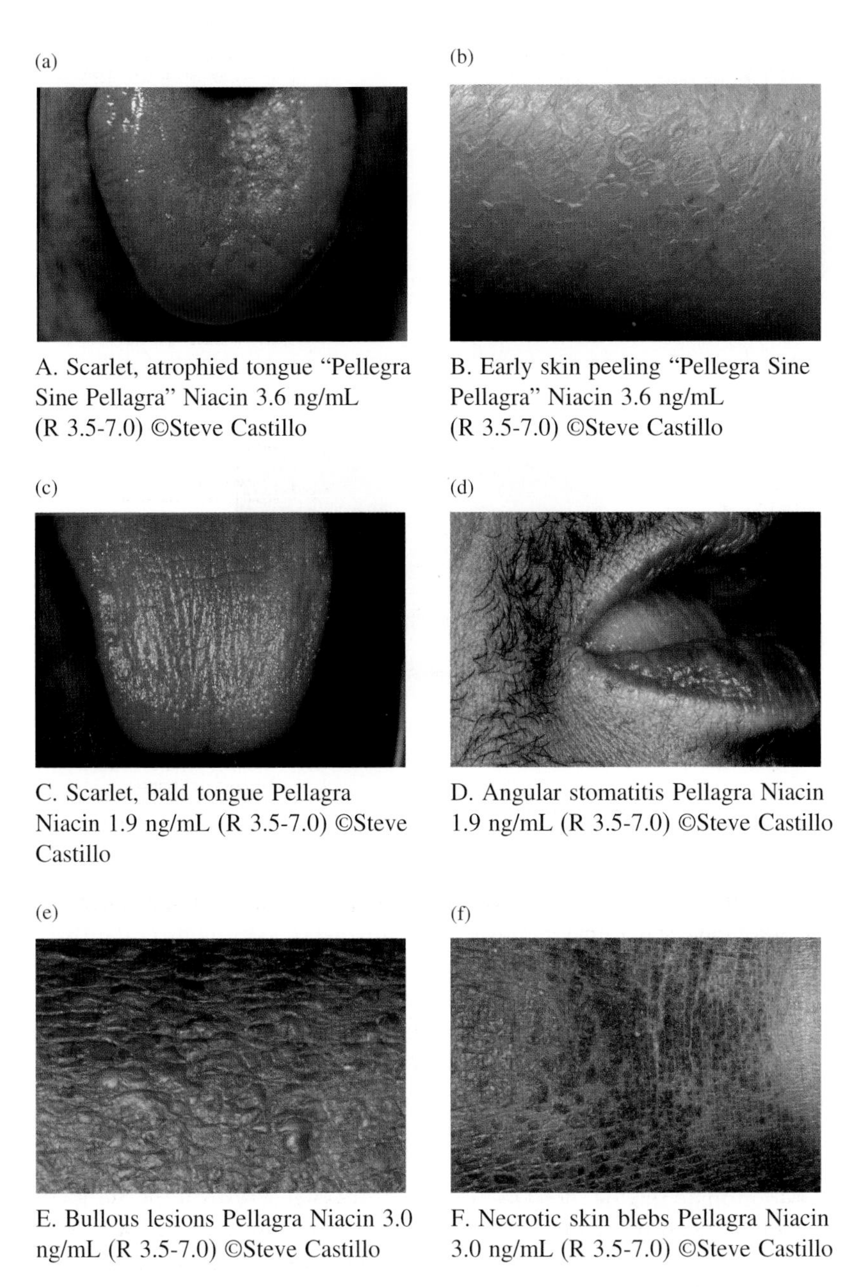

(a) A. Scarlet, atrophied tongue "Pellegra Sine Pellagra" Niacin 3.6 ng/mL (R 3.5-7.0) ©Steve Castillo

(b) B. Early skin peeling "Pellegra Sine Pellagra" Niacin 3.6 ng/mL (R 3.5-7.0) ©Steve Castillo

(c) C. Scarlet, bald tongue Pellagra Niacin 1.9 ng/mL (R 3.5-7.0) ©Steve Castillo

(d) D. Angular stomatitis Pellagra Niacin 1.9 ng/mL (R 3.5-7.0) ©Steve Castillo

(e) E. Bullous lesions Pellagra Niacin 3.0 ng/mL (R 3.5-7.0) ©Steve Castillo

(f) F. Necrotic skin blebs Pellagra Niacin 3.0 ng/mL (R 3.5-7.0) ©Steve Castillo

Fig. 6. Nutrient-based lesions associated with laboratory validated niacin deficit. Photographs may not be reproduced, copied, projected, televised, digitized, or used in any manner without photographer's express written permission.

(g)

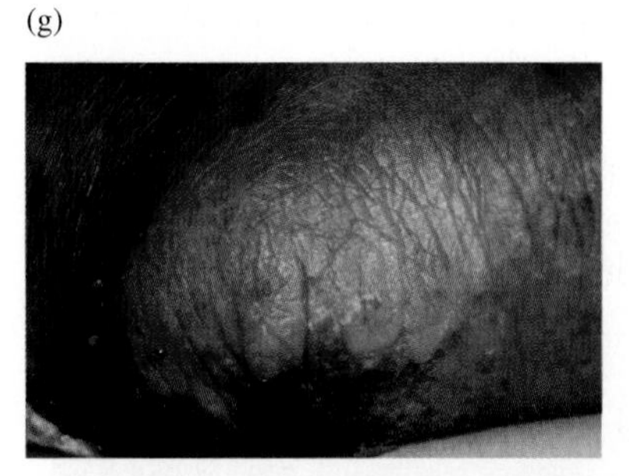

G. Exfoliated elbow Pellagra Niacin 1.9 ng/mL (R 3.5-.70) ©Steve Castillo

(h)

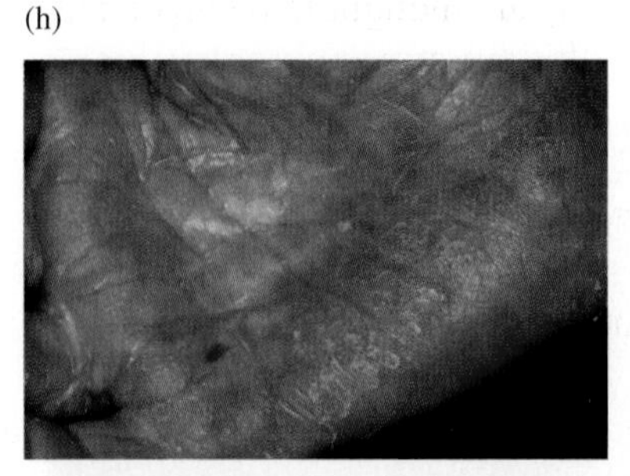

H. Exfoliated palm Pellagra Niacin 1.9 ng/mL (R 3.5-7.0) ©Steve Castillo

(i)

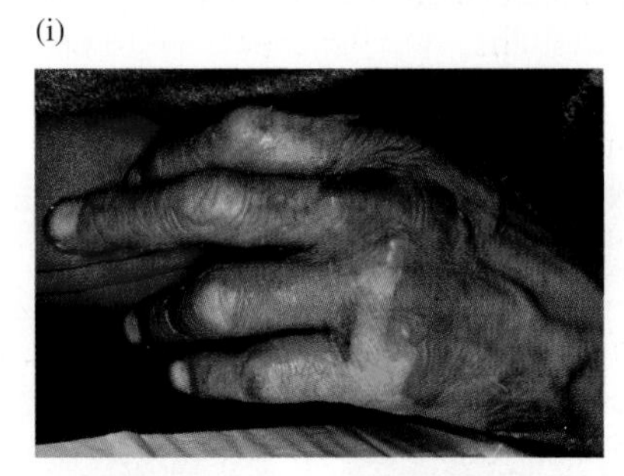

I. Exfoliated knuckles Pellagra Niacin 1.9 ng/mL (R3.5-7.0) ©Steve Castillo

Fig. 6. (Continued)

neuropathology from 74 chronic alcoholics who presented only mental, neurological, and GI symptoms at death *(185)*. Although pellagra is frequently related to alcohol or malabsorption *(186,187)*, use of non-fortified grains like rice and corn as primary energy sources, combined with low protein diets, can expedite niacin deficiency. Pre-formed niacin is derived primarily from fortified breads and cereals *(188)*. Food intake data from rice-eating PD patients *(189)* tallied thiamin, niacin, and riboflavin intakes between 50 and 83% of the Recommended Dietary Allowances (RDA).

Functional niacin deficits include diarrhea, with and without bleeding, dysphagia, nausea, vomiting, and anorexia *(190)*. Autopsy reports reveal scarlet, exfoliated oral surface, extending throughout the GI tract *(191)*. Diffuse inflammation is also found in vaginal mucosal tissues *(182,192,193)*; amenorrhea is common *(182,191)*. Long-term neuropsychological manifestations of photophobia, asthenia, depression, hallucinations, confusion, memory loss, and psychosis

(194) are thought to be mediated by impaired conversion of tryptophan to the neurotransmitter serotonin *(193)*. Early unresolved cases resulted in dementia and death *(191,195)*.

A comprehensive niacin assessment is shown in Fig. 5. Pellagrous lesions observed in dialysis patients at San Francisco General Hospital (Fig. 6) include sore, scarlet, atrophied tongue, angular stomatitis, bullous skin vesicles, necrotic blebs, and exfoliated elbows, palms, and knuckles. Pellagra sine Pellagra was also observed with severe glossitis and mild, tissue-like peeling of the skin.

5.5.2. Vitamin B_6

5.5.2.1. Vitamin B_6 Deficiency. Although V-B_6 deficiency was identified following the development of seizures and anemia in infants given purified diets *(196)*, experimental depletion in adults was achieved using an anti-metabolite, desoxypyridoxine. Vilter et al. *(197)* described seborrheic dermatitis of the nasolabial folds, eyebrows, angles of the mouth, retroauricular spaces, and scalp as the primary lesion, evolving after 19–21 days of desoxypyridoxine treatment. Dermatitis was followed by glossitis, glossodynia, and red, hypertrophied filiform papillae on the lateral aspects of the tongue. Sensory neuritis began as tingling, numbness in the hands and feet, and ascended quickly as extremities became hyperesthetic to pinprick.

Microcytic anemia accompanies V-B_6 deficiency resulting from the role of plasma pyridoxal phosphate (PLP) in synthesizing aminolevulinate synthase, an essential enzyme in the first step for heme synthesis *(198)*. Without heme to sequester iron, differential smears may reveal sideroblasts, immature erythroid cells with mitochondrial granules of non-heme iron *(199)*.

Functional deficits of deficiency include depression, confusion *(200)*, and hypoactive deep tendon reflexes *(197)*. Neurological sequelae are not well understood, but may result from altered PLP function in metabolism of synaptic transmitter amines, noradrenalin, adrenaline, tyramine, dopamine, and serotonin *(201)*. Accumulation of abnormal tryptophan metabolites within the brain is thought to exacerbate neuropsychiatric symptoms *(202)*, as V-B_6 is required in tryptophan conversion to nicotinic acid *(198)*. PLP is also necessary to form the singular inhibitory neurotransmitter, gamma-aminobutyric acid, whose absence may predispose to seizure activity, recalcitrant to anti-seizure medications, in severe V-B_6 deficiency *(203,204)*.

Drug therapies can contribute to V-B_6 deficit. Loop diuretics deplete V-B_6 in early kidney failure as they decrease tubular resorption of water. V-B_6, vitamin C, and oxalate losses have been shown to

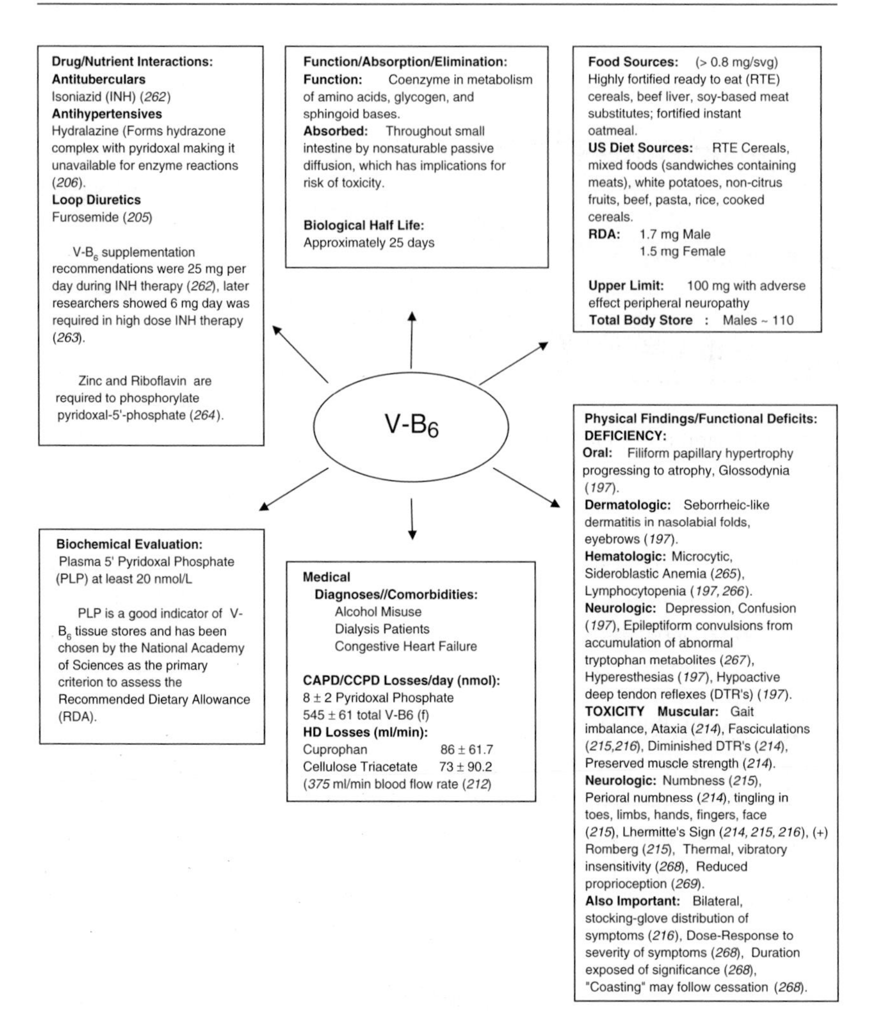

Fig. 7. Comprehensive vitamin B_6 (V-B_6) assessment in renal failure. If not annotated otherwise, data cited from refs *(198,150–195)*.

correlate with water diuresis *(205)*. The antihypertensive medication, hydralazine, forms hydrazone with PLP, compromising V-B_6 decarboxylase activity *(206)*. Isoniazid (INH), used as an anti-tubercular, combines with pyridoxine to form isonicotinyl hydrazine, which can result in PLP-mediated INH neurotoxicity in low V-B_6 intake, that is, in alcoholic, malnourished, *(207)*, or CKD populations *(208–211)*.

V-B_6 losses through the dialyzer are significant, which double with high efficiency, high flux cellulose triacetate membranes, compared to cuprophan membranes *(212)*. Losses through the peritoneal membrane

(a)

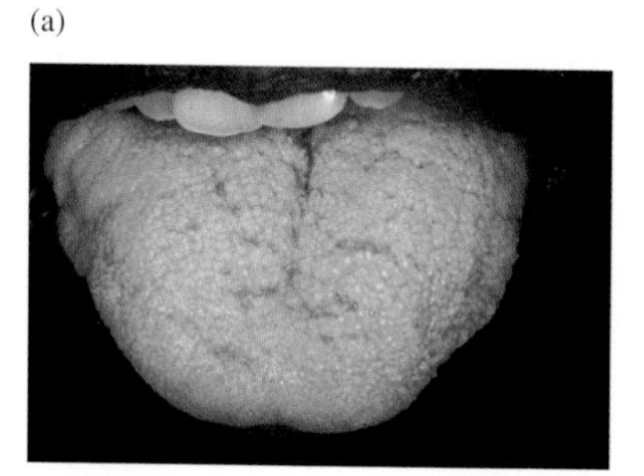

A. Filiform papillary hypertrophy with fused papilla Plasma Pyridoxal Phosphate (PLP) 1.3 ng/mL (R 3.6-18) ©Steve Castillo

(b)

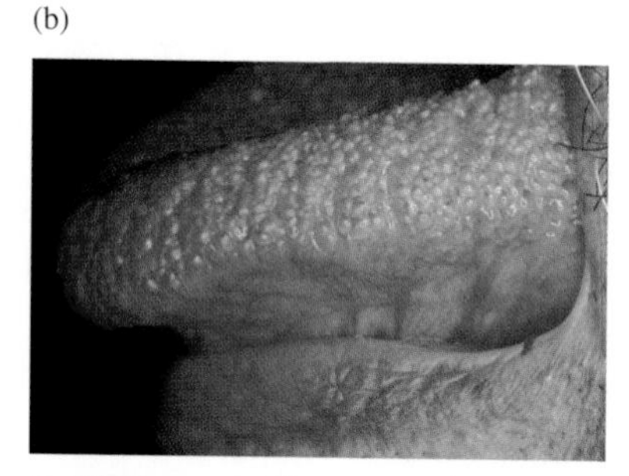

B. Filiform papillary hypertrophy Lateral view PLP 1.3 ng/mL (R 3.6-18) ©Steve Castillo

(c)

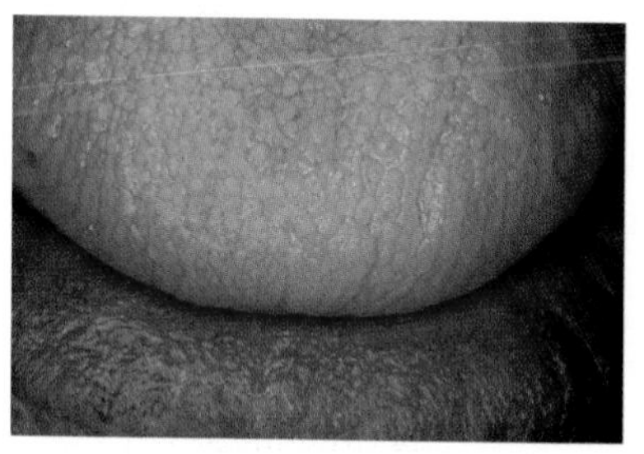

C. Erosion of papilla with longstanding V-B_6 deficit PLP 3.1 ng/mL (R 3.6-18) ©Steve Castillo

(d)

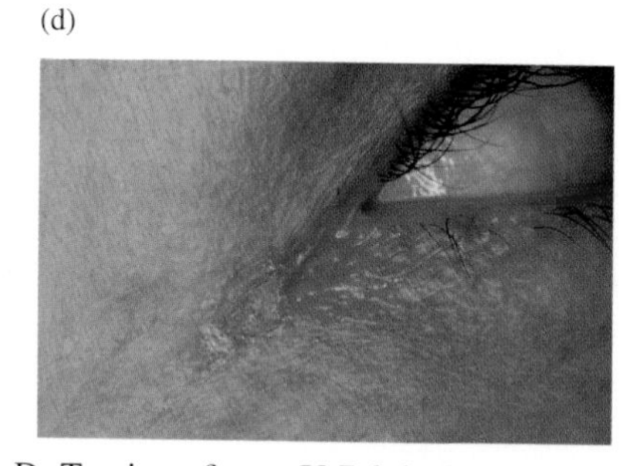

D. Tearing of eyes V-B6 depletion PLP 6.4 ng/mL (R 3.6-18) ©Steve Castillo

(e)

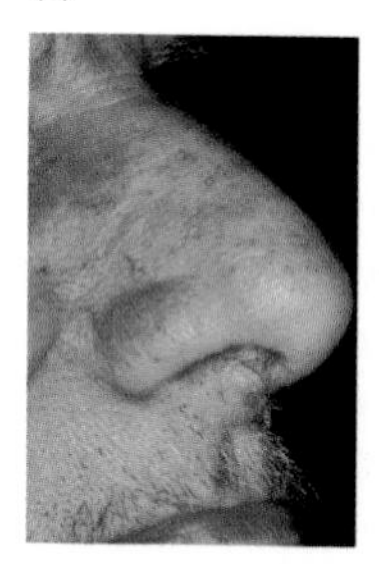

E. Seborrheic-like dermatitis Nasolabial fold, nares PLP 1.3 ng/mL (R 3.6-18) ©Steve Castillo

Fig. 8. Nutrient-based lesions associated with laboratory validated vitamin B_6 (V-B_6) deficit.

are lower than HD membranes, but dietary intake alone has not been able to sustain PLP concentrations *(213)*.

A comprehensive V-B_6 assessment is shown in Fig. 7. The primary lesion photographed by Castillo in deficient HD patients was filiform papillary hypertrophy (Fig. 8). Mild seborrheic-like dermatitis and increased tearing were also identified in patients with PLP concentrations between 1.3 and 6.4 ng/mL (laboratory range 3.6–18.0). PLP is believed to reflect tissue stores of the nutrient, which was used by the NAS to estimate the RDA for V-B_6.

5.5.2.2. Vitamin B_6 Toxicity. High doses of V-B_6 present in foods do not cause toxicity, but V-B_6 accumulates with supplementation due to its non-saturable passive absorption process. Toxicity is generally relegated to the dorsal root ganglia with its increased blood vessel permeability, unlike the blood-brain barrier that insulates the brain *(214)*. Clinical symptoms of V-B_6 toxicity include numbness, paresthesias, ataxia, Lhermitte's sign, and pain. Symptoms documented are sensory deficit, sensory ataxia, (+) Romberg's sign, and loss of Achilles reflexes *(215)*. Neuropathy associated with V-B_6 toxicity presents in a bilateral, stocking-glove distribution barring injury, beginning at the distal digits *(216)*. Limited clearance of V-B_6 in dialysis patients has been associated with elevated PLP concentrations with supplementation as low as 50 mg daily *(217)*. With its potential for both deficiency and toxicity, it is good nutrition practice to document PLP concentrations when renal vitamin intake is uncertain or when supplemental V-B_6 is prescribed with anti-tubercular or cardiovascular treatments.

5.5.3. Zinc

5.5.3.1. Zinc Deficiency. Prasad et al. *(218)* first described zinc deficiency as "adolescent nutritional dwarfism," documenting the nutrient's role in growth and sexual maturity. While the early syndrome occurred with iron deficit, identification of an autosomal recessive defect causing acrodermatitis enteropathica *(219)* produced a clear phenotypic presentation of zinc deficiency alone. Skin lesions with predictable distribution around body orifices and extremities predominate, with functional deficits of diarrhea, compromised T-cell function, and altered central nervous function. The pivotal role of zinc in gene expression, cellular growth, and differentiation is the underlying cause of generalized metabolic impairment in deficiency *(220)*.

Subjects without kidney disease enrolled in well-controlled, induced zinc depletion studies show classic dermatitis around the mouth, nose, inflammation of the mucosal membranes *(221)*, and structural changes in hair bulb formation *(222)*—signs that appear as plasma zinc falls.

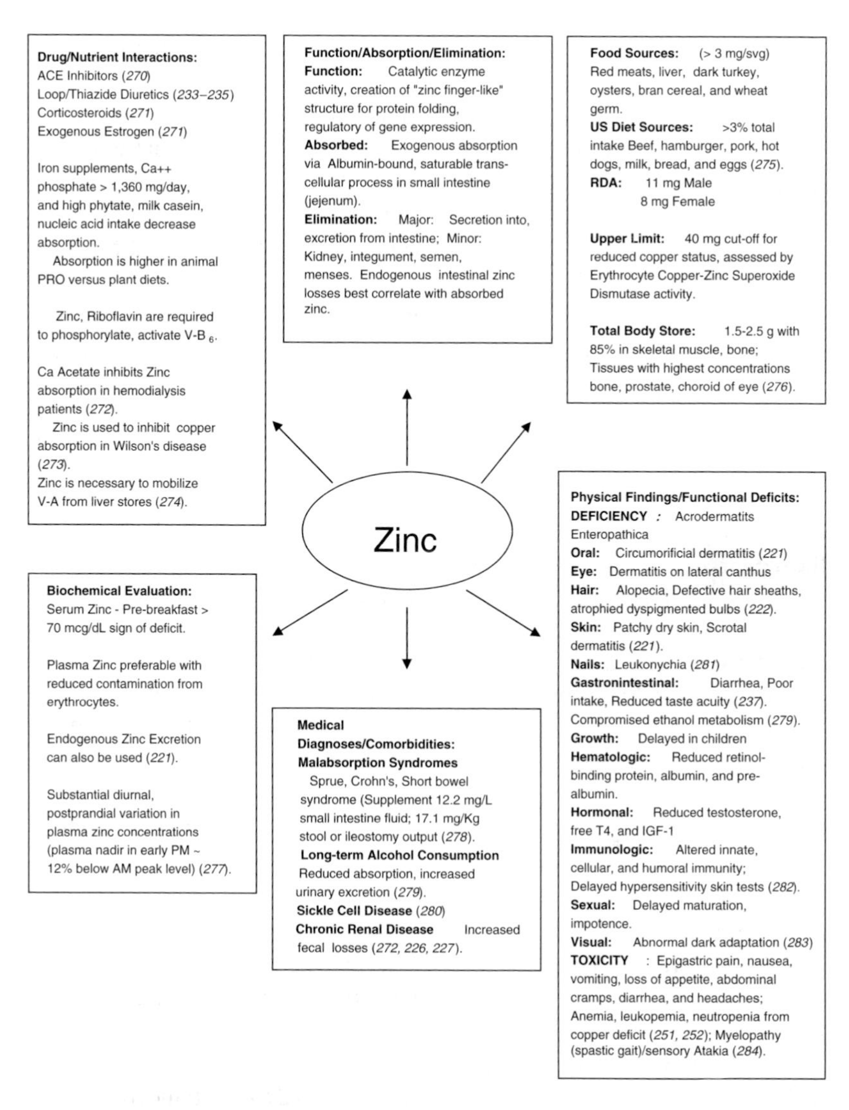

Fig. 9. Comprehensive zinc assessment in renal failure. If not annotated otherwise, data cited from ref. *(230)*.

Use of plasma/serum zinc as an indicator of zinc status was criticized in early studies, but more recent kinetic research suggests that plasma zinc concentration may be a valid indicator of whole-body zinc status in the absence of infection or stress *(221)*, particularly when levels are low.

Zinc studies in kidney failure consistently reveal reduced serum concentrations *(223–225)*. Metabolic research in patients receiving

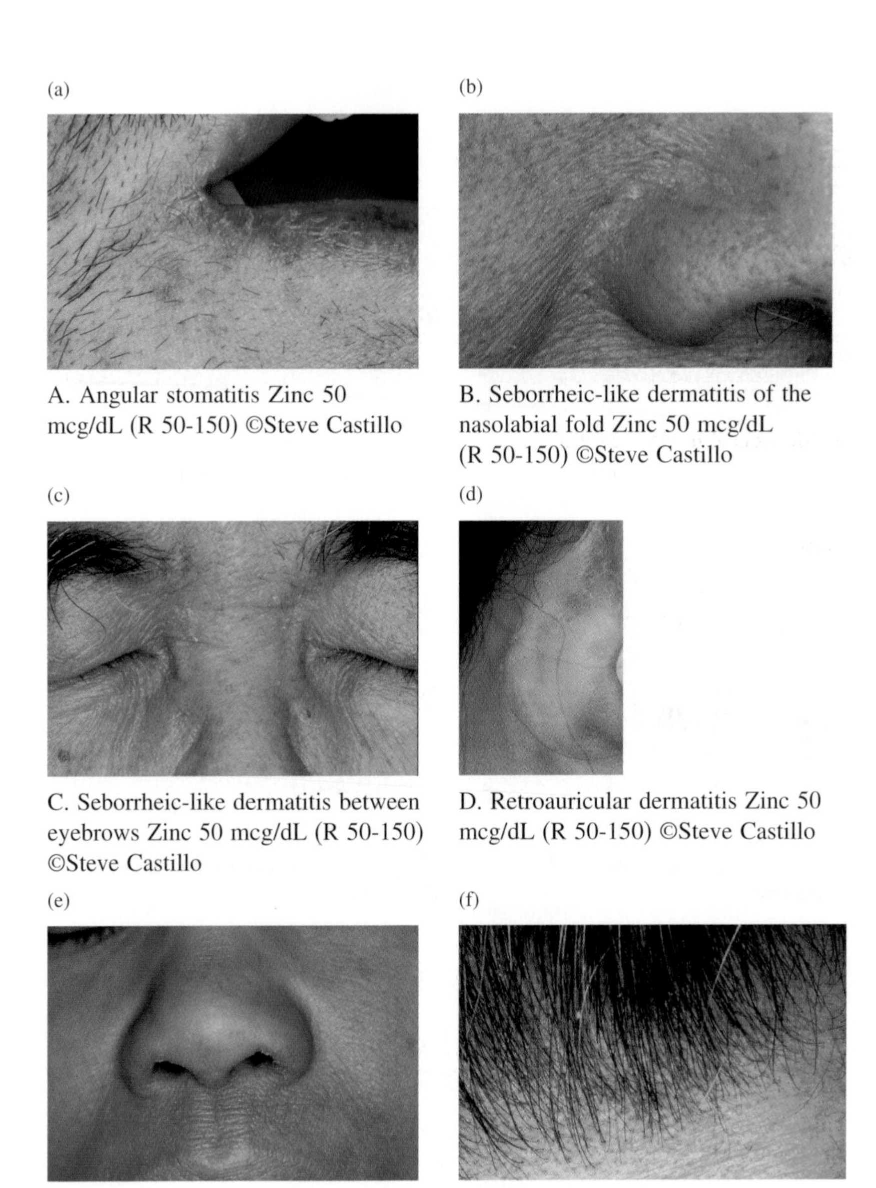

A. Angular stomatitis Zinc 50 mcg/dL (R 50-150) ©Steve Castillo

B. Seborrheic-like dermatitis of the nasolabial fold Zinc 50 mcg/dL (R 50-150) ©Steve Castillo

C. Seborrheic-like dermatitis between eyebrows Zinc 50 mcg/dL (R 50-150) ©Steve Castillo

D. Retroauricular dermatitis Zinc 50 mcg/dL (R 50-150) ©Steve Castillo

E. Circumorificial dermatitis Zinc 52 mcg/dL (R 50-150) ©David Giacalone

F. Seborrheic-like dermatitis of the scalp Zinc 50 mcg/dL (R 50-150) ©Steve Castillo

Fig. 10. Nutrient-based lesions associated with laboratory validated zinc deficit. Photographs may not be reproduced, copied, projected, televised, digitized, or used in any manner without photographer's express written permission.

(g)

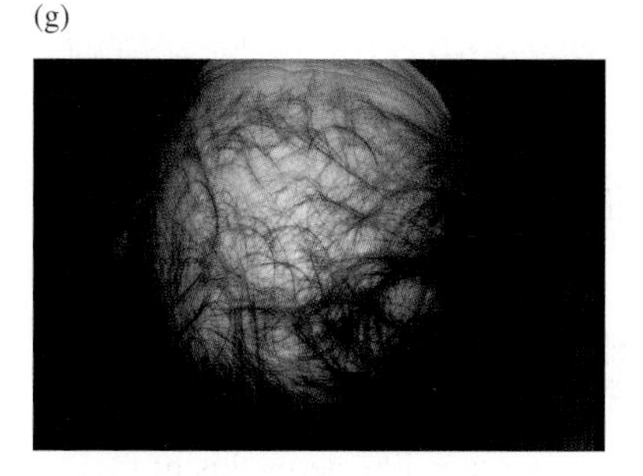

G. Alopecia Zinc 45 mcg/dL (R 50-150) ©Steve Castillo

(h)

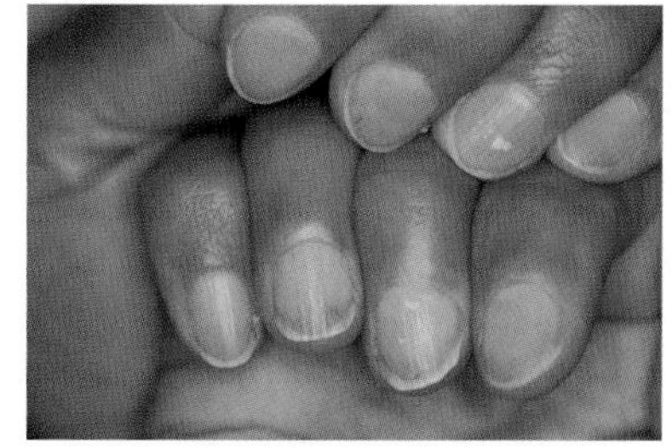

H. Leukonychia Zinc 45 mcg/dL (R 50-150) ©David Giacalone

(i)

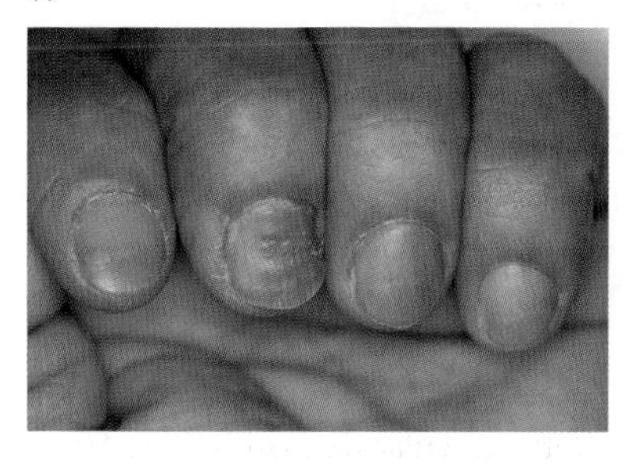

I. Beau's Lines R hand Hospitalization, bacteremia ©David Giacalone

(j)

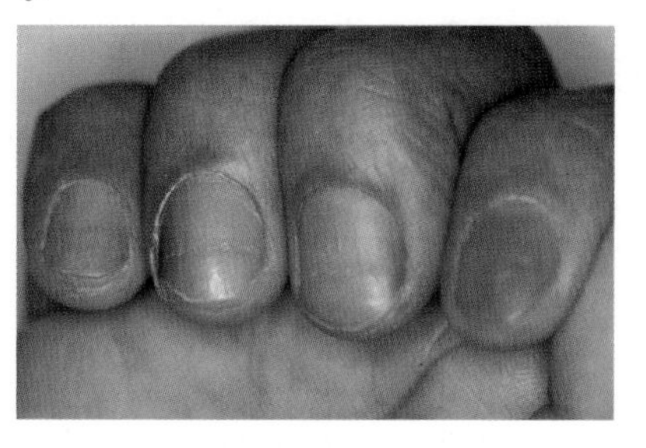

J. Beau's Lines L hand Hospitalization, bacteremia ©David Giacalone

(k)

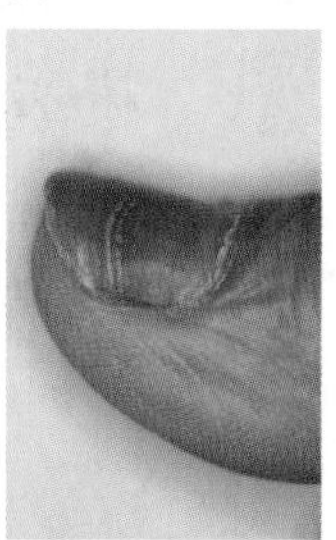

K. Beau's Lines Thumb Hospitalization, bacteremia ©David Giacalone

Fig. 10. *(Continued)*

maintenance HD on fixed zinc intakes indicate increased fecal loss is a major contributor *(226)*. Aluminum hydroxide gels and ferrous sulfate were shown to increase zinc losses *(227)*. Calcium carbonate also binds zinc, but to a lesser degree than calcium acetate *(228)*. Other factors that may contribute to zinc depletion in predialysis patients include reduced zinc intake with low protein diets *(229–231)* and increased urinary losses with angiotensin enzyme inhibitors *(232)* and loop/thiazide diuretics *(233–235)*.

Functional studies in HD patients present conflicting reports on the ability of zinc to improve taste acuity, impotence, and immunity. Supplementation of 120 mg elemental zinc as sulfate post-HD treatment for 6 weeks was shown to improve taste acuity in 95% of cases *(236)*; 50 mg zinc as acetate over a 6- to 12-week period was associated with significant improvement in taste thresholds for salt, sweet, and bitter *(237)*. Matson et al. *(238)* failed to demonstrate improved taste acuity or increased serum zinc in HD patients or controls following 60 mg elemental zinc as sulfate, but poor response may have been related to the inability to normalize plasma zinc concentrations in this study. Indeed, research by Henkin et al. *(239,240)* in non-renal patients suggests salivary zinc concentration best correlates with reversal of taste dysfunction.

A double blind study by Mahajan et al. *(241)* showed improved potency, libido, and frequency of sexual intercourse by HD patient report, following a 6-month course of 50 mg elemental zinc. Later work by Rodger et al. *(242)* with supplementation up to 100 mg elemental zinc daily did not demonstrate improved potency using subjective measures of sperm count, nocturnal penile tumescence, testosterone, sex hormone binding globulin, and gonadotrophin concentration.

Ribeiro et al. *(243)* showed improved delayed hypersensitivity skin tests with Escherichia coli and phytohemagglutinin (PHA)-stimulated lymphocyte blastogenesis after 100 days of 15 mg elemental zinc as acetate twice daily, consistent with improved skin tests by Brodersen et al. *(244)*, Bonomini et al. *(245)*, and Antoniou et al. *(246)*. Briggs et al. *(247)* did not find increased cellular lymphocyte blast formation in vitro with cells isolated from zinc-supplemented HD patients exposed to PHA, pokeweed, conconavalin A mitogens, or streptokinase/streptodornase soluble antigen.

A comprehensive zinc assessment is shown in Fig. 9; lesions observed in hypozincemic dialysis patients are shown in Fig. 10. Seborrheic-like dermatitis in the scalp, between the eyebrows, behind the ears, and along the nasolabial folds predominates; angular stomatitis, circumorificial dermatitis, alopecia, and leukonychia are also seen. Beau's Lines *(248)* were observed in the nails of a HD patient hospitalized with bacteremia complete probable zinc-related lesions.

5.5.3.2. Zinc Toxicity. Given research using zinc supplementation 3–10 times the RDA, concerns about zinc-induced copper deficiency is appropriate. Literature supports increased risk for copper deficiency with modest zinc supplementation, when the zinc to copper ratio is high *(249)*. Monitored zinc to copper intake at 15:1 has shown limited effects on copper absorption *(250)* and is frequently used as the industry standard for nutritional supplements. Classic copper deficiency symptoms of anemia (normocytic and hypochromic), leukopenia, and neutropenia have been observed in patients receiving total parenteral nutrition *(251)* and jejunostomy feedings *(252)*. Idiopathic myelopathy is recently described as a functional copper deficit that curiously mimics sensory ataxia and gait disturbances observed in V-B_{12}deficiency *(253)*. With consistent evidence that zinc deficit exists in kidney failure, the best advice may be to "start low and go slow." Copper indicators should be monitored if zinc supplementation continues for more than 90 days.

5.6. Placing Physical Findings Within the Clinical Context of the Patient

Lesions and functional deficits must be placed within the clinical context of the patient. Their presence fits within a time frame, consistent with the patient's nutritional history, disease state, and laboratory findings. These data enable construction of a nutrient-focused hypothesis, explaining why a particular nutrient imbalance would occur in a particular patient, at a particular time. Findings are shared with the interdisciplinary team, and the RD develops the nutrient intake plan with the patient.

Advanced practice RDs must estimate the patient's nutrient store and availability from food and supplements against the nutritional "cost" *(254)* of dialysis therapy, hospitalization, and medical diagnoses with their prerequisite drug/nutrient profiles. Assiduous physical assessment aimed at early identification of physical findings, functional deficits consistent with nutrient imbalance, with a prompt nutrient replacement response, is crucial in end-organ failure. It is the author's belief that nutrient-focused physical examinations will ultimately direct the nutritional management required by Medicare guidelines *(162)* to achieve "effective nutritional status."

5.7. Summary of the Nutrition Physical Assessment

RDs in today's interdisciplinary team must master comprehensive nutrient assessments that consider tissue stores, dietary intake, nutrient disposition in health and disease, interaction with phamaceuticals/nutraceuticals, and laboratory assessment. Professional and regulatory statutes have strengthened the RD's role by mandating a physical assessment component. Physical assessment techniques, historical studies describing lesions/functional deficits in selected nutrient imbalance, and physiological mechanisms responsible for altered tissue integrity have been explored. Conscientious nutrition-focused physical examinations are complementary to the renal nutrition specialist's disciplined reasoning, hypothesizing nutrient imbalance. Observable lesions and functional deficits emerge as the ultimate expression of clinically significant nutritional disease.

REFERENCES

1. Kopple JD, Greene T, Chumlea WC et al. Relationship between nutritional status and the glomerular filtration rate: results from the MDRD study. Kidney Int 2000;57:1688–703.
2. Rocco MV, Dwyer JT, Larive B et al. The effect of dialysis dose and membrane flux on nutritional parameters in hemodialysis patients: results of the HEMO Study. Kidney Int 2004;65:2321–34.
3. Kopple JD, Wolfson M, Chertow GM et al. Clinical practice guidelines for nutrition in chronic renal failure: a report from the Nutrition Work Group of the National Kidney Foundation Dialysis Outcomes Quality Initiative. Am J Kidney Dis 2000;35:S1–104.
4. Leavey SF, Strawderman RL, Young EW et al. Cross-sectional and longitudinal predictors of serum albumin in hemodialysis patients. Kidney Int 2000;58: 2119–28.
5. Kaysen GA, Don BR. Factors that affect albumin concentration in dialysis patients and their relationship to vascular disease. Kidney Int Suppl 2003;S94–7.
6. Kaysen GA, Chertow GM, Adhikarla R, Young B, Ronco C, Levin NW. Inflammation and dietary protein intake exert competing effects on serum albumin and creatinine in hemodialysis patients. Kidney Int 2001;60:333–40.
7. Lowrie EG, Lew NL. Death risk in hemodialysis patients: the predictive value of commonly measured variables and an evaluation of death rate differences between facilities. Am J Kidney Dis 1990;15:458–82.
8. Kaysen GA. Serum albumin concentration in dialysis patients: why does it remain resistant to therapy? Kidney Int Suppl 2003;S92–8.
9. Pupim LB, Caglar K, Hakim RM, Shyr Y, Ikizler TA. Uremic malnutrition is a predictor of death independent of inflammatory status. Kidney Int 2004;66: 2054–60.

10. Chertow GM, Ackert K, Lew NL, Lazarus JM, Lowrie EG. Prealbumin is as important as albumin in the nutritional assessment of hemodialysis patients. Kidney Int 2000;58:2512–7.
11. Ikizler TA, Wingard RL, Harvell J, Shyr Y, Hakim RM. Association of morbidity with markers of nutrition and inflammation in chronic hemodialysis patients: a prospective study. Kidney Int 1999;55:1945–51.
12. Kopple JD, Mehrotra R, Suppasyndh O, Kalantar-Zadeh K. Observations with regard to the National Kidney Foundation KDOQI clinical practice guidelines concerning serum transthyretin in chronic renal failure. Clin Chem Lab Med 2002;40:1308–12.
13. Iseki K, Yamazato M, Tozawa M, Takishita S. Hypocholesterolemia is a significant predictor of death in a cohort of chronic hemodialysis patients. Kidney Int 2002;61:1887–93.
14. Kopple JD, Greene T, Chumlea WC et al. Relationship between nutritional status and the glomerular filtration rate: results from the MDRD study. Kidney Int 2000;57:1688–703.
15. Kaysen GA. The microinflammatory state in uremia: causes and potential consequences. J Am Soc Nephrol 2001;12:1549–57.
16. Utaka S, Avesani CM, Draibe SA, Kamimura MA, Andreoni S, Cuppari L. Inflammation is associated with increased energy expenditure in patients with chronic kidney disease. Am J Clin Nutr 2005;82:801–5.
17. Caglar K, Peng Y, Pupim LB et al. Inflammatory signals associated with hemodialysis. Kidney Int 2002;62:1408–16.
18. Stenvinkel P, Lindholm B, Heimburger O. Novel approaches in an integrated therapy of inflammatory-associated wasting in end-stage renal disease. Semin Dial 2004;17:505–15.
19. Kalantar-Zadeh K, Stenvinkel P, Pillon L, Kopple JD. Inflammation and nutrition in renal insufficiency. Adv Ren Replace Ther 2003;10:155–69.
20. Eustace JA, Astor B, Muntner PM, Ikizler TA, Coresh J. Prevalence of acidosis and inflammation and their association with low serum albumin in chronic kidney disease. Kidney Int 2004;65:1031–40.
21. Menon V, Greene T, Wang X et al. C-reactive protein and albumin as predictors of all-cause and cardiovascular mortality in chronic kidney disease. Kidney Int 2005;68:766–72.
22. Rao M, Sharma M, Juneja R, Jacob S, Jacob CK. Calculated nitrogen balance in hemodialysis patients: influence of protein intake. Kidney Int 2000;58: 336–45.
23. Chauveau P, Fouque D, Combe C et al. Acidosis and nutritional status in hemodialyzed patients. French Study Group for Nutrition in Dialysis. Semin Dial 2000;13:241–6.
24. Ahuja TS, Mitch WE. The evidence against malnutrition as a prominent problem for chronic dialysis patients. Semin Dial 2004;17:427–31.
25. National Kidney Foundation. Clinical practice guidelines for chronic kidney disease: evaluation, classification and stratification. Am J Kidney Dis 2002;39(2 Suppl 1):S1–S266.
26. Kopple JD and Masssry SG. Nutrition Management of Renal Disease. Baltimore, MD: Williams & Wilkins, 1997.

27. Kopple JD, Greene T, Chumlea WC et al. Relationship between nutritional status and the glomerular filtration rate: results from the MDRD study. Kidney Int 2000;57(4):1688–1703.
28. Laws RA, Tapsell KC, Kelly J. Nutritional status and its relationship to quality of life in a sample of chronic hemodialysis patients. J Ren Nutr 2000;10:139–47.
29. Valderrabano F, Jofre R, Lopez-Gomez JM. Quality of life in end-stage renal disease patients. Am J Kidney Dis 2001;38:443–64.
30. Leung J, Dwyer J, Miller J, Patrick SW, Rocco M, Uhlin L. The role of the dietitian in a multicenter clinical trial of dialysis therapy: the Hemodialysis (HEMO) Study. J Ren Nutr 2001;11(2):101–8.
31. Kalantar-Zadeh K, Unruh M. Health related quality of life in patients with chronic kidney disease. Int Urol Nephrol 2005;37(2):367–78.
32. Owen WF, Jr. Patterns of care for patients with chronic kidney disease in the United States: dying for improvement. J Am Soc Nephrol 2003;14(7 Suppl 2):S76–80.
33. Avesani CM, Draibe SA, Kamimura MA, Dalboni MA, Colugnati FA, Cuppari L. Decreased resting energy expenditure in non-dialysed chronic kidney disease patients. Nephrol Dial Transplant 2004;19(12):3091–7.
34. Avesani CM, Draibe SA, Kamimura MA et al. Assessment of body composition by dual energy X-ray absorptiometry, skinfold thickness and creatinine kinetics in chronic kidney disease patients. Nephrol Dial Transplant 2004;19(9):2289–95.
35. Beto JA, Bansal VK. Nutrition interventions to address cardiovascular outcomes in chronic kidney disease. Adv Chronic Kidney Dis 2004;11(4):391–7.
36. Panesar A, Agarwal R. Resting energy expenditure in chronic kidney disease: relationship with glomerular filtration rate. Clin Nephrol 2003;59(5):360–6.
37. Rigalleau V, Lasseur C, Chauveau P et al. Body composition in diabetic subjects with chronic kidney disease: interest of bio-impedance analysis, and anthropometry. Ann Nutr Metab 2004;48(6):409–13.
38. Snively CS, Gutierrez C. Chronic kidney disease: prevention and treatment of common complications. Am Fam Physician 2004;70(10):1921–8.
39. Rocco MV, Dwyer JT, Larive B et al. The effect of dialysis dose and membrane flux on nutritional parameters in hemodialysis patients: results of the HEMO Study. Kidney Int 2004;65(6):2321–34.
40. Steiber AL, Handu DJ, Cataline DR, Deighton TR, Weatherspoon LJ. The impact of nutrition intervention on a reliable morbidity and mortality indicator: the hemodialysis-prognostic nutrition index. J Ren Nutr 2003;13(3):186–190.
41. Steiber AL, Kalantar-Zadeh K, Secker D, McCarthy M, Sehgal A, McCann L. Subjective Global Assessment in chronic kidney disease: a review. J Ren Nutr 2004;14(4):191–200.
42. Utaka S, Avesani CM, Draibe SA, Kamimura MA, Andreoni S, Cuppari L. Inflammation is associated with increased energy expenditure in patients with chronic kidney disease. Am J Clin Nutr 2005;82(4):801–5.
43. Foster BJ, Leonard MB. Measuring nutritional status in children with chronic kidney disease. Am J Clin Nutr 2004;80(4):801-814.
44. Foster BJ, Leonard MB. Nutrition in children with kidney disease: pitfalls of popular assessment methods. Perit Dial Int 2005;25(Suppl 3):S143–6.
45. Buckner S, Dwyer JT. Do we need a nutrition-specific quality of life questionnaire for dialysis patients? J Ren Nutr 2003;13(4):295–302.

46. Gibson RS. Principles of Nutritional Assessment Measuring Food Consumption of Individuals, 2nd edn. New York: Oxford University Press, 2005, pp. 41–64.
47. Dwyer J, Picciano MF, Raiten DJ. Estimation of usual intakes: what we eat in America-NHANES. J Nutr 2003; 133(2):609S–23S.
48. Bialostosky K, Wright JD, Kennedy-Stephenson J, McDowell M, Johnson CL. Dietary intake of macronutrients, micronutrients, and other dietary constituents: United States 1988–94. Vital Health Stat 11 2002;(245):1–158.
49. Blair D. Dietary methodology issues related to energy balance measurement for NHANES III. Vital Health Stat 4 1992;(27):43–50.
50. Liu T, Wilson NP, Craig CB et al. Evaluation of three nutritional assessment methods in a group of women. Epidemiology 1992;3(6):496–502.
51. Posner BE, Smigelski CG, Krachenfels MM. Dietary characteristics and nutrient intake in an urban homebound population. J Am Diet Assoc 1987;87(4): 452–6.
52. Iannotti RJ, Zuckerman AE, Blyer EM, O'Brien RW, Finn J, Spillman DM. Comparison of dietary intake methods with young children. Psychol Rep 1994;74(3 Pt 1):883–9.
53. Koebnick C, Wagner K, Thielecke F et al. An easy-to-use semiquantitative food record validated for energy intake by using doubly labelled water technique. Eur J Clin Nutr 2005;59(9):989–95.
54. Nelson M, Atkinson M, Darbyshire S. Food photography II: use of food photographs for estimating portion size and the nutrient content of meals. Br J Nutr 1996;76(1):31–49.
55. Nelson M, Haraldsdottir J. Food photographs: practical guidelines II. Development and use of photographic atlases for assessing food portion size. Public Health Nutr 1998;1(4):231–7.
56. Nelson M, Haraldsdottir J. Food photographs: practical guidelines I. Design and analysis of studies to validate portion size estimates. Public Health Nutr 1998;1(4):219–30.
57. Barnard JA, Tapsell LC, Davies PS, Brenninger VL, Storlien LH. Relationship of high energy expenditure and variation in dietary intake with reporting accuracy on 7 day food records and diet histories in a group of healthy adult volunteers. Eur J Clin Nutr 2002;56(4):358–67.
58. Payette H, Gray-Donald K. Dietary intake and biochemical indices of nutritional status in an elderly population, with estimates of the precision of the 7-d food record. Am J Clin Nutr 1991;54(3):478–88.
59. Pedersen AN, Fagt S, Ovesen L, Schroll M. Quality control including validation in dietary surveys of elderly subjects. The validation of a dietary history method (the SENECA-method) used in the 1914-population study in Glostrup of Danish men and women aged 80 years. J Nutr Health Aging 2001;5(4):208–16.
60. Rosell MS, Hellenius ML, de Faire UH, Johansson GK. Associations between diet and the metabolic syndrome vary with the validity of dietary intake data. Am J Clin Nutr 2003;78(1):84–90.
61. Westerterp KR, Goris AH. Validity of the assessment of dietary intake: problems of misreporting. Curr Opin Clin Nutr Metab Care 2002;5(5):489–93.
62. Hankin JH, Wilkens LR, Kolonel LN, Yoshizawa CN. Validation of a quantitative diet history method in Hawaii. Am J Epidemiol 1991;133(6):616–28.

63. Hankin JH, Wilkens LR. Development and validation of dietary assessment methods for culturally diverse populations. Am J Clin Nutr 1994;59(1 Suppl):198S–200S.
64. Hartman AM, Brown CC, Palmgren J et al. Variability in nutrient and food intakes among older middle-aged men. Implications for design of epidemiologic and validation studies using food recording. Am J Epidemiol 1990;132(5):999–1012.
65. Osler M, Schroll M. A dietary study of the elderly in the city of Roskilde 1988/89. Methodological aspects of the relative validity of the dietary history method. Dan Med Bull 1990;37(5):462–6.
66. Willett W. Nutritional Epidemiology. Food Frequency Methods, 2nd edn. New York: Oxford University Press, 1998, pp. 74–94.
67. Kaskoun MC, Johnson RK, Goran MI. Comparison of energy intake by semiquantitative food-frequency questionnaire with total energy expenditure by the doubly labeled water method in young children. Am J Clin Nutr 1994;60(1):43–7.
68. Cade J, Thompson R, Burley V, Warm D. Development, validation and utilization of food-frequency questionnaires - a review. Public Health Nutr 2002;5(4): 567–87.
69. Freedman LS, Carroll RJ, Wax Y. Estimating the relation between dietary intake obtained from a food frequency questionnaire and true average intake. Am J Epidemiol 1991;134(3):310–20.
70. Haraldsdottir J. Minimizing error in the field: quality control in dietary surveys. Eur J Clin Nutr 1993;47(Suppl 2):S19–24.
71. Block G, Woods M, Potosky A, Clifford C. Validation of a self-administered diet history questionnaire using multiple diet records. J Clin Epidemiol 1990;43(12):1327–35.
72. Block G, Hartman AM, Naughton D. A reduced dietary questionnaire: development and validation. Epidemiology 1990;1(1):58–64.
73. Hankin JH, Stram DO, Arakawa K et al. Singapore Chinese Health Study: development, validation, and calibration of the quantitative food frequency questionnaire. Nutr Cancer 2001;39(2):187–95.
74. Mitch WE, Klahr S, ed. Handbook of Nutrition and the Kidney, 4th edn. Baltimore, MD: Lippincott Williams & Wilkins, 2002.
75. Ohri-Vachapati P, Seghal AR. Qualify of life implications in inadequate protein nutrition among hemodialysis patients. J Ren Nutr 1999;9(1):9–13.
76. Diaz-Buxo JA, Lowrie EG, Lew NL. Quality-of life evaluation using short form 36: comparison in hemodialysis and peritoneal dialysis patients. Am J Kideny Dis 2000;35(2): 293–300.
77. Edgell ET, Cooms SJ, Carter WB. A review of health-related quality of life measures used in end-stage renal disease. Clin Ther 1996;18:887–938.
78. Carmichael P, Papoola J, John I et al. Assessment of quality of life in a single center dialysis population using the KDQOL-SF36 questionnaire. Qual Life Res 2000;9:195–205.
79. Osawa Y, Aoike I, Sakurabayashi T, Miyazaki S, Yuasa Y, Suzuki M. [Serum calcium levels at first renal replacement therapy and patients survival]. Clin Calcium 2005;15(Suppl 1):167–72.
80. NKF KDOQI Clinical practice guidelines in chronic renal failure. Am J Kidney Dis 2000;35(6 Suppl 2):31–5140.

81. Combe C, McCullough KP, Asano Y, Ginsberg N, Maroni BJ, Pifer TB. Kidney Disease Outcomes Quality Initiative (KDOQI) and the Dialysis Outcomes and Practice Patterns Study (DOPPS): nutrition guidelines, indicators, and practices. Am J Kidney Dis 2004;44(5 Suppl 3):39–46.
82. Gibson RS, Sazawal S, Peerson JM. Design and quality control issues related to dietary assessment, randomized clinical trials and meta-analysis of field-based studies in developing countries. J Nutr 2003;133(5 Suppl 1):1569S–73S.
83. Byham-Gray LD. Outcomes research in nutrition and chronic kidney disease: perspectives, issues in practice, and processes for improvement. Adv Chronic Kidney Dis 2005;12(1):96–106.
84. Sarkar SR, Kuhlmann MK, Khilnani R et al. Assessment of body composition in long-term hemodialysis patients: rationale and methodology. J Ren Nutr 2005;15(1):152–8.
85. Woteki CE. Measuring dietary patterns in surveys. Vital Health Stat 4 1992;(27):101–8.
86. Burrowes J, Levin N. Morbidity and mortality in dialysis patients. Diet Curr 1992;19:10–14.
87. Owen WFJ, Lew NL, Liu Y, Lowrie EG, Lazarus JM. The urea reduction ratio and serum albumin concentration as predictors of mortality in patients undergoing hemodialysis. N Engl J Med 1993;329:1001–06.
88. Chumlea WC, Dwyer JT, Paranandi L et al. Nutritional status assessed from anthropometric measures in the HEMO Study. J Ren Nutr 2003;13:31–8.
89. Collins AJ, Hanson G, Umen A, Kjellstrand C, Keshaviah P. Changing risk factor demographics in end-stage renal disease patients entering hemodialysis and the impact on long-term mortality. Am J Kidney Dis 1990;15:422–32.
90. Blumenkrantz MJ, Kopple JD, Gutman RA et al. Methods for assessing nutritional status of patients with renal failure. Am J Clin Nutr 1980;33:1567–85.
91. Chumlea WC. Anthropometric and body composition assessment in dialysis patients. Semin Dial 2004;17:466–70.
92. Ellis KJ, Shypailo RJ. Nuclear-based methods for the measurement of body composition: applications in pediatric research. In: Kral JG, Van Itallie TB, eds. Recent Developments in Body Composition Analysis: Methods and Applications. London: Smith-Gordon, 1993, pp. 75–86.
93. Seidell JC, Cigolini M, Charzewska J et al. Indicators of fat distribution, serum lipids, and blood pressure in European women born in 1948–the European Fat Distribution Study. Am J Epidemiol 1989;130:53–65.
94. Kvist H, Sjostrom L, Tylen U. Adipose tissue volume determinations in women by computed tomography: technical considerations. Int J Obes 1986;10: 53–67.
95. Baumgartner RN, Rhyne RL, Garry PJ, Chumlea WC. Body composition in the elderly from MRI: Associations with cardiovascular disease risk factors. In: Ellis KJ, Eastman JD, eds. Human Body Composition: In Vivo Methods, Models and Assessment. New York: Plenum Press, 1993, pp. 35–8.
96. Wang J, Dilmanian FA, Thornton J. In vivo neuteron activation analysis for body fat: comparisions by seven methods. In: Ellis KJ, Eastman JD, eds. Human Body Composition: InVivo Methods, Models and Assessment. New York: Plenum Press, 1992, pp. 31–4.

97. Chertow GM, Lazarus JM, Lew NL, Ma L, Lowrie EG. Development of a population-specific regression equation to estimate total body water in hemodialysis patients. Kidney Int 1997;51:1578–82.
98. Stall S, Ginsberg NS, DeVita MV et al. Percentage body fat determination in hemodialysis and peritoneal dialysis patients: a comparison. J Ren Nutr 1998;8:132–6.
99. Ikizler TA, Sezer MT, Flakoll PJ et al. Urea space and total body water measurements by stable isotopes in patients with acute renal failure. Kidney Int 2004;65:725–32.
100. Dumler F, Schmidt R, Kilates C, Faber M, Lubkowski T, Frinak S. Use of bioelectrical impedance for the nutritional assessment of chronic hemodialysis patients. Miner Electrolyte Metab 1992;18:284–7.
101. Haarbo J, Gotfredsen A, Hassager C, Christiansen C. Validation of body composition by dual energy X-ray absorptiometry (DEXA). Clin Physiol 1991;11: 331–41.
102. Johnson J, Dawson-Hughes B. Precision and stability of dual-energy X-ray absorptiometry measurements. Calcif Tissue Int 1991;49:174–8.
103. Orwoll ES, Oviatt SK. Longitudinal precision of dual-energy x-ray absorptiometry in a multicenter study. The Nafarelin/Bone Study Group. J Bone Miner Res 1991;6:191–7.
104. DeVita MV, Stall SH. Dual-energy X-ray absorptiometry: a review. J Ren Nutr 1999;9:178–81.
105. Lukaski HC. Soft tissue composition and bone mineral status: evaluation by dual-energy X-ray absorptiometry. J Nutr 1993;123:438–43.
106. Roubenoff R, Kehayias J, Dawsonhughes B, Heymsfield S. Use of dual-energy X-ray absorptiometry in body-composition studies - not yet a gold standard. Am J Clin Nutr 1993;58:589–91.
107. Kohrt WM. Body composition by DXA: tried and true? Med Sci Sports Exerc 1995;27:1349–53.
108. Binkley N, Kiebzak GM, Lewiecki EM et al. Recalculation of the NHANES database SD improves T-score agreement and reduces osteoporosis prevalence. J Bone Miner Res 2005;20:195–201.
109. Tataranni PA, Ravussin E. Use of dual-energy X-ray absorptiometry in obese individuals. Am J Clin Nutr 1995;62:730–4.
110. Guo SS, Wisemandle WA, Tyleshevski F et al. Inter-machine and inter-method differences in body composition measures from dual energy x-ray absorptiometry. J Nutr Health Aging 1997;1:29–37.
111. Schoeller D, Tylavsky FA, Baer DJ et al. QDR 4500A dual X-ray absorptiometer underestimates fat mass in comparison with criterion methods in adults. Am J Clin Nutr 2005;8:1018–25.
112. Khaled MA, Lukaski HC, Watkins CL. Determination of total body water by deuterium NMR. Am J Clin Nutr 1987;45:1–6.
113. Rebouche C, Pearson G, Serfass R, Roth C, Finley J. Evaluation of nuclear magnetic resonance spectroscopy for determination of deuterium in body fluids: application to measurement of total body water in human infants. Am J Clin Nutr 1987;45:373–80.
114. Chumlea WC, Guo SS, Zeller CM et al. Total body water reference values and prediction equations for adults. Kidney Int 2001;59:2250–8.

115. Chumlea WC, Guo SS, Zeller CM, Reo NV, Siervogel RM. Total body water data for white adults 18 to 64 years of age: The Fels Longitudinal Study. Kidney Int 1999;56:244–52.
116. Cheek D. Extra-cellular volume: its structure and measurement and influence of age and disease. J Pediatr 1961;58:103–25.
117. Vaisman N, Pencharz P, Koren G, Johnson J. Comparison of oral and intravenous administration of sodium bromide for extra-cellular water measurements. Am J Clin Nutr 1987;46:1–4.
118. Chumlea WC, Guo SS, Kuczmarski RJ et al. Body composition estimates from NHANES III bioelectrical impedance data. Int J Obes Relat Metab Disord 2002; 26:1596–1609.
119. Siri W. Body composition from fluid spaces and density analysis of methods. In: Henschel A, ed. Techniques for Measuring Body Composition. Vol. 61. Washington, DC: National Academy Press, 1961, pp. 223–44.
120. Moore F, Haley H, Brooks L. Changes of body composition in disease. Surg Gynecol Obstet 1952;95:155–80.
121. Depner TA. Quantifying hemodialysis. Am J Nephrol 1996;16:17–28.
122. Arkouche W, Fouque D, Pachiaudi C et al. Total body water and body composition in chronic peritoneal dialysis patients. J Am Soc Nephrol 1997;8: 1906–14.
123. Woodrow G, Oldroyd B, Turney JH, Davies PS, Day JM, Smith MA. Measurement of total body water and urea kinetic modelling in peritoneal dialysis. Clin Nephrol 1997;47:52–7.
124. Watson PE, Watson ID, Batt RD. Total body water volumes for adult males and females estimated from simple anthropometric measurements. Am J Clin Nutr 1980;33:27–39.
125. Sun SS, Chumlea WC, Heymsfield SB et al. Development of bioelectrical impedance analysis prediction equations for body composition with the use of a multicomponent model for use in epidemiological surveys. Am J Clin Nutr 2003;77:331–40.
126. Chumlea WC, Schubert CM, Reo N, Sun SS, Siervogel RM. Total body water volume for White children and adolescents and anthropometric prediction equations: the Fels Longitudinal Study. Kidney Int 2005: 68:2317–22.
127. Chumlea WC, Sun SS. Bioelectrical impedance analysis. In: Heymsfield SB, Lohman TG, Wang Z, Going SB, eds. Human Body Composition. Champaign, IL: Human Kinetics Books, 2005, pp. 79–88.
128. Lee SW, Song JH, Kim GA, Lee KJ, Kim MJ. Assessment of total body water from anthropometry-based equations using bioelectrical impedance as reference in Korean adult control and haemodialysis subjects. Nephrol Dial Transplant 2001;16:91–7.
129. Konings CJ, Kooman JP, Schonck M et al. Influence of fluid status on techniques used to assess body composition in peritoneal dialysis patients. Perit Dial Int 2003;23:184–90.
130. Chumlea WC, Guo SS. Bioelectrical impedance and body composition: present status and future directions. Nutr Rev 1994;52:123–31.
131. Hanai T. Electrical properties of emulsions. In: Sherman P, ed. Emulsion Science. London: Academic Press, 1968, p. 354.

132. Van Loan M, Koop L, King J, Wong W, Mayclin P. Fluid changes during pregnancy: use of bioimpedance spectroscopy. J Appl Physiol 1995;78: 1037–42.
133. Van Loan M, Withers P, Mattie J, Mayclin P. Use of Bio-impedance spectroscopy (BIS) to determine extracellular fluid (ECF), intracellular fluid (ICF), total body water (TBW), and fat-free mass (FFM). In: Ellis K, Eastman J, eds. Human Body Composition: In Vivo Methods, Models, and Assessment. New York: Plenum Publishers, 1993, p. 67.
134. Van Loan MD, Strawford A, Jacob M, Hellerstein M. Monitoring changes in fat-free mass in HIV-positive men with hypotestosteronemia and AIDS wasting syndrome treated with gonadal hormone replacement therapy. AIDS 1999;13:241–8.
135. De Lorenzo A, Andreoli A, Matthie J, Withers P. Predicting body cell mass with bioimpedance by using theoretical methods: a technological review. J Appl Physiol 1997;82:1542–58.
136. Klahr S, Levey A, Beck G. Modification of Diet in Renal Disease Study Group: the effects of dietary protein restriction and blood pressure control on the progression of chronic renal disease: The Modification of Diet in Renal Disease Study. N Engl J Med 1994;330:377–884.
137. Eknoyan G, Levey A, Beck G, Agodoa L, Daugirdas J, Kusek J. The hemodialysis (HEMO) study: rationale for selection of interventions. Semin Dial 1996;9: 24–33.
138. Lohman TG, Roche AF, Martorell R, eds. Anthropometric Standardization Reference Manual. Champaign, IL: Human Kinetics Publishers, 1988:vi + 177.
139. Chumlea WC. Anthropometric assessment of nutritional status in renal disease. J Ren Nutr 1997;7:176–81.
140. USDHHS. NHANES III Anthropometric Procedures (videotape). Washington, DC, 1996.
141. WHO. Physical Status: The Use and Interpretation of Anthropometry. Geneva. WHO, 1995 (854).
142. Donini LM, Chumlea WC, Vellas B, Del Balzo V, Cannella C. Obesity in the Elderly. J Nutr Health Aging 2006;10:52–4.
143. Despres J, Prudhomme D, Pouliot M, Tremblay A, Bouchard C. Estimation of deep abdominal adipose-tissue accumulation from simple anthropometric measurements in men. Am J Clin Nutr 1991;54:471–7.
144. Patrick JM, Bassey EJ, Fentem PH. Changes in body fat and muscle in manual workers at and after retirement. Eur J Appl Physiol Occup Physiol 1982;49: 187–96.
145. Chumlea WC, Wisemandle WA, Guo SS, Siervogel RM. Relations between frame size and body composition and bone mineral status. Am J Clin Nutr 2002;75:1012–6.
146. Chumlea WC, Guo SS, Steinbaugh ML. The prediction of stature from knee height for black and white adults and children, with application to the mobility-impaired. J Am Diet Assoc 1994;94:1385–8, 1391.
147. Chumlea WC, Guo SS, Wholihan K, Cockram DB, Kuczmarski RJ, Johnson CL. Stature prediction equations for elderly non-Hispanic White, non-Hispanic Black, and Mexican American persons: from NHANES III (1988–94). J Am Diet Assoc 1998;98:137–42.

148. Najjar M, Rowland M. Anthropometric reference data and prevalence of overweight. United States, 1976–80. Vital and Health Statistics, Series 11, No 238, National Center for Health Statistics 1987, pp. 1–73.
149. Nelson EE, Hong CD, Pesce AL, Peterson DW, Singh S, Pollak VE. Anthropometric norms for the dialysis population. Am J Kidney Dis 1990;16:32–7.
150. Sun SS, Chumlea WC. Statistical Methods. In: Going SB, ed. Human Body Composition. Champaign, IL; Human Kinetics, 2005, pp. 151–60.
151. Lohman TG, Caballero B, Himes JH et al. Estimation of body fat from anthropometry and bioelectrical impedance in Native American children. Int J Obes Relat Metab Disord 2000;24:982–8.
152. Wagner DR, Heyward VH, Kocina PS, Stolarczyk LM, Wilson WL. Predictive accuracy of BIA equations for estimating fat-free mass of black men. Med Sci Sports Exerc 1997;29:969–74.
153. Lukaski H. Methods for the assessment of human body composition: traditional and new. Am J Clin Nutr 1987;46:537–56.
154. Chumlea WC, Baumgartner RN. Body composition, morbidity and mortality. In: Marriot BM, Grumpstrup-Scott J, eds. Body Composition and Physical Performance, Applications for the Military Services. Washington, DC: National Academy Press, 1992, pp. 175–84.
155. Woteki CE, Briefel RR, Kuczmarski RM. Contributions of the National Center for Health Statistics. Am J Clin Nutr 1988;47:320–8.
156. Kuczmarski R, Flegal K, Campbell S, Johnson C. Increasing prevalence of overweight among US adults. JAMA 1994;272:205–11.
157. Federal Register (June 3, 1976) Fed Register (41 FR 22501).
158. 1995 Comprehensive Accreditation Manual for Hospitals. Oakbrook Terrace, IL: Joint Commission on Accreditation of Healthcare Organizations, 1994, pp. 151–2.
159. Gilmore GA, Maillet JO, Mitchell BE. Determining educational preparation based on job competencies of entry-level dietetics practitioners. J Am Diet Assoc 1997;97:306–16.
160. American Dietetic Association Standards of Practice in Nutrition Care for the Registered Dietitian, Scope of dietetics practice framework. American Dietetics Association 2005: Retrieved March 10, 2006. Accessed at http://www.eatright.org:80/cps/rde/xchg/ada/hs.xsl/career_1917_ENU_HTML.htm.
161. KDOQI National Kidney Foundation: Clinical practice guidelines for nutrition in chronic renal failure. Am J Kidney Dis 2000;35(Suppl 2):S1–S140.
162. Department of Health and Human Services. Medicare Program; Conditions for Coverage for End Stage Renal Disease Facilities; Proposed Rule. Federal Register 2005;70(23):6184–232.
163. Marieb EN, Mallatt J. Human Anatomy, 3rd edn. San Francisco: Addison Wesley Longman, Inc., 2001.
164. Hammond KA, Hillhouse J. Nutrition-Focused Physical Assessment Skills for Dietitians, Study Guide, 2nd edn. Chicago, IL: Dietitians in Nutrition Support, 2000.
165. Hammond KA. Physical assessment: a nutritional perspective. Nurs Clin North Am 1997;32:779–90.
166. Evans-Stoner N. Nutrition assessment: a practical approach. Nurs Clin North Am 1997;32:637–50.

167. Manning E, Shenkin A. Nutrition assessment of the critically ill. Crit Care Clin 1995;11:603–34.
168. Kight MA. The Nutrition Physical Examination. CRNQ 1987;11(3):9–12.
169. Kight MA, Kelly MP, Castillo ST, Migliore V. Conducting physical examination rounds for manifestations of nutrient deficiency or excess. Nutr Clin Pract 1999;14:93–8.
170. Kelly MP, Kight MA, Rodriguez R, Castillo ST. A diagnostically reasoned case study with particular emphasis on B_6 and zinc imbalance directed by clinical history and nutrition physical examination findings. Nutr Clin Pract 1998;13: 32–9.
171. Detsky AS, McLaughlin JR, Baker JP. What is subjective global assessment? J Parenter Enteral Nutr 1987;11:8–13.
172. McCann L. Subjective global assessment as it pertains to nutrition status. Dial Transplant 1996;25:190–225.
173. Steiber AL, Kalantar-Zadeh K, Secker D, McCarthy M, Sehgal A, McCann L. Subjective global assessment in chronic kidney disease: a review. J Ren Nutr 2004;14:191–200.
174. LeBlond RF, DeGowin RL, Brown DD, eds. DeGowin's Diagnostic Examination: The Complete Guide to Assessment, Examination, Differential Diagnosis, 8th edn. San Francisco: The McGraw-Hill Companies, 2004.
175. Osaki T, Ohshima M, Tomita Y, Matsugi N, Nomura Y. Clinical and physiological investigations in patients with taste abnormality. J Oral Pathol Med 1996;25:38–43.
176. Bean WB. An analysis of subjectivity in the clinical examination in nutrition. J Appl Physiol 1948;1:458–68.
177. Sheth TN, Choudhry NK, Bowes M, Detsky A. The relation of conjunctival pallor to the presence of anemia. J Gen Intern Med 1997;12:102–06.
178. Walsh FB, Howard JE. Conjunctival and corneal lesions in hypercalcemia. J Clin Endocrinol 1947;7:644–52.
179. Tosti A, Baran R, Dawber P. The nails in systemic disease and drug-induced changes. In: DeBerker DAR, Baran R, Dawber RPR, eds. Handbook of Diseases of the Nails and their Management. Osney Mead, Oxford: Blackwell Science Ltd., 1995, p. 91.
180. DeBerker D. What do Beau's lines mean? Int J Dermatol 1994;33:545–6.
181. Goldberger J, Waring CH, Willets DG. The prevention of pellagra. Public Health Rep 1915;30:3117.
182. Goldsmith GA, Sarett, HP, Register UD, Gibbons J. Studies of niacin requirements in man. I. Experimental pellagra in subjects on corn diets low in niacin and tryptophan. J Clin Invest 1952;31:533–42.
183. Rose WC. Amino acid requirements of man. Fed Proc 1949;8:546.
184. Harris HF. Pellagra. New York: McMillan Company, 1919, p. 234.
185. Ishii N, Nishihara Y. Pellagra among chronic alcoholics: clinical and pathological study of 20 necroscopy cases. J Neurol Neurosurg Psychiatry 1981;44:209–15.
186. Hawn LJ, Guldan GJ, Chillag S, Klein L. A case of pellagra and a South Carolina history of the disorder. J S C Med Assoc 2003;99:220–3.
187. Pitsavas S, Andreou C, Bascialla F, Bozikas VP, Karavatos A. Pellagra encephalopathy following B-complex vitamin treatment without niacin. Int J Psychiatry Med 2004;34:91–5.

188. Standing committee on the scientific evaluation of dietary reference intakes Niacin. In: Dietary Reference Intakes for Thiamin, Riboflavin, Niacin, Vitamin B_6, Folate, Vitamin B_{12}, Pantothenic Acid, Biotin, and Choline. Washington, DC: National Academy Press, 1998, pp. 123–49.
189. Wang AY-M, Sea MM-M, Ip T, Law MC, Chow KM, Lui SF, Li PK-T, Woo J. Independent effects of residual renal function and dialysis adequacy on dietary micronutrient intakes in patients receiving continuous ambulatory peritoneal dialysis. Am J Clin Nutr 2002;76:569–76.
190. Goldsmith GA. Experimental niacin deficiency. J Am Diet Assoc 1956;32: 312–6.
191. Saunders EB. The gynecological, obstetrical, and surgical aspects of pellagra. Am J Insanity 1911;67:541–51.
192. Goldsmith GA, Rosenthal HL, Gibbens J, Unglaub WG. Studies of niacin requirement in man. II. Requirement on wheat and corn diets low in tryptophan. J Nutr 1955;56:371–86.
193. Daya SM. Vitamin A abnormalities, pellagra, Vitamin C deficiency. In: Mannis MJ, Macsai MS, Huntley AC, eds. Eye and Skin Disease. Philadelphia: Lippincott-Raven, 1996, pp. 124–5.
194. Pitche PT. Pellagra. Sante 2005;15:205–8.
195. Perry ML. A report of two cases of pellagra. Am J Insanity 1911;67:553–7.
196. Snyderman SE, Carretteto R, Holt LE. Pyridoxine deficiency in the human being. Fed Proc 1950;9:371–84.
197. Vilter RW, Mueller JF, Glazer HS, Jarrold T, Abraham J, Thompson C, Hawkins VR. The effect of V-B_6 deficiency induced by desoxypyridoxine in human beings. J Lab Clin Med 1953;42:335–57.
198. Standing committee on the scientific evaluation of dietary reference intakes Vitamin B_6. In: Dietary Reference Intakes for Thiamin, Riboflavin, Niacin, Vitamin B_6, Folate, Vitamin B_{12}, Pantothenic Acid, Biotin, and Choline. Washington, DC: National Academy Press, 1998, pp. 150–188.
199. Hoffman LM, Ross J. The role of heme in the maturation of erythroblasts: The effects of inhibition of pyridoxine metabolism. Blood 1980;55:762–71.
200. Hawkins WW, Barsky J. An experiment on human V-B_6 deprivation. Science 1948;108:284–6.
201. Snider D. Pyridoxine supplementation during isoniazid therapy. Tubercle 1980;61:191–96.
202. Guilarte TR. Vitamin B_6 and cognitive development. Nutr Rev 1993;51:193–8.
203. Roberts E. Gamma-Aminobutyric Acid-V-B_6 relationships in the brain. Am J Clin Nutr 1963;12:291–307.
204. Wood JD, Peesker SJ. The effect on GABA metabolism in brain of isonicotinic acid hydrazide and pyridoxine as a function of time after administration. J Neurochem 1972;19:1527–37.
205. Mydlik M, Derzsiova K, Zemberova E. Influence of water and sodium diuresis and furosemide on urinary excretion of V-B_6, oxalic acid, and V-C in chronic renal failure. Miner Electrolyte Metab 1999;25:352–6.
206. Vidrio J. Interaction with pyridoxal as a possible mechanism of hydralazine hypotension. J Cardiovasc Pharmacol 1990;15:150–6.
207. Caselli MA, Prywes HC, Sharnoff DG. Isoniazid toxicity as a cause of peripheral neuropathy. J Am Podiatry Assoc 1981;71:691–3.

208. Wood ER. Isoniazid toxicity: pyridoxine controlled seizures in a dialysis patient. J Kans Med Soc 1981;82:551–2.
209. Blumberg EA, Gil RA. Cerebellar syndrome caused by isoniazid. Ann Pharmacother 1990;24:829–31.
210. Siskind MS, Thienemann D, Kirlin L. Isoniazid-induced neurotoxicity in chronic dialysis patients: report of three cases and a review of the literature. Nephron 1993;64:303–06.
211. Asnis DS, Bhat G, Melchert AF. Reversible seizures and mental status changes in a dialysis patient on isoniazid preventive therapy. Ann Pharmacother 1993;27:444–6.
212. Kasama R, Koch T, Canals-Navas C, Pitone JM. V-B_6 and hemodialysis: impact of high-flux/high efficiency dialysis and review of the literature. Am J Kidney Dis 1996;27:680–6.
213. Ross EA, Shah GM, Reynolds RD, Sabo A, Pichon M. V-B_6 requirements of patients on chronic peritoneal dialysis. Kidney Int 1989;36:702–06.
214. Schaumberg H, Kaplan J, Windebank A, Vick N, Rasmus S, Pleasure D, Brown MJ. Sensory neuropathy from pyridoxine abuse. N Engl J Med 1983;309:445–8.
215. Parry GJ, Bredesen DE. Sensory neuropathy with low-dose pyridoxine. Neurology 1985;35:1466–8.
216. Dalton K, Dalton MJT. Characteristics of pyridoxine overdose neuropathy syndrome. Acta Neurol Scand 1987;76:8–11.
217. Kelly MP, Kight MA, Rodriguez R, Castillo S. A diagnostically reasoned case study with particular emphasis on B_6 and zinc imbalance directed by clinical history and nutrition physical examination findings. Nutr Clin Pract 1998;13: 32–9.
218. Prasad AS, Halsted JA, Nadimi M. Syndrome of iron deficiency anemia, hepatosplenomegaly, hypogonadism, dwarfism, and geophagia. Am J Med 1961;31:532.
219. Moynahan EJ. Acrodermatitis enteropathica: a lethal inherited human zinc deficiency disorder. Lancet 1974;2:399–400.
220. Hambidge M. Human zinc deficiency. J Nutr 2000;130:1344S–9S.
221. Lowe NM, Woodhouse LR, Sutherland B, Shames DM, Burri BJ, Abrams AS, Turnlund JR, Jackson MJ, King JC. Kinetic parameters and plasma zinc concentrations correlate well with loss and gain of zinc from man. J Nutr 2004;134: 2178–81.
222. Baer MT, King JC. Tissue zinc levels and zinc excretion during experimental zinc depletion in young men. Am J Clin Nutr 1984;39:556–70.
223. Hsieh YY, Shen WS, Lee LY, Wu TL, Ning HC, Sun CF. Long term changes in trace elements in patients undergoing chronic hemodialysis. Biol Trace Elem Res 2006;109:115–21.
224. Krachler M, Scharfetter H, Wirnsberger GH. Kinetics of the metal cations magnesium, calcium, copper, zinc, strontium, barium, and lead in chronic hemodialysis patients. Clin Nephrol 2000;54:35–44.
225. Hosokawa S, Oyamaguchi A, Yoshida O. Trace elements and complications in patients undergoing chronic hemodialysis. Nephron 1990;55:375–9.
226. Mahajan SK, Bowersox EM, Rye DL, Abu-Hamdan DK, Prasad AS, McDonald FD, Biersack KL. Factors underlying abnormal zinc metabolism in uremia. Kidney Int 1989;36(Suppl 27):S269–73.

227. Abu-Hamdan DK, Mahajan SK, Migdal SD, Prasad AS, McDonald FD. Zinc tolerance test in uremia: effect of ferrous sulfate and aluminum hydroxide. Ann Intern Med 1986;104:50–2.
228. Hwang SJ, Lai YH, Chen HC, Tsai JH. Comparisons of the effects of calcium carbonate and calcium acetate on zinc tolerance test in hemodialysis patients. Am J Kidney Dis 1992;19:57–60.
229. Standing committee on the scientific evaluation of dietary reference intakes Zinc. In: Dietary Reference Intakes for V-A, V-K, Arsenic, Boron, Chromium, Copper, Iodine, Iron, Manganese, Molybdenum, Nickel, Silicon, Vanadium, and Zinc. Washington, DC: National Academy Press, 2001, pp. 442–501.
230. Gilli P, Fagioli F, Paoli E, Vitali E, Farinelli A. Is zinc status a problem in dietary treatment of chronic renal failure? (letter). Nephron 1985;40:382.
231. Blendis LM, Ampil M, Wilson DR, Kiwan J, Labranche J, Johnson M, Williams C. The importance of dietary protein in the zinc deficiency of uremia. Am J Clin Nutr 1981;34:2658–61.
232. Golik A, Modai D, Averbukh M, Sheffy M, Shamis A, Cohen N, Shaked U, Dolev E. Zinc metabolism in patients treated with captopril versus enalapril. Metabolism 1990;39:665–7.
233. Wester PO. Urinary zinc excretion during treatment with different diuretics. Acta Med Scand 1980;208:209–12.
234. Leary WP, Reyes AJ, Wynne RD, Van der Byl K. Renal excretory actions of furosemide, of hydrochlorothiazide and vasodilator flosequinan. J Int Med Res 1990;18:120–41.
235. Cohen N, Golik A, Dishi V, Zaidenstein R, Weissgarten J, Averbukh Z, Modao D. Effect of furosemide oral solution versus furosemide tablets on diuresis and electrolytes in patients with moderate congestive heart failure. Miner Electrolyte Metab 1996;22:248–52.
236. Atkin-Thor E, Goddard BW, O'Nion J, Stephen RL, Kolff WJ. Hypoguesia and zinc depletion in chronic dialysis patients. Am J Clin Nutr 1978;31:1948–51.
237. Mahajan SK, Prasad AS, Lanbujon J, Abbasi AA, Briggs WA, McDonald FD. Improvement of uremic hypogeusia by zinc: a double blind study. Am J Clin Nutr 1980;33:1517–21.
238. Matson A, Wright M, Oliver A, Woodrow G, King N, Dye L, Blundell J, Brownjohn A, Turney J. Zinc supplementation at conventional doses does not improve the disturbance of taste perception in HD patients. J Ren Nutr 2003;13:224–8.
239. Henkin RI, Martin BM, Agarwal RP. Decreased parotid saliva gustin/carbonic anhydrase VI secretion: an enzyme disorder manifested by gustatory and olfactory dysfunction. Am J Med Sci 1999;318:380–91.
240. Henkin RI, Martin BM, Agarwal RP. Efficacy of exogenous oral zinc in treatment of patients with carbonic anhydrase VI deficiency. Am J Med Sci 1999;318: 392–404.
241. Mahajan SK, Abbasi AA, Prasad AS, Rabbaoni P, Briggs WA, McDonald FD. Effect of oral zinc therapy on gonadal function in hemodialysis patients. Ann Intern Med 1982;97:357–61.
242. Rodger RSC, Sheldon WL, Watson MJ, Dewar JH, Wilkinson R, Ward MK, Kerr DNS. Zinc deficiency and hyperprolactinaemia are not reversible causes of sexual dysfunction in uraemia. Nephrol Dial Transplant 1989;4:888–92.

243. Ribeiro RCJ, Sales VSF, Neves FDAR. Draibe S, Brandao-Neto J. Effects of zinc on cell mediated immunity in chronic hemodialysis patients. Biol Trace Element Res 2004;98:209–17.
244. Brodersen HP, Heltkamp W, Larbig D (Letter). Zinc supplementation and cellular and humoral immune function in chronic hemodialysis patients. Nephrol Dial Transplant 1994;9:736–7.
245. Bonomini M, DiPaolo B, DeRiso F. Effects of zinc supplementation on chronic hemodialysis patients. Nephrol Dial Transplant 1993;8:1166–8.
246. Antoniou LD, Shalhoub RJ, Schechter GP. The effect of zinc on cellular immunity in chronic uremia. Am J Clin Nutr 1981;34:1912–7.
247. Briggs WA, Pedersen MM, Mahajan SK, Silliz DH, Prasad AS, McDonald F. Lymphocyte and granulocyte function in zinc-treated and zinc-deficient hemodialysis patients. Kidney Int 1982;21:827–32.
248. Weismann K. Lines of Beau: possible markers of zinc deficiency. Acta Derm Venereol 1977;57:88–90.
249. Sandstead HH. Requirements and toxicity of essential trace elements, illustrated by zinc and copper. Am J Clin Nutr 1995;61:621S–4S.
250. August D, Jangbhorbani M, Young VR. Determination of zinc and copper absorption at three dietary Zinc:Copper ratios by using stable isotopic methods in young adult and elderly subjects. Am J Clin Nutr 1989;50:1457–63.
251. Fujita M, Itakura R, Takagi Y, Okada A. Copper deficiency during TPN: clinical analysis of three cases. J Parenter Enteral Nutr 1989;13:421–5.
252. Jayakumar S, Micallif-Eynaud PD, Lyon TDB, Cramb R, Jilaihawi AN, Prakash D. Acquired copper deficiency following prolonged jejunostomy feeds. Ann Clin Biochem 2005;42:227–31.
253. Kumar N, Gross JB, Ahlskog JE. Copper deficiency myelopathy produces a clinical picture like subacute combined degeneration. Neurol 2004;63:33–9.
254. Kelly MP, Kight MA, Torres M, Migliore V. Nutritional cost of hospitalization for HD patients and time to achieve nutritional resiliency. J Ren Nutr 1994; 4: 1–11.
255. Leikin JB, Paloucek FP, eds. Poisoning and Toxicology Handbook, 2nd edn. Cleveland: Lexi-Comp Inc, 1996–1997, pp. 555–556.
256. Brown RB. Nutrition. In: Tierney LM, McPhee SJ, Papadakis MA, eds. Current Medical Diagnosis & Treatment 2005, 44th edn. San Francisco: Lange Medical Books/McGraw-Hill, 2005, p. 1231.
257. Etchason JA, Miller TD, Squires RW, Allison TG, Gau GT, Marttila JK, Kottke BA. Niacin-induced hepatitis: A potential side effect with low dose, time-release niacin. Mayo Clin Proc 1991;66:23–28.
258. Henkin Y, Johnson KC, Segrest JP. Rechallenge with crystalline niacin after drug-induced hepatitis from sustained released niacin. J Am Med Assoc 1990;264: 241–243.
259. Stern RH, Freeman D, Spence JD. Differences in metabolism of time-release and unmodified nicotinic acid: Explanation of the differences in hypolipidemic action? Metabolism 1992;41:879–881.
260. Dastur DK, Santhadevi N, Quadros EV, Avari FCR. The B-vitamins in malnutrition with alcoholism: A model of intervitamin relationships. Br J Nutr 1976;36:143–159.

261. Spies TD, Bean WB, Ashe WF. Recent advances in the treatment of pellagra and associated deficiencies. Ann Intern Med 1939;12:1830–1844.
262. Oestreicher R, Dressler SH, Middlebrook G. Peripheral neuritis in tuberculous patients treated with isoniazid. Am Rev Tuberc 1954;70:504–508.
263. Tuberculosis Chemotherapy Centre, Madras. The prevention and treatment of INH toxicity in the therapy of pulmonary TB: An assessment of the prophylactic effect of pyridoxine in low dosage. Bull World Health Organ 1963;29:457–481.
264. Zinc and the regulation of V-B_6 metabolism. Nutr Rev 1990;48:255–258.
265. Bottomley SS. Sideroblastic anemias. In: Lee GR, Bethell TC, Foerster J, Athens JW, Lukens JN, eds. Wintrobe's Clinical Hematology, 9th edn, Vol 1. Philadelphia: Lea & Febiger, 1993, pp. 852–871.
266. Cheslock KE, McCully MT. Response of human beings to a low V-B_6 diet. J Nutr 1960;70:507–513.
267. Guilarte TR. Vitamin B6 and cognitive development: Recent rresearch findings from human and animal studies. Nutr Rev 1993;51:193–198.
268. Berger AR, Schaumburg HH, Schroeder C, Apfel S, Reynolds R. Dose response, coasting, and differential fiber vulnerability in human toxic neuropathy: A prospective study of pyridoxine neurotoxicity. Neurology 1992;42:1367–1370.
269. Ludolph AC, Masur H, Oberwittler C, Koch HG, Ullrich, K. Sensory neuropathy and V-B_6 treatment for homocystinuria. Eur J Pediatr 1993;152:271.
270. Golik A, Modai D, Averbukh Z, Sheffy M, Shamis A, Cohen N, Shaked U, Dolev E. Zinc metabolism in patients treated with captopril versus enalapril. Metabolism 1996;39:665–667.
271. Pelton R, LaVelle JB, Hawkins EB, Krinsky DL, eds. Drug-Induced Nutrient Depletion Handbook, 2dn edn. Hudson, OH: Lexi-Comp, Inc., 2001
272. Hwang SJ, Lai YH, Chen HC, Tsai JH. Comparison of the effect of calcium carbonate and calcium acetate on zinc tolerance test in hemodialysis patients. Am J Kidney Dis 1992;19:57–60.
273. Brewer GJ, Yuzbasiyan-Gurkan V, Lee DY. Use of zinc-copper metabolic interactions in the treatment of Wilson's disease. J Am Coll Nutr 1990;9:487–491.
274. Solomons NW, Russell RM. The interaction of vitamin A and zinc: Implications for human nutrition. Am J Clin Nutr 1980;33:2031–2040.
275. Mares-Perlman JA, Subar AF, Block G, Greger JL, Luby MH. Zinc intake and sources in the US adult population: 1976–1980. J Am Coll Nutr 1995;14:349–357.
276. Dibley MJ. Zinc. In: Bowman BA, Russell RM, eds. Present Knowledge in Nutrition, 8th edn. Washington, DC: ILSE Press, 2001, pp. 329–343.
277. King JC, Hambidge M, Westcott, Kern DL, Marshal G. Daily variation in plasma zinc concentrations in women fed meals at six-hour intervals. J Nutr 1994;124:508–516.
278. American Medical Association Department of Foods and Nutrition. Guidelines for essential trace element preparations for parenteral use. J Am Med Assoc 1979;241:2051–2054.
279. Das I, Hahn HKJ. Effects of zinc deficiency on ethanol metabolism and ethanol aldehyde dehydrogenase activities. J Lab Clin Med 1984;104:610–617.
280. Prasad AS, Beck FWJ, Kaplan J. Effect of zinc supplementation on incidence of infections and hospital admissions in sickle cell disease. Am J Hematol 1999;61:194–202.

281. Pfeiffer CC, Jenney EH. Fingernail white spots: Possible zinc deficiency. J Am Med Assoc 1974;228:157.
282. Ribeiro RC, Sales, VSF, Neves FdAR, Draibe S, Brandao-Neto J. Effects on cell-mediated immunity in chronic hemodialysis patients. Biol Trace Elem Res 2004:98:209–218.
283. Morrison SA, Russell RM, Carney EA, Oaks EV. Zinc deficiency: A cause of abnormal dark adaptation in cirrhotics. Am J Clin Nutr 1978;31:276–281.
284. Fosmire GJ. Zinc toxicity. Am J Clin Nutr 1990;51:225–227.

II Chronic Kidney Disease During Stages 1–4 in Adults

1 Prevention

5 Hypertension

Kristie J. Lancaster

LEARNING OBJECTIVES

1. To identify the nutrients and foods most commonly believed to be associated with blood pressure.
2. To evaluate the evidence for or against the effect of certain nutrients and food groups on blood pressure.
3. To discuss the current dietary recommendations for the prevention and treatment of hypertension.
4. To describe how current dietary recommendations for hypertension apply to people with chronic kidney disease.

Summary

Hypertension is a major cause of chronic kidney disease. A key component of treatment is lifestyle modification, especially dietary change. A number of nutrients and food groups have been touted to reduce blood pressure, including sodium, potassium, calcium and dairy foods, magnesium, protein, certain types of fatty acids, vitamin C, and fruits and vegetables. However, the studies examining the associations between these nutrients and foods and blood pressure are of varying quality. The Dietary Approaches to Stop Hypertension studies have successfully lowered blood pressure with a diet high in fruits, vegetables, and low-fat dairy foods.

Key Words: Blood pressure; sodium; potassium; DASH diet; calcium; omega-3 fatty acids; monounsaturated fatty acids; vitamin C; protein; magnesium.

From: *Nutrition and Health: Nutrition in Kidney Disease*
Edited by: L. D. Byham-Gray, J. D. Burrowes, and G. M. Chertow

1. INTRODUCTION

According to national data, approximately 65.2 million U.S. adults (31.3%) had hypertension in 1999–2000, about a 30% increase since 1988–1994 *(1)*. A significant number of patients with chronic kidney disease (CKD) have high blood pressure. Hypertension can be treated and sometimes prevented with the proper dietary intake. Some studies suggest that certain nutrients, including sodium, potassium, calcium, magnesium, protein, unsaturated fats, and vitamin C, may have an effect on blood pressure level. Research has also found associations of blood pressure with food groups such as fruits, vegetables, and dairy products. This chapter will review the evidence for treating hypertension with diet modification.

2. NUTRIENTS AND BLOOD PRESSURE

2.1. Sodium

Although the debate over the extent of dietary sodium's influence on blood pressure has raged for decades, Intersalt, the cross-sectional epidemiological study conducted in 32 countries in the mid-1980s found that higher 24-hours urine sodium excretion (a marker for sodium intake) was positively associated with higher blood pressure *(2,3)*. The association was stronger for the older participants (40–59 years old). Other cross-sectional studies have also found an association between sodium intake and blood pressure *(4)*.

Randomized, controlled trials that reduce sodium intake have also found significantly lower blood pressure in the sodium-restricted group *(5–7)*. Pooled estimates from a meta-analysis of randomized trials to reduce sodium intake for at least 4 weeks found a significant reduction in blood pressure for the treatment groups *(8)*. The second trial of Dietary Approaches to Stop Hypertension (DASH) tested both the DASH diet—a diet high in fruits, vegetables, and low-fat dairy foods—and sodium intake *(9)*. Participants with prehypertension (systolic blood pressure of 120–139 mm Hg or diastolic of 80–89) or stage 1 hypertension (systolic of 140–159 mm Hg or diastolic of 90-99 mm Hg) were randomized into the typical U.S. diet (control) or the DASH diet in a parallel group design. Within each diet arm, they spent a month consuming that diet at each of three sodium levels: high, intermediate, and low (150, 100, and 50 mmol/day of sodium, respectively, for a 2000-kcal diet). In both diet groups, blood pressure was significantly lower for the lower sodium levels, showing an independent effect of sodium on blood pressure.

One 2-year prospective study did not find a significant association between sodium excretion and blood pressure, but found that fewer than 20% of 296 healthy participants experienced a significant increase in blood pressure at high sodium intakes *(10)*. Indeed, some people seem to be more sensitive than others to the effects of sodium on blood pressure. Despite this fact, most recent studies show that sodium restriction does reduce blood pressure. The effect is more pronounced in African Americans, older people, and persons diagnosed with hypertension. As a result, the prevailing recommendation is to reduce sodium intake in hypertension.

The 2004 Dietary Reference Intake for sodium recommends consuming no more than 1500 mg of sodium per day, 1300 mg for those over age 50 *(11)*, considerably lower than the previous recommendation of 2400 mg per day. The 2005 Dietary Guidelines for Americans recommends that most people consume less than 2300 mg per day, but highlights the 1500 mg limit for people with hypertension, African Americans, and older adults *(12)*.

2.2. *Potassium*

Both observational and experimental studies have shown that potassium intake is inversely associated with blood pressure. A meta-analysis of 33 randomized controlled trials conducted from 1981 to 1995 found that potassium supplementation (from the diet in six of the studies) significantly reduced systolic blood pressure by 3.11 mm Hg and diastolic blood pressure by 1.97 mm Hg *(13)*. A more recent meta-analysis also found that low potassium significantly increased risk of hypertension *(14)*. Even a low dose (600 mg) of potassium can significantly reduce blood pressure *(15)*. In addition to interventions, cross-sectional epidemiological studies suggest the same result *(4,16)*.

Increasing potassium intake can have a greater effect on blood pressure when sodium intake is high *(17)*. This association may be due to the increase in sodium excretion that results from increasing potassium intake. Intersalt and other studies have also found a significant association of blood pressure with the sodium to potassium ratio, the lower the sodium : potassium ratio, the lower the blood pressure *(2,4,13)*. However, the literature does not mention an ideal ratio.

The Institute of Medicine and the 2005 Dietary Guidelines for Americans have raised the recommended intake of potassium to 4700 mg per day *(11,12)*. This increase was instituted in large part in an effort to prevent and treat hypertension.

2.3. Calcium and Dairy Foods

Results from cross-sectional studies examining dietary calcium and blood pressure have been mixed *(4,18)*. Most trials examining the effect of calcium supplementation on blood pressure have found no change or only small reductions that are inconsistently significant *(19–23)*. A pooled analysis of 42 randomized controlled trials found a small blood pressure reduction from calcium supplements and dietary calcium *(24)*. Dietary calcium had almost twice the reduction in blood pressure compared to supplemental calcium, but the difference was not significant.

Some studies have shown that diets high in calcium, mostly from dairy products, have a blood pressure lowering effect. Of greatest note, the DASH trials found that increasing low-fat dairy intake in a diet high in fruits and vegetables reduced blood pressure even further than high fruit and vegetable intake alone *(25)*. However, a study that examined blood pressure and milk intake included three treatment groups: (i) skim milk, (ii) high calcium skim milk, and (iii) high calcium skim milk enriched with potassium. Only the group consuming the high potassium milk significantly reduced blood pressure *(26)*. Any association of dairy consumption and blood pressure may be due to calcium, but may also involve other nutrients present in milk. More studies are needed to determine the extent of the influence of calcium intake on blood pressure.

2.4. Magnesium

Trials examining the effect of magnesium intake on blood pressure have yielded inconsistent results. A 2002 meta-analysis of 20 trials supplementing magnesium suggest small, dose-dependent reductions in blood pressure *(27)*. However, most of the trials that have included magnesium to date have had small sample sizes, which limit the conclusions that can be drawn from the data. As with calcium, more carefully controlled studies are needed to further elucidate any association between magnesium and blood pressure.

2.5. Protein

A 1996 review of studies examining the association of dietary protein with blood pressure found little or no effect of protein intake on blood pressure among intervention studies *(28)*. However, some observational studies did find decreased blood pressure with higher protein intake. Since that review, the cross-sectional Intersalt study found that total and urea nitrogen excretion, as markers of protein intake, were associated with lower blood pressure when adjusting for age, sex, alcohol intake, BMI, and urinary sodium, potassium, calcium, and magnesium excretion *(29)*. The INTERMAP study suggests that the effect of protein on diet may vary depending on the type of protein. The cross-sectional study collected dietary intake data and found an inverse association between vegetable protein intake and blood pressure, but a positive association between animal protein intake and blood pressure *(30)*. In addition, three randomized trials that found an inverse association between protein and blood pressure increased the protein levels in the diet by adding vegetable protein *(31–33)*. These studies suggest that vegetable protein may be beneficial in controlling blood pressure.

2.6. Fatty Acids

Three meta-analyses of randomized controlled trials administering large doses of omega-3 oils to people with hypertension found an inverse association with blood pressure *(34–36)*. There was a greater effect on blood pressure for persons 45 years or older *(34)*. However, there was little effect of omega-3 fatty acids on healthy adults or in trials using small doses of the oil. Although the evidence for this association is relatively strong, the potential side effects associated with consuming large doses of fish oil may be prohibitive.

Some studies suggest that monounsaturated fat can also reduce blood pressure in people with hypertension. A randomized crossover study that examined the effect of a high monounsaturated fatty acid (MUFA) oil (olive oil) and a high polyunsaturated fatty acid (PUFA) oil (sunflower oil) on blood pressure *(37)*. The olive oil significantly reduced blood pressure, but the sunflower oil did not. The OmniHeart Study examined the effects of a high carbohydrate diet, a high protein diet, and a diet rich in MUFAs in the form of olive, canola, and safflower oils and also in nuts and seeds *(28)*. The MUFA diet lowered blood pressure more than the high carbohydrate and had a similar effect to the high protein diet. Another crossover study of 42 healthy participants tested four diets enriched with (i) saturated fatty acids, (ii) MUFAs, (iii) n-3 PUFAs, and (iv) n-6 PUFAs. The MUFA group had the lowest blood pressure and the saturated fatty acid group had the highest *(38)*. These studies suggest that omega-3 and MUFAs may have a beneficial effect on blood pressure. However, most of these studies had small sample sizes and involved people with hypertension, limiting the generalizability of the results.

2.7. Vitamin C

The effect of vitamin C on blood pressure is unclear. Some cross-sectional studies from Europe have found an inverse association *(39,40)*. However, one study from Germany divided survey respondents into four groups, high vitamin C-high fruits and vegetables; high vitamin C-low fruits and vegetables; low vitamin C-high fruits and vegetables; and low vitamin C-low fruits and vegetables (control) *(41)*. The women in the two high fruits and vegetables groups had lower systolic blood pressure compared to the control group. The high vitamin C-low fruits and vegetables group did not have improved blood pressure compared to the control group. This suggests that any differences in blood pressure may largely be due to fruit and vegetable intake, not dietary vitamin C intake. Additionally, two interventions found little difference in blood pressure between vitamin C and placebo groups *(42,43)*.

3. THE DASH DIET

The key to modifying diet to reduce blood pressure may rest in overall dietary patterns as opposed to consumption of a single nutrient or food. The DASH studies illustrate how a dietary pattern can successfully affect blood pressure level. The original DASH study featured three treatment arms—(i) a usual intake that mirrored the typical U.S. diet; (ii) the typical diet modified to include higher amounts of fruits, vegetables, and fiber, and lower in snacks and sweets; and (iii) a diet high in fruits, vegetables, and low-fat dairy foods, and lower in total fat, saturated fat, and cholesterol *(25)*. Table 1 summarizes the components of the DASH (combination) diet. Participants ate the assigned diet for 8 weeks. At the end of the 8 weeks, participants eating the fruits and vegetables diet had systolic and diastolic blood pressures 2.8 and 1.1 mm Hg lower than the mean blood pressure in the control group (p <0.001 and p = 0.07, respectively). The combination fruit, vegetable, and low-fat dairy diet group had systolic blood pressure 5.5 mm Hg and diastolic blood pressure 3.0 mm Hg lower than the control group (p < 0.001 for both). The effect of the diet was greater for African Americans and for those with hypertension.

The second investigation into this diet compared the control and combination (DASH) diets at three different sodium levels as described above *(7)*. In the DASH-Sodium trial, blood pressure decreased as sodium intake decreased. Consuming the DASH diet at the lowest sodium level (50 mmol/day) resulted in lowest blood pressure. A behavioral intervention using the DASH diet (Premier) found that

Table 1
The DASH Eating Plan (Based on 2000 Calorie Diet)

Food group	*Servings per day*	*Nutrient sources*
Grains and grain products	7–8	Major sources of energy and fiber
Vegetables	4–5	Rich sources of potassium, magnesium, and fiber
Fruits	4–5	Important sources of potassium, magnesium, and fiber
Low-fat or fat-free dairy foods	2–3	Major sources of calcium and protein
Meats, poultry, and fish	2 or less	Rich sources of magnesium and protein
Nuts, seeds, and dry beans	4–5 per week	Rich sources of magnesium, potassium, protein, and fiber
Fats and oils	2–3	27% calories as fat
Sweets	5 per week	

Taken from ref. *(47)*.

it is possible to successfully incorporate the DASH dietary pattern and reduce blood pressure in a community setting *(44)*. The results of these carefully controlled studies and the subsequent behavioral study show that an eating pattern can be an effective tool in reducing blood pressure.

4. RECOMMENDATIONS FOR DIET IN HYPERTENSION

4.1. The Seventh Report of the Joint National Committee (JNC) on Prevention, Detection, Evaluation, and Treatment of High Blood Pressure Recommendations

The importance of lifestyle modification, especially dietary change, in prevention and treatment of hypertension is well recognized. JNC 7 was written in 2003 to incorporate results from the latest hypertension studies and trials into useful guidelines for preventing and treating hypertension *(45)*. JNC 7 sees lifestyle modification as a critical component of hypertension management. Its dietary recommendations (Table 2) include losing weight if overweight or obese, reducing sodium intake to less than 100 mEq/l (2.4 g) of sodium, limiting alcohol consumption to no more than two drinks per day, and adopting the eating pattern shown to effectively lower blood pressure in the DASH studies. The report also includes an approximate amount of reduction in systolic blood pressure that can be achieved for each of these dietary changes. It advocates adopting more than one strategy to achieve even greater results.

Table 2
JNC 7 Diet-Related Lifestyle Modifications

Modification	*Recommendation*	*Approximate systolic BP reduction, range*
Weight reduction	Maintain normal body weight (BMI, 18.5–24.9)	5–20 mm Hg/10-kg weight loss
Adopt DASH eating plan	Consume a diet rich in fruits, vegetables, and low-fat dairy products with a reduced content of saturated and total fat	8–14 mm Hg
Dietary sodium reduction	Reduce dietary sodium intake to no more than 100 mEq/l (2.4 g sodium or 6 g sodium chloride)	2–8 mm Hg
Moderation of alcohol consumption	Limit consumption to no more than 2 drinks per day [1 oz. or 30 ml ethanol (e.g. 24 oz. beer, 10 oz. wine, or 3 oz. 80-proof whiskey)] in most men and no more than 1 drink per day in women and lighter-weight persons	2–4 mm Hg

BMI, body mass index; DASH, Dietary Approaches to Stop Hypertension.
Taken from ref. *(42)*.

4.2. KDOQI Guidelines

Dietary modification is a critical component of managing CKD. The National Kidney Foundation Kidney Disease Outcomes Quality Initiative (KDOQI) has guidelines for diet in hypertension in CKD (Table 3) *(46)*. With the exception of protein intake, the KDOQI recommendations for macronutrient intake in CKD largely mirror the recommendations for the general population. The sodium recommendations are in keeping with the sodium recommendations described above. However, because of potassium retention in kidney disease, the recommended intake of 4700 mg per day is not advisable for CKD patients, especially for those with glomerular filtration rate (GFR) <60 ml/min/1.73 m^2. Similarly, protein intake should be lower in CKD (about 10% of energy), especially in stages 3 and 4, to reduce the production of nitrogenous wastes and to try to slow the progression of the disease. Although the DASH diet is beneficial in lowering blood pressure, the higher potassium and phosphorus intakes that result from the diet may lead to hyperkalemia and hyperphosphatemia respectively, especially in patients with GFR <60 ml/min/1.73 m^2.

Table 3
KDOQI Nutritional Recommendations

	Stage of CKD		
Nutrient	*Stages 1–2*	*Stages 1–4*	*Stages 3–4*
Sodium (mg/day)		<2400	
Total fat (% of energy)		<30	
Saturated fat (% of energy)		<10	
Cholesterol (mg/day)		<200	
Carbohydrate (% of energy)		50–60	
Protein (g/kg/day % of energy)	1.4 (~18)		0.6–0.8 (~10)
Phosphorus (g/day)	1.7		0.8–1.0
Potassium (g/day)	>4		2–4

CKD, Chronic Kidney Disease; KDOQI, Kidney Disease Outcomes Quality Initiative.

Reprinted with permission from ref. *(45)*.

5. SUMMARY

In summary, current evidence suggests that lowering sodium intake and increasing intake of potassium, vegetable protein, omega-3 and MUFAs, fruits, and vegetables, can help lower blood pressure. In CKD, however, increasing some of these nutrients and food groups is contraindicated.

REFERENCES

1. Fields LE, Burt VL, Cutler JA, Hughes J, Roccella EJ, Sorlie P. The burden of adult hypertension in the United States 1999 to 2000: a rising tide. Hypertension 2004;44:398–404.
2. Intersalt Cooperative Research Group. Intersalt: an international study of electrolyte excretion and blood pressure. Results for 24 hour urinary sodium and potassium excretion. BMJ 1988;297:319–328.
3. Elliott P, Stamler J, Nichols R, Dyer AR, Stamler R, Kesteloot H, Marmot M. Intersalt revisited: further analyses of 24 hour sodium excretion and blood pressure within and across populations. Intersalt Cooperative Research Group. BMJ 1996;312:1249–1253.
4. Hajjar IM, Grim CE, George V, Kotchen TA. Impact of diet on blood pressure and age-related changes in blood pressure in the US population: analysis of NHANES III. Arch Intern Med 2001;161:589–593.
5. Appel LJ, Espeland MA, Easter L, Wilson AC, Folmar S, Lacy CR. Effects of reduced sodium intake on hypertension control in older individuals: results from the Trial of Nonpharmacologic Interventions in the Elderly (TONE). Arch Intern Med 2001;161:685–693.
6. Espeland MA, Kumanyika S, Yunis C, Zheng B, Brown WM, Jackson S, Wilson AC, Bahnson J. Electrolyte intake and nonpharmacologic blood pressure control. Ann Epidemiol 2002;12:587–595.
7. TOHP Group. The effects of nonpharmacologic interventions on blood pressure of persons with high normal levels. Results of the Trials of Hypertension Prevention, Phase I. JAMA 1992;267:1213–1220.
8. He FJ, MacGregor GA. Effect of modest salt reduction on blood pressure: a meta-analysis of randomized trials. Implications for public health. J Hum Hypertens 2002;16:761–770.
9. Sacks FM, Svetkey LP, Vollmer WM, Appel LJ, Bray GA, Harsha D, Obarzanek E, Conlin PR, Miller ER III, Simons-Morton DG, Karanja N, Lin PH; DASH-Sodium Collaborative Research Group. Effects on blood pressure of reduced dietary sodium and the Dietary Approaches to Stop Hypertension (DASH) diet. DASH-Sodium Collaborative Research Group. N Engl J Med 2001;344:3–10.
10. Ducher M, Fauvel JP, Maurin M, Laville M, Maire P, Paultre CZ, Cerutti C. Sodium intake and blood pressure in healthy individuals. J Hypertens 2003;21:289–294.
11. Institute of Medicine. Dietary Reference Intakes for Water, Potassium, Sodium, Chloride, and Sulfate. National Academy Press, Washington, DC, 2004.

12. U.S. Department of Health and Human Services and U.S. Department of Agriculture. Dietary Guidelines for Americans, 2005. 6th Edition, U.S. Government Printing Office, Washington, DC, January 2005.
13. Whelton PK, He J, Cutler JA, Brancati FL, Appel LJ, Follmann D, Klag MJ. Effects of oral potassium on blood pressure. Meta-analysis of randomized controlled clinical trials. JAMA 1997;277:1624–1632.
14. Geleijnse JM, Kok FJ, Grobbee DE. Impact of dietary and lifestyle factors on the prevalence of hypertension in Western populations. Eur J Public Health 2004;14:235–239.
15. Naismith DJ, Braschi A. The effect of low-dose potassium supplementation on blood pressure in apparently healthy volunteers. Br J Nutr 2003;90:53–60.
16. Elliott P, Stamler J, Dyer AR, Appel L, Dennis B, Kesteloot H, Ueshima H, Okayama A, Chan Q, Garside DB, Zhou B. Association between protein intake and blood pressure: the INTERMAP Study. Arch Intern Med 2006;166:79–87.
17. Gallen IW, Rosa RM, Esparaz DY, Young JB, Robertson GL, Batlle D, Epstein FH, Landsberg L. On the mechanism of the effects of potassium restriction on blood pressure and renal sodium retention. Am J Kidney Dis 1998;31:19–27.
18. Ruidavets JB, Bongard V, Simon C, Dallongeville J, Ducimetiere P, Arveiler D, Amouyel P, Bingham A, Ferrieres J. Independent contribution of dairy products and calcium intake to blood pressure variations at a population level. J Hypertens 2006;24:671–681.
19. Dwyer JH, Dwyer KM, Scribner RA, Sun P, Li L, Nicholson LM, Davis IJ, Hohn AR. Dietary calcium, calcium supplementation, and blood pressure in African American adolescents. Am J Clin Nutr 1998;68:648–655.
20. Kawano Y, Yoshimi H, Matsuoka H, Takishita S, Omae T. Calcium supplementation in patients with essential hypertension: assessment by office, home and ambulatory blood pressure. J Hypertens 1998;16:1693–1699.
21. Reid IR, Horne A, Mason B, Ames R, Bava U, Gamble GD. Effects of calcium supplementation on body weight and blood pressure in normal older women: a randomized controlled trial. J Clin Endocrinol Metab 2005;90:3824–3829.
22. Sacks FM, Brown LE, Appel L, Borhani NO, Evans D, Whelton P. Combinations of potassium, calcium, and magnesium supplements in hypertension. Hypertension 1995;26(6 Pt 1):950–956.
23. Sacks FM, Willett WC, Smith A, Brown LE, Rosner B, Moore TJ. Effect on blood pressure of potassium, calcium, and magnesium in women with low habitual intake. Hypertension 1998;31:131–138.
24. Griffith LE, Guyatt GH, Cook RJ, Bucher HC, Cook DJ. The influence of dietary and nondietary calcium supplementation on blood pressure: an updated metaanalysis of randomized controlled trials. Am J Hypertens 1999;12(1 Pt 1): 84–92.
25. Appel LJ, Moore TJ, Obarzanek E, Vollmer WM, Svetkey LP, Sacks FM, Bray GA, Vogt TM, Cutler JA, Windhauser MM, Lin PH, Karanja N. A clinical trial on the effects of dietary patterns on blood pressure. N Engl J Med 1997; 336:1117–1124.
26. Green JH, Richards JK, Bunning RL. Blood pressure responses to high-calcium skim milk and potassium-enriched high-calcium skim milk. J Hypertens 2000; 18:1331–1339.

27. Jee SH, Miller ER III, Guallar E, Singh VK, Appel LJ, Klag MJ. The effect of magnesium supplementation on blood pressure: a meta-analysis of randomized clinical trials. Am J Hypertens 2002;15:691–696.
28. Obarzanek E, Velletri PA, Cutler JA. Dietary protein and blood pressure. JAMA 1996;275:1598–1603.
29. Stamler J, Elliott P, Kesteloot H, Nichols R, Claeys G, Dyer AR, Stamler R. Inverse relation of dietary protein markers with blood pressure. Findings for 10,020 men and women in the INTERSALT Study. INTERSALT Cooperative Research Group. INTERnational study of SALT and blood pressure. Circulation 1996;94:1629–1634.
30. Elliott P, Stamler J, Dyer AR, Appel L, Dennis B, Kesteloot H, Ueshima H, Okayama A, Chan Q, Garside DB, Zhou B. Association between protein intake and blood pressure: the INTERMAP Study. Arch Intern Med 2006;166:79–87.
31. Appel LJ, Sacks FM, Carey VJ, Obarzanek E, Swain JF, Miller ER III, Conlin PR, Erlinger TP, Rosner BA, Laranjo NM, Charleston J, McCarron P, Bishop LM; OmniHeart Collaborative Research Group. Effects of protein, monounsaturated fat, and carbohydrate intake on blood pressure and serum lipids: results of the OmniHeart randomized trial. JAMA 2005;294:2455–2464.
32. He J, Gu D, Wu X, Chen J, Duan X, Chen J, Whelton PK. Effect of soybean protein on blood pressure: a randomized, controlled trial. Ann Intern Med 2005;143:1–9.
33. Burke V, Hodgson JM, Beilin LJ, Giangiulioi N, Rogers P, Puddey IB. Dietary protein and soluble fiber reduce ambulatory blood pressure in treated hypertensives. Hypertension 2001;38:821–826.
34. Geleijnse JM, Giltay EJ, Grobbee DE, Donders AR, Kok FJ. Blood pressure response to fish oil supplementation: metaregression analysis of randomized trials. J Hypertens 2002;20:1493–1499.
35. Morris MC, Sacks F, Rosner B. Does fish oil lower blood pressure? A meta-analysis of controlled trials. Circulation 1993; 88:523–533.
36. Dickinson HO, Mason JM, Nicolson DJ, Campbell F, Beyer FR, Cook JV, Williams B, Ford GA. Lifestyle interventions to reduce raised blood pressure: a systematic review of randomized controlled trials. J Hypertens 2006;24:215–233.
37. Ferrara LA, Raimondi AS, d'Episcopo L, Guida L, Dello Russo A, Marotta T. Olive oil and reduced need for antihypertensive medications. Arch Intern Med 2000;160:837–842.
38. Lahoz C, Alonso R, Ordovas JM, Lopez-Farre A, de Oya M, Mata P. Effects of dietary fat saturation on eicosanoid production, platelet aggregation and blood pressure. Eur J Clin Invest 1997;27:780–787.
39. Bates CJ, Walmsley CM, Prentice A, Finch S. Does vitamin C reduce blood pressure? Results of a large study of people aged 65 or older. J Hypertens 1998;16:925–932.
40. Ness AR, Khaw KT, Bingham S, Day NE. Vitamin C status and blood pressure. J Hypertens 1996;14:503–508.
41. Beitz R, Mensink GB, Fischer B. Blood pressure and vitamin C and fruit and vegetable intake. Ann Nutr Metab 2003;47:214–220.
42. Fotherby MD, Williams JC, Forster LA, Craner P, Ferns GA. Effect of vitamin C on ambulatory blood pressure and plasma lipids in older persons. J Hypertens 2000;18:411–415.

43. Block G, Mangels AR, Norkus EP, Patterson BH, Levander OA, Taylor PR. Ascorbic acid status and subsequent diastolic and systolic blood pressure. Hypertension 2001;37:261–267.
44. Elmer PJ, Obarzanek E, Vollmer WM, Simons-Morton D, Stevens VJ, Young DR, Lin PH, Champagne C, Harsha DW, Svetkey LP, Ard J, Brantley PJ, Proschan MA, Erlinger TP, Appel LJ; PREMIER Collaborative Research Group. Effects of comprehensive lifestyle modification on diet, weight, physical fitness, and blood pressure control: 18-month results of a randomized trial. Ann Intern Med 2006;144:485–495.
45. Chobanian AV, Bakris GL, Black HR, Cushman WC, Green LA, Izzo JL Jr, Jones DW, Materson BJ, Oparil S, Wright JT Jr, Roccella EJ; National Heart, Lung, and Blood Institute Joint National Committee on Prevention, Detection, Evaluation, and Treatment of High Blood Pressure; National High Blood Pressure Education Program Coordinating Committee. The Seventh Report of the Joint National Committee on Prevention, Detection, Evaluation, and Treatment of High Blood Pressure: the JNC 7 report. JAMA 2003;289:2560–2572.
46. KDOQI clinical practice guidelines on hypertension and antihypertensive agents in chronic kidney disease. Guideline 6: Dietary and other therapeutic lifestyle changes in adults. Am J Kidney Dis 2004;43(Suppl 1):S115–S119.
47. Facts about the DASH Eating Plan. National Institutes of Health, Department of Health and Human Services. NIH Publication No. 03-4082. May 2003.

6 Diabetes Mellitus

Joni J. Pagenkemper

LEARNING OBJECTIVES

1. To describe techniques for early intervention and screening in those at risk of or with diabetes to reduce microvascular and macrovascular complications.
2. To integrate the goals of therapy in diabetes and chronic kidney disease and their comorbidities regarding glycemic control, hypertension, and lipid management.
3. To identify the nutrition recommendations for patients with diabetes and chronic kidney disease.
4. To discuss the current glucoregulatory medications for diabetes.

Summary

Diabetic nephropathy is a complication of diabetes mellitus (DM) that can be prevented or slowed by medical and nutritional interventions. Glycemic control coupled with the management of hypertension and dyslipidemia can alter the progression of chronic kidney disease (CKD) in stages 1 through 4. Careful management of carbohydrate (CHO) intake is essential to diabetes meal planning, emphasizing how much and when to consume CHO. At all stages of CKD *(1–4)*, persons with DM should achieve a dietary protein intake that meets, but does not exceed, the Recommended Dietary Allowance of 0.8 g/kg body weight/day, or 10% of total calories.

Key Words: Diabetes; diabetic nephropathy; chronic kidney disease; medical nutrition therapy.

From: *Nutrition and Health: Nutrition in Kidney Disease*
Edited by: L. D. Byham-Gray, J. D. Burrowes, and G. M. Chertow

1. INTRODUCTION

The development of diabetic nephropathy, a complication of diabetes, can be prevented or slowed down by medical and nutritional interventions. This chapter provides an overview as to how glycemic control coupled with the management of hypertension (HTN) and dyslipidemia can alter the progression of chronic kidney disease (CKD) in some stages 1 through 4. There are a number of references cited for additional depth and overview of diabetes management which is beyond the scope of this chapter *(1–4)*.

1.1. Facts and Statistics about Diabetes

Diabetes is a group of diseases marked by high levels of blood glucose resulting from defects in insulin production, insulin action, or both. Type 2 diabetes mellitus (T2DM), the most common form of diabetes, is due to a progressive insulin secretory defect with insulin resistance, whereas type 1 diabetes mellitus (T1DM) develops from an autoimmune β-cell destruction that usually leads to absolute insulin deficiency *(1)*. Although diabetes is a chronic, progressive disease with no known cure, steps can be taken to control the disease and lower the risk of complications associated with the disease.

One in three Americans and one in two minorities born in 2000 will develop diabetes in their lifetime *(5)*. Today, almost 21 million people (7% of the population) in the United States have diabetes, with nearly one-third unaware that they have the disease. An estimated 41 million people have pre-diabetes. Prevalence increases with age as 21% of people 60 years of age or older have diabetes. It is estimated that one dollar out of every 10 health care dollars is spent on diabetes and its complications *(5)*.

Diabetes is the leading cause of kidney failure, accounting for 45% of new cases of stage 5 CKD in 2003 *(6)*. Risk of nephropathy is disproportionately found among certain ethnic groups, with a greater risk, 2.6–5.6 times, in non-Hispanic blacks, 4.5–6.6 times in Mexican-Americans, and 6 times in American Indians *(5)*. These rates are troublesome, with access to affordable quality healthcare often being a factor. Most individuals are not seen by a nephrologist until they are diagnosed with stage 3 or 4 CKD, although earlier referral has been found to improve quality of care while reducing costs by prolonging the initiation of dialysis *(7,8)*. Additionally, early intervention with medical nutrition therapy (MNT) has an important role in slowing the progression of CKD while maintaining optimal nutritional status *(9)*.

2. EARLY INTERVENTION AND SCREENING

Early screening and detection, along with tight glucose and blood pressure control, can delay the onset and course of diabetic nephropathy *(1,7)*. Persistent hyperglycemia leads to nephropathy via several mechanisms, including hypertrophy, increased endothelial cell permeability, and increased matrix protein synthesis. Hyperglycemia also causes an increase in vasodilatory prostaglandins, which promotes an increase in renal perfusion and intraglomerular pressure, resulting in hyperfiltration *(10)*.

There is a direct correlation between elevated blood pressure and the degree of decline in glomerular filtration rate (GFR). Alteration of hemodynamic factors produces increases in mesangial matrix formation and basement membrane thickening. Higher systolic blood pressure specifically leads to extracellular matrix accumulation, increased glomerular permeability, proteinuria, and glomerulosclerosis *(10)*. Elevated total cholesterol and low-density lipoprotein cholesterol (LDL-C) have been found to be independent risk factors for the progression of renal disease. Nephropathy in smokers tends to progress more rapidly than in non-smokers, with adverse effects on blood pressure; therefore, smoking cessation should always be encouraged *(1)*.

In general, every 10 mm Hg reduction in systolic blood pressure lowers the risk for any diabetes-related complication by 12% *(5)*. Studies have also shown that each percentage drop in hemoglobin A1C (HbA1C) blood test value (e.g., from 8 to 7%) can reduce the risk of microvascular complications by up to 40% *(11–13)*. Better control of blood lipids [LDL, high density lipoprotein (HDL), and triglyceride (TG)] can decrease complications of cardiovascular disease (CVD) by 20 to 50% *(5,14,15)*. Additionally, treatment with angiotensin-converting enzyme inhibitors (ACEI) or angiotensin II receptor blockers (ARBs) effectively reduces the decline in kidney function, independent of any blood pressure lowering effect *(16)*.

The earliest clinical evidence of nephropathy is the appearance of microalbuminuria (≥30 mg/day). Persistent microalbuminuria in the range of 30–300 mg/day is the earliest stage of nephropathy in T1DM and a marker for the development of nephropathy in T2DM *(1)*. At this point, without specific interventions, approximately 80% of those with T1DM will progress to overt nephropathy or clinical albuminuria (>300 mg/day) with HTN within 10–15 years. Once this occurs, the GFR will gradually decline, with stage 5 CKD developing in 50% of persons having T1DM within 10 years and >75% by 20 years. A higher proportion of patients with T2DM have microalbuminuria and overt nephropathy shortly after diagnosis due to being previously undiagnosed with diabetes for several years. Ironically, after 20 years, only about 20% of patients with T2DM and overt or clinical nephropathy will progress to stage 5 CKD *(1,7)*.

The American Diabetes Association (ADA) recommends an annual test for the presence of microalbuminuria in patients with T1DM after at least 5 years of the disease and starting at diagnosis for patients with T2DM, since they often go undiagnosed for many years *(1,7)*. Screening for microalbuminuria is most commonly performed in an office setting by obtaining a random spot urine collection for measurement of an albumin-to-creatinine ratio *(1,7)*. In addition, microalbuminuria is also used as a marker of increased cardiovascular morbidity and mortality for patients with either T1DM or T2DM *(1,17)*. The presence of microalbuminuria doubles the risk of cardiovascular morbidity and mortality in T2DM *(16)*.

According to the National Kidney Foundation (NKF), urinary albumin excretion (UAE) in addition to the level of GFR may be used to determine the stage of CKD. The current NKF classification is based on GFR levels, as a significant decrease in GFR may be found in the absence of increased UAE in a substantial number of adults with T1DM and T2DM, particularly those with stage 3 CKD *(1,17)*. Elevations in UAE over baseline levels can be due to exercise (within 24 h of the test), infection, fever, heart failure, marked hyperglycemia, and/or HTN *(1)*.

Serum creatinine should be measured at least annually to estimate GFR and the stage of CKD in all adults with diabetes, regardless of the degree of UAE. The GFR can be easily estimated using the newer prediction equation developed by Levey et al. *(18)* based on data collected from the Modification of Diet in Renal Disease study:

$$186 \times S_{cr}^{-1.154} \times A^{-0.23} \times (0.742 \text{ if female}) \times (1.21 \text{ if African American}),$$

where S_{cr} = serum creatinine, and A = age.

Estimated GFR can be calculated by using the equation at the website http://www.kidney.org/professionals/kdoqi/gfr_calculator.cfm.

Monitoring microalbuminuria annually can help assess response to therapy after instituting ACEI or ARB therapy and achieving blood pressure and glycemic control. Although unconfirmed by prospective trials, reduction of microalbuminura to the near normal range is suggested to improve renal and cardiovascular prognosis *(1)*. Table 1 summarizes additional metabolic consequences common to each stage of CKD.

Table 1
Stages of Diabetic Nephropathy/Chronic Kidney Disease and Metabolic Consequences

Stage 1:	• GFR ≥90 ml/min/1.73 m^2; normal or increased GFR with enlarged kidneys • Hyperglycemia leads to increased kidney filtration due to osmotic load and toxic effects of high blood glucose
Stage 2:	• Mildly decreased GFR 60–89 ml/min/1.73 m^2 • Clinically silent phase, but continued kidney hyperfiltration and hypertrophy • Concentration of PTH starts to rise
Stage 3:	• Moderately decreased GFR 30–59 ml/min/1.73 m^2. • Microalbuminuria (defined as 30–300 mg/day); loss of albumin in urine • Within 5 years 20% develop nephropathy on standard care; 50% do not progress • Microalbuminuria is a better predictor of progression in type 1 than type 2 diabetes • Microalbuminuria is a predictor of increased cardiovascular disease, particularly in type 2 diabetes • Microalbuminuria is associated with higher hemoglobin A1Clevels (>8.1) • Decrease in calcium absorption (GFR <50 ml/min/1.73 m^2) • Onset of anemia (erythropoietin deficiency) • Lipoprotein activity falls
Stage 4:	• GFR severely decreased 15–29 ml/min/1.73 m^2 • Overt nephropathy • Almost always with hypertension, >300 mg/day albumin in urine, about 10% have nephrotic range proteinuria • TGL concentrations start to rise • Hyperphosphatemia • Metabolic acidosis • Tendency toward hyperkalemia
Stage 5:	• Kidney failure (GFR <15 ml/min/1.73 m^2) needing renal replacement therapy (dialysis or transplantation) • Over 30 years, >24% of patients with T1DM progress to this stage • Over 25 years, about 8% of patients with T2DM progress to this stage

GFR, glomerular filtration rate: PTH, parathyroid hormone; TGL, triglyceride; T1DM, type 1 diabetes mellitus; T2DM, type 2 diabetes mellitus.

Adapted from refs *(19,17,2)*.

3. GOALS OF THERAPY

3.1. Glycemic Control

Glycemic control is fundamental to the management of diabetes and for preventing complications. The Diabetes Control and Complications Trial (DCCT) showed that intensive therapy to control blood glucose in T1DM resulted in fewer incidences of diabetic nephropathy *(11)*. Microalbuminuria was reduced by one-third, and in those with microalbuminuria, the risk of progressing to proteinuria was about half in those with tight glucose control *(1,7,11)*. The United Kingdom Prospective Diabetes Study found that patients with T2DM demonstrated a 25% reduction in nephropathy with intensive diabetes therapy and, for every percentage point decrease in HbA1C, there was a 35% reduction in the risk of complications *(1,12,13)*.

Because tight glycemic control has been shown to reduce the risk of diabetic complications, a number of organizations have created guidelines for glycemic control *(1,20,21)* (Table 2). The American College of Endocrinology and the International Diabetes Federation make more stringent recommendations for glycemic control than the ADA. The ADA advocates that glycemic control be individualized. Patients with special considerations, such as children, pregnant women, the elderly, or those with advanced chronic diseases, may require less or more intense goals for glycemic control *(1)*.

Treatment efficacy can be assessed by performing a HbA1C test, which measures a patient's average glycemia over the preceeding 2–3 months. The HbA1C test should be done quarterly in patients whose therapy has changed or who are not meeting blood glucose goals. In patients who are meeting treatment goals and who have stable blood

Table 2
Glycemic Targets

Glycemic target	*ADA*	*ACE*	*IDF*
A1C %	<7.0	≤6.5	<6.5
Fasting glucose, mg/dl	90–130*	<110*	<100(t)
Postprandial glucose, mg/dl	<180*T	<140*(T)	<145(t)

A1C, hemoglobin A1C; ADA, American Diabetes Association *(1)*; ACE, American College of Endocrinology *(20)*; IDF, International Diabetes Federation; t, Self-monitored blood glucose; T, 2h.

* Plasma equivalent.

glucose control, the HbA1C test should be performed at least twice yearly *(1)*. The limitations of HbA1C should be taken into consideration when monitoring glycemic control in clients with CKD, as this value may be reduced due to the shortened lifespan of erythrocytes *(3)*. However, the Kidney Disease Outcomes Quality Initiative (KDOQI) Clinical Practice Guidelines for Diabetes in CKD state that "Target HbA1C for all persons with diabetes should be $\leq$7%, irrespective of the presence or absence of CKD" *(3)*. Table 3 illustrates how HbA1C levels correlate with mean plasma glucose concentrations, which can help patients better understand this relationship.

It is now recommended that intensive diabetes management with the goal of achieving near normal glycemia be implemented as early as possible and in as many patients with diabetes as is safely possible using a physician-coordinated collaborative and integrated health-care team approach *(1)*. Diet is often the most complex aspect of intensive diabetes therapy, so it is important to simplify and streamline nutrition priorities for the patient with CKD *(22)*.

Currently, it is estimated that among adults diagnosed with diabetes, 16% take insulin only, 12% take insulin and oral medications, 57% take oral medications only, and 15% follow lifestyle management with no insulin or oral medications *(5)*. Careful attention to the relationship between diet, insulin, and physical activity is necessary to achieve HbA1C goals without undesired hypoglycemia and weight gain. Food variations often explain erratic blood glucose results and

Table 3
Correlation Between A1C Level and Mean Plasma Glucose Levels

	Mean plasma glucose	
A1C (%)	*mg/dl*	*mmol/L*
6	135	7.5
7	170	9.5
8	205	11.5
9	240	13.5
10	275	15.5
11	310	17.5
12	345	19.5

Adapted from table in ref. *(1)*. Based on data from the DCCT ref. *(84)*.

episodes of hypoglycemia and hyperglycemia *(23)*. However, fluctuations in blood glucose levels can also occur due to alterations in insulin metabolism with changing kidney function. Because insulin is degraded by the kidneys, patients with progressive renal impairment often begin to require less exogenous insulin due to increased exposure and reduced clearances and need to be alert to the signs and symptoms of hypoglycemia and how to treat it *(24)*. Studies have shown that insulin reductions of up to 25–40% may be necessary with moderate to severe kidney impairment *(25)*. With reduced kidney function, there is also decreased clearance of some oral agents used to treat diabetes. Reduced kidney mass also decreases gluconeogenesis, which can compromise the ability of a patient to defend against hypoglycemia *(3)*.

3.1.1. Treating Hypoglycemia

The 15/15 rule should be used to treat hypoglycemia; namely, take 15 g of carbohydrate (CHO), wait 15 mins and retest the blood sugar or evaluate relief of symptoms. If necessary, repeat with another 15 g of CHO and continue the process until the blood glucose level normalizes. *(4,26)*. The usual methods for treating hypoglycemia may result in fluid overload and/or hyperkalemia in the oliguric patient with diabetes. Better choices that will provide 15 g of CHO are 10 jelly beans, 6 lifesavers, or 3–4 commercial glucose tablets. Once the blood glucose returns to normal, if the next regular meal will be delayed, then the patient should eat a snack with additional CHO to stabilize the glucose level 1 h later *(4,26)*.

3.2. Hypertension Management

Tight blood pressure control plays an important role in the treatment and prevention of diabetic nephropathy *(7)*. In patients with T1DM, the onset of HTN is frequently associated with the development of diabetic nephropathy. It is estimated that 65% of patients with T2DM also have HTN, which is often present as part of the metabolic syndrome. Microvascular and macrovascular complications of diabetes are exacerbated when HTN is present as a comorbidity *(27)*. The Seventh Report of the Joint National Committee (JNC VII) and the ADA set a target blood pressure goal of <130/80 mmHg for patients with diabetes and kidney disease based on the findings of several large, randomized trials *(1,28)* (refer to Chapter 5 for more details).

3.3. Reduce Cardiovascular Morbidity

Cardiovascular death rates are two to four times higher in those with diabetes than in those without diabetes *(1,29)*, with heart disease and stroke responsible for about 65% of deaths in people with diabetes. Consensus panels evaluated the increased morbidity and mortality in the diabetes population, and in the Third Report of the National Cholesterol Education Program (NCEP), diabetes was reclassified as a risk factor for coronary heart disease (CHD) *(30)*. According to the NCEP, patients with diabetes should achieve LDL-C levels less than 100 mg/dl *(15)*. Typically, statin drug therapy and lifestyle changes are necessary to achieve LDL-C goals. Attention is then focused on achieving the secondary targets, namely TG less than 150 mg/dl, HDL-C greater that 40 mg/dl in men and greater than 50 mg/dl in women, and non-HDL less than 130 mg/dl *(30)*.

People with T1DM tend to have a normal lipid profile if glycemia is controlled. On the other hand, people with T2DM usually have elevated TG levels and decreased HDL-C with normal total cholesterol and LDL-C levels *(14)*. Those with diabetes tend to have smaller and more dense LDL-C particles which are more atherogenic. Patients with diabetes and CVD also have excess thromboxane production in vivo *(14)*. Thromboxane is a potent vasoconstrictor and platelet aggregant. Aspirin blocks thromboxane synthesis and is recommended at doses of 81–325 mg daily for primary and secondary prevention of CVD in patients with diabetes *(1)* (refer to Chapter 7 for further information).

4. MEDICAL NUTRITION THERAPY (MNT)

MNT is an integral component of diabetes and CKD management and should emphasize the role of the patient in problem solving as much as possible, using a variety of education strategies and techniques. General goals include the following *(1,24)*:

- maintenance of near normal blood glucose levels by balancing food intake, exercise, and available insulin with stage of kidney function;
- achievement of optimal serum lipids and blood pressure to reduce the risk of CVD;
- adequate energy intake to attain or maintain a reasonable body weight;
- achievement of acceptable biochemical parameters and fluid status;
- prevention and treatment of acute and long-term complications of diabetes and CKD;
- improvement in overall health through appropriate food choices and physical activity; and

- consideration of lifestyle, personal and cultural preferences, and financial situation while respecting the individual's willingness to make changes.

Monitoring blood glucose, HbA1C, lipids, blood pressure, and kidney function are essential to evaluate nutrition-related outcomes. If goals are not met, changes must be made in the overall management plan. The role of the dietitian is to help patients learn a problem-solving approach that evaluates diet as one of many factors that impact these goals *(22)*.

4.1. Dietary Strategies for Carbohydrate Management

Careful management of CHO intake is essential to any diabetes meal planning approach and emphasizes knowing how much and when to consume CHO. This component of the diet has the greatest impact as both the amount (in grams) and type of CHO in food influences blood glucose levels *(31)*. Many dietary strategies can be used to achieve consistency, and the registered dietitian is well qualified to match the appropriate meal planning strategy to each patient's lifestyle and capabilities.

A "constant carbohydrate" meal plan suits patients who use diet alone to control their blood glucose levels, those on fixed doses of insulin or oral diabetes medications, or patients with low literacy. Emphasis is placed on keeping the amount of CHO relatively constant for each meal and eating meals at the same time every day. Insulin would be adjusted around usual CHO intake as much as possible rather than CHO being altered to meet the insulin regimens. Some modifications in food choices may be needed to accommodate sodium, potassium, and phosphorus restrictions *(22,24)*.

The "carbohydrate counting" method provides the most flexibility and is the preferred meal-planning approach for adjusting insulin around usual dietary intake. Initially, careful record keeping by reading food labels and measuring portion sizes for CHO foods is essential until an accurate ratio of insulin to grams of CHO (I/C ratio) is established. The patient counts the grams of total CHO to be eaten and then matches it with the proper amount of insulin. This requires motivation and a higher literacy level. The final insulin dose must also take into account exercise and current blood glucose levels (correction factor adjustment). It is important to keep protein and fat intake relatively constant and not vary widely in CHO consumption to order to avoid excess energy intake and undesired weight gain at the expense of blood glucose control *(22,24)*.

Individuals who have premeal glucose values within target range but who are not meeting HbA1C targets should consider monitoring 2-h postprandial glucose (PPG) after the start of the meal. Treatment should be aimed at reducing PPG values to <180 mg/dl and thereby comparably reducing HbA1C *(1)*. There are now pharmacologic agents that primarily modify PPG levels.

The Exchange List for Meal Planning is familiar to many people with diabetes. Foods are categorized into six food groups with similar amounts of CHO, protein, fats, and calories. Foods within each group can be traded, depending on the amounts consumed *(22,24)*. Although the exchange lists can provide structure and a higher degree of metabolic control, they also require a higher level of literacy due to their complexity.

Low CHO diets, defined by the Institute of Medicine (IOM) as restricting total CHO to <130 gm/day, are specifically not recommended in the management of diabetes *(1,32)*. CHO are an essential source of energy for optimal functioning of the brain and central nervous system and for exercise performance. CHO are the only source of dietary fiber and they supply important water-soluble vitamins, minerals, and other phytochemicals and antioxidants important for good health. The ADA recommends that CHO be derived primarily from whole grains, fruits and vegetables, and non-fat or low-fat dairy products *(1)*. However, with decreasing kidney function, food choices and portions may need to be adjusted based on serum potassium and phosphorus levels. Dietary guidelines for diabetes are often liberalized on the renal diet to provide adequate calories from increased amounts of unsaturated fats and simple CHO, especially with low protein diets.

4.1.1. Glycemic Index

Controlling high postprandial blood glucose levels is an ongoing challenge in diabetes management. Both the quantity and the type or source of CHO found in foods influence PPG levels. Most experts agree, and the ADA position is, that the evidence for total CHO intake from a meal or snack is a more reliable predictor of PPG *(4,31,33,34)*. The impact and relative importance that type or source of CHO has on PPG levels continues to be an area of debate. Two methods that categorize CHO-containing foods based on their glycemic response are the glycemic index (GI) and the glycemic load (GL) *(4,31)*.

The GI ranks CHO foods based on how they affect postprandial glycemia. Some foods result in a marked increase followed by a more or less rapid fall in blood glucose, whereas others produce a smaller peak along with a more gradual decline in plasma glucose *(4,31)*. The specific type of CHO (starch vs. sucrose) present in a particular food does not always predict its effect on blood glucose *(35)*. Foods ingested in 50 g portions are quantified in comparison to a reference food, either glucose or white bread, and the increase in blood glucose (over the fasting level) that is observed after 2 h determines the GI value for that food.

The GL of a particular food is the product of the GI of the food and the amount of CHO in a serving, so it primarily reflects CHO intake. No intervention studies have been done to show a benefit from using the GL technique in persons with diabetes to determine the effect of a mixed meal on PPG and insulin levels, as well as the effects on all day glucose and insulin levels. In fact, no studies to date have found that the amount of CHO eaten per day is significantly associated with the development of type 2 diabetes. This greatly diminishes the importance of GL as a contributor to the onset or development of diabetes *(36–37)*.

There are a number of confounding issues regarding the GI that make its usefulness complicated *(4,31,35,38,39)*. Some of these include the following:

- The GI of a food varies substantially depending on the kind of food, its ripeness, the length of time it was stored, how it was cooked, how it was processed, and its variety (e.g., types of potatoes or rice).
- The GI does not predict PPG response in individuals with diabetes as accurately as it does in healthy persons.
- The GI of a food varies from person to person (inter-individual variation of 10%) and even in a single individual (intra-individual coefficient of variation ranges 23–54%) from day to day, depending on blood glucose levels, insulin resistance, and prior food intake.
- The GI of a food might be one value when eaten alone and another value when it is eaten with other foods as part of a complete meal.
- The GI value is based on a portion that contains 50 g of CHO, which is rarely the total amount eaten.
- Most GI values reflect the blood glucose response to food for only 2 h, whereas glucose levels after eating some foods remain elevated for up to 4 h or longer in people with diabetes because of their altered hormonal responses.
- Another concern is that many high fat foods have a low GI. If food manufacturers begin lowering the GI of processed foods by adding high fat ingredients or high fructose corn syrup (which has a low GI), then consumers once again face the dilemma of altered snack or convenience foods being perceived as healthy.

In summary, there appears to be a small effect from a low GI diet over a high GI diet, primarily on PPG. Use of a low GI diet does not affect fasting plasma glucose values. The GI concept can be used as an adjunct to help "fine-tune" glycemic control. However, there is still insufficient evidence for the ADA to recommend use of low GI diets as a primary strategy in food/meal planning *(4,37)*.

4.1.2. Fiber

Fiber intake should be encouraged within the constraints of the renal diet. High fiber diets (50 g fiber/day) have been shown to reduce glycemia in subjects with T1DM and T2DM and to reduce hyperinsulinemia and lipemia in subjects with T2DM *(4,26)*. Because there are potential barriers to achieving such a high fiber intake, the ADA position as a first priority encourages people with diabetes to aim for the same fiber intake goals set for the general population (14 g / 1000 kcal/day).

4.2. Protein Guidelines for Diabetes and CKD

Although glucose is the primary stimulus for insulin release, protein/amino acids enhance insulin release when ingested with CHO, and thus aid the clearance of glucose from the blood *(4,26)*. Dietary protein intake appears to have an important impact in those with evidence of diabetic nephropathy. The KDOQI Diabetes and CKD Nutrition Management Guidelines and the ADA position statement for Nutrition Recommendations and Interventions for Diabetes state:

- At all Stages of CKD *(1–4)*, persons with diabetes should achieve a dietary protein intake that meets, but does not exceed, the Recommended Dietary Allowance (RDA) of 0.8 g/kg body weight/day *(3)*.
- Reduction of protein intake to 0.8–1 g/kg/day in the early stages of CKD and to 0.8 g/kg/day in the later stages of CKD may improve measures of kidney function and is recommended *(4)*.

Further protein restriction to 0.6 g/kg body weight/day should GFR begin to drop more rapidly than 4 ml/min/year at CKD stage 4 has been shown to slow the decline in selected patients who are stable and closely monitored *(11–13)*. However, this needs to be balanced with the possibility of undernutrition in patients with diabetes who are prescribed a low protein diet, as insulin deficiency stimulates gluconeogenesis and increases protein degradation. Metabolic acidosis also needs to be corrected to maintain positive nitrogen balance. Based on the available evidence, the KDOQI Diabetes Work Group concluded that "limiting dietary protein will slow the decline in kidney function and progression of albuminuria, and it may prevent Stage 5 CKD" *(3)*. The benefits of limiting dietary protein intake are more evident in T1DM than in T2DM, but that may be due to fewer studies having been done in the latter population. Based on two meta-analyses, low protein diets had more pronounced benefits in diabetic than non-diabetic kidney disease *(40,41)*. Even modest limitations of dietary protein (0.89 g/kg body weight/day vs. 1.02 g/kg body weight/day) substantially reduced the risk of stage 5 CKD or death in persons with T1DM and stage 2 CKD *(42)*.

At the other end of the spectrum, high protein diets are a particular concern in diabetes because they appear to have more pronounced effects on renal hemodynamics (including glomerular hyperfiltration and increased intraglomerular pressure) and kidney damage (increased albuminuria and accelerated loss of kidney function based on GFR) *(3)*. Therefore, the KDOQI Diabetes Work Group state that diabetic persons with CKD should avoid high protein diets ($\geq$20% of total daily calories) that are recommended in common popular diets such as Atkins, Protein Power, the Zone, South Beach, and Sugar Busters *(3)*.

The KDOQI Diabetes Work Group suggests that the Dietary Approaches to Stop Hypertension (DASH) and DASH-Sodium diets *(43)* that emphasize sources of protein other than red meat may be an alternative to lower total protein intake in persons with HTN, diabetes, and in CKD stages 1–2. Diets that emphasize proteins from plant sources (vegetables, soy, whole grains, legumes, and nuts) instead of animal sources (particularly red meat) may be renal-sparing *(3,43–47)*.

The NKF KDOQI Guidelines on Hypertension in CKD recommend a version of the DASH diet with modifications for CKD stages 3–4 *(43,48)*. These modifications decreased dietary protein from 1.4 g/kg body weight/day to 0.6–0.8 g/kg body weight/day as well as restricted phosphorus (0.8–1 g/day) and potassium (2–4 g/day) intake *(3,48)*. Because most persons with diabetes and CKD have HTN, which is characterized by enhanced sodium retention, a dietary sodium restriction to 2.4 g/day (100 mmol/day), as recommended by the DASH-Sodium diets and the 2005 Dietary Guidelines, would apply to this population *(3,43)*.

Individuals will remain well nourished as long as an adequate energy intake is maintained for those who achieve the RDA for protein (0.8 g/kg body weight/day). Regardless of the level of protein intake, greater than 50% of the protein should be of high biologic value (HBV), coming predominantly from lean poultry, fish and soy- and vegetable-based proteins *(3)* [see Chapter 25, for the biologic values of selected animal and vegetable food sources, noting that anything >60% is considered HBV for adults *(49,50)*]. The ADA Nutrition position statement defines good quality protein sources as having a high protein digestibility-corrected amino acid score (PDCAAS) and which provide all nine indispensible amino acids *(4)*. Examples are meat, fish, poultry, eggs, milk, cheese, and soy *(32)*.

4.3. Dietary Fat Recommendations

When protein intake is limited, an increase in CHO and/or fats is required for adequate energy intake. According to the National Academy of Sciences, Institute of Medicine (NAS/IOM), non-protein calories (90% of total) should be distributed as 30% or less from dietary fats and up to 60% from complex CHO *(3,32)*. The metabolic profile and need for weight loss should be considered when determining the fat content of the diet *(32)*.

Dietary fat, when eaten with CHO, slows glucose absorption and delays the peak glycemic response *(31)*. Saturated fat and trans fatty acids are the principal dietary determinants of plasma LDL-C, the major risk factor for CVD. Because individuals with diabetes are considered to be at similar risk for CVD as those with a history of CVD, the most recent guidelines for dietary fat intake (amount and type) from the NCEP and the American Heart Association (AHA) would apply. These recommendations state that total fat should be 25–35% of total calories, with saturated fat <7% and trans fat <1% of calories *(30,51)* (refer to Chapter 7 for more detail).

The KDOQI Diabetes Work Group suggests that increasing the intake of omega-3 and monounsaturated fatty acids may provide favorable effects on progression of CKD, although the level of evidence is weak and based on the opinion of the Work Group. Few studies have examined the effects of fatty acid intake or supplements on the markers of kidney disease and the risk factors in patients with diabetes, and those studies have been short term and in small numbers *(3,52–53)*.

The IOM established guidelines for the intake of omega-3 fatty acids. Adequate intake (AI) of alpha-linolenic acid was established at 1.6 g/day for men and 1.1 g/day for women. The more physiologically potent eicosapentaenoic acid (EPA) and docosahexaenoic acid (DHA) can be substituted for up to 10% of these amounts *(32)*. The AHA and the KDOQI Clinical Practice Guidelines for Cardiovascular Disease in Dialysis Patients recommends 12–16 ounces of oily cold water fish weekly, which would provide EPA and DHA well in excess of the 10% AI amounts recommended *(51,54)*. Patients with documented CHD are advised to consume 1 g of EPA and DHA daily, either from fish or supplements. Under a physician's care, supplements which provide 2–4 g of EPA + DHA daily are recommended for individuals with hypertriglyceridemia *(51)*.

4.4. Energy Needs and Weight Management

In the early stages of CKD, the primary goal for MNT is to prevent protein and energy malnutrition (PEM), which increases the risk of poor clinical outcomes and morbidity and mortality *(55)*. The DCCT recommended the usual level of 25 kcal/kg body weight/day be increased to 35 kcal/kg body weight/day for people with diabetes *(11)*. The NKF KDOQI Nutrition guidelines for stages 1–3 CKD are based on energy expenditure *(17,55)*. The ADA MNT Guides for nondialysis use the basal energy expenditures equation (considering stress, protein intake, and weight goals) *(2,9)*.

The energy needs of patients with CKD parallel those of normal healthy individuals *(55)*. However, a recent review of evidence-based guidelines for MNT in CKD recommends a range of calorie levels, depending on age and level of kidney function *(56)*. The NKF KDOQI Nutrition guidelines for stage 4 CKD recommend 30–35 kcal/kg/day for individuals $\geq$60 years of age and 35 kcal/kg/day for those $<$60 years of age, using standard body weight based on the second National Health and Nutrition Examination Survey (NHANES II) *(58)*. The ADA MNT Guides for nondialysis patients recommends 35–45 kcal/kg/ideal body weight (to achieve positive or neutral nitrogen balance) *(9)*. In a review of emerging research since the year 2000, and after evidence-based guidelines were published, overall, nondialyzed patients with CKD seem to have lower energy requirements in comparison to healthy controls and maintenance dialysis patients *(56)*.

Only one study, by Avesani et al. (2001) *(57)*, tried to determine whether the measured resting energy expenditure (MREE) of nondialyzed patients with CKD ($n = 24$) differed from that of nondialyzed patients with CKD and diabetes ($n = 24$, with only 3 T1DM). The MREE was based on indirect calorimetry using a metabolic cart. Outcomes showed that CKD patients with diabetes had higher MREE (12.5% higher) than those without diabetes, with significant correlations to lean body mass and creatinine clearance. Multiple regression analysis showed that having diabetes added 182 kcal to the REE. Dietary energy intakes were not significantly different between those with diabetes (mean 23.4 $\pm$ 5.4 kcal/kg/day) and those CKD patients without diabetes (mean 24.8 $\pm$ 6.9 kcal/kg/day). However, there was no evidence of PEM [normal level of serum albumin and mean body mass index (BMI) 26 kg/m^2].

4.4.1. Weight Management and Exercise

For overweight individuals with T2DM in stages 1–3 CKD, moderate weight loss of 5–10% with maintenance improves insulin sensitivity, glycemic, and blood pressure control and reduces CVD risk *(2,58)*. The primary approach for achieving weight loss is therapeutic lifestyle change, which includes a moderate reduction in energy intake (500–1000 kcal/day) and an increase in moderate physical activity (contributing approximately 200 kcal/day), which should result in a slow but progressive weight loss (1–2 lb/week) *(1)*.

Recommendations for physical activity should be modest and based on the patient's willingness and ability. The ADA guidelines recommend a goal of at least 150 min/week of moderate intensity activity (50–70% maximum heart rate) over at least 3 days/week. Resistance exercises three times/week are also encouraged *(1,4)*. It is important to monitor blood glucose levels before exercising in people with T1DM because vigorous exercise could lead to hypoglycemia or hyperglycemia, depending on the initial blood glucose levels. For planned exercise, a reduction in insulin dosage is the preferred method to prevent hypoglycemia. However, for unplanned exercise, an additional 10–15 g of CHO may be needed for every 60 min of moderate intensity exercise *(4)*.

Pharmacologic treatments for obesity such as sibutramine (Meridia®, Abbott Laboratories) and orlistat (Xenical®, Roche US Pharmaceuticals) are the most studied drugs, and they are the only two approved for long-term treatment of obesity. Most patients who take these drugs, including those with diabetes, have sustained weight loss of up to 10% of their body weight compared with patients on placebo *(59)*.

Sibutramine has a dual action as an appetite suppressant and as a norephineprhine and serotonin re-uptake inhibitor. Its utility is limited in patients with CKD patients because it can cause mild tachycardia and an increase in blood pressure *(59)*. Orlistat is a potent inhibitor of pancreatic lipase. It acts in the lower gut to prevent the breakdown of long-chain fatty acids inhibiting the absorption of fat in the intestine resulting in excess fat removal in the stool. Unfortunately, insurance does not typically cover weight loss therapies and monthly out-of-pocket costs exceeding $100–$200 can be prohibitive for most patients.

4.5. MNT Summary

Table 4 summarizes the nutrition recommendations for patients with diabetes and CKD *(3)*. There is no evidence for additional vitamin and mineral supplements in persons with diabetes who do not have

Table 4
Daily Nutrition Recommendations for Diabetes and Chronic Kidney Disease

	Recommendation		
Nutrient	*Stages 1–2 CKD*	*Stages 1–4 CKD*	*Stages 3–4 CKD*
Calories		30–35 kcal/kg	
Obese		(–) 500–1000 kcal	
Carbohydrate (% kcals)		60%	
Fiber		14 g/1000 kcal	
Protein g/kg/day (% kcals)		0.8 g/kg	
Fat (% kcals)		30%	
Saturated fat (% kcals)		7%	
Cholesterol, mg		<200 mg	
Sodium, mg		<2400 mg	
Phosphorus (mg/day)	1700 mg		800–1000 mg
Potassium (mg/day)	>4000 mg		2400 mg

Adapted from refs. *(3,4)*. Reprinted with permission from the National Kidney Foundation, Inc.

underlying deficiencies. Routine supplementation with antioxidants is not advised because of uncertainties related to long-term efficacy and safety *(26)*. A detailed MNT protocol for CKD (non-dialysis) from the American Dietetic Association is presented in Chapter 25 *(9)*.

4.5.1. Medicare Reimbursement for MNT

Effective on January 1, 2002, Medicare beneficiaries diagnosed with diabetes and non-dialysis kidney disease inclusive of postrenal transplant were eligible for reimbursement of MNT. Beneficiaries must have a fasting glucose level $\geq$126 mg/dl to meet the diagnostic criterion for diabetes. Chronic renal insufficiency or non-dialysis kidney disease is defined as a GFR of 13–50 ml/min/1.73 m^2, not severe enough to require dialysis or renal transplantation. Effective on February 27, 2001, Diabetes Self Management Training (DSMT) services were eligible for Medicare reimbursement. Refer to the Centers for Medicare and Medicaid Services (CMS) guidelines published elsewhere for more detail *(2,60–62)*.

5. MEDICAL MANAGEMENT OF DIABETES

5.1. Insulin Therapies and Oral Agents

Until 1995, insulin and sulfonylureas were the only classes of drugs available in the United States for the treatment of diabetes. Since then, new insulin analog and insulin delivery devices, four new classes of oral medications, and a new class of "incretin mimetics" have been approved for diabetes management *(63)*.

Patients with T1DM require exogenous insulin in contrast to patients with T2DM who do not require exogenous insulin for survival, at least initially *(1)*. Patients with T2DM continue to lose beta-cell function despite treatment. At diagnosis, they usually have less than 50% of their normal insulin secretion and, after 6 years, less than 25%. This progressive decline in beta-cell function is the reason many fail oral therapy and require insulin *(64)*. Some studies have shown that beta-cell function may be preserved with early insulin therapy that reduces the initial "glucose toxicity" *(65)*. Simply introducing a once-daily injection of a basal insulin at bedtime, such as insulin glargine, detemir, or NPH, to an existing oral regimen can help reach glycemic goals *(66)*.

Insulin dosing is very patient-specific, and there is no one way that works best for all patients. When a basal and bolus insulin combination is used, it increases the patient's flexibility and may allow the patient to skip or change meal times or adjust the bolus dose based on the meal content. However, the newer "physiologic" approach, which tries to mimic normal insulin secretion, requires multiple daily injections or a pump. A peakless basal insulin is usually given once or twice daily, and a short-acting bolus insulin is usually given before each meal *(65)*. Fasting hyperglycemia is controlled mainly by basal insulin secretion, which regulates hepatic glucose production; bolus insulin is for postprandial glycemic control.

Most patients with T2DM prefer the most simple insulin regimens possible in combination with oral agents such as using insulin glargine and insulin pens or premixed insulins *(64)*. A comparison of current insulin therapies used in the treatment of T1DM and T2DM *(64,67)* and current oral agents used in the treatment of T2DM are presented in Tables 5 and 6 *(63,66,68–71)*, respectively. The first generation sulfonylureas should be avoided in patients with CKD because they rely on the kidney for elimination of the drug and active metabolites, which could result in prolonged action and a greater risk of hypoglycemia *(3)*.

5.2. Postprandial Glycemic Control and Incretin Mimetics

High postprandial blood glucose levels are a major problem that impact glycemic control in most people with T2DM. This cycle is triggered when the release of insulin is both delayed and blunted in response to blood glucose levels. The alpha cells of the pancreas respond by increasing the excretion of the hormone glucagon, which stimulates the liver to release glucose in the blood and causes further increases in postprandial blood glucose levels *(72)*. To achieve the currently recommended evidence-based glycemic target for HbA1C of 7% or less, better control of postprandial hyperglycemia is needed. This is because as blood glucose control approaches target levels, postprandial blood glucose makes a proportionally greater contribution to overall glycemic exposure *(72)*.

Incretins are hormones that enhance insulin secretion in response to elevated blood glucose levels from oral intake by stimulating the beta cells of the pancreas to produce the right amount of insulin at the right time. They also suppress the secretion of glucagon in the presence of hyperglycemia (which helps decrease hepatic glucose output) and slows gastric emptying, both of which help improve blood glucose regulation *(73)*.

5.2.1. Glucagon-Like Peptide-1

Glucagon-like peptide-1 (GLP-1), secreted by intestinal L cells, appears to be the major mediator of the incretin effect as it binds to receptors in the pancreas, stomach, lung, and brain, and it also stimulates insulin secretion from the pancreas in response to high blood glucose levels *(74)*. Both insulin and GLP-1 are normally present and work together in healthy individuals without diabetes. Secretion of these hormones is deficient in patients with T2DM. GLP-1 is estimated to account for 70–80% of the endogenous incretin effect that has a role in maintaining normoglycemia in nondiabetic individuals following a caloric load *(75)*.

Table 5
Comparison of Insulins Used in the Treatment of Type 1 and Type 2 Diabetes

Type	*Insulin name/ (manufacturer)*	*Onset/peak/duration*	*Insulin delivery devices/comments*
Rapid-acting analogs (Bolus/meal insulins)	Humalog-Lispro(Lilly)	Onset: 5–10 min	Clear; pre-filled pen; used in pumps
	Novolog-Aspart (Novo Nordisk)	Peak: 45–90 min	Clear; pre-filled pen; insulin cartridge; used in pumps
	Glulisine-Apidra (Sanofi Aventis)	Duration: 3–5 h	Clear; insulin cartridge for Opticlick pen
			All prescription only
Short-acting regular (Bolus/meal insulins)	Humulin R (Lilly)	Onset: 30 min	Clear; no significant changes in disposition at any stage of renal failure
		Peak: 2–4 h	Clear; pre-filled pen; insulin cartridge; Innolet device

(Continued)

Table 5
(Continued)

Type	*Insulin name/ (manufacturer)*	*Onset/peak/duration*	*Insulin delivery devices/comments*
	Novolin R (Novo Nordisk)	Duration: 6–8 h	Both can be mixed with NPH All available without a prescription
Intermediate-acting NPH	Humulin N (Lilly)	Onset: 2 h	Cloudy; pre-filled pen
(Basal/background insulins)	Novolin N (Novo Nordisk)	Peak: 6–8 h	Cloudy; pre-filled pen; insulin cartridge; Innolet device
		Duration: 10–12 h (NPH)	All available without prescription
Lente	Novolin L (Novo Nordisk)	14–16 h (Lente)	Cloudy
Long-acting (Basal/background insulins)	Glargine-Lantus (Sanofi Aventis)	Onset: 1.1 h(Lantus) 2 h (Levemir)	Clear; pre-filled SoloStar® pen Clear; pre-filled FlexPen;
		Peak: nearly flat (Lantus)	Reduced incidence of nocturnal hypoglycemia
	Detemir-Levemir (Novo Nordisk)	slight peak 5–14 h (Levemir)	Lantus taken once daily morning or bedtime or BID

		Duration: 22–24 h (Lantus) 14–24 h (Levemir)	Levemir requires BID dosing q 12 h. All prescription only; Not to be mixed with short-acting analogues
Mixtures (pre-mixed at factory)	Humalog 75/25 (Lilly)	Varies according to type of insulin combination	75% NPH/25% Humalog; pre-filled pen
			70% NPH/30% Novolog; pre-filled pen; insulin cartridge
	Novolog 70/30 (Novo Nordisk)		
	Humulin 70/30 (Lilly)		*70% NPH/30% Regular
	Novolin 70/30 (Novo Nordisk)		*70% NPH/30% Regular; pre-filled pen; InnoLet device
	Humulin 50/50 (Lilly)		*50% NPH/50% Regular
	Humalog 50/50 (Lilly)		50% NPH/50% Humalog
			All are cloudy in appearance *available without prescription; others by prescription only

Adapted from refs. *(64,67)*.

Note: All insulins are from human recombinant DNA sources; U-100 (100 units insulin per ml); 10-ml vials or 1.5-ml or 3-ml pen cartridge; open vials to be used in 28 days at refrigerator or room temperature. In 2005 Eli Lilly discontinued four insulin products: Humulin® U Ultralente; Humulin® L Lente; Ilentin® Pork insulin; Porkinsulin (regular and NPH formulations). Careful glucose monitoring and dose adjustments of all insulins may be necessary in patients with renal impairment due to increased circulating levels and sensitivity to insulin as renal function declines.

* Pertains to available without prescription.

Table 6
Comparison of Oral Agents Used in the Treatment of Type 2 Diabetes

Class	*Oral agent generic (trade)*	*Target organ action*	*Glycemic reduction*	*Benefits*	*Side effects*	*Precautions*
Secretagogues: ↓Sulfonylureas (2nd generation)	Glyburide (Diabeta, Micronase, Glynase) Glipizide (Glucotrol/XL)[a] Glimepiride (Amaryl)	Pancreas: stimulate phase II insulin release in β-cells to reduce postprandial BG	↓FBG 50–60 mg/dl ↓A1C 1.5–2	Reduces fasting or postprandial BG (requires residual β-cell function)	Hypoglycemia (risk less with meglitinides); weight gain; hyperinsulinemia	Use with caution in elderly patients, hepatic or renal impairment; Take before meals (Glucotrol 30 min before meals) Avoid glyburide at GFR <60 ml/min/1.73 m^2
↓Meglitinides (shorter onset and duration of action)	Repaglinide (Prandin)[a] Nataglinide (Starlix)		↓PPG 40–50 mg/dl ↓A1C 0.5–1.5	Most helpful in ↓ PPG	Headaches	Note: Avoid 1st generation sulfonylureas in CKD

Biguanides	Metformin Riomet – liquid (Glucophage) Fortamet (Glucophage XR)	Liver; Decrease hepatic glucose production (and some ↑ in insulin sensitivity in hepatic and peripheral tissues)	↓FBG 50–60 mg/dl ↓A1C 1.5-2	Does not cause weight gain; improves lipid profile (slight ↑ in HDL, 10–15% ↓ LDL and TGL)	Nausea; diarrhea; metallic taste lactic acidosis ↓vit B12 levels (in 7% of patients in clinical trials)	Use caution in patients with heart failure (class III/IV), renal or hepatic or severe pulmonary disease; Contraindicated if CrCl <60 ml/min or if sCr >1.5 in men or sCr >1.4 in women; CrCl 30–60 ml/min, limit metformin dose to 1000–1500 mg/day; CrCl <30 ml/min, avoid metformin; hold 48 h for any radiocontrast dye procedure or surgery; take with meals and titrate dose up slowly to minimize GI effects (subsides after 3 weeks)

(Continued)

Table 6
(Continued)

Class	*Oral agent generic (trade)*	*Target organ action*	*Glycemic reduction*	*Benefits*	*Side effects*	*Precautions*
Thiazolidinediones (insulin sensitizers)	Pioglitazone (Actos) Rosiglitazone (Avandia)	Peripheral tissue: Improves insulin sensitivity (and some ↑ in hepatic glucose production)	↓FBG 30–60 mg/dl ↓A1C 0.8–1.8	Improve lipid abnormalities (esp. Actos): converts small dense LDLs to less atherogenic larger particles; ↑ in HDL 5-10%, ↓ TGL 10-20%; Improves β-cell function	Weight gain; edema; anemia ↑ fracture risk in women	Contraindicated if ALT > 3× the upper limit of normal; use with caution in patients with heart failure (class III/IV) or hepatic disease; Monitor LFTs before starting and every 2 months for 1 year and periodically after that; Takes 4–6 weeks to see effect on BG; induces ovulation; Take with or without meals; no dose adjustment for CKD

						Avandia: possible ↑ heart attack risk
Alpha-glucosidase inhibitor	Acarbose Miglitol	Intestine: slows down CHO absorption in intestine	↓PPG 50 mg/dl ↓A1C 0.5–1.0	Does not cause weight gain or hypoglycemia	Flatulence; diarrhea; abdominal discomfort	Avoid if sCr > 2.0 mg/dl; avoid in patients with GI disorders; Take with first bite of food; Start with low dose and titrate up slowly to minimize GI effects; If used with certain other DB medications, treat hypoglycemia with pure glucose (gels/tabs)

(Continued)

Table 6
(Continued)

Class	*Oral agent generic (trade)*	*Target organ action*	*Glycemic reduction*	*Benefits*	*Side effects*	*Precautions*
Combination pills	Glyburide/ Metformin (Glucovance)	Pancreas/liver			Side effects same as drug in combination	Precautions same as drug in combination; Take with meals
	Glipizide/ Metformin (Metaglip)					
	Rosiglitizone/ Metformin (Avandamet)	Peripheral tissue/liver				
	Pioglitazone/ Metformin (Acto*plus*met)	Peripheral tissue/pancreas				
	Avandia/Amaryl (Avandaryl)					

FBG, fasting blood glucose; PPG, postprandial glucose; sCr, serum creatinine; CrCl, creatinine clearance; CHF, congestive heart failure; CHO, carbohydrate; ALT, alanine transaminate; TGL, triglyceride.

CKD, chronic kidney disease; GFR, glomerular filtration rate; GI, glycemic index; HDL, high density lipoprotein; LDL, low density lipoprotein;

Adapted from refs. *(63,66,68,69,3)*.

[a] Safest for kidney disease with no dose adjustment necessary.

5.2.2. Exenatide (Byetta®)

Exenatide injection, Byetta® (Amylin Pharmaceuticals/Eli Lilly), is the first of a new class of drugs known as "incretin mimetics" approved to treat people with T2DM who are unable to control their blood glucose levels with oral agents and to specifically help improve postprandial hyperglycemia. Studies consistently show a sustained 1% reduction in HbA1C levels with initial weight reductions averaging 4–5 lb. Glycemic control has been maintained for up to 2 years with long-term use of exenatide, resulting in continued weight loss of up to 10–15 lb, an effect rarely seen with other diabetes medications. Additionally, fasting glucose was reduced by 28 mg/dl *(63,75)*.

5.2.3. Pramlintide (Symlin®)

Another injectable drug, pramlintide acetate (Symlin®, also marketed by Amylin Pharmaceuticals) has many of the same incretin effects but with different mechanisms of action. Pramlintide is a synthetic version of the hormone amylin, which is co-produced and co-secreted with insulin in the beta cells of the pancreas and then binds to receptors in the stomach and brain. It acts to suppress appetite, slow gastric emptying, and suppress glucagon secretion, which reduces hepatic glucose output postprandially. Amylin secretion has been shown to be delayed and diminished in more advanced cases of T2DM and markedly reduced or absent in people with T1DM *(75,76)*. Without sufficient amylin, glucose enters the blood rapidly following meals, producing glycemic peaks. Without amylin, it is harder for insulin to lower these peaks, making glycemic control more challenging. In contrast to exenatide, pramlintide has no stimulation on the beta cells *(76–77)*. In cohort studies, the addition of amylin to insulin resulted in significant reductions in A1C (between 0.5 and 0.7%) with a mean weight loss of 2 kg over 6 months *(78)*. These findings have been sustained over 2 years in uncontrolled open-label extension studies with additional weight loss up to 4 kg *(79)*. Patients with moderate or severe renal impairment (CrCl 20–50 ml/min) did not show increased exposure or reduced clearance of pramlintide compared to subjects with normal renal function *(75)*.

It should be emphasized that current research is exploring new medications to assist with glycemic control, and availability of pharmacological agents changes rapidly. A comparison of pramlintide (Symlin) and exenatide (Byetta) *(80–81)* is presented in Table 7 and Table 8 provides a summary of the current glucoregulatory options for T2DM *(74,82)*.

Table 7
Comparison of Pramlintide (Symlin®) and Exenatide (Byetta®)

	Pramlintide (Symlin®)	*Exenatide (Byetta®)*
Description	Analog of amylin, a naturally occurring pancreatic hormone	Mimics the action of gut incretin hormone, glucagon-like peptide-1
Class	Amylin mimetic	Incretin mimetic
Mechanism of action	Improves postprandial glycemic control by: • Suppresing glucagon secretion[a] • Inducing satiety and reduces food intake • Slowing gastric emptying	Improves postprandial glycemic control by: • Enhancing glucose-dependent insulin secretion • Restoring first-phase insulin response to food • Suppresing glucagon secretion[a] • Reducing food intake • Slowing gastric emptying
Adverse effects	GI: mild and transient nausea and vomiting	GI: mild and transient nausea and vomiting; others include diarrhea, dyspepsia, headache, dizziness, nervousness, risk of pancreatitis
Use	• Insulin using (basal & bolus) patients with type 1 or type 2 diabetes • Can be used in CKD patients with CrCl >20 ml/min	• Patients with type 2 diabetes using sulfonylurea, metformin, or combination of both • Can be used in CKD patients with CrCl >30 ml/min

Contraindications	Should not be used in patients who have: • poor compliance with current insulin regimen • presence of hypoglycemia unawareness • recurrent severe hypoglycemia requiring assistance in the past 6 months • poor compliance with prescribed self-blood glucose monitoring • an A1C >9% • confirmed diagnosis of gastroparesis • require the use of drugs that stimulate gastric motility • pediatric patients • end-stage renal disease or on dialysis (not studied in these patients)	Should not be used in patients who have: • severe renal impairment CrCl <30 ml/min or on dialysis • severe gastrointestinal impairment • type 1 diabetes • type 2 diabetes taking insulin or other oral diabetes medications other than those above • pregnancy

(Continued)

Table 7
(Continued)

	Pramlintide (Symlin®)	*Exenatide (Byetta®)*
Administration	• Injected subcutaneously up to three times daily prior to meals or snacks that contain at least 250 kcals and 30 g CHO; eat within 15 min. • Type 2 diabetes: Start with 60 mcg (10 units) x 1 week; advance to maintenance dose of 120 mcg (20 units) as tolerated (no nausea) • Type 1 diabetes: Start with 15 mcg (2.5 units) and titrate by 15 mcg increments every 3–7 days as tolerated up to maintenance dose of 60 mcg (10 units)	• Injected subcutaneously twice daily before morning and evening meal using pre-filled fixed-dose disposable injector pens (5 mcg or 10 mcg); eat within 60 min. • Starting dose is 5 mcg twice daily for 30 days • Advance to maintenance dose of 10 mcg twice daily as tolerated (no nausea)
Precautions	• Not to be mixed with insulin; requires separate injection • Recommend initial 50% reduction in mealtime insulin dose to reduce risk of hypoglycemia • Titrate dose upward slowly to minimize nausea	• Initially reduce sulfonylurea dose by half to minimize risk of hypoglycemia

CHO, carbohydrate; CKD, chronic kidney disease; CrCl, creatinnine clearance; GI, glycemic index.
Adapted from refs. *(77,82)*. Pramlintide acetate/Symlin® (Amylin Pharmaceuticals, Inc.) and Exenatide/Byetta® (Amylin and Eli Lilly).
[a] Not in the presence of hypoglycemia.

Table 8
Current Glucoregulatory Options for Type 2 Diabetes

Agent	*Regulates food intake*	*Regulated gastric emptying*	*Modulates glucose absorption*	*Enhances insulin secretion*	*Reduces hepatic glucose production*	*Enhances insulin sensitivity*	*Enhances glucose uptake*
Insulin	No	No	No	Yes	Yes	Yes	Yes
Insulin secretagogues	No	No	No	Yes	Yes	Possibly	Yes
Biguanides	No	No	No	No	Yes	Yes	Yes
Alpha-Glucosidase Inhibitors	No	No	Yes	No	No	No	No
Thiazoladinediones	No	No	No	Possibly	Yes	Yes	Yes
Amylin mimetics	Yes	Yes	No	No	Yes	No	No
Incretin mimetics	Yes	Yes	No	Yes	Yes	No	No

Adapted from refs. *(82,74)*.

5.3. *Self-Monitoring of Blood Glucose (SMBG)*

In recent years there has been an impressive increase in new self-monitoring of blood glucose (SMBG) systems to improve glycemic control. The ADA recommends a minimum of once-daily monitoring for patients on insulin or insulin secretagogues to assist in the prevention of hypoglycemia. However, to obtain optimal glucose control, it is necessary for a patient using insulin therapy to test a minimum of three times/day to detect variations in blood glucose levels that may require adjustments in insulin dosages *(1)*.

Results of SMBG can be useful in preventing hypoglycemia and for adjusting medications, food intake, and physical activity to achieve glycemic goals. The ADA acknowledges that the benefits of tight control may not be indicated for the elderly (≥65 years of age) or patients with advanced complications. The optimal frequency and timing of SMBG for patients with T2DM is not known, but should be sufficient to reach glucose goals *(1)*. A reasonable goal would be to monitor fasting blood glucose and then vary a second blood check by rotating different days and times to see if a pattern emerges prior to certain meals, 2 h postprandial, or at bedtime. It is especially important to test 2 h postprandially in a person with T2DM, as this time period tends to be most problematic in this population. It is also important for health care providers to evaluate each person's monitoring technique for accuracy, both initially and at regular intervals thereafter. Despite education, few patients self-monitor their blood glucose levels on a regular basis or adhere to the SMBG regimen prescribed by their health care providers *(83)*.

6. CONCLUSION/SUMMARY

The incidence of both diabetes and CKD is on the rise. Living with either diabetes or CKD is challenging for anyone. Managing diabetes in the presence of CKD requires additional effort on the part of the patient and the health care team to improve outcomes and to decrease morbidity and mortality in this population. Because of the complexity of diabetes and CKD and their comorbidities, multiple drug therapies, in conjunction with MNT, are necessary to achieve the goals of therapy; therefore, patient adherence becomes a serious concern. The nutrition needs of patients with diabetes and CKD change as the progression and treatment of the disease changes. Understanding the treatment goals at each stage is crucial to optimize the nutritional status of patients throughout the course of their disease and to individualize their education and meal plans accordingly *(22)*.

7. CASE STUDY

G.O. is a 58-year-old Caucasian male who is morbidly obese and is referred for MNT with interest in starting Symlin to aid weight loss and improve glycemic control. Patient is not interested in gastric bypass surgery. His medical history reveals T2DM × 19 years, HTN, hyperlipidemia, CKD, anemia, gout, arthritis; morbid obesity; retinopathy. Current medications include Novolin 70/30, takes 80 u. a.m. and 90 u. with supper; Zocor 40 mg; Tricor 48 mg; Lasix 80 mg; Vasotec 40 mg bid; Atenelol 100 mg; Losartin 100 mg; Lasix 80 mg; Norvasc 10 mg; Allopurinol 100 mg; Cardura 4 mg; Colchicine 0.6 mg; Procrit 8000 units SC q wk; MVI daily. His physical exam presents with height: 5′11″; weight: 370# (BMI 52 kg/m^2); BP 140/60 mmHg; pedal edema 2+. Laboratory data included HbA1C 8.5, Glucose 193 mg/dl (postprandial); SCr. 2.6 mg/dl; MAC 123 mm; Hgb 10.1 g/dl; TC 201 mg/dl; TG 671 mg/dl; HDL 32 mg/dl; LDL unable to calculate. He routinely only checks blood sugar (BS) two times/day at a.m. and HS; sporadically will check four times/day including lunch and dinner; reports three to four episodes of hypoglycemia monthly (usually from meal delay). He did not bring meter or BS log with him. He is married; works 7 a.m. to 3 p.m. as a computer analyst for a large company; quit smoking 18 years ago; no alcohol consumed. He complains of swelling in feet and does not regularly exercise; reports balance/stability problems with gout and excess weight. Dietary history revealed that he has used the Atkins Diet in the past and lost up to 40 lb only to regain the weight. Most recently on Atkins diet 9 months ago but developed kidney failure requiring hospitalization (admission labs indicated: Gluc 102, BUN 197, Cr 6.6, Ca++ 13.9, K+ 4.8, C02 25, Hgb 11.4). Kidney function stabilized and no longer follows Atkins diet. Eats out weekly at Olive Garden restaurant. Reports no previous diabetes classes or education from a dietitian. Desires weight loss to 300#. G.O., in attempt to lose weight, has altered his meal pattern recently and reports that breakfast at work 8:30 a.m. (1 c. cereal with 1/2 c. 1% milk and banana), a.m. snack (snacks on bag of microwave popcorn throughout morning), noon lunch (frozen Weight Watchers meal or 2 c. homemade navy bean/corn chowder soup and 1.5 oz chips with diet Coke), p.m. snack (apple, hard candy, and diet Coke), supper 5:30 p.m. (6 oz. meat, 2 starch servings, green salad or vegetable, diet Coke), HS snack (ice cream bar or cookies).

7.1. Case Study Questions

1. What is the patient's estimated GFR and CKD stage?
2. Because patient did not bring meter or SBGM records with him, what is his average blood sugar based on his HbA1C?

3. Is patient a good candidate for Symlin? Why or why not?
4. If Symlin initiated, what precautions/guidelines need to be followed?
5. What dietary protein intake would you recommend?

REFERENCES

1. American Diabetes Association: Standards of Medical Care in Diabetes - 2007. Diabetes Care 2007; 30 (Suppl 1): S4–S41.
2. Ross TA, Boucher JL, O'Connell BS, ed. American Dietetic Association Guide to Diabetes Medical Nutrition Therapy and Education. Diabetes Care and Education Dietetic Practice Group. Chicago, IL: American Dietetic Association, 2005.
3. National Kidney Foundation. K/DOQI clinical practice guidelines for diabetes and chronic kidney disease. Am J Kidney Dis 2007; 49 (Suppl 2): S1–S179.
4. Nutrition Recommendations and Interventions for Diabetes – 2006. A position statement of the American Diabetes Association. Diabetes Care 2006; 29(9): 2140–2157.
5. American Diabetes Association Web site. Available at: http://www.diabetes.org. Diabetes Statistics and National Diabetes Fact Sheet, 2005. Accessed November 10, 2005.
6. United States Renal Data System. USRDS Annual Data Report at www.usrds.org. Accessed August 14, 2006.
7. Molitch ME, DeFronzo RA, Franz MJ, Keane WF, Mogensen CE, Parving HH, Steffes MW. American Diabetes Association: Nephropathy in Diabetes (Position Statement). Diabetes Care 2004; 27 (Suppl 1): S79–S83.
8. Levinsky NG. Specialist evaluation in chronic kidney disease: too little, too late. Ann Intern Med 2002; 137: 542–543.
9. American Dietetic Association Medical Nutrition Therapy Evidence-Based Guide for Practice. Chronic Kidney Disease (non-dialysis) Medical Nutrition Protocol [CD-ROM]. Chicago, IL: American Dietetic Association, 2002.
10. Nesbitt KN. An overview of diabetic nephropathy. J Pharmacy Practice 2004; 17(1): 75–79.
11. Diabetes Control and Complications Trial Research Group. The effect of intensive treatment of diabetes on the development and progression of long-term complications in insulin-dependent diabetes mellitus. N Engl J Med 1993; 329: 977–986.
12. UK Prospective Diabetes Study Group. Intensive blood-glucose control with sulphonylureas or insulin compared with conventional treatment and risk of complications in patients with type 2 diabetes (UKPDS 33). Lancet 1998; 352: 837–853.
13. UK Prospective Diabetes Study Group. Effect of intensive blood-glucose control with metformin on complications in overweight patients with type 2 diabetes (UKPDS 34). Lancet 1998; 352: 854–865.
14. Schering D, Kasten S. The link between diabetes and cardiovascular disease. J Pharmacy Practice 2004; 17(1): 61–65.
15. American Diabetes Association. Dyslipidemia management in adults with diabetes (position statement). Diabetes Care 2004; 27 (Suppl 1): S68–S71.
16. Namyi Y, Milite CP, Inzucchi SE. Evidence-based treatment of diabetic nephropathy. Practical Diabetology 2005; December: 36–41.

17. National Kidney Foundation. KDOQI clinical practice guidelines for chronic kidney disease: evaluation, classification, and stratification. Am J Kidney Dis 2002; 39 (Suppl 1): S1–S266.
18. Levey AS, Bosch JP, Lewis JB, Greene T, Rogers N, Roth D. A more accurate method to estimate glomerular filtration rate from serum creatinine: a new prediction equation: Modification of Diet Renal Disease Study Group. Ann Intern Med 1999;130:461–470.
19. McCann L, ed. Pocket Guide to the Nutrition Assessment of the Patient with Chronic Kidney Disease (CKD), 3rd edn. New York, NY: National Kidney Foundation, 2002.
20. American College of Endocrinology. American College of Endocrinology Consensus Statement on Guidelines for Glycemic Control. Endocr Pract 2002; 8 (Suppl 1): 5–11.
21. International Diabetes Federation Clinical Guidelines Task Force. Global guidelines for type 2 diabetes. Available at: http://www.idf.org/webdata/docs/IDF%20GGT2D.pdf. Accessed April 3, 2006.
22. Pagenkemper JJ. Nutrition management of diabetes in chronic kidney disease. In: A Clinical Guide to Nutrition Care in Kidney Disease. Byham-Gray L, Wiesen K (ed.). Chicago IL: American Dietetic Association, 2004.
23. Delahanty LM. Implications of the Diabetes Control and Complications Trial for renal outcomes and medical nutrition therapy. J Ren Nutr 1998; 8: 59–63.
24. Renal Dietitians Dietetic Practice Group of the American Dietetic Association. National Renal Diet Professional Guide, 2nd edn. Chicago, IL: American Dietetic Association, 2002.
25. Apidra® (Insulin Glulisine Injection), Prescribing Information, Package Insert. Bridgewater, NJ: Sanofi-Aventis Pharmaceuticals, Inc., 2005.
26. Franz MJ, Bantle JP, Beebe CA, Brunzell JD, Chiasson JL, Garg A, Holzmeister L, Hoogwerf BJ, Mayer-Davis E, Mooradian AD, Purnell JQ, Wheeler M. Evidence-based nutrition principles and recommendations for the treatment and prevention of diabetes and related complications (Technical Review). Diabetes Care 2002; 25: 148–198.
27. Durst SW, Schering D. Hypertension management in the diabetes patient. J Pharmacy Practice 2004; 17(1): 55–60.
28. National High Blood Pressure Education Program. The seventh report of the Joint National Committee on Detection, Evaluation, and Treatment. JAMA 2003; 289: 2560–2572.
29. American Diabetes Association. Consensus development conference on the diagnosis of coronary heart disease in people with diabetes. Diabetes Care 1998; 21: 1551–1559.
30. Expert Panel on Detection, Evaluation, and Treatment of High Blood Cholesterol in Adults. Executive summary of the Third Report of the National Cholesterol Education Program (NCEP) expert panel on detection, evaluation, and treatment of high blood cholesterol in adults (adult treatment panel III). JAMA 2001; 285: 2486–2497.
31. Sheard NF, Clark NG, Brand-Miller JC, Franz MJ, Pi-Sunyer FX, Mayer-Davis E, Kulkarni K, Geil P. Dietary carbohydrate (amount and type) in the prevention and management of diabetes: a statement of the American Diabetes Association. Diabetes Care 2004; 27: 2226–2271.

32. National Academy of Sciences, Institute of Medicine. Dietary Reference Intakes: Energy, Carbohydrate, Fiber, Fat, Fatty Acids Cholesterol, Protein, and Amino Acids. Washington, DC: National Academy Press, 2002.
33. Wolever TM, Bolognesi C. Source and amount of carbohydrate affect postprandial glucose and insulin in normal subjects. J Nutr 1996; 126: 2798–2806.
34. Wolever TM, Bolognesi C. Prediction of glucose and insulin responses of normal subjects after consuming mixed meals varying in energy, protein, fat, carbohydrate and glycemic index. J Nutr 1996; 126: 2807–2812.
35. Foster-Powell K, Miller JB. International tables of glycemic index. Am J Clin Nutr 1995; 62: 871S–890S.
36. Pi-Sunyer FX. Do glycemic index, glycemic load, and fiber play a role in insulin sensitivity, disposition index, and type 2 diabetes? Diabetes Care 2005; 28: 2978–2979.
37. Franz MJ. The glycemic index: not the most effective nutrition therapy intervention. Diabetes Care 2003; 26(8): 2466–2468.
38. Freeman J. The glycemic index debate: does the type of carbohydrate really matter? Diabetes Forecast 2005; September: 11.
39. Pi-Sunyer FX. Glycemic index and disease. Am J Clin Nutr 2002; 76: S290–S298.
40. Kasiske BL, Lakatua JD, Ma JZ, Louis TA. A meta-analysis of the effects of dietary protein restriction on the rate of decline in renal function. Am J Kidney Dis 1998; 31: 954–961.
41. Pedrini MT, Levey AS, Lau J, Chalmers TC, Wang PH. The effect of dietary protein restriction on the progression of diabetic and nondiabetic renal disease: a meta-analysis. Ann Intern Med 1996; 124: 627–632.
42. Hansen HP, Tauber-Lassen E, Jensen BR, Parving HH. Effect of dietary protein restriction on prognosis in patients with diabetic nephropathy. Kidney Int 2002; 62: 220–228.
43. Sacks FM, Svetkey LP, Vollmer WM, Appel LJ, Bray GA, Harsha D, Obarzanek E, Conlin PR, Miller ER III, Simons-Morton DG, Karanja N, Lin PH. Effects on blood pressure of reduced dietary sodium and the Dietary Approaches to Stop Hypertension (DASH) diet. DASH-Sodium Collaborative Research Group. N Engl J Med 2001; 344: 3–10.
44. He J, Gu D, Wu X, Chen J, Duan X, Chen J, Whelton PK. Effect of soybean protein on blood pressure: a randomized, controlled trial. Ann Int Med 2005; 143: 1–9.
45. Azadbakht L, Shakerhosseini R, Atabak S, Jamshidian M, Mehrabi Y, Esmaill-Zedah A. Beneficiary effect of dietary soy protein on lowering plasma levels of lipid and improving kidney function in type 2 diabetes with nephropathy. Eur J Clin Nutr 2003; 57: 1292–1294.
46. Jibani MM, Bloodworth LL, Foden E, Griffiths KD, Galpin OP. Predominantly vegetarian diet in patients with incipient and early clinical diabetic nephropathy: effects of albumin excretion rate and nutritional status. Diabet Med 1991; 8: 949–953.
47. Knight EL, Stampfer MJ, Hankinson SE, Spiegelman D, Curhan GC. The impact of protein intake on renal function decline in women with normal renal function or mild renal insufficiency. Ann Intern Med 2003; 138: 460–467.
48. National Kidney Foundation. KDOQI clinical practice guidelines on hypertension and antihypertensive agents in chronic kidney disease. Am J Kidney Dis 2004 (suppl 1); 43: S115–S119.

49. Food Policy and Food Science, Nutrition Division. Amino-Acid Content of Foods and Biological Data on Proteins. Rome: Food and Agriculture Organization of the United Nations, 1970.
50. Food and Nutrition Board. Evaluation of Protein Nutrition. Washington, D.C.: National Academy of Sciences. National Research Council, Publ 711, 1959.
51. Diet and lifestyle recommendations revision 2006. A scientific statement from the American Heart Association Nutrition Committee. Circulation 2006; 114: 82–96.
52. Hamazaki T, Takazakura E, Osawa K, Urakaze M, Yano S. Reduction in microalbuminuria in diabetics by eicosapentaenoic acid ethyl ester. Lipids 1990; 9: 541–545.
53. Rossing P, Hansen BV, Nielsen FS, Myrup B, Holmer G, Parving HH. Fish oil in diabetic nephropathy. Diabetes Care 1996; 19: 1214–1219.
54. National Kidney Foundation. KDOQI clinical practice guidelines for cardiovascular disease in dialysis patients. Am J Kidney Dis 2005 (Suppl 3); 45: S90–S95.
55. National Kidney Foundation. KDOQI clinical practice guidelines for nutrition in chronic renal failure. Am J Kidney Dis 2000; 35: S40–S45, S58–S61.
56. Byham-Gray LD. Weighing the evidence: energy determinations across the spectrum of kidney disease. J Ren Nutr 2006; 16(1): 17–26.
57. Avesani CM, Cuppari L, Silva AC, Sigulem DM, Cendoroglo M, Sesso R, Draibe SA. Resting energy expenditure in pre-dialysis diabetic patients. Nephrol Dial Transplant 2001; 16: 556–560.
58. Maggio CA, Pi-Sunyer FX. Obesity and type 2 diabetes. Endocrinol Metab Clin North Am 2003; 32: 805–822.
59. Hollander P. Pharmacologic treatment of obesity. On the Cutting Edge: Diabetes Care and Education. Spring 2006; 27(2): 22–26.
60. Medicare program; revisions to payment policies and five-year review of and adjustments to the relative value units under the physician fee schedule for calendar year 2002; final rule, 66 Federal Register 55246-55332 (2001) (codified at 42 CFR 405).
61. Additional clarification for medical nutrition therapy (MNT) services. Available at: http://www.cms.hhs.gov/manuals/. Accessed November 4, 2007.
62. Medicare program, expanded coverage for outpatient diabetes self-management training and diabetes outcome measurements; final rule and notice. 65 Federal Register 83130-83154 (2000) (codified at 42 CFR 410, 414,424,480, 498).
63. Drucker DJ. The evidence for achieving glycemic control with incretin mimetics. The Diabetes Educator March/April 2006; 32(Suppl 2): S72–S81.
64. Daugherty KK. Review of insulin therapy. J Pharmacy Practice. 2004; 17(1): 10–19.
65. DeWitt DE, Hirsch IB. Outpatient insulin therapy in type 1 and type 2 diabetes mellitus: scientific review. JAMA 2003; 289: 2254–2264.
66. Cornell S, Briggs A. Newer treatment strategies for the management of type 2 diabetes mellitus. J Pharmacy Practice 2004; 17(1): 49–54.
67. Saenz M. Insulin analogs: tools for better blood-glucose control. On the Cutting Edge: Diabetes Care and Education. Spring 2006; 27(2): 12–14.
68. Koski RR. Oral antidiabetic agents: a comparative review. J Pharmacy Practice 2004; 17(1): 39–48.

69. Hamrick J. Oral medications for diabetes: a review. On the Cutting Edge: Diabetes Care and Education. Spring 2006; 27(2): 15–18.
70. Nissen SE, Wolski K. Effect of rosiglitazone on the risk of myocardial infarction and death from cardiovascular courses. N Engl J Med. 2007; 356: 2457–2471.
71. Strotmeyer ES, Schwarte AV, Newman AB. Thiazolidinediones and the risk of nontraumatic fractures in patients with diabetes: In reply. Arch Intern Med. 2006; 166(9): 1043.
72. Monnier L, Lapinski H, Colette C. Contributions of fasting and postprandial plasma glucose increments to the overall diurnal hyperglycemia of type 2 diabetic patients: variations with increasing levels of HbA1C. Diabetes Care 2003; 26: 881–885.
73. Drucker DJ. Incretin-based therapies: a clinical need filled by unique metabolic effects. Diabetes Educ March/April 2006; 32(Suppl 2): S65–S71.
74. Kruger DF, Martin CL, Sadler CE. New insights into glucose regulation. The Diabetes Educator March/April 2006; 32: 221–228.
75. Uwaifo GI, Ratner RE. Novel pharmacologic agents for type 2 diabetes. Endocrinol Metab Clin N Am 2005; 34: 155–197.
76. Blair EM. Pramlintide: a new tool for glycemic control. On the Cutting Edge: Diabetes Care and Education. Spring 2006; 27(2): 6–8.
77. Kruger D. Symlin and Byetta: two new antihyperglycemic medications. Practical Diabetology 2006; March: 49–52.
78. Hollander PA, Levy P, Fineman MS, Maggs DG, Shen LZ, Stroebel SA, Weyer C, Kolterman OG. Pramlintide as an adjunct to insulin therapy improves long-term glycemic and weight control in patients with type 2 diabetes. Diabetes Care 2003; 26: 784–790.
79. Karl DM. Learning to use pramlintide. Practical Diabetology 2006; March: 43–46.
80. Clark WL. Exenatide. Diabetes Self-Management 2006; Jan/Feb: 36–40.
81. Blair EM. Exenatide: glycemic control and weight loss. On the Cutting Edge: Diabetes Care and Education. Spring 2006; 27(2): 4-6.
82. DeFronzo RA. Pharmacologic therapy for type 2 diabetes mellitus. Ann Intern Med 1999; 131: 281–303.
83. Briggs A, Cornell S. Self-monitoring blood glucose (SMBG): now and the future. J Pharmacy Practice 2004; 17(1): 29–38.
84. Rohlfing CL, Wiedmeyer H-M, Little RR, England JD, Tennill A, Goldstein DE. Defining the relationship between plasma glucose and HbA1C: analysis of glucose profiles and HbA1C in the Diabetes Control and Complications Trial. Diabetes Care 2002; 25: 275–278.

7 Dyslipidemias

Judith A. Beto
and Vinod K. Bansal

LEARNING OBJECTIVES

1. Define dyslipidemia in chronic kidney disease from a pathophysiological and metabolic perspective.
2. Review current clinical practice guidelines and peer-reviewed recommendations.
3. Identify key assessment and intervention strategies.
4. Outline parameters of dietary and non-dietary treatment to reduce lipidemia.

Summary

Chronic kidney disease (CKD) patients are at higher risk for cardiovascular disease (CVD) than the non-CKD population. The Kidney Disease Outcome Quality Initiative guidelines support intervention to reduce and manage CVD risk factors including dyslipidemia in this population. The achievement and reduction of dyslipidemia to optimal levels in CKD stages 1–4 can be most successfully accomplished using a multifactorial health care team approach that includes comprehensive assessment and intervention using lifestyle, dietary, and pharmacological strategies.

Key Words: Dyslipidemia; nutrition; diet; chronic kidney disease.

From: *Nutrition and Health: Nutrition in Kidney Disease*
Edited by: L. D. Byham-Gray, J. D. Burrowes, and G. M. Chertow

1. INTRODUCTION

The National Kidney Foundation's Kidney Disease Outcome Quality Initiative (KDOQI) has produced a cohort of clinical practice guidelines directed toward improving the quality and breadth of care given to patients with chronic kidney disease (CKD) *(1)*. The KDOQI Clinical Practice Guidelines for Chronic Kidney Disease: Evaluation, Classification, and Stratification created a clinical action plan that defined five stages of kidney function by glomerular filtration rate levels *(2)*. This classification system re-directed the existing focus when patients were near or at kidney failure (stage 5 requiring renal replacement therapy such as hemodialysis or transplantation) to earlier clinical intervention during stages 1–4 which might delay or retard progression. By applying the KDOQI classification system to existing population surveys, it is now conservatively estimated that more than 20 million Americans (one out of nine adults) have some risk factors for CKD *(2)*.

Recent analyses have suggested an increase in cardiovascular disease (CVD) as high as 100% greater in CKD compared to the general population even when matching for gender, race, age, and other confounding risk factors *(3)*. Control of dyslipidemia in earlier stages of CVD may help lower or prevent cardiovascular risk *(4)*. Lipid screening has been added to the expanded activities for the National Kidney Foundation's Kidney Early Evaluation Program beginning in 2006 *(5)*. The American Kidney Fund's Minority Intervention Kidney Evaluation and Take Charge: Protect Your Kidneys public education programs both include lipid levels as part of their CKD risk factor evaluation *(6)*. This chapter will focus on the definition and treatment of dyslipidemia in stages 1–4 of CKD.

2. PATHOPHYSIOLOGY

Dyslipidemia is defined as elevated serum levels of lipid components in the blood: total cholesterol (TC), high-density lipoproteins (HDL), low-density lipoproteins (LDL), and other lipid particles. Abnormal lipid profiles are seen in kidney function impairment and particularly in protein-losing nephropathies such as nephrotic syndrome *(4,7)*.

Lipid metabolism, specifically cholesterol synthesis, takes place in the liver. The liver produces bile which is the primary lipid-reducing agent stored in the gallbladder. Cholesterol can be synthesized in the liver. It can be removed from circulating lipoproteins and be directly absorbed from the small intestine from cholesterol secreted in the bile or from dietary cholesterol. Excess circulating cholesterol results in hyperlipidemia. Individual variation in lipid response comprises differences in absorption or biosynthesis of primary and receptor-mediated lipid products that may or may not be CKD related. Chronic elevation may lead to deposits on inner arterial walls (fatty plaque) resulting in accumulation and atherosclerosis.

The process of reverse cholesterol transport also exists whereby cholesterol may be removed from areas of lipid accumulation and returned to circulation. The exact mechanisms responsible, however, are still being understood. HDL, as the primary carrier, transports cholesterol back to the liver where it is either reused or excreted. This metabolic process uses a cohort of enzymes and protein pathways to decrease monocyte penetration of wall sites. Macrophages attract oxidized LDL which appear to increase inflammatory effects *(8,9)*.

3. EXISTING CLINICAL PRACTICE GUIDELINES AND PEER-REVIEWED RECOMMENDATIONS

KDOQI suggests that all CKD patients should be managed as high-risk using existing lipid guidelines for non-CKD high-risk patients. Specific KDOQI guidelines have been published for dyslipidemias in CKD and are discussed in this chapter *(4,10)*. Currently, there are no long-term studies of lipid management in CKD patients that provide any additional information to direct care. A meta-analysis of the general population supported the beneficial, low-risk effect of dietary and pharmaceutical modifications to reduce level of serum lipids *(11)*.

4. ASSESSMENT

4.1. Biochemical

Serum lipids should be assessed at the first office visit. Patients may have been screened for TC with a non-fasting sample as part of general risk assessment. Ideally, a fasting lipid profile should be obtained. The lipid profile should include TC, HDL, LDL, and triglycerides. Values reported outside of reasonable laboratory parameters should be repeated for reliability. Instructions for fasting should be reinforced with the patient prior to the subsequent blood draws *(4,10)*.

Serum lipids should be drawn annually or whenever a treatment change warrants re-assessment of effect. Serum lipid patterns may change during CKD stages 4 and 5. The ramifications of the duration of lipid abnormalities and their relationship to later CVD risk are unknown. When hyperlipidemia is particularly resistant to standard treatment, the clinician may consider additional biochemical testing for contributory inflammatory markers (c-reactive protein and newer novel biomarkers such as monocyte chemoattractant protein or plasma selectin) *(4,12)*. Hidden sources of infection should be investigated (i.e., foot and nail infection particularly in diabetics) and advanced periodontal disease.

4.2. Physical

Measured, rather than self-reported, height and weight should be recorded. The body mass index should be calculated and compared to standardized tables for baseline assessment. Patients with fluid accumulation such as edema, amputees, or other body composition imbalances need special adaptations to standardized formula. *(13)*. Physical assessment should include an evaluation of recommended cardiac activity level and intensity based on American Heart Association guidelines in preparation for lifestyle intervention *(14)*.

Patterns of body composition have not received much attention in the literature. Ideally, individuals should have optimal lean muscle mass in proper proportion to adipose tissue to achieve a body mass index comparable to normal body weight *(14,15)*. There are several hand-held instruments that can be used to estimate lean body mass with individual strengths and weaknesses on reliability and validity of data over time. Employing a single instrument to track changes of an individual over time using their own baseline to measure progress may be more consistent and reliable rather than comparing to a heterogeneous group mean or trend. Adipose tissue location, particularly abdominal fat stores estimated by waist circumference, has also been used in cardiovascular risk factor evaluation.

A detailed medical history should be taken including, but not limited to prior laboratory values, family history of associated lipid or vascular disorders, comorbid conditions, and current medications *(4,10,14,15)*.

4.3. Nutritional

A registered dietitian will be most able to evaluate the dietary intake and recommend specific food changes to promote lower serum lipids. Dietary intake patterns can be assessed by one of several methods (see Chapter 4). Specific attention should be given to the type of fat consumed by saturation level, pattern of fat consumption throughout day, and use of fat in food preparation. Detailed information from a computer analysis of nutrient content should guide dietary changes *(4,10,13–15)*.

A single educational session will be insufficient to enact dietary change. Routinely scheduled long-term monitoring is required. The services of a registered dietitian delivered to Medicare-eligible patients diagnosed with diabetes and kidney disease are reimbursable with a referral from their treating physician *(16)*. This professional assessment and on-going monitoring are essential to attain the dietary goals.

5. INTERVENTION

5.1. Lifestyle

The American Heart Association recommends establishing a consistent daily physical activity pattern for all adults. Sustained cardiac activity will promote the use of circulating lipids for energy, rather than storage as atherosclerotic deposits. Also, regular exercise to achieve a sustained cardiovascular benefit heart rate level may reduce serum fat particle size *(14)*. Patients can visually understand this exercise principle if a bottle of salad oil combined with red vinegar is shown first in a resting state of separated layers and then shaken (as in exercise) to distribute and reduce the fat particle size.

The increased cardiac output and corresponding muscle strength generally have minimal risk when undertaken within the context of daily activities found in the home (climbing stairs, walking, vacuuming, and carrying groceries). CKD patients may have anemia which can reduce oxygen-carrying capacity of the blood and exhibit symptoms of fatigue. Anemia can be treated with multifaceted protocols that include iron supplementation in conjunction with injectable erthyropoeitin, the kidney hormone decreased in CKD.

The American Heart Association recommends a minimum of 10,000 steps per day to achieve basic cardiac health *(14)*. A simple pedometer can be used to record steps taken per day. Complicated models monitoring stride distances are not necessary. Although the accuracy between pedometers has been shown to be variable, the use of the same pedometer by the same patient on a daily basis minimizes variability and provides a consistent baseline measurement upon which to monitor physical activity. Physical activity can be increased by small increments of as few as 100–250 steps per day until the minimal goal has been reached or as higher goals are attained. The use of a pedometer should become routine over time. The placement of the pedometer in relation to the hip flex movement is important to obtain accurate and consistent results.

To sustain motivation, many individuals benefit from pairing with a "walking buddy" or walking group to provide continuous support and activity opportunities. Recent lifestyle studies have shown individuals who owned dogs increased their daily walking frequency and distance when compared to individuals who did not. In addition, social support generally has been shown to decrease relative risks of death in both the general and chronic disease populations *(17)*.

5.2. Dietary

The general dietary principles to treat dyslipidemia by diet in CKD stages 1–4 are summarized in Table 1. The key component of dietary intervention is type and amount of fat consumed with emphasis on reducing saturated and *trans*-fatty acid content. This is a general population goal for both healthy and CKD individuals *(4,15)*. As such,

Table 1
Summary of Dietary Recommendations to Address Dyslipidemia in CKD Stages 1–4

Dietary modification	*Intervention method*	*Anticipated change*
Match energy intake to energy output	Calculate and implement amount of calories required using goal body weight	Attain and maintain healthy body weight
Match caloric distribution among diet components	Provide adequate dietary protein while providing sufficient total calories Emphasize quality of protein when quantity limited to potentially retard progression	Maintain normal serum albumin Decrease risk of protein-calorie malnutrition
Decrease total fat calories to ≤30% of total calories	Decrease total fat calories consumed from all dietary sources	Normalize serum lipids
Decrease total cholesterol intake <300 mg/day	Reduce intake of dietary cholesterol (i.e., egg yolks and animal fats); substitute whole milk dairy products with skim and low fat alternatives; replace egg yolks with egg substitute products	Normalize serum cholesterol

Decrease saturated fat calories <7% of total calories	Decrease intake or avoid saturated fats (i.e., animal fats, butter, full fat dairy products, mayonnaise, avocado and tropical oils such as palm and coconut)	Decrease LDL
Change type of fats used in food preparation	Promote intake of non-hydrogenated vegetable oils (peanut, canola and olive) or nut oils (walnut and flaxseed)	Decrease LDL
Increase use of monounsaturated fats within total fat intake amount	Promote use of olive oil, sunflower oil and canola oil	Increase HDL
Increase use of omega-3 fatty acids within total fat intake amount	Increase consumption of green leafy vegetables, flaxseed, nuts (almonds, walnuts), and use of fatty fish 1–2 servings/week	Increase HDL
Avoid use of *trans*-unsaturated fatty acids	Avoid commercially fried foods, hydrogenated fats, partially hydrogenated vegetable oils and margarines and processed foods containing these fats	Decrease LDL Increase HDL

CKD, chronic kidney disease; HDL, high density lipoprotein; LDL, low density lipoprotein.

the integration of diet modifications can be beneficial to both the individual as well as other individuals living within that household, potentially maximizing the benefit and compliance to everyone.

5.2.1. Determination of Body Weight and Total Daily Calories

General guidelines suggest reducing total fat calories but also assume individuals will not consume more calories per day than they need to attain or sustain a healthy weight. The estimation of the amount of daily dietary fat to be consumed should be based on a reasonable body weight. Obesity will be promoted or sustained using the current body weight if presently at obese or overweight levels. Body mass index of 25–29.9 kg/m^2 is considered overweight by government guidelines. The consumption of "empty" calories will contribute to overall non-lean body mass. The definition of "healthy" weight in many patients within the context of chronic disease has challenges of its own, and the use of formulas contained in KDOQI guidelines has been shown to be used inconsistently in practice *(18,19)*.

Recent literature from the chronic dialysis population (CKD stage 5) has shown a trend toward greater survival at higher body weight levels compared to normal and underweight levels. Although the exact mechanism is not fully understood, this "J-curve" observation may be related to the cushion of additional body fat stores available for energy during concurrent hospitalization or stress periods. It is important to maintain a reasonable body weight during CKD stages 1–4 and avoid malnutrition when progressing to CKD stage 5. The exact benefit or need for a "cushion" of fat stores and muscle mass, however, has yet to be determined *(20)*.

5.2.2. Amount and Type of Dietary Fat

A typical 2000 calorie per day diet should contain ≤30% total fat. This is calculated as 600 calories (30% of 2000 calories and 600 calories/9 calories per gram of fat) which equals approximately 66 g or less of total fat per day. No more than 7% of total calories (140 calories or 16 g of fat) should come from saturated fat *(14)*.

Saturated fat has more hydrogen bonds than polyunsaturated and monounsaturated fat. Typically, saturated fat remains solid at room temperature (such as animal fat from meat, lard, and butter) whereas unsaturated fat is softer or liquid at room temperature (such as vegetables oils, or tub compared to stick margarine). These hydrogen bonds are more difficult to break down and metabolize thus circulating as larger fat particles in the serum. The composition of the diet should be changed to encourage the intake of predominantly polyunsaturated and monounsaturated sources within the total daily fat intake.

Trans-fatty acids have been altered during processing to change the natural "cis" configuration (the most unsaturated version) to the "trans" configuration (primarily to increase shelf life). *Trans*-fatty acids function as saturated fatty acids during metabolism and have been implicated in reducing HDL and increasing LDL cholesterol *(8)*. They are found predominantly in processed foods (cookies, crackers, and baked goods). Many fast food restaurants use frying oils that contain a high content of *trans*-fatty acids.

Nutrition labels allow for rounding of fat grams on a label to 0.5 g or zero if that food contains less than 5 g of fat per portion and rounding to the nearest 1 g if containing more than 5 g of fat per portion. If a consumer uses the nutrition label as the general guide to counting fat grams per day, the accuracy of their estimate can be proportionate to the number of food items they consume each day. Nutrition labels in the United States must now also contain *trans*-fatty acid composition.

5.2.3. Other Beneficial Dietary Modifications

Other factors that may help reduce hyperlipidemia include fiber and use of plant sterols. The level of dietary fiber has decreased with the increased consumption of refined foods. The Westernized diet promotes the use of animal protein compared to plant sources.

Increasing the consumption of dietary fiber to levels of 20–30 g/day has been linked to lower LDL levels. Soluble fiber binds to bile acids which may decrease the absorption of cholesterol. Both soluble and insoluble fiber may also decrease gastrointestinal transit time which may improve insulin sensitivity by slowing carbohydrate absorption. Insoluble fiber is found primarily in wheat products but has shown less LDL effect than soluble. Soluble fiber is found in a wide variety of foods including barley, bran, raw or partially cooked fruits and vegetables, nuts and seeds, oats and oatmeal. A wide variety of over-the-counter psyillum capsules and soft fiber equivalents are available as alternatives or supplements to dietary modification. Fiber intake should be increased gradually over time in conjunction with liberal fluid intake to decrease gastrointestinal symptoms until gut adapts to higher load *(4,10,14,15)*. Dietary fluid and potassium restrictions may require adaptations when progression to CKD stage 5 is eminent.

Plant sterols and their stanol esters are naturally present in small quantities in plant sources. Most research has been done in soybean derivatives where they have been chemically concentrated to produce commercial products marketed as butter or margarine substitutes. Plant sterol esters in this new format which exceed what can be consumed by diet alone have been shown to potentially lower LDL in the general population. Clinical trials have included more than 1800 people with doses of up to 25 g/day. No clinical studies have been done in CKD patients but they are rated as a safe food-grade additive. A daily intake of approximately two tablespoons consumed as part of two separate meals per day (total 2–3 g/day) is the recommended dose with no evidence that higher levels produce a greater effect *(21)*.

6. PHARMACOLOGICAL

There are several classes of drugs used for reducing serum lipids. Each drug has a unique mechanism by which it changes lipid metabolism or absorption. Most use the gastrointestinal tract as an exchange or as a means to excrete excess lipid through the feces.

Regular bowel habits will promote the efficacy of many of these drugs. The gastrointestinal tract can increase absorption of specific dietary components such as potassium in CKD as a compensatory mechanism to decreased absorption-reabsorption by the kidneys. Constipation may be a problem due to lower fluid intake and binding features of concurrent drugs such as phosphate binders in later stages of CKD.

The type of pharmacological intervention used is not as important as the attainment of the overall goal of lipid reduction. A variety of options may be necessary to achieve compliance within financial and administration issues while avoiding side-effects and potential complications. Dietary and pharmacological intervention should be used together to maximize lipid reduction effect as their mechanisms of action are complementary, not competitive *(4,10,11)*.

7. SUMMARY

The achievement and reduction of dyslipidemia to optimal levels in CKD stages 1–4 is a multifactorial approach. Practitioners need to use comprehensive assessment techniques to evaluate the status of the individual. Intervention strategies that include physical activity, dietary changes, and pharmacological options need to be continually monitored to achieve goals. The use of a registered dietitian is necessary to provide the support and continuous monitoring of daily food intake to modify existing habits and promote new patterns. The individual should attain a healthy body weight while maintaining optimal nutrition status in preparation for progression to stage 5 CKD. A strong integrated health team approach is necessary to diagnose, evaluate, and treat dyslipidemia throughout the CKD stages. KDOQI guidelines provide a template upon which to plan and coordinate quality care.

REFERENCES

1. Eknoyan G, Levin NW. An overview of the National Kidney Foundation-Dialysis Outcomes Quality Initiative implementation. Adv Ren Replace Ther 1999;6:3–6.
2. National Kidney Foundation. KDOQI clinical practice guidelines for chronic kidney disease: evaluation, classification, and stratification. Am J Kidney Dis 2002;39(S1):S1–S266.

3. Snyder S, Pendergraph B. Detection and evaluation of chronic kidney disease. Am Fam Physician 2005;72:1723–1732.
4. National Kidney Foundation. KDOQI clinical practice guidelines for managing dyslipidemias in chronic kidney disease. Am J Kidney Dis 2003;41:S1–S92.
5. National Kidney Foundation Kidney Early Evaluation Program (KEEP) Annual Data Report 2005. Am J Kidney Dis 2005;46(S3):S1–S158
6. American Kidney Fund. Take Charge: Protect Your Kidneys. Available at http:www.kidneyfund.org. Accessed November 12, 2007.
7. Expert Panel on Detection, Evaluation, and Treatment of High Blood Cholesterol in Adults. Executive Summary of the Third Report of the National Cholesterol Education Program (NCEP). JAMA 2001;285:2486–2497.
8. Gropper SS, Smith JL, Groff JL. Advanced Nutrition and Human Metabolism. 4th Edn. Belmont CA: Thomson Wadsworth, 2005, pp. 128–167.
9. Mozaffarian D, Rimm EB, King IB, Lawler RL, McDonald GB, Levy WC. *trans* fatty acids and systemic inflammation in heart disease. Am J Clin Nutr 2004; 80:1521–1525.
10. National Kidney Foundation. KDOQI clinical practice guidelines for cardiovascular disease in dialysis patients. Am J Kidney Dis 2002;39(S3):S1–S266.
11. Cholesterol Treatment Trialists (CTT) Collaborators. Efficacy and safety of cholesterol-lowering treatment: prospective meta-analysis of data from 90 056 participants in 14 randomised trials of statins. Lancet 2005;366:1267–1278.
12. Catilla P, Echarri E, Ddvalos A et al. Concentrated red grape juice exerts antioxidant, hypolipidemic, and anti-inflammatory effects in both hemodialysis and healthy subjects. Am J Clin Nutr 2006;84:252–262.
13. National Kidney Foundation. KDOQI clinical practice guidelines for nutrition in chronic renal failure. Am J Kidney Dis 2000;35(S2):S1–S140.
14. Lichtenstein AH, Appel LJ, Brands J et al. Diet and lifestyle recommendations revison 2006: A scientific statement from the American Heart Assocation Nutrition Committee. Circulation 2006;114:82–96.
15. Olendzki B, Speed C, Domino FJ. Nutritional assessment and counseling for prevention and treatment of cardiovascular disease. Am Fam Physician 2006;73:257–264.
16. Medical nutrition services—An overview. Available at http:www.cms.hhs.gov/ medicalnutritiontherapy. Accessed November 12, 2007.
17. Evidence for improving patient care and outcomes: The Dialysis Outcomes and Practice Patterns Study (DOPPS) and Kidney Disease Outcomes Quality Initiative (KDOQI). Kidney Int 2004;44(S2):S1–S67.
18. Harvey KS. Methods for determining health body weight in end stage renal disease. J Ren Nutr 2006;16:269–276.
19. Schatz S, Pagenkemper J, Beto J. Body weight estimation in chronic kidney disease: Results of a practitioner survey. J Ren Nutr 2006;16:283–286.
20. Kopple JD. The phenomenon of altered risk factor patterns or reverse epidemiology in persons with advanced chronic kidney failure. Am J Clin Nutr 2005;81:1257–1266.
21. Functional foods fact sheets: Plant stanols and sterols. International Food Information Council Foundation, Washington DC, April 2003.

2 Treatment

8 Nutrition and Pharmacologic Approaches

Kathy Schiro Harvey

LEARNING OBJECTIVES

1. Describe nutrition risks associated with chronic kidney disease, including protein energy malnutrition, renal osteodystrophy, cardiovascular complications, and anemia.
2. Identify appropriate interventions to prevent and/or treat nutrition risks associated with chronic kidney disease, including estimating protein, calorie, sodium, and phosphorus needs.
3. Discuss how to treat chronic kidney disease complications such as cardiovascular disease, hypertension, hyperlipidemia, anemia, and diabetes.
4. Describe nutrition strategies that may be appropriate for treating IgA nephropathy.
5. Describe how self-management techniques can be used to enhance nutrition knowledge and promote behavior changes in people with chronic kidney disease.

Summary

Chronic kidney disease (CKD) is a worldwide public health problem and a progressive, debilitating condition. People with CKD are at high risk of malnutrition, and the first priority of nutrition therapy must focus on the prevention and/or treatment of protein energy malnutrition. Additional nutrition therapy goals include slowing the progression of CKD and its uremic complications and

From: *Nutrition and Health: Nutrition in Kidney Disease*
Edited by: L. D. Byham-Gray, J. D. Burrowes, and G. M. Chertow

preventing renal osteodystrophy, cardiovascular disease, and diabetic complications. Nutrition and pharmaceutical therapies which address these issues are complex, but with appropriate intervention patients can maintain nutrition health, slow the progression to stage 5 CKD, stabilize cardiac status, and reduce diabetic complications. Successful nutrition therapy includes the active involvement of a qualified renal dietitian.

Key Words: Chronic kidney disease stages 1–4; protein restriction; malnutrition; energy needs; hyperphosphatemia.

1. INTRODUCTION

Chronic kidney disease (CKD) is a worldwide public health problem and a progressive, debilitating condition. In 2002, the National Kidney Foundation (NKF) Kidney Disease Outcomes Quality Initiative (KDOQI) published Clinical Practice Guidelines for Chronic Kidney Disease: Evaluation, Classification, and Stratification *(1)* (Tables 1 and 2). Using the KDOQI definition, the NKF estimates that 20 million Americans have CKD *(2)*. The prevalence of stage 5 CKD in the

Table 1
Chronic Kidney Diseases: A Clinical Action Plan

Stage	*Description*	*GFR (ml/ min/1.73 m^2)*	*Action*[a]
1	Kidney damage with normal of ↑ GFR	≥90	Diagnosis and treatment, Treatment of comorbid conditions, stowing progression, CVD risk reduction
2	Kidney damage with mild ↓ GFR	60–89	Estimating progression
3	Moderate ↓ GFR	30–59	Evaluating and treating complications
4	Severe ↓ GFR	15–29	Preparation for kidney replacement therapy
5	Kidney failure	<15 (or dialysis)	Replacement (if uremia present)

CVD, cardiovascular disease, GFR, glomerular filtration rate. Chronic kidney disease is defined as either kidney damage or GFR <60 ml/min/1.73 m^2 for ≥3 months. Kidney damage is defined as pathologic abnormalities or markers of damage, including abnormalities in blood or urine tests or imaging studies.

[a]Includes actions from preceding stages.

Reprinted with permission of the National Kidney Foundation *(1)*.

Table 2
Abbreviated MDRD and Cockroft-Gault Equations for Estimating GFR *(1)*

Estimated GFR ml/min/1.73 m^2 =

1. Abbreviated MDRD equation:
   $\text{Exp}(5.228\text{-}1.154 \times \ln(\text{Scr}) - 0.203 \times \ln(\text{Age}) - (0.299 \text{ if female}) + (0.192 \text{ if African American})$
2. Cockcroft-Gault Equation:
   $$\frac{(140 - \text{Age}) \times \text{Weight}}{72 \times \text{SCr}} \times (0.85 \text{ if female})$$

United States, defined as those receiving dialysis or kidney transplantation, is it much less than 472,000 *(3)*. Decreased glomerular filtration rate (GFR) is associated with complications in virtually all organ systems. In general, the severity of complications worsens as the level of GFR declines. Among the most serious complications are high blood pressure, cardiovascular disease (CVD), hyperlipidemia, anemia, malnutrition, and bone disease. The incidence and effect of these complications or comorbid conditions may explain the lower occurrence of stage 5 CKD compared to stages 1–4; many people with CKD die before they reach stage 5 *(4–6)*.

This chapter will discuss the nutrition-related risk factors that are associated with the progression and outcomes of CKD. It is not surprising that nutrition plays a significant role in CKD considering that most chronic diseases in America today are correlated with diet and lifestyle behaviors. CVD, hypertension, obesity, and diabetes are all common to CKD and are related to a sedentary lifestyle, a high fat and a high sodium diet, which are common in Western societies. This chapter will suggest interventions that can improve or stabilize overall health and nutrition status and possibly prevent the progression of CKD to stage 5.

2. MALNUTRITION IN CKD

According to the KDOQI Clinical Practice Guidelines for CKD, protein energy malnutrition develops during the course of the disease *(1)*. As kidney function decreases, there is a spontaneous drop in protein and energy intake. Kopple *(7)* evaluated clinically stable patients with moderate to advanced CKD and found that the nutrition parameters—protein intake, serum albumin, cholesterol, transferrin, body weight, body fat and muscle, and body mass index

(BMI)—declined with GFR. Energy and protein intakes were strongly associated with many of the nutrition parameters. Ikizler assessed nutrition status and protein intake in patients with CKD and found that as kidney function declined, protein intake decreased as did nutrition markers of serum transferrin and cholesterol *(8)*. Other researchers have evaluated the incidence of malnutrition and potential factors associated with it. There seems to be agreement that inadequate intake is a significant cause, related to anorexia caused by an accumulation of uremic toxins, metabolic and hormonal derangements, depression, and gastrointestinal abnormalities *(1,8)*. Additional factors related to malnutrition include acidosis and inflammation.

Metabolic acidosis is common in CKD and is associated with increased protein catabolism. The degradation of essential branched chain amino acids and muscle protein is stimulated during metabolic acidosis, resulting in muscle catabolism and suppression of albumin synthesis *(9)*. Evidence strongly suggests a chronic inflammatory state in CKD, especially as GFR drops below 60 ml/min/1.73 m^2. Inflammation is associated with anorexia, increased skeletal muscle protein breakdown, increased whole body protein catabolism, cytokine-mediated hypermetabolism, and disruption of growth hormone and IFG-1 axis leading to decreased anabolism *(1,10)*.

2.1. Protein Needs

Considering the strong risk of malnutrition in CKD, calculating appropriate energy and protein needs is critical. Protein-restricted diets for CKD have been used for over 30 years, mainly to alleviate uremic symptoms, but also to blunt the glomerular hyperfiltration in residual nephrons of diseased kidneys, and thus slow the progression to renal failure. Studies in humans indicate that an intake of animal protein can induce an increase in GFR. The mechanisms for this effect on renal dynamics are not completely understood, but two major theories have been proposed. One theory is that a high protein diet can increase the release of hormones such as glucagon, insulin-like growth factor-1 and kinins, which all stimulate an increased GFR. Additionally, intrarenal mechanisms may play a role in protein-induced hyperfiltration. The increased filtered load of amino acids can enhance proximal sodium reabsorption, ultimately leading to an elevation in GFR *(11)*.

Studies in humans have shown positive effects of reduced protein intake on kidney function in people with CKD. Smaller studies seem to have the most positive results, suggesting that a low protein diet may protect against CKD progression in at least some conditions, such as diabetic nephropathy and chronic glomerular disease. Ihle and Walser *(12–14)* followed patients on very low protein diets and found a reduction in GFR decline and progression to stage 5 CKD. Larger controlled trials have shown less impact of restricted protein intake, including the largest trial to date, the Modification of Diet in Renal Disease Study *(15,16)*. This study followed over 800 patients with CKD for 2–3 years. Initial results showed only a modest effect of a low protein intake (~0.6 g/kg) on rate of progression of CKD, but after secondary analysis of the data, the study group concluded that the balance of evidence is more consistent with a beneficial effect of protein restriction. In two meta-analyses, Fouque et al. *(17,18)* evaluated 43 randomized controlled trials and concluded that reducing protein intake in patients with CKD reduces the occurrence of renal death by about 40% compared with higher or unrestricted protein intakes.

Discrepancies between researchers and study findings can be explained by many factors.

- Methods of assessing kidney function varied and included creatinine clearance by urinary creatinine, serum creatinine, GFR by estimation equations, and GFR by iothalamate. Measurement of kidney function and rate of decline will vary depending on the assessment tool used.
- Blood pressure at baseline and during the study period will affect outcome, in the same manner as the type of treatment used to control blood pressure. Studies included patients with different blood pressure ranges and treatments.
- The underlying disease may also impact the rate of decline of kidney function. Some studies included people with diabetes treated either with or without exogenous insulin, polycystic disease, glomerular nephropathies, and so on. Protein intake may affect kidney function differently depending on the primary disease.
- The level of kidney function at the start of the intervention may have affected outcomes. Protein restriction studies included patients with mild-to-severe CKD, and it is possible that protein restriction may have a greater impact at different stages of CKD.
- Outcome measurements varied from study to study: change in creatinine clearance vs. rate of decline of GFR vs. stage 5 CKD vs. death. Clinical interpretation will be affected by the measurement tools.

Studies have been done to evaluate the effects of different protein sources on kidney function in CKD. There is some evidence that plant proteins have less effect on GFR in humans compared to animal proteins. Elliott et al. evaluated epidemiological data as part of the International Collaborative Study of Macronutrients and Micronutrients and Blood Pressure (INTERMAP) Study and found that subjects who consumed a diet higher in vegetable protein had lower systolic and diastolic blood pressure compared to those who ate high animal protein diets *(19)*. In studies with CKD patients following vegetable protein diets, results are inconsistent. Soroka et al., Anderson et al., and Wheeler et al. *(20–22)* found no difference in renal function in subjects following animal-based low protein diets vs. plant-based low protein diets. In most cases, both diets resulted in slowing the deterioration of renal function and decreasing proteinuria. However, Stephenson et al. and Teixeira et al. *(23,24)* studied soy protein diets in patients with diabetes and concluded that substitution of soy for animal protein resulted in decreased albumin excretion and improved GFR and lipid profiles. In all studies, subjects maintained adequate nutrition status and improved clinical profiles compared to the usual uncontrolled diets. Plant-based low protein diets appear to provide the same or more benefits as animal-based low protein diets for those with CKD and offer an alternative to those who prefer a vegetarian diet.

Because of the risk of muscle wasting and malnutrition in CKD, and the possibility that it can be exacerbated by a low protein intake, Castaneda et al. *(25)* evaluated mechanisms for preventing catabolism and muscle wasting in a group of older adults with stage 4 CKD. Subjects on low protein diets who practiced resistance training had stable body weights, improved muscle strength, and decreased muscle catabolism compared to those who did not participate in resistance training.

Given that a reduced protein intake is most likely a positive intervention in CKD, the question remains, what level of restriction is optimal and at what stage of CKD should restriction begin? The Recommended Dietary Allowances of the Food and Nutrition Board for protein is 0.8 g/kg/day. This level is believed to provide adequate protein for most healthy adult men and women, although it is well below the average protein intake in the United States of 90–100 gm/day *(16)*. KDOQI Guidelines have evolved over the past several years with regard to protein recommendations, from 0.6 g/kg/day to 1.4 gm/kg/day depending on the stage of CKD *(1,26,27)*. General recommendations for stages 1–2 CKD suggest a protein intake of 0.8–1.4 gm/kg/day and 0.6–0.8 gm/kg/day for stages 3–4 CKD. For stage 5 (kidney failure), 0.6 gm/kg/day would be acceptable to treat or ameliorate uremic symptoms, prior to initiating dialysis or transplant. The KDOQI Guidelines for Nutrition in Chronic Renal Failure and those of the American Dietetic Association (ADA) state that at least 50% of the protein must be of high biological value (Table 3) *(25,26,28)* (refer to Chapter 25 for additional information concerning high biological value proteins).

Table 3
Low Protein Meal Plan

Morning
1 cup rice krispies with ½ cup non-dairy creamer and 1 tsp sugar
½ grapefruit
Coffee with cream
Noon
Grilled cheese sandwich with
1½ ounce cheese and 2–3 Tbs margarine
1 cup carrot and cucumber slices with
4 Tbs ranch dressing
1 apple with caramel dip
Fruit drink
Evening
Chicken stirfry with 3 oz chicken and
1 cup sliced vegetables
1½ cup rice with margarine
1 cup grapes
Iced tea

Approximately 2300 calories, 52 gm protein, 1700 mg sodium, 1700 mg potassium, 870 mg phosphorus.

In patients with nephrotic syndrome or proteinuria, a low protein diet can result in decreased protein excretion. Several researchers followed nephrotic subjects at various stages of CKD and found that low protein diets result in decreased proteinuria, while nutrition markers increased or remained stable *(29–32)*. These findings suggest that protein restriction can be safely used in patients with nephrotic syndrome who maintain an adequate energy intake, and whose urinary protein losses are replaced.

2.2. Energy Needs

When protein intake is restricted, it is imperative that energy intake remain adequate to spare protein for essential anabolic functions and to maintain nitrogen balance. Various studies have evaluated energy needs in patients with CKD to determine if they are equal to, greater than, or less than those of healthy people. Most controlled studies involve small groups of patients. The overall conclusions are that energy expenditure in CKD is close to that of healthy controls (30–35 kcal/kg body weight); however, energy expenditure may increase with progressive chronic kidney failure *(33–35)*. Unfortunately, numerous studies have shown that people with CKD consistently eat less than recommended. Passey et al., Chauveau et al. and Guarniera et al. followed patients with CKD on low protein diets which included specific instructions to consume >30 kcal/kg/day. Although protein intake was maintained, energy intake declined over the study periods to below recommended levels, as did indicators of lean tissue mass *(36–38)*. Byham-Gray *(39)* conducted an evidence-based review of energy needs in patients with CKD, concluding that definitive evidence is lacking and that practitioners must implement a reasonable approach which follows practice guidelines, as well as being mindful of patient values and preferences.

The KDOQI Guidelines recommend an energy intake of 30–35 kcal/kg/day based on the stage of CKD, the individual's age, and overall nutrition health *(1,26)*. The ADA also recommends 30–35 kcal/kg/day or Basal Energy Expenditure considering activity and stress factors, protein intake, and weight goals *(28)*. Although it is possible that some people with CKD can maintain nitrogen balance and nutrition health with <30 kcal/kg/day, the risk of malnutrition is high and there is a strong tendency to eat less than recommended. A calorie recommendation of at least 30–35 kcal/kg may be prudent, with close monitoring to see that nutrition status is adequate.

2.3. Vitamin Supplements

Due to the restricted protein recommendation in stages 3 and 4, and because it is common for people with CKD to eat less calories than prescribed, supplemental water-soluble vitamins at approximately the level of the Dietary Reference Intakes is advised *(28)*. Folate, pyridoxine (B6), and cobalamin (B12) needs may be higher to help prevent hyperhomocysteinemia. Biotin needs may also be higher due to decreased biotin consumption with the low protein diet. Regarding fat-soluble vitamins, research indicates that routine supplementation is not advised. Serum levels of vitamin A increase as kidney function declines, including elevations in liver and plasma. Studies of vitamin E have been inconclusive, showing serum levels in CKD being lower, normal, and higher than standard. Vitamin E supplements are occasionally recommended by physicians for their antioxidant properties. Adequate vitamin K levels are essential for maintaining bone health, but there is little research on supplementing in the CKD population. Vitamin D status must be monitored regularly in CKD and supplemented as indicated *(40)* (Table 4).

Table 4
Vitamin Supplementation Recommendations in Chronic Kidney Disease (CKD) Stages 1–4

Vitamin	*US DRI (adults >18 years)*	*CKD Stages 1–4*
Vitamin C	75–90 mg/day	60–100 mg/day
Thiamin (B1)	1.1–1.2 mg/day	1.5 mg/day
Riboflavin (B2)	1.1–1.3 mg/day	1.8 mg/day
Niacin	14–16 mg/day	14–20 mg/day
Folate	0.4 mg/day	>1.0 mg/day
Pyridoxine (B6)	1.3–1.7 mg/day	>5 mg/day
Cobalamin (B12)	2.4 microgram/day	>2–3 microgram/day
Biotin	30 microgram/day	30–100 microgram/day
Pantothenic acid	5 mg/day	5 mg/day

Dietary Reference Intake (DRI) levels represent Recommended Dietary Allowances (RDA) and Adequate Intakes (AI) *(40)*.

3. CALCIUM, PHOSPHORUS AND VITAMIN D

CKD is associated with a variety of bone disorders and abnormalities in calcium, phosphorus, vitamin D, and parathyroid hormone (PTH) metabolism, which begin to occur in the early stages of CKD. Although it is unclear which factor begins the damaging cycle, it is most likely that combinations of derangements at sites throughout the body are occurring simultaneously. Reduction in kidney function even at stages 2 and 3 will result in phosphate retention *(41)*. However, hyperphosphatemia at this stage is rarely present, and patients are more likely to present with normal or low serum phosphorus levels. It has been postulated that a transient and possibly undetectable increase in serum phosphorus due to decreased kidney function leads to a decrease in blood levels of calcium, which then stimulates the parathyroid glands to release more PTH. The elevation in blood levels of PTH would decrease the tubular reabsorption of phosphate and increase urine phosphates. The overall goal of this mechanism is to maintain serum phosphorus and calcium levels at normal, but at the expense of elevated blood levels of PTH. Serum PTH levels will slowly rise and reach above normal range as GFR falls below 60 ml/min/1.73 m^2 *(42,43)*. Restriction of dietary phosphorus in conjunction with a low protein diet has been shown to have a direct effect on lowering serum PTH *(41,44,45)*. It appears that the calcimetic response to PTH becomes blunted during the course of CKD, as skeletal resistance to the calcium-mobilizing action of PTH takes place. There appears to be down-regulation of PTH receptors throughout the body resulting in less mobilization of calcium, which helps prevent hypercalcemia *(42,43,46,47)*.

Studies have shown that the skeletal resistance to PTH is also due to a deficiency of dihydroxycholecalciferol [1,25(OH)2D3] or vitamin D *(42,47,48)*. Despite the presence of adequate functioning kidney mass in early CKD (stage 2), the production of 1,25(OH)2D3 does not increase adequately to meet the needs of the target organs. Absolute deficiency of and/or resistance to vitamin D develops. The number of vitamin D receptors decreases as the loss of kidney function progresses, resulting in resistance to vitamin D action. By stage 4 CKD, blood levels of 1,25(OH)2D3 are definitely low. Restriction of dietary phosphorus has been associated with a significant increase in blood levels of 1,25(OH)2D3, and treatment with vitamin D has been shown to prevent bone loss in patients at stages 3 and 4 *(42,47)*.

Because naturally occurring phosphorus is found in high protein foods, a restricted protein diet, as is recommended in CKD, tends to be lower in phosphorus than a typical American diet (Table 3). However, the phosphorus content of convenience, processed and fast foods must also be considered in planning for a low phosphorus intake. Because nutrition labels do not usually contain information on the phosphorus content of foods and processed foods change often, frequent consultation with food manufacturers is needed to obtain accurate nutrient amounts (see Chapter 25).

In addition to a phosphorus restriction, use of phosphate binders such as calcium acetate or calcium carbonate is usually required as kidney function declines. Phosphate-binding compounds render dietary phosphate and phosphate contained in swallowed saliva and intestinal secretions unabsorbable. When taken with meals and snacks, these compounds bind phosphate in the intestine before it can be absorbed (Table 5). They are most effective when dietary intake of phosphate is below 1000 mg/day. As intake increases, binder effectiveness is reduced and hyperphosphatemia may persist. Additionally binder compliance is often poor, as patients complain of the inconvenience and distaste in taking potentially large doses of pills every time they eat. Noncompliance can also be related to gastrointestinal side effects of nausea, stomach or intestinal discomfort, diarrhea and constipation.

Table 5
Phosphate Binder Equivalents

Calcium based binders	*Non-calcium based binders*
1 Phoslo (167 mg calcium)	1 Renagel 800 mg
2 Tums (400 mg calcium)	2 Renagel 400 mg
1½ Tums EX (450 mg calcium)	1 Fosrenol 250 mg
1 Tums Ultra (400 mg calcium)	
2 CaCO3 500 mg (400 mg calcium)	

Binding doses are approximately equal in phosphate binding capacity. (Note: Renagel and Fosrenol have not been approved for CKD Stages 1–4).

Bone, PTH, and Vitamin D abnormalities such as high turnover bone disease, bone resorption, osteomalacia, adynamic bone disease, overt fractures, microfractures and bone pain are widespread in CKD *(42)*. The high mortality seen in CKD may be more directly related to soft tissue calcification resulting from hyperphosphatemia, high calcium-phosphorus product, high PTH, or high calcium load. Vascular calcification has been implicated in the high rate of atherosclerosis and cardiac dysfunction seen in the CKD population *(47,49)*. As patients with CKD progress toward stages 4 and 5, abnormalities with hyperphosphatemia, hypocalcemia, PTH secretion, and low 1,25(OH)2D3 worsen, impacting greater bone loss, cardiovascular events, and the high prevalence of death, which occurs prior to stage 5 CKD. The pathogenesis and progression of abnormal bone metabolism in CKD is complex, involving a combination of minerals and hormones which act on multiple organ systems throughout the body. More may be learned by referring to Chapter 14.

The KDOQI Clinical Practice Guidelines for Bone Metabolism and Disease in CKD recommend basic starting points for identification and treatment of renal bone abnormalities based on evidence and expert opinion (see Chapter 25) *(42)*. However, due to a lack of strong evidence for many of the guideline topics, controversy continues about the best clinical practices for preventing and treating CKD bone metabolism disorders. For example, which vitamin D is best to use at which level of kidney disease; ergocalciferol, calcitriol, alfacalcidol, or doxercalciferol? Maximum level of calcium consumption is also controversial with practitioners recommending intakes from 1500 to 2500 mg/day. Since most scientific research in this field is funded by pharmaceutical companies with a large financial interest in study conclusions, it is a challenge for clinicians to accurately determine which drug treatment is most effective. Treatment guidelines must be considered with an open mind, accepting that a variety of recommendations may be appropriate given that patients are monitored closely and treatment plans are adjusted as outcomes warrant. A low protein meal plan as recommended for CKD Stages 3 and 4 is also low phosphorus (Table 3).

4. ANEMIA

Anemia develops during the course of CKD and the incidence of anemia increases as GFR declines, primarily due to insufficient production of erythropoietin (EPO) by the diseased kidneys *(1)*. Additional causes may be blood loss from repeated laboratory testing, gastrointestinal bleeding, severe hyperparathyroidism, acute and chronic inflammation, or deficiency of iron, folate, or vitamin B12. It is recommended that all patients with GFR <60 ml/min/1.73 m^2 be evaluated for anemia. If hemoglobin (Hgb) falls below 12.5 g/dL in males or 11.0 g/dL in females, an anemia workup is warranted. This workup should include assessment of red blood cells, reticulocyte count, serum iron, total iron binding capacity, percent transferrin saturation (TSAT), serum ferritin, and test for occult blood in stool *(1)*.

Anemia in CKD is generally normocytic and normochromic. Microcytosis may reflect iron deficiency; macrocytosis may indicate B12 or folate deficiency. Elevated reticulocyte count may suggest active hemolysis, such as hemolytic uremic syndrome. An abnormal white blood cell count and/or platelet count may reflect a more generalized bone marrow dysfunction such as malignancy or vasculitis. Low levels of serum iron, percent TSAT, or ferritin may indicate iron deficiency. If iron deficiency is present, a stool occult blood test is recommended to test for gastrointestinal bleeding. If a reversible cause of anemia is not present or has been corrected, then EPO deficiency is the most likely primary cause of anemia (Fig. 1).

4.1. Iron Deficiency

Iron deficiency is indicated if TSAT is <20% and/or serum ferritin level is <100 ng/mL *(49)*. Due to the quantity of iron required, increasing iron intake from foods will not adequately replenish iron stores. Oral iron supplementation is needed, at least 200 mg elemental iron in two to three divided doses in the form of ionic iron salts such as iron sulfate, fumarate, or gluconate. These are less costly and provide known amounts of elemental iron. Oral iron is best absorbed when taken without food or other medications and supplemental ascorbic acid

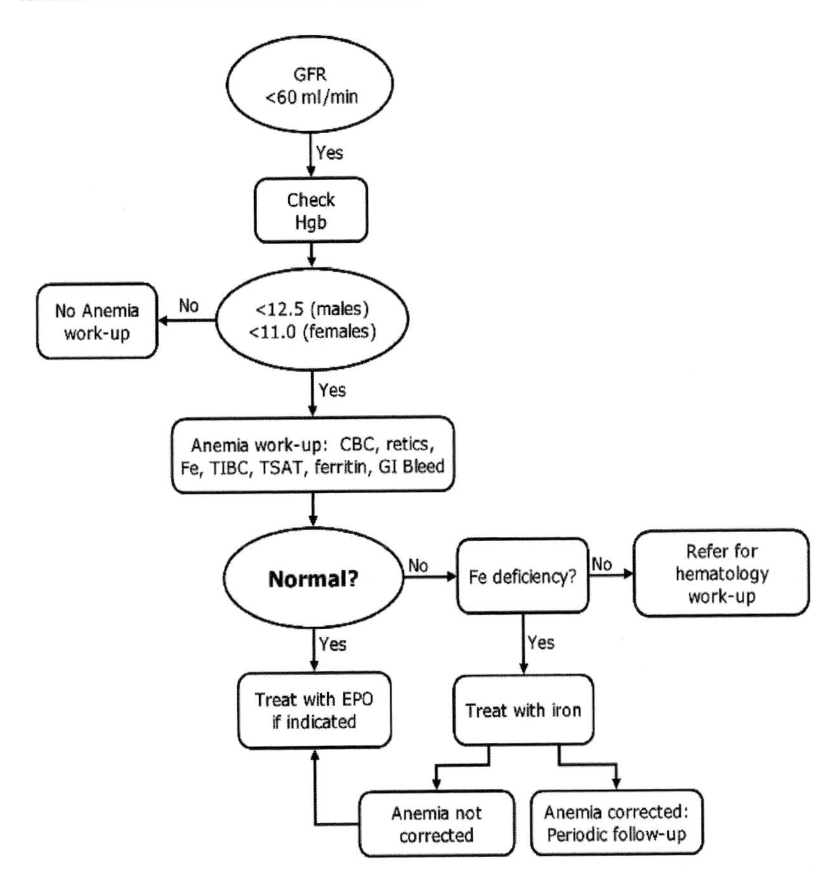

Fig. 1. Anemia work-up for chronic kidney disease stages 1–4. Retics, reticulocyte count; CBC, complete blood count; EPO, erythropoietin; Fe, serum iron; GFR, glomerular filtration rate; GI, gastrointestinal; TIBC, total iron binding capacity; TSAT, transferrin saturation. Adapted from the National Kidney Foundation *(1)*.

does not seem to improve iron absorption. Intestinal iron absorption is inversely related to iron stores and increases as erythropoiesis improves with EPO therapy. If iron status does not improve after 1–3 months of oral iron, IV iron therapy may be needed. Iron dextran or sodium ferric gluconate (500–1000 mg) may be administered and repeated as needed to obtain and maintain adequate iron stores. While receiving iron therapy, iron status should be checked every 1–3 months *(50)*.

4.2. EPO Therapy

Once adequate iron status is verified (TSAT >20%, serum ferritin >100 ng/mL) and other causes of anemia have been ruled out and/or corrected, EPO therapy may begin in doses of 80–120 units/kg/week. The goal is to induce a slow, steady increase in Hgb and to achieve a target level (11–12 gm/dL) within a 2- to 4-month period. If the increase in Hgb is <0.5–0.6 gm/dL over 2–4 weeks, the dose of EPO may be increased by 50%. If Hgb increases to >3 gm/dL/month, the dose of EPO may be decreased by 25%. If the target Hgb is surpassed, the EPO dose can be reduced by 25% or maintained at the same dose, but administered at a reduced frequency *(50)*.

The most common cause of an inadequate response to EPO therapy for the treatment of anemia is iron deficiency. If iron-replete patients are not achieving their Hgb goal, they should be evaluated for conditions such as infection or inflammation, chronic blood loss, osteitis fibrosa, aluminum toxicity, hemoglobinopathies, folate, or vitamin B12 deficiency, multiple myeloma, or hemolysis. Additionally, severe malnutrition can result in the unavailability of needed substrate for protein syntheseis in hematopoietic cells *(50)*. It is unclear whether the presence of anemia in CKD directly worsens prognosis or progression to stage 5 CKD. However, evidence shows that low Hgb levels in CKD are associated with higher rates of hospitalizations, CVD, cognitive impairment, and mortality *(1)*.

5. CARDIOVASCULAR RISKS

People with CKD are at increased risk for CVD, including coronary heart disease, cerebrovascular disease, peripheral vascular disease, and heart failure. CVD is the leading cause of death regardless of the stage of CKD, accounting for 40–50% of all deaths. In persons diagnosed with diabetes and kidney disease, CVD is related to the level of albuminuria, with many studies showing increased CVD mortality in people with diabetes, at all levels of albuminuria and proteinuria *(1)*.

The Framingham Heart Study showed that the prevalence of CVD, left ventricular hypertrophy (LVH) and cardiac dysfunctions was higher in those with elevated serum creatinine levels compared to those with normal levels of serum creatinine *(51)*. In a Canadian study, patients with renal insufficiency were more likely to die after a cardiac event compared to those with normal kidney function *(52)*. Brugts et al. *(53)* followed healthy elderly subjects and found that kidney function was an independent predictor of myocardial infarction. De Nicola et al. *(54)* evaluated the treatment of CVD risk factors in patients with CKD and found it to be largely inadequate or omitted, and those with diabetes and more advanced CKD received the worse treatment.

Most persons diagnosed with CKD do not develop kidney failure, dying before stage 5, and the cause of death is likely CVD. Risk factors for CVD are more prevalent in the CKD population, including the traditional risk factors of older age, hypertension, lipid disorders, diabetes, and physical inactivity, and non-traditional risk factors such as anemia and inflammation. Shlipak et al. *(55)* evaluated risk factors associated with CV deaths in elderly Americans and found that in those with CKD the traditional risks of diabetes, hypertension, smoking, low physical activity, LVH, and moderate alcohol use were associated with elevated risk of CV death. It appears that traditional risk factors may be the optimal targets for CVD risk reduction in elderly patients with CKD. The following sections explore the interventions to prevent and treat the progression of CVD in patients with CKD.

5.1. Hypertension

Hypertension is both a cause and complication of CKD, with 50–75% of patients with CKD having blood pressure >140/90 mmHg. It is generally recommended that people with kidney disease should maintain blood pressure levels less than those suggested for the general population. The NKF KDOQI Workgroup on Hypertension and Antihypertensive Agents in Chronic Kidney Disease published guidelines for identifying levels of risk and treatment options (Table 6). These guidelines include coordination with the seventh report of the Joint National Committee (JNC 7) for Prevention, Detection, Evaluation and Treatment of High Blood Pressure *(27)*. Refer to Chapter 5 for a more detailed discussion on hypertension and CKD.

For the earliest stages of hypertension (120–130 mmHg systolic/ 80–89 mmHg diastolic) dietary and lifestyle modifications are the first line of therapy. Studies have identified dietary patterns that are effective in reducing hypertension and CVD. Low fat diets that include nonhydrogenated and unsaturated fats as the predominant form of dietary fat appear to provide some protection against heart disease. Diets containing an abundance of whole grains, fruits, vegetables, and omega-3 fatty acids also offer significant protection *(27,56,57,58,59)*. In 2000, the Dietary Approaches to Stop Hypertension (DASH) trial showed that combining these interventions successfully reduced blood pressure in adults *(60)* (refer to Chapter 5). In 2001, the DASH Research Group showed that adding sodium restriction to the DASH dietary plan further reduced blood pressure *(61)*. Compared to the typical American diet, the DASH plan is lower in fat and sodium, and higher in potassium, magnesium, calcium, fiber and antioxidants. These results indicate that a whole foods approach which may include

Table 6
Risk Stratification and Indication for Antihypertensive Therapy and Target Blood Pressure from JNC7, Modified by KDOQI Work Group

Blood pressure stage (blood pressure, mm Hg)	*No risk factors, no TOD or CVD*	*One or more risk factors (except diabetes or CKD), no TOD or CVD*	*TOD or CVD, and/or diabetes, and/or CKD*
Prehypertension (120–139/80–89)	Lifestyle modification	Lifestyle modification	Drug therapy and simultaneous lifestyle modification for CHF, diabetes or CKD; lifestyle modification for other TOD or CVD; target blood pressure is <130/80 mm Hg; consider modifications for proteinuria.
Stage 1 (140–159/90–99)	Drug therapy after 12 months of lifestyle modification; target blood pressure is <140/90 mm Hg	Drug therapy after 6 months of lifestyle modification; target blood pressure is <140/90 mm Hg	Drug therapy and simultaneous lifestyle modification; target blood pressure is <130/80 mm Hg; consider modifications for proteinuria.

Stage 2 (>160/>100)	Drug therapy and simultaneous lifestyle modification; target blood pressure is <140/90 mm Hg	Drug therapy and simultaneous lifestyle modification; target blood pressure is <140/90 mm Hg	Drug therapy and simultaneous lifestyle modification; target blood pressure is <130/80 mm Hg; consider modifications for proteinuria.

JNC7, Joint National Committee for Prevention, Detection, Evaluation and Treatment of High Blood Pressure; CHF, congestive heart failure; CVD, cardiovascular disease; TOD, target organ damage; CKD, chronic kidney disease.

Adapted from the National Kidney Foundation *(27)*.

interactions between nutrients may be more effective in treating hypertension than simply limiting or increasing one nutrient.

The DASH trials demonstrated the short-term efficacy and safety of these diets, but did not include hypertensive adults with CKD. Sodium handling by the kidney is altered in CKD and that sodium retention has a major role in hypertension, primarily through expansion in extracellular fluid volume. Therefore, it is generally recommended that sodium intake be limited to less than 2.4 g/day in people with CKD. Sodium restriction of less than 1.2 g/day may lower blood pressure further. In those patients with "salt-wasting" nephropathy such as in some tubulointerstitial diseases, dietary sodium should not be restricted, but may require treatment with antihypertensive agents if hypertension develops *(27)*.

While the DASH diet is effective in lowering blood pressure in the general population, the macronutrient and mineral components may not be appropriate for all stages of CKD. For stages 1 and 2 CKD, the DASH diet may be safe and effective for treating hypertension and preventing progression of CVD. However, for CKD stages 3 and 4, the diet may contain higher protein and phosphorus intakes than recommended (refer to Chapter 5). Additionally, patients with CKD are at risk for hyperkalemia due to a reduced potassium excretion as kidney function declines, and the potassium-sparing effects of angiotensin converting enzyme inhibitors (ACEi) and angiotensin receptor blockers (ARBs) or potassium-sparing diuretics, used for treating hypertension. The DASH plan includes about 4500 mg/day of potassium which is generally higher than recommended for people with CKD stage 3 or 4. Therefore, the DASH dietary plan should not be routinely recommended for people with CKD stage 3 or 4. However, sodium limits to less than 2.4 g/day would be appropriate. The low protein meal plan in Table 3 meets the sodium limits of CKD stages 3 and 4.

Additional lifestyle changes may impact hypertension (refer to Chapter 5). It is generally accepted that smoking cessation and moderation of alcohol intake will improve heart health and help treat hypertension; these recommendations are appropriate for people with CKD. But the beneficial effects of weight loss and exercise on hypertension, CVD, and progression of CKD are unknown. Weight loss to obtain and maintain an appropriate body weight for height (BMI $<25\ kg/m^2$) is most likely beneficial for CKD stages 1 and 2, as well as fitness programs to improve muscle mass, flexibility, and strength. However, considering the high risk of malnutrition as kidney function declines, weight loss programs should be approached with caution in people with CKD stages 3 and 4. Due to the high risk of protein energy malnutrition in this population, maintaining an adequate calorie and protein intake is first priority. In obese subjects, controlled and gradual weight loss might be appropriate if patients are monitored closely to assure that they are able to consume an adequate protein intake, and that they are not losing muscle mass. This becomes especially critical as appetite and food intake declines spontaneously with decreasing kidney function. Exercise programs which include resistance training have been shown to help reduce catabolism and help maintain muscle mass in people with stage 4 CKD *(25)*.

5.1.1. Drug Therapies for Hypertension

In addition to dietary therapies, pharmacologic treatments for hypertension will impact the progression of CKD. All antihypertensive agents can be used to lower blood pressure in CKD, but multidrug regimens are usually required to reach blood pressure goals. Drug therapies should be targeted to specific CKD causes and CVD risk. Preferred agents may be indicated for patients with or without hypertension, such as nondiabetic kidney diseases associated with proteinuria. These are especially affected by treatment with angiotensin converting enzyme inhibitors (ACEi) and ARBs, which will reduce proteinuria and slow the progression of CKD. Concomitant use of diuretics and/or nondihydropyridine calcium channel blockers may increase the antiproteinuric effects of ACEi and ARBs. Dihydropyridine calcium channel blockers seem to be less effective at slowing kidney diseases with proteinuria. For patients with CVD, preferred antihypertensive agents will depend on diagnosis, symptoms, and comorbid conditions. Because fluid retention is one of the major causes of hypertension in CKD, almost all patients will require diuretic therapy in order to achieve blood pressure goal. Combinations of diuretics, ACE inhibitors and ARBs, plus beta blockers, calcium channel blockers, and aldosterone antagonists may be appropriate, depending on cardiac and comorbid conditions. The NKF KDOQI Guidelines on Hypertension and Antihypertensive Agents (Table 7) in CKD provide an extensive review of appropriate pharmacologic agents for specific conditions in CKD *(27)*.

5.2. Dyslipidemia

Dyslipidemias (abnormal lipid levels, lipoprotein composition, or both) are common in people with CKD. Considering the high incidence of atherosclerotic cardiovascular disease (ACVD) in this population, it has been recommended that all people with CKD should be evaluated for lipid disorders *(62)* (refer to Chapter 7 for further details). Although there have been no randomized controlled intervention trials which

Table 7
Preferred Antihypertensive Agents for Cardiovascular Disease in CKD

Cardiovascular Disease Type	*Thiazide or Loop Diuretics*	*ACE Inhibitors or ARBs*	*Beta Blockers*	*Calcium Channel Blockers*	*Aldosterone Antagonists*
Heart Failure with Systolic Dysfunction	X	X	X (carvedilol, bisoprolol, metoprolol succinate		X
Post MI with Systolic Dysfunction		X	X		X
Post MI			X		
Chronic Stable Angina			X	X	
High Risk for Coronary Artery Disease	X	X	X	X	
Recurrent Stroke Prevention	X	X			
Supraventricular Tachycardias			X	X (Non-dihydropyridines)	

Adapted from the National Kidney Foundation *(27)*.

show that dyslipidemias cause ACVD in CKD, a large observational study has shown that using statins to improve lipid profiles was associated with lower all-cause mortality and reduction in CVD in stage 5 CKD patients *(63)*. In the general population, the relationship between dyslipidemias and ACVD is significant in both men and women, old and middle age, smokers and non-smokers, hypertensives and normotensives, diabetics and nondiabetics, and in conditions of both higher and lower low density lipoprotein (LDL), total cholesterol (TC), high density lipoprotein (HDL) and TG. Considering the evidence, there is reason to assume that dyslipidemias can contribute to ACVD in patients with CKD also; therefore, it is prudent to evaluate lipid profiles in people with CKD to detect abnormalities that can be treated to reduce the incidence of ACVD.

It is unclear whether dyslipidemias cause or are the result of decreased kidney function, or if conditions such as proteinuria cause both reduced kidney function and dyslipidemias. A meta-analysis concluded that the rate of decline in kidney function was significantly less in those treated with cholesterol-lowering agents compared to placebo *(64)*. The prevalence of dyslipidemias in CKD can be influenced by many factors, including changes in proteinuria, GFR, malnutrition, diet therapy, and treatments of CKD. The KDOQI Clinical Practice Guidelines for Managing Dyslipidemias in CKD recommends an evaluation of lipid profiles more frequently than in the general population, at least annually and more often after therapies which may affect lipids such as treatment with diet, lipid lowering agents, or immunosuppressive agents *(62)*. Treatments for dyslipidemias in CKD are similar to those recommended by the National Cholesterol Education Program (NCEP) Adult Treatment Panel III (ATP III) Guidelines (Table 8) *(62,65)*.

First step therapy involves therapeutic lifestyle changes (TLC) including diet modifications, exercise, moderate alcohol intake and smoking cessation (Table 9). There is ample evidence in the general population that lipid-lowering diets can improve lipid profiles, but no randomized trials have examined the safety and efficacy of a low-fat, low cholesterol diet in CKD. Few studies have evaluated the effect of exercise and weight loss on abnormal lipid profiles in CKD although one small study by Goldberg et al. found a decrease in TG with exercise *(67)*.

The effect of weight reduction on dyslipidemia in obese subjects with CKD is unknown. Nutrition therapy as recommended by ATP III is similar to the DASH diet for treating hypertension and is mostly likely safe for CKD stages 1 and 2. In CKD stages 3 and 4, the risk

Table 8
Management of Dyslipidemias in Adults with Chronic Kidney Disease

Dyslipidemia	*Goal range*	*Begin with*	*If goal not met add*	*Alternative therapy*
TG≥500 mg/dL	TG <500 mg/dL	TLC	TLC + fibrate or niacin	Fibrate or niacin
LDL 100–129 gm/dL	LDL <100 mg/dL	TLC	TLC + low dose statin	Bile acid sequestrant or niacin
LDL≥130 gm/dL	LDL <100 mg/dL	TLC + low dose statin	TLC + maximum dose statin	Bile acid sequestrant or niacin
TG≥200 mg/dL and non-HDL ≥130 mg/dL	Non-HDL <130 mg/dL	TLC + low dose statin	TLC + maximum dose statin	Fibrate or niacin

LDL, low density lipoprotein; HDL, high density lipoprotein; TG, triglyceride; TLC, therapeutic life style changes. Adapted from the National Kidney Foundation *(62)*.

of malnutrition is high and the priority of nutrition therapy should be adequate calories and protein, reduced sodium intake to control hypertension, low phosphorus intake for controlling serum phosphorus and PTH, consistent meals, and carbohydrate intake to maintain goal blood glucose levels in people with diabetes. If these goals are being met, it may be reasonable to implement dietary changes to improve lipid profiles for 2–3 months before beginning drug treatment (Table 10). For others, drug treatment of lipid disorders may be first-line therapy.

For those patients with nephrotic range proteinuria, increases in total and LDL cholesterol and even TG are common. Treatment of the underlying glomerular disease may induce a remission in nephrotic syndrome and improve lipid profiles. Some studies have shown reduced urine protein excretion and improved cholesterol levels after treatment with ACE inhibitors and some show little impact of this treatment on lipid levels *(62)*. However, there is substantial evidence that treatments with these drugs reduce the rate of kidney disease progression in people with proteinuria, regardless of their effect on plasma lipids.

Table 9
Therapeutic Lifestyle Changes (TLC) for Adults with Chronic Kidney Disease

Diet

Emphasize reduced saturated fat:
- Saturated fat <7% total calories
- Polyunsaturated fat < 10% total calories
- Monounsaturated fat < 20% total calories
- Total fat 25–35% total calories
- Cholesterol <200 mg per day
- Carbohydrate 50-6-% total calories

Emphasize components that may reduce dyslipidemia
- Fiber 20-30 gm/day, emphasize 5-10 gm/day viscous (soluble) fiber
- Consider plant stanols/sterols, 2 gm/day
- Improve glycemic control

Emphasize total calories to attain/maintain BMI 22-26
- Match intake of overall calories to overall energy needs
- Goal waist circumference: men <40 inches, women <35 inches
- Goal Waist-hip ratio: men <1.0, women < 0.8

Physical Activity

Moderate daily lifestyle activities
- Use pedometer to attain/maintain 10,000 steps/day
- Emphasize regular daily motion and distance

Moderate planned physical activity
- 3–4 times/week, 20–30 minute periods of activity
- Choose walking, swimming, supervised exercise as tolerated
- Include resistance exercise to build lean muscle mass, reduce excess body fat

Lifestyle Habits

Drink alcohol in moderation, limit one drink/day with physician approval
Smoking cessation

Adapted from the National Kidney Foundation *(63)*.

5.2.1. Drug Therapies for Dyslipidemia

For patients who do not improve LDL and TG levels with TLC, drug therapies are appropriate, including use of statins, fibrates, niacin, and bile acid sequestrants. The KDOQI Clinical Practice Guidelines for Managing Dyslipidemias in CKD contain detailed recommendations for when and how to implement drug therapies, including combining

Table 10
Dietary Modifications to Improve Lipid Profiles in Chronic Kidney Disease

Food choices	*Choose more of*	*Choose less of*
Eggs	• Limit 2 eggs/week, or use egg whites or cholesterol-free egg substitutes	• Egg yolks and whole eggs (including in baked goods)
Meats, poultry, meat alternatives	• Lean meats, trimmed of fat • Poultry without skin • Low-fat tofu, soy protein products	• High-fat meats (sausage, bacon, liver, sweetbreads) • Deli meats such as ham, cold cuts, processed meats
Fish, shellfish	• Fish or shellfish baked or broiled without added fat	• Fried fish or shellfish • Avoid eating bones of fish (sardines, anchovies, etc.) due to high phosphorus content
Fats and oils	• Unsaturated oils (safflower, corn, soybean, cottonseed, canola, olive, peanut) • Margarine made from oils above, especially soft and liquid forms • Salad dressings made from oils above	• Hydrogenated and partially hydrogenated fats • Trans fats • Coconut, palm kernel and palm oils • Butter, lard, shortening, bacon fat, stick margarine • Salad dressings or dips made with egg yolk, cheese, sour cream, whole milk

Breads and grains	• Breads without toppings or cheese • Bagels, English muffins • Cereals–oat, wheat, corn, multigrain • Crackers – low-fat animal crackers, unsalted soda crackers and bread sticks, melba toast • Homemade breads made with recommended fats and oils	• High fat content breads (croissants, flaky dinner rolls) • Granola that contain coconut or hydrogenated fats • High-fat crackers (>3 gm fat/serving) • Commercially baked pastries biscuits, muffins
Fruits and vegetables	• Fresh, frozen or low-sodium canned (avoid high potassium choices per serum potassium level)	• Fried fruits or vegetables or served with butter or cream sauces
Sweets	• Sugar, syrup, honey, jam, preserves, candy made without fat • Low-fat and non-fat sherbet, sorbet, fruit ice • Angel food cake, fig and fruit bar cookies • Other cookies, cakes and pies made with egg whites or egg substitutes • Non-dairy regular and frozen whipped toppings in moderation	• Candy made with chocolate, cream, butter, frostings • Ice cream and regular frozen desserts • Commercially baked cookies, cakes, pies • Commercially fried pastries such as doughnuts • Whipped cream

Diet decisions should be made in consultation with a qualified Renal Dietitian, to adapt food choices to individual medical and nutritional conditions. Adapted from the National Kidney Foundation *(62)*.

therapies with stage of CKD and contraindications for combination treatments (Table 8) *(62)*.

6. DIABETES IN CKD

Diabetic kidney disease (nephropathy) is a leading cause of CKD, with estimates of greater than 40% of all CKD due to diabetes. Prevention of diabetic nephropathy should be a major public health goal as all people with diabetes are at risk for CKD. Without interventions, approximately 80% of those with type 1 diabetes and 20–40% of those with type 2 diabetes will progress to overt nephropathy (albuminuria >300 mg/day) and hypertension. Refer to Chapter 6 for an in-depth discussion of evaluations and therapies to identify, treat, and prevent CKD in people with diabetes.

7. IGA NEPHROPATHY

IgA nephropathy (IgAN) is the most common primary glomerulonephritis in the world. It is an immune complex-mediated glomerulonephritis characterized by the deposition of IgA within the mesangial regions of the glomeruli. Many patients develop a chronic, slowly progressive decline in kidney function over 10–20 years, leading to stage 5 CKD in 20–40% of cases *(66,67)*. Treatment often includes corticosteroids and/or ACE inhibitors. Two additional therapies involving nutrition interventions have been examined.

7.1. Omega-3 (n-3) Polyunsaturated Fatty Acids

The omega-3 polyunsaturated fatty acids (PUFA) eicosapentanoic acid (EPA) and docosahexanoic acid (DHA) (found in cold water fish, flaxseed oil, and canola oil) have anti-inflammatory effects in humans. EPA and DHA undergo biologic transformation into trienoic eicosanoids, which lead to decreased production of proinflammatory mediators, resulting in vasodilatory rather than vasoconstriction properties. In the kidney, EPA is used as substrate which results in reduced manifestations of mesangial cell activation *(68)*. Prospective studies have evaluated the effect of EPA and DHA supplements on the progression of IgAN and found that in long-range follow-up, the rate of loss of kidney function can be lowered with doses of 1.9 g EPA and 1.4 g DHA *(69)*. As a person would need to consume approximately two servings of salmon per day to obtain that amount of EPA and DHA, supplements provide a convenient and easy way to increase noninflammatory PUFA. An additional benefit may come from reducing intake of pro-inflammatory PUFAs, which are found in vegetable oils such as safflower, sunflower, and corn oils.

7.2. Low Antigen Content Diet

The rationale for use of a low antigen content diet for the treatment of IgAN involves the pathogenesis of IgA renal lesions. IgA is secreted by β-cells as part of the mucosal immune system in the intestinal tract, as triggered by viral and dietary antigens crossing the intestinal mucosal barrier *(68)*. Animal studies have shown IgA that is produced in response to dietary antigens can form glomerular deposits *(70)*. Studies in humans have shown that a low antigen content diet was able to reduce nephritogenic dietary antigens *(71)*. Long-term effects and consequences of following the low antigen content diet have not been reported, although one study recommends following the restricted diet for 1 month initially, then 10–15 days every month thereafter *(72)*. The low antigen content diet includes avoidance of foods containing common allergens such as gluten, nuts, dairy foods, citrus foods such as oranges and grapefruit, cantaloupe, honeydew, berries, chocolate, shellfish, eggs, and sulfites. Although the diet is complicated and restrictive in many foods and nutrients, a qualified renal dietitian could combine the low antigen components with an appropriate renal diet in those patients with IgAN. Regular nutrition assessments would be necessary to assure that the patient was consuming adequate nutrients.

8. ADVANCED GLYCATION END PRODUCTS

Advanced glycation end products (AGEs) are a heterogeneous group of compounds with significant prooxidant and proinflammatory actions. They form in the body under physiologic conditions produced by nonenzymatic glucose–protein interactions, producing highly reactive dicarbonyls which contribute to oxidative stress, inflammatory response, and tissue injury. Endogenous AGE formation is enhanced in diabetes, possibly due to the presence of elevated serum glucose levels. AGEs have been implicated in the vascular damage seen in diabetes and CVD, and, in animal studies, vascular disease has been suppressed by restricting food AGE content. High levels of AGEs are also seen in kidney failure, presumably due to accumulation and secondary to decreased renal clearance *(73–76)*.

AGEs form externally during the heat processing of food [i.e., the Maillard (or Browning) reaction], which produce fluorescent compounds, brown pigments, and flavoring agents *(77)*. Maillard reactions are responsible for the rich color and flavor, which forms on roasted and broiled meats and baked breads. Studies have shown that in CKD, higher AGE intake correlates with both higher serum markers of AGEs and greater markers of inflammation. Reducing dietary intake of AGEs will decrease serum levels and some inflammatory markers and may impact inflammatory vascular disease *(78,79)*. Dietary changes to reduce AGE consumption include avoiding cooking methods which promote Maillard reactions (frying, broiling, and grilling) and instead cooking meat, fish, poultry, vegetables, and so on by boiling, poaching, stewing, or steaming. These changes appear to be low risk, as long as patients are able to maintain an adequate protein and energy intake and it may be appropriate for those wishing to reduce accumulation of toxic AGEs.

9. SELF-MANAGEMENT BEHAVIOR

Nutrition management of CKD involves one of the most complex diet therapies prescribed. Patients with CKD must learn to balance intake of protein and calories, as well as several micronutrients. If the patient has diabetes, he/she must include carbohydrate control as well. Depending on the individual's overall health, clinicians may advise modifying lipids, smoking cessation, increased exercise, and so on. These are not behavior changes that patients can learn to do in brief office visits with their clinicians, but involve intense education in new lifestyles.

Self-management is a health care education approach, which recognizes the central role that patients play in health promotion, disease prevention, and successful disease management. Patients and clinicians work as partners to learn problem-solving skills and appropriate interventions and resources. Patients develop skills through modeling and practices to be self-supporting and in order to make successful behavior changes. This approach is significantly different from traditional knowledge-based teaching, where patients are provided with basic knowledge of why and how to change with no opportunity for practicing or experiencing the change as part of the learning process.

The self-management approach has been effective in changing behavior in patients with CKD *(80,81)*. In hypertension treatment studies, self-management has been shown to be more effective than information-only education at teaching people how to follow a low sodium diet. In addition, this approach has been shown to be more effective at improving clinical outcomes *(82,83)*.

10. CONCLUSION/SUMMARY

People with CKD require nutrition therapy and lifestyle changes, and it is unreasonable to expect the physician's office to provide effective nutrition education in a monthly clinic visit. Qualified registered dietitians are the cornerstone of all successful dietary interventions, being experts at assessing nutrient needs, conducting comprehensive nutritional assessments, identifying foods to be limited on the diet, recipe modification, shopping and cooking methods, and modifying diets for cultural and religious practices and beliefs. Dietitians are trained to determine the patients' readiness to learn and to make dietary changes and can assess what and how to teach effectively to promote healthful lifestyles. Teaching self-management is time-consuming, requires frequent contacts with clinicians for education, goal setting, teaching self-monitoring and problem-solving skills, evaluating progress, and providing feedback. Incorporating the skills of a qualified dietitian will result in the most successful outcomes for the patient with CKD.

11. CASE STUDY

BH has been referred to a nephrologist by the local Community Health Center. Last month, BH went to the Health Center, complaining of "flu symptoms" like fatigue, chronic cough, nausea, headaches, and fever. BH is 66 years old, height 182 cm, body weight 83 kg, with a previous diagnosis of diabetes. He quit smoking last year but drinks alcohol occasionally. Medications included Glyburide 2.5 mg everyday. His blood pressure was 163/89 mm Hg with trace peripheral edema in his lower extremities. His lungs were clear. Laboratory test results showed: creatinine 2.3 mg/dL, BUN 51 mg/dL, HbA1C 6.1%, potassium 4.6 meq/L, bicarbonate 26 meq/L, TC 219 mg/dL. The Health Center prescribed Lasix 20 mg once daily, Lisinopril 5 mg once daily, and referred BH to the nephrologist. When the nephrologist sees BH a month later his blood pressure is 150/80 mm Hg. Trace edema is still present below the knees. Cough, headache, and fever have resolved, but he complains of ongoing fatigue and low energy. Current laboratory tests show creatinine 1.8 mg/dL, BUN 52 mg/dL, Hgb 10.6 gm/dL, potassium 5.3 meq/L, calcium 9.2 meq/L, phosphorus 5.0 mg/dL, Intact PTH 79 pg/mL, albumin 4.9 g/dL, glucose 95 mg/dL. Calculated GFR is 40 ml/min/1.73 m^2, CKD stage 3. The nephrologist orders an anemia workup including iron studies. Ferritin is 83 ng/mL and TSAT is 18%. The nephrologist discusses CKD with him the importance of achieving a lower blood pressure and good blood sugar control. BH is told to add Norvasc 5 mg and FeSO4 325 mg (four tablets daily) to his medications and is referred to a dietitian for nutrition assessment and therapy.

11.1. Case Study Questions

1. Why is EPO therapy not prescribed, considering BH's low Hgb?

11.2. Additional Follow-Up

One month later, BH and his wife meet with the dietitian. She questions BH about lifestyle and usual eating patterns. He eats two meals per day at restaurants, plus an afternoon snack at work. Exercise includes housework and light yard work on weekends.

2. What are nutrition priorities for BH?

11.3. Additional Follow-Up

Three weeks later, BH and his wife return for follow-up, after seeing the nephrologist. Current laboratory values include serum phosphorus 4.8 mg/dL, potassium 4.6 meq/L, Hgb 10.8 gm/dL, ferritin 110 ng/dL, TSAT 23% and blood pressure 132/79. The nephrologist has instructed BH to take Tums 1 with each meal and has begun EPO injections 8500 units/week. BH reviews his food records with the dietitian, showing the changes he has made over the past several weeks.

3. BH's Hgb is increasing with oral iron. Why is BH started on EPO therapy?
4. Why are phosphate binders prescribed (Tums) when the serum phosphorus level is decreasing?
5. What additional nutrition therapy should be presented to BH at this clinic visit?

11.4. Additional Follow-Up

One month later, BH returns for follow-up. Current laboratory values (from his nephrology visit) include serum phosphorus 4.3 mg/dL, Hgb 11.7 gm/dL, blood pressure 129/80, blood sugars range 90–130 mg/dL, and body weight has decreased to 80 kg. No medication changes are noted. BH reports that with his smaller meat portions, he often feels like he is not getting enough to eat.

6. Is BH's recent weight loss a benefit or risk, considering his usual weight, diabetes, hypertension, and CKD?

REFERENCES

1. NKF KDOQI clinical practice guidelines for chronic kidney disease: evaluation, classification, and stratification. Am J Kidney Dis 2002;39:S37–S75, S112–S155, S170–S212.
2. National Kidney Foundation at http://www.kidney.org/kidneyDisease/. Accessed January 28, 2006.
3. U.S. Renal Data System. USRDS 2006 Annual Data Report: atlas of chronic kidney disease and end-stage renal disease in the United States, National Institutes of Health, National Institute of Diabetes and Digestive and Kidney Diseases, Bethesda, MD, 2006.
4. Shulman, NB, Ford CE, Hall WD, Blaufox MD, Simon D, Langford HG, Schneider KA. Prognostic value of serum creatinine and effect of treatment of hypertension on renal function. Results from the Hypertension Detection and Follow-up Program (HDFP). The Hypertension Detection and Follow-Up Program Cooperative Group. Hypertension 1989;13:S180–S193.
5. Keith DS, Nichols GA, Gullion CM, Brown JB, Smith DH. Longitudinal follow-up and outcomes among a population with chronic kidney disease in a large managed care organization. Arch Intern Med 2004;164:659–663.
6. Go AS, Chertow GM, Fan D, McCulloch CD, Hsu CY. Chronic kidney disease and the risks of death, cardiovascular events, and hospitalization. N Engl J Med 2004;351:1296–1305.
7. Kopple JD, Greene T, Chumlea WC, Hollinger D, Maroni BJ, Merrill D, Scherch LK, Schulman G, Wang S, Zimmer GS. Relationship between nutritional status and the glomerular filtration rate: Results from the MDRD Study. Kidney Int 2000;57:1688–1703.
8. Ikizler TA, Green JH, Wingard RL, Parker RA, Hakim RM. Spontaneous dietary protein intake during progression of chronic renal failure. J Am Soc of Nephrol 1995;6:1386–1391.
9. Mitch WE, Maroni BJ. Factors causing malnutrition in patients with chronic uremia. Am J Kidney Dis 1999;33:176–179.
10. Stenvinkel P, Heimburger O, Paultre F, Diczfalusy U, Wang T, Berglund L, Jogestrand T. Strong association between malnutrition, inflammation, and atherosclerosis in chronic renal failure. Kidney Int 1999;55:1899–1911.
11. Rose BD. Protein restriction and progression of chronic kidney disease. www.uptodate.com 2005;13.2:1–12.
12. Ihle BU, Becher GJ, Whitworth JA, Charlwood RA, Kincaid-Smith PS. The effect of protein restriction on the progression of renal insufficiency. N Engl J Med 1989;321:1773–1777.
13. Walser M, Hill SB, Ward L, Magder L. A crossover comparison of progression of chronic renal failure: Ketoacids versus amino acids. Kidney Int 1993;43:933–939.
14. Walser M, Hill S. Can renal replacement be deferred by a supplemented very low protein diet? J Am Soc Nephrol 1999;10:110–116.
15. Klahr S, Levey AS, Beck GJ, Caggiula AW, Hunsicker L, Kusek JW, Striker G. The effects of dietary protein restriction and blood-pressure control on the progression of chronic renal disease. N Engl J Med 1994;330:877–884.
16. Levey AS, Green T, Beck GJ, Caggiula AW, Kusek JW, Hunsicker LG, Klahr S. Dietary protein restriction and the progression of chronic renal disease: what

have all of the results of the MDRD study shown? J Am Soc Nephrol 1999;10: 2426–2439.
17. Fouque D, Laville M, Boissel JP, Chifflet R, Labeeuw M, Zech PY. Controlled low protein diets in chronic renal insufficiency: meta-analysis. BMJ 1992;304:216–220.
18. Fouque D, Wang P, Laville M, Boissel JP. Low protein diets for chronic renal failure in non diabetic adults (Cochran Review). In: The Cochran Library, 2000. Oxford: Update Software, p. 4.
19. Elliot P, Stamler J, Dyer AR, Appel L, Dennis B, Kesteloot H, Ueshima H, Okayama A, Chan Q, Garside DB, Zhoe B. Association between protein intake and blood pressure. Arch Intern Med 2006;166:79–87.
20. Soroka N, Silverberg DS, Greemland M, Birk Y, Blum M, Peer G, Iaina A. Comparison of a vegetable-based (soya) and an animal-based low-protein diet in predialysis chronic renal failure patients. Nephron 1998;79:173–180.
21. Anderson JW, Blake JE, Turner J, Smith BM. Effects of soy protein on renal function and proteinuria in patients with type 2 diabetes. Am J Clin Nutr 1998;68:1348S–1353S.
22. Wheeler ML, Fineberg SE, Fineberg NS, Gibson RG, Hackward LL. Animal versus plant protein meals in individuals with type 2 diabetes and microalbuminuria, effects on renal, glycemic and lipid parameters. Diabetes Care 2002;25:1277–1282.
23. Stephenson TJ, Setchell KD, Kendall CW, Jenkins DJ, Anderson JW, Fanti P. Effect of soy protein-rich diet on renal function in young adults with insulin-dependent diabetes mellitus. Clin Nephrol 2005;64:1–11.
24. Teixeira SR, Tappenden KA, Carson LA, Jones R, Prabhudesai M, Marchall WP, Erdman JW. Isolated soy protein consumption reduces urinary albumin excretion and improves the serum lipid profile in men with type 2 diabetes mellitus and nephropathy. J Nutr 2004;134:1874–1880.
25. Castaneda C, Gordon PL, Uhlin KL, Levey AS, Kehayias JL, Dwyer JT, Fielding RA, Roubenoff R, Singh MF. Resistance training to counteract the catabolism of a low-protein diet in patients with chronic renal insufficiency. Ann Intern Med 2001;135:965–976.
26. NKF KDOQI clinical practice guidelines for nutrition in chronic renal failure. Am J Kidney Dis 2000;35:S58–S61.
27. NKF KDOQI clinical practice guidelines for hypertension and antihypertensive agents in chronic kidney disease. Am J Kidney Dis 2004;43:S16–230 (Guideline 1,5,6,7,9).
28. Wiggins KL. Guideline 2, Nutrition Care of Adult Dialysis Patients. In: Guidelines for Nutrition Care of Renal Patients, 3rd ed. Chicago: American Dietetic Association, 2002, pp. 5–18.
29. Maroni BJ, Staffield C, Young VR, Manatunga A, Tom K. Mechanisms permitting nephrotic patients to achieve nitrogen equilibrium with a protein-restricted diet. J Clin Invest 1997;99:2479–2487.
30. Walser M, Hill S, Tomalis EA. Treatment of nephrotic adults with a supplemented, very low-protein diet. Am J Kidney Dis 1996;28:354–364.
31. D'Amico G, Gentile MG, Manna G, Fellin G, Ciceri R, Cofano CR, Petrini C, Lavarda F, Perolini S, Porrini M. Effect of vegetarian soy diet on hyperlipidaemia in nephrotic syndrome. Lancet 1992;339:1131–1134.

32. Gentile MG, Fellin G, Cofano F, Delle Fave A, Manna G, Ciceri R, Petrini C, Lavarda F, Possi F, D'Amico G. Treatment of proteinuric patients with a vegetarian soy diet and fish oil. Clin Nephrol 1993;40:315–320.
33. Monteon FJ, Laidlaw SA, Shaib JK, Kopple JD. Energy expenditure in patients with chronic renal failure. Kidney Int 1986;30:741–747.
34. Kopple JD, Monteon FJ, Shaib JK. Effect of energy intake on nitrogen metabolism in nondialyzed patients with chronic renal failure. Kidney Int 1986;29:734–742.
35. Kuhlmann U, Schwidkardi M, Trebst R, Lange H. Resting metabolic rate in chronic renal failure. J Ren Nutr 2001;11:202–206.
36. Passey C, Bunker V, Jackson A, Lee H. Energy balance in predialysis patients on a low-protein diet. J Ren Nutr 2003;13:120–125.
37. Chauveau P, Barthe N, Rigalleau V, Ozenne S, Castaing F, Delclaux C, de Precigout V, Combe C, Aparicio M. Outcome of Nutritional Status and Body Composition of Uremic Patients on a Very Low Protein Diet. Am J Kidney Dis 1999;34:500–507.
38. Guarnieri GF, Toigo G, Situlin R, Carraro M, Tamaro G, Lucchesli A, Oldrizzi L, Rugiu C Maschio G. Nutritional state in patients on long-term low-protein diet or with nephrotic syndrome. Kidney Int 1989;36:S195–S199.
39. Byham-Gray LD. Weighing the evidence: energy determinations across the spectrum of kidney disease. J Ren Nutr 2006;16:17–26.
40. Wiggins KL. Vitamins and minerals in chronic kidney disease. In: Renal Care: Resources and Practical Applications. Chicago: American Dietetic Association, 2004, pp. 39–60.
41. Martinez I, Saracho R, Montenegro J, Llach F. The importance of dietary calcium and phosphorous in the secondary hyperparathyroidism of patients with early renal failure. Am J Kidney Dis 1997;29:496–502.
42. NKF KDOQI clinical practice guidelines for bone metabolism and disease in chronic kidney disease. Am J Kidney Dis 2003;42: S29–S91.
43. Reichel H, Deibert B, Schmidt-Gayk H, Ritz E. Calcium metabolism in early chronic renal failure: implications for the pathogenesis of hyperparathyroidism. Nephrol Dial Transplant 1991;6:162–169.
44. Barsotti G, Morelli E, Guiducci A, Ciardella F, Giannoni A, Lupetti S, Giovannetti S. Reversal of hyperparathyroidism in severe uremics following very low-protein and low-phosphorus diet. Nephron 1982;30:310–313.
45. Barsotti G, Cupisti A. The role of dietary phosphorus restriction in the conservative management of chronic renal disease. J Ren Nutr 2005;15: 189–192.
46. Cochran M, Coates PT, Morris HA. The effect of calcitriol on fasting urine calcium loss and renal tubular reabsorption of calcium in patients with mild renal failure–actions of a permissive hormone. Clin Nephrol 2005;64:98–102.
47. Andress DL. Vitamin D in chronic kidney disease: A systemic role for selective vitamin D receptor activation. Kidney Int 2006;69:33–43.
48. Yumita S, Suzuki M, Akiba T, Akizawa T, Seino Y, Kurokawa K. Levels of serum 1,25(OH)2D in patients with pre-dialysis chronic renal failure. Tohoku J Exp Med 1996;180:45–56.
49. Kramer H, Toto R, Peshock R, Cooper R, Victor R. Association between chronic kidney disease and coronary artery calcification: the Dallas Heart Study. J Am Soc Nephrol 2005;16:507–513.

50. NKF KDOQI clinical practice guidelines for anemia of chronic kidney disease update 2000. Am J Kidney Dis 2001;37:S182–238. (Guideline 1–12,15,16,20).
51. Culleton BF, Larson MG, Wilson PWF, Evans JC, Parfrey PS, Levy D. Cardiovasculare disease and mortality in a community-based cohort with mild renal insufficiency. Kidney Int 1999;56:2214–2219.
52. Keough-Ryan TM, Kiberd BA, Dipchand CS, Cox JL, Rose CL, Thompson KJ, Clase CM. Outcomes of acute coronary syndrome in a large Canadian cohort: impact of chronic renal insufficiency, cardiac, interventions, and anemia. Am J Kidney Dis 2005;46:845–855.
53. Brugts JJ, Knetsch AM, Mattace-Raso FUS, Hofman A, Witteman JCM. Renal function and risk of myocardial infarction in an elderly population, The Rotterdam Study. Arch Intern Med 2005;165:2659–2665.
54. De Nicola L, Minutolo R, Chiodini P, Zoccali C, Castellino P, Donadio C, Strippoli M, Casino F, Giannattasio M, Petrarulo F, Virgilio M, Llaraia E, Di Iorio BR, Savica V, Conte G. Global approach to cardiovascular risk in chronic kidney disease: Realty and opportunities for intervention. Kidney Int 2006;69:538–545.
55. Shlipak MG, Fried LF, Cushman M, Manolio TA, Peterson D, Stehman-Breen C, Bleyer A, Newman A, Siscovick D, Psaty B. Cardiovascular mortality risk in chronic kidney disease, comparison of traditional and novel risk factors. JAMA 2005;293:1737–1745.
56. Toto RD. Treatment of hypertension in chronic kidney disease. Semin Nephrol 2005;25:435–439.
57. Sica D, Carl D. Pathologic basis and treatment considerations in chronic kidney disease-related hypertension. Semin Nephrol 2005;25:246–251.
58. De Nicola L, Minutolo R, Zamboli P, Cestaro R, Marzano L, Giannattasio P, Cristofano C, Chimienti S, Savica V, Bellinghieri G, Rapisarda F, Fatuzzo P, Conte G. Italian audit on therapy of hypertension in chronic kidney disease: The TABLE-CKD Study. Semin Nephrol 2005;25:425–430.
59. Appel LJ, Brands MW, Daniels SR, Karanja N, Elmer PJ, Sacks FM. Dietary Approaches to Prevent and Treat Hypertension: A Scientific Statement From the American Heart Association. Hypertension 2006;47:296–308.
60. Conlin PR, Chow D, Miller ER, Svetkey LA, Lin PH, Harsha DW, Moore TJ, Sacks FM, Appel LJ. The effect of dietary patterns on blood pressure control in hypertensive patients: Results from the Dietary Approaches to Stop Hypertension (DASH) Trial. Am J Hypertens 2000;13:949–955.
61. Sacks FM, Svetkey LP, Vollmer WM, Appel LJ, Bray GA, Harsha D, Obarzanek E, Conlin PR, Miller ER, Simons-Morton DG, Karanja N, Lin PH. Effects on blood pressure of reduced Dietary Sodium and the Dietary Approaches to Stop Hypertension (DASH) diet. N Engl J Med 2001;344:3–10.
62. NKF KDOQI clinical practice guidelines for managing dyslipidemias in chronic kidney disease. Am J Kidney Dis 2003;41:S22–S58. (Guideline 1–4).
63. Seliger SL, Weiss NS, Gillen DL, Kestenbaum B, Ball A, Sherrard DJ, Stehman-Breen CO. HMG-CoA reductase inhibitors are associated with reduced mortality in ESRD patients. Kidney Int 2002;61:297–304.
64. Fried LF, Orchard TJ, Kasiske BL. The effect of lipid reduction on renal disease progression: A meta-analysis. Kidney Int 2001;59:260–269.
65. Executive Summary of the Third Report of the National Cholesterol Education Program (NCEP) Expert Panel on Detection, Evaluation, and Treatment of High

Blood Cholesterol in Adults (Adult Treatment Panel III). JAMA 2001;285: 2486–2497.
66. Donadio JV. The emerging role of omega-3 polyunsaturated fatty acids in the management of patients with IgA nephropathy. J Ren Nutr 2001;11:122–128.
67. Pettersson EE, Rekola S, Berglund L, Sundqvist KG, Angelin B, Diczfalusy U, Bjorkhem I, Bergstrom J. Treatment of IgA nephropathy with omega-3-polyunsaturated fatty acids: a prospective, double-blind, randomized study. Clin Nephrol 1994;41:183–190.
68. Souba WW, Wilmore DW. Diet and nutrition in the care of the patient with surgery, trauma, and sepsis. In: Shils ME, Olson JA, Shike M, Ross AC, ed. Modern Nutrition in Health and Disease, 9th edn. Philadelphia: Lippincott Williams & Wilkins, 1999, p. 1609.
69. Donadio JV, Larson TS, Bergstralh EJ, Grande JP. A randomized trial of high-dose compared with low-dose omega-3 fatty acids in severe IgA nephropathy. J Am Soc Nephrol 2001;12:791–799.
70. Genin G, Laurent B, Sabatier JC, Colon S, Berthoux FC. IgA mesangial deposits in C3H/He mice after oral immunization with ferritin or bovine serum albumin. Clin Exp Immunol 1986;63:385–391.
71. Ferri C, Puccini R, Longombardo G, Paleologo G, Migliorini P, Moriconi L, Pasero G, Cioni L. Low-antigen-content diet in the treatment of patients with IgA nephropathy. Nephrol Dial Transplant 1993;8:1193–1198.
72. Ferri C, Puccini R. Low-Antigen-Content Diet. http://www.do.med.unipi.it/apher/write/diet.htm. Accessed April 1, 2006
73. Vlassara H. Advanced glycation in health and disease, role of the modern environment. Ann N Y Acad Sci 2005;1043:452–460.
74. Uribarri J, Cai W, Sandu O, Peppa M, Goldberg T, Vlassara H. Diet-derived advanced glycation end products are major contributors to the body's AGE pool and induce inflammation in healthy subjects. Ann N Y Acad Sci 2005;1043: 461–466.
75. Vlassara H, Cai W, Crandall J, Goldberg T, Oberstein R, Dardaine V, Peppa M, Rayfield EJ. Inflammatory mediators are induced by dietary glycotoxins, a major risk factor for diabetic angiopathy. Proc Natl Acad Sci USA 2002:99: 15596–15601.
76. Lapolla A, Fedele D, Reitano R, Bonfante L, Pastori G, Seraglia R, Tubaaro M, Traldi P. Advanced glycation end products/peptides. Ann N Y Acad Sci 2005;1043:267–275.
77. Stipanuk MH. Biochemical and Physiological Aspects of Human Nutrition. Philadelphia: Saunders, 2000, pp. 9–11.
78. Uribarri J, Peppa M, Cai W, Goldberg T, Lu M, He C, Vlassara H. Restriction of dietary glycotoxins reduces excessive advanced glycation end products in renal failure patients. J Am Soc Nephrol 2003;14:728–731.
79. Uribarri J, Peppa M, Cai W, Goldberg T, Lu M, Suresh B, Vassolotti J, Vlassara H. Dietary glycotoxins correlate with circulating advanced glycation end product levels in renal failure patients. Am J Kidney Dis 2003;42:532–538.
80. Gillis BP, Caggiula AW, Chiavacci AT, Coyne T, Doroshendo L, Milas NC, Nowalk MP, Scherch LK. Nutrition intervention program of the Modification of Diet in Renal Disease Study: A self-management approach. J Am Diet Assoc 1995;96:1288–1294.

81. Milas NC, Nowalk MP, Akpele L, Castaldo L, Coyne T, Doroshenko L, Kigawa L, Korzec-Ramirez D, Scherch LK, Snetselaar L. Factors associated with adherence to the dietary protein intervention in the Modification of Diet in Renal Disease Study. J Am Diet Assoc 1995;96:1295–1300.
82. Midgley JP, Matthew AG, Greenwood CM, Logan AG. Effect of reduced dietary sodium on blood pressure: A meta-analysis of randomized controlled trials. JAMA 1996;275:1590–1597.
83. Bodenheimer T, Lorig K, Holman H, Grumback K. Patient self-management of chronic disease in primary care. JAMA 2002;288:2469–2475.

III Chronic Kidney Disease (Stage 5) in Adults

1 Treatment

9 Dialysis

Karen Wiesen and Graeme Mindel

LEARNING OBJECTIVES

1. Describe the goals of medical nutrition therapy for patients receiving hemodialysis, peritoneal dialysis and nocturnal home dialysis.
2. Describe the differences in nutrition therapy between the different types of renal replacement therapies.
3. Identify factors that may impact the nutritional status of patients receiving maintenance dialysis.

Summary

Medical nutrition therapy plays an integral role in the health of patients with stage 5 chronic kidney disease. This chapter reviews the different types of renal replacement therapies and corresponding dietary recommendations for calories, protein, sodium, fluid, potassium, calcium, phosphorus, lipids, vitamins and trace minerals. It also briefly discusses factors that may influence nutritional status in patients receiving maintenance dialysis.

Key Words: Malnutrition; medical nutrition therapy; hemodialysis; peritoneal dialysis; nocturnal home hemodialysis.

1. INTRODUCTION

Medical nutrition therapy (MNT) plays an integral role in the health of the patient with stage 5 chronic kidney disease (CKD) receiving maintenance dialysis. The health professional must understand the role of nutrition in stage 5 CKD, the factors affecting assessment and maintenance of adequate nutritional status and the nutritional implications

From: *Nutrition and Health: Nutrition in Kidney Disease*
Edited by: L. D. Byham-Gray, J. D. Burrowes, and G. M. Chertow

associated with the different types of renal replacement therapies (RRT). Currently there are over 350,000 patients receiving RRT in the United States with those numbers expected to steadily increase *(1,2)*. The average age of the patient starting dialysis is increasing with a significant number of patients presenting to dialysis with numerous comorbid conditions that may have already negatively impacted their nutritional status. The nutritional status of the patient at the initiation of RRT is an important risk factor for future outcomes, and malnutrition is associated with increased mortality in patients receiving maintenance dialysis *(3)*. Nutritional management should include ongoing diet education, nutrition assessment, individualized interventions and monitoring of nutritional status. The goals of MNT in stage 5 CKD are (i) to achieve and maintain neutral or positive nitrogen balance; (ii) to achieve and maintain good nutritional status; (iii) to prevent the accumulation of electrolytes and minimize fluid imbalance; and (iv) to minimize the effect of metabolic disorders associated with stage 5 CKD.

Patients beginning RRT now have a choice of modalities that include hemodialysis (HD), peritoneal dialysis (PD), and nocturnal home hemodialysis (NHD) and renal transplantation. This chapter will review the MNT for each modality and discuss factors that may impact nutrition assessment and overall nutritional status. Renal transplantation is another treatment option and is discussed in Chapter 10.

2. FACTORS INFLUENCING NUTRTIONAL STATUS

Protein energy malnutrition (PEM) in the maintenance dialysis population is associated with increased morbidity and mortality *(4–7)*. Between 10 and 50% of HD patients and 18 and 56% of PD patients have some degree of malnutrition *(8,9)*. Routine monitoring of the patients' nutritional status through the use of anthropometric measures, biochemical parameters, and diet histories and interviews is important in the early detection and prevention of PEM. The National Kidney Foundation Kidney Disease Outcomes Quality Initiative (KDOQI) Clinical Practice Guidelines for Nutrition recommends that a panel of measures be used to routinely assess nutritional status *(10)*. Refer to Chapter 4 for more information on individual parameters used in nutrition assessment. Use of a variety of assessment tools is important because some of the traditional anthropometric and biochemical measures employed to assess nutritional status can be influenced by multiple catabolic factors such as anorexia, inflammation, acidosis and dialysis-related losses. This section will provide a brief overview of some of these factors.

Serum albumin has long been used as a marker of nutritional status in the maintenance dialysis population and has been shown to independently correlate with an increased risk of mortality *(5–7)*. Serum albumin between 3.5 and 4 g/dL is associated with an increased risk of death twofold, and a low albumin at the initiation of dialysis increases the risk of hospitalization as well as length of stay in the first year of dialysis *(3,5)*. Emerging evidence now indicates that there are two types of malnutrition present in the dialysis population. Classic or type 1 malnutrition is defined by loss of lean body mass, inadequate oral intake, normal to mildly depleted albumin and normal C-reactive protein levels (CRP) and is responsive to nutrition intervention *(8,11)*. Type 2 malnutrition is caused by inflammation and characterized by markedly low albumin despite adequate oral intake, increased oxidative stress, elevated CRP and other pro-inflammatory markers and is not reversible with nutrition intervention *(11)*. Type 2 is also referred to as Malnutrition-Inflammation Complex Syndrome because of the interrelationship between malnutrition and inflammation in the dialysis patient. Both conditions have been found to frequently coexist and there may be a cause and effect relationship between these two factors, presenting a challenge for the clinician trying to assess which type of malnutrition is present.

Anorexia or poor appetite is a subjective factor in nutrition assessment; however, recent studies have shown it to be predictive of poor clinical outcomes as well as being associated with inflammation *(12)*. Anorexia is estimated to be present in one-third of patients receiving maintenance dialysis *(13)*. The etiology of anorexia is multifactorial and includes uremia, inflammation, infection, delayed gastric emptying, comorbid conditions, medications, and psychosocial and socioeconomic factors, absorption of glucose in PD and early satiety and age *(13–15)*. Refer to Chapter 11 for further information on PEM and inflammation.

Metabolic acidosis impacts nutritional status by increasing protein catabolism and possibly decreasing protein synthesis leading to negative nitrogen balance and loss of lean body mass *(15,16)*. Correction of acidosis with sodium bicarbonate has been shown to correct the protein catabolism *(17)*. Metabolic acidosis may induce insulin resistance and chronic inflammation, both of which may also increase protein catabolism but more research is needed *(16,18,19)*. KDOQI recommends that pre-dialysis bicarbonate levels be $\geq$ 22 mEq/L and that levels be monitored monthly *(10)*.

HD has long been considered a catabolic procedure. Approximately 2–3 g of peptides and 4–9 g net amino acids are lost during HD and 5–15 g protein during PD *(20–22)*. The use of bio-incompatible membranes and reuse of high flux dialyzers have both been shown to increase amino acid loss *(23,24)*. Dialysate protein losses are also higher with polysulphone dialyzers processed with bleach and losses significantly increase after the sixteenth use *(25)*. HD has been shown to induce a protein catabolic state that can stimulate muscle and whole body protein loss, decrease protein synthesis and increase energy expenditure with the effects lasting up to 2 hours post-dialysis *(26,27)*. PD is not as catabolic unless the patient has peritonitis. A mild inflammatory response may be triggered by peritoneal dialysate bioincompatability, endotoxin transfer from the dialysate, or the PD catheter itself that can lead to protein catabolism *(22,28)*. Patients receiving PD who are classified as high transporters by their peritoneal equilibration test have a higher incidence of poor nutrition due to the loss of larger amounts of protein into the dialysate. Serum albumin levels are usually low in these patients and they may require nutritional supplementation.

Assessment of dialysis adequacy should be part of the routine evaluation of nutrition in patients receiving maintenance dialysis. The relationship between Kt/V (a marker for dialysis adequacy where K = clearance, t = time, and v = volume) and the protein equivalent of nitrogen appearance (PNA), however, may be confounded by mathematical coupling. Using Kt alone, the non-normalized dose of dialysis may be more closely associated with serum albumin levels *(29)*. The Hemodialysis (HEMO) Study looked at the optimal dose of dialysis for patients receiving maintenance HD in order to determine the best parameters needed for achieving dialysis adequacy. Patients were randomly assigned to standard dose of dialysis (single pool Kt/V 1.25) or high dose dialysis (single pool Kt/V 1.65) and to low or high flux dialyzers. When the study group looked at the effect of dialysis dose and membrane flux on various nutritional parameters, they found no significant differences between the various groups *(30)*.

The relationship between nutritional intake and dose of dialysis in patients receiving continuous PD was evaluated as part of the CANUSA (Canada-USA) PD study group. A number of different nutritional markers were evaluated, and in the first 6 months, there was a positive correlation between the PD dose and all of the nutritional markers except for serum albumin levels. After 6 months, reduction in overall clearance because of loss of residual renal function was associated with a trend toward declining nutritional parameters *(31)*. Current minimum recommendations are a Kt/V of 1.3–1.4 for maintenance HD and a weekly total Kt/V of 1.7 for continuous PD *(32)*.

The use of short daily hemodialysis and NHD has been shown to reduce anorexia and improve a number of nutritional markers *(20)*. Both of these treatment modalities are not available to all dialysis patients because of reimbursement obstacles and limited data to support their use. Although the observations have been reproduced in a number of studies, further investigation is needed.

3. NUTRIENT RECOMMENDATIONS IN DIALYSIS

3.1. Energy

KDOQI recommends a daily energy intake of 35 kcal/kg standard or adjusted body weight/day for stable maintenance dialysis patients who are less than 60 years old and 30–35 kcal/kg standard or adjusted body weight/day for those age 60 and older *(10)*. These recommendations are based on metabolic studies that showed 35 kcal/kg was necessary to maintain neutral nitrogen balance and stable body composition. As patients 60 years or older may be more sedentary and have less lean body mass, a lower energy intake of 30–35 kcal/kg body weight is thought to be acceptable. Energy intakes should be adjusted if the patient is involved in heavy physical exercise, underweight, or catabolic *(10,33)*. Energy intakes for acutely ill maintenance dialysis patients are the same.

3.1.1. Hemodialysis

Many HD patients are unable to consume the recommended energy intake, resulting in a low body weight and body mass index, both of which are associated with increased mortality in the HD patient *(10,20)*. The average energy intake has been reported at 24–27 kcal/kg/day *(13,34)*. In the HEMO Study, patients averaged 23 kcal/kg/day, significantly less than the NKF-KDOQI guidelines for calories and less than the HEMO guidelines of 28 kcal/kg/day *(34)*. Energy intake was less and patient reported appetite was sub-optimal on dialysis days when compared to a non-dialysis day. Contributing factors were thought to be fatigue after dialysis or the catabolic, physiologic and metabolic effects of dialysis on the body *(35)*. Other factors such as taste disturbances, medications, missed meals on dialysis days, overly restricted diet, delayed gastric emptying, repeated hospitalizations and psychosocial concerns may also contribute to inadequate energy intake *(8,14,36)*. A standard dialysate solution containing 200 mg/dL glucose contributes only a small amount of calories during thrice weekly dialysis and does not significantly contribute to energy intake *(20)*. Assessment and counseling by the dietitian is important in helping the patient achieve an adequate intake. The use of nutritional supplements, may need to be considered (refer to Chapter 12).

3.1.2. Peritoneal Dialysis

For patients receiving PD, energy levels should include calories from both diet and the dialysate as calories absorbed during dialysis can be significant and lead to weight gain. Several formulas have been published for determining caloric load from PD and these are found in Table 1 *(37,38)*. The most accurate method is to compare the grams of glucose infused with the grams of glucose in the effluent *(39)*. Glucose absorption differs between therapies with patients on continuous cyclic PD (CCPD) absorbing approximately 40% of calories due to shorter dwell times, whereas patients on continuous ambulatory PD (CAPD) absorb approximately 60% of calories *(40,41)*. It may be difficult to restrict calories for weight reduction in PD patients without compromising protein intake and nutritional status. To help with weight

Table 1
Estimate of Calories Absorbed from Peritoneal Dialysis

Formula	*Example*	*Comment*
Glucose absorbed (kcal) = (11.3X–10.0) × L inflow × 3.4 Where X = average glucose concentration infused. *(37)*	4 L of 1.5% (1.36) + 4 L of 4.25% (3.8) dextrose = average 2.6%. [(11.3 × 2.6) –10.9] × 8L = (29.4–10.9) × 8 L = 18.5 × 8L = 148 g glucose. 148 × 3.4 = 503 kcal	Does not account for differences in membranes. Gives a rough estimate of glucose and kcal absorption
Glucose absorbed (kcal) = $(1–D/D_0)\ x_1$ Where D/D_0 is the fraction of glucose remaining and the x_1 is the initial glucose *(38)*	Infused = 4 L of 38 g/L + 4 L of 13.6 g/L = 152 + 54.4 = 206 g 206 g = × 1. Measured remaining glucose = 1200 mg/dL in 10 L effluent. = 1200 × (10,000 mL/100,000 = 1200 × 0.1 = 120 g D/D_0 = 120/206 = 0.58 (1–0.58) × 206 = 86.5 86.5 × 3.4 = 294 kcal.	Considers dialysis modality and membrane transport type

Simple Estimate: Glucose absorbed (kcal) = glucose infused (mL) × % absorption. G glucose absorbed × 3.4 = kcal absorbed % Absorption: APD: 40% CAPD: 60% Icodextrin = 40% G glucose /L: 1.5 % = 15 g 2.5% = 25 g 4.25% = 42.5 g icodextrin = similar to 2.5% solution	4 L (1.5%) + 4 L (2.5%), on CAPD. 60 g + 100 g × 60% = 160 × 60% = 96 g × 3.4 = 326 kcal	Does not consider membrane transport type or type of PD modality. Provides a rough estimate

APD, automated peritoneal dialysis; CAPD, continuous ambulatory peritoneal dialysis; PD, peritoneal dialysis.
Adapted from ref. *(39)*

control, patients should be encouraged to limit excessive sugars and fats and exercise if possible.

Icodextrin is an alternative polyglucose PD solution produced from the hydrolysis of cornstarch. Because it is a macromolecule, it is absorbed more slowly by the peritoneal membrane resulting in a sustained ultrafiltration (UF) and lower glucose absorption *(40–43)*. Icodextrin provides a caloric load similar to a 2.5% dextrose over the longer dwell. One of the metabolites of icodextrin is maltose which can interfere with certain blood glucose monitors and strips causing a falsely elevated reading *(42)*. People with diabetes who use icodextrin need to check with the manufacturer of their glucose meter and strips to assure they are using a system not affected by the maltose.

3.1.3. Nocturnal Hemodialysis

There have been limited studies examining the nutritional needs of patients receiving NHD. The current recommendations are extrapolated from the needs of the conventional HD patient. The diet should be individualized to the patient based on lab data, weight gains and length of dialysis. Because of the frequency and length of dialysis, the diet for NHD tends to be more liberal than conventional HD. There are no established energy guidelines for NHD, so current recommendations follow the KDOQI Nutrition guidelines for conventional HD, and energy needs can be adjusted as needed to maintain weight (see Table 2). With NHD, appetite has been reported to improve leading to an increase in both protein and calorie intake *(20,44)*. Weight gain may occur, so energy needs should be adjusted and exercise encouraged if the patient is physically able to participate *(20,44)*.

3.2. Protein

3.2.1. Hemodialysis

Adequate protein intake is important to ensure that the patient maintains positive or neutral nitrogen balance. The KDOQI Nutrition Guidelines recommend 1.2 g/kg standard or adjusted body weight for the clinically stable HD patient with at least 50% from high biological value (HBV) sources. HBV protein or animal protein is used more efficiently and provides the required essential amino acids *(10)*. Vegetarian patients will require ongoing counseling by the dietitian to help ensure they consume adequate protein from legumes or soy products without excess mineral load. Protein recommendations are based on a small number of nitrogen balance studies and do not differentiate for age. While it is possible that protein needs of the elderly patient receiving maintenance HD may be slightly reduced, the catabolic effects of dialysis along with other comorbid conditions may outweigh this reduction. A small number of studies have shown that the risk level for malnutrition in elderly HD patients is higher than in younger patients *(45–48)*.

Protein intake is often inadequate in HD patients and contributes to PEM. In the HEMO Study, less than 20% of the patients at baseline met the current KDOQI nutrition guidelines for protein with the average intake being 0.93 g/kg/day *(34)*. Protein needs are also influenced by metabolic acidosis, infection, inflammation or surgical procedures associated with increased catabolism. There is limited data on the protein needs of the acutely ill maintenance HD patient, and KDOQI recommends that these patients receive at least 1.2 g protein/kg/day *(10)*. Hospitalized HD patients generally consume less than the recommended amount of protein and may need intensive nutrition counseling, monitoring and possibly nutritional support to provide adequate protein to meet their needs.

Table 2
Daily Nutrient Recommendations for Adult Dialysis Patients

Nutrient	*Hemodialysis*	*Peritoneal dialysis*	*Nocturnal hemodialysis*
Energy (kcal/kg SBW or adj IBW)	30–35 $\geq$ 60 years, 35 < 60 years	Same and include dialysate kcal	None established. Use KDOQI for HD
Protein (g/kg SBW or adj IBW)	1.2 ($\geq$50% HBV)	1.2–1.3 ($\geq$50% HBV)	Use KDOQI for HD and individualize
Sodium	2–3 g/day	2–4 g/day. Monitor BP control, fluid balance	2.4–4 g/day. Monitor BP control, fluid balance
Potassium	2–3 g/day	3–4 g/day, unrestricted. Monitor serum levels	Dependent on serum levels
Calcium	$\geq$2000 mg total elemental	Same	Same
Phosphorus	800–1000 mg or <17 mg/kg weight	Same. Adjust to meet protein needs	Mild to unrestricted. Adjust to control serum levels
Fluid	750–1000 mL + UO	Individualize for fluid balance	Individualize for BP and fluid balance
Thiamine	1.2–1.5 mg/day	Same	Same
Riboflavin	1.1–1.3 mg/day	Same	Same
Niacin	20 mg/day	Same	Same
Biotin	30 μg/day	Same	Same
Pantothenic acid	5–10 mg/day	Same	Same
Cobalamin	2–3 μg/day	Same	Same
Pyridoxine	10 mg/day	Same	Same

(Continued)

Table 2
(Continued)

Nutrient	*Hemodialysis*	*Peritoneal dialysis*	*Nocturnal hemodialysis*
Folate	1–10 mg/day[a]	Same[a]	Same[a]
Vitamin C	60–100 mg/day	Same	Same
Vitamin A	None	None	None
Vitamin D	Individual	Individual	Individual
Vitamin E	Optional[b]	Optional[b]	Optional[b]
Vitamin K	None[c]	None[c]	None[c]
Zinc	If needed[d]	If needed[d]	If needed[d]
Copper	None	None	None
Iron	Individualize[e]	Individualize[e]	Individualize[e]
Selenium	None	None	None
Magnesium	None	None	None
Aluminum	None	None	None

SBW: standard body weight; adj IBW: adjusted ideal body weight; HD: hemodialysis; BP: blood pressure; UO: urine output; HBV, high biological value; KDOQI, Kidney Disease Outcomes Quality Initiative.

Source: Data from refs. *(20,44,59,89,91,92)*.

[a] 1 mg is standard recommendation but higher amounts may be needed (See text).

[b] 400 IU may be indicated (See text).

[c] May need supplement if on antibiotics and have poor oral intake.

[d] May be supplemented up to 15 mg elemental zinc.

[e] Varies based on erythropoietin dose.

3.2.2. Peritoneal Dialysis

Protein requirements in PD are higher than for HD due to increased losses during dialysis. PD patients lose between 5 and 15 g protein/day through the peritoneum. During episodes of peritonitis, this loss can increase by 50–100% and may remain elevated for 2–3 weeks after resolution of the infection *(40,49,50)*. Metabolic balance studies found that in clinically stable patients, 1.2–1.3 g protein/kg/day was required to maintain neutral or positive nitrogen balance *(51,52)*. Based on these studies, KDOQI recommends that PD patients consume no less than 1.2 g protein/kg/day and that 1.3 g/kg/day be prescribed with 50% from HBV sources *(10)*. Achieving this intake is sometimes difficult for patients, and the dietitian needs to assess protein intake for adequacy.

3.2.3. PNA—Hemodialysis and Peritoneal Dialysis

Protein equivalent of nitrogen appearance (PNA) is used to estimate protein intake in the clinically stable dialysis patient. Protein is metabolized to nitrogenous products, and in the stable patient, nitrogen waste products removed are equal to protein intake. PNA is calculated from the urea appearance rate. Urea nitrogen appearance rate is calculated from 24-h collections of urea dialysate and urine concentrations in the PD patient or estimated from the interdialytic changes in serum urea nitrogen in the HD patient. The equations can be found in the KDOQI Nutrition Guidelines *(10)*. After PNA is calculated, it can be normalized (nPNA) to body size. PNA may also be normalized to actual edema free body weight; however, this method will give a higher nPNA in malnourished patients with low body weights than in the overweight patient in good nutritional status *(10,49)*. In these cases, normalizing to ideal body weight may be more appropriate *(10)*. If the patient is catabolic, the PNA will be high in proportion to the actual dietary protein intake. If the patient is anabolic, then the reverse will occur *(22)*.

3.2.4. Nocturnal Hemodialysis

There are no established guidelines for protein requirements in NHD, so current recommendations are to implement the KDOQI guidelines for protein and adjust to maintain adequate serum albumin levels (see Table 2) *(44)*. Patients receiving NHD usually have good protein intakes possibly due to an increased clearance of middle and larger molecular weight substances during the longer dialysis *(44)*.

3.2.5. Nutrition Support

Patients who are malnourished at the initiation of dialysis, later develop malnutrition, or have peritonitis will have increased dietary protein and energy requirements *(49)*. Achieving adequate intake may be difficult as oral intake may have spontaneously declined. It may be necessary to liberalize the diet and encourage the use of oral nutritional supplements in the form of modular protein powders or nutritionally complete liquid products. Tube feeding or intradialytic parenteral nutrition (IDPN) may be considered for the HD patient who needs significant nutrition support. The use of intraperitoneal amino acids has been shown to improve nutritional status in malnourished PD patients and is as an option in the patient who has failed to achieve/maintain an adequate protein intake *(53–55)*. Both IDPN and IPN are limited by cost and insurance reimbursement. For more information, refer to Chapter 12.

3.3. Sodium and Fluid

3.3.1. Hemodialysis

Sodium and fluid control are very important in patients receiving maintenance HD. When the glomerular filtration rate (GFR) falls below 15 ml/min/1.73 m^2, the kidneys ability to compensate and excrete sodium declines, leading to sodium retention. Because a patient's GFR declines within the first few months on HD to 1–2 ml/min/1.73 $m^2$and the patient becomes oliguric or anuric, diet and dialysis become the two controlling factors in sodium and fluid balance *(20)*. Excessive fluid and sodium intake between treatments can result in sodium and fluid overload leading to hypertension and cardiac problems such as congestive heart failure (CHF) *(56,57)*. In addition, removal of large interdialytic fluid weight gains to achieve a dry weight may not be possible during one treatment and may cause intradialytic hypotension, cramping, angina, arrhythmias and malaise *(20,58)*. The recommended sodium intake for HD patients is 2–3 g/day while the recommended fluid intake is 750–1000 mL plus urine output and, in general, should not exceed 1500 mL/ day including that in food *(20,58)*. The goal is to minimize interdialytic weight gains and control blood pressure. Ideally, interdialytic weight gains between treatments should not exceed 2–3 kg or 3–5% of the patient's dry weight *(39,58,59)*. The accessibility of high sodium convenience and fast foods makes patient compliance difficult. Patients should be counseled on avoiding high sodium foods, label reading, making appropriate choices when eating out and ways to help control thirst. While high interdialytic weight gains can indicate excessive consumption of sodium and fluid, very low interdialytic weight gains, especially in the elderly, may be an early indicator of poor oral intake *(60)*. The elderly dialysis patient, already at risk for malnutrition, may over restrict both their food and fluid intake. A thorough diet history, along with a review of other nutritional parameters, should be obtained to assess the nutritional adequacy of the diet.

3.3.2. Peritoneal Dialysis

The recommended sodium restriction for patients receiving PD is 2–4 g/day and should be individualized depending on cardiac status, blood pressure control and fluid balance. Sodium is usually easily cleared in PD, with the majority of patients clearing 3–4 g of sodium daily depending on their dialysis prescription *(61,62)*. Excess dietary sodium intake will affect volume retention and blood pressure control *(56,62,63)*. Patients receiving PD who are volume overloaded, hypertensive and unresponsive to management by medication may require a stricter sodium and fluid restriction *(64)*. To correct volume overload, it becomes necessary to use a hypertonic dialysate solution. The frequent use of high dextrose concentrations can damage the peritoneum leading to alterations in and possible loss of ultrafiltration (UF) by the peritoneal membrane *(43,65,66)*. Hypertonic dextrose solutions can also aggravate hypertriglyceridemia, hyperglycemia and hyperinsulinemia and promote weight gain *(40)*. Patients may initially start out with a liberal sodium intake but it may become necessary to reassess the patient for declining residual renal function and uncontrolled blood pressure to determine whether a reduction in sodium intake is indicated.

Fluid removal in PD is regulated by the strength of the dialysate concentration used, that is, the higher the concentration, the more fluid removed *(40,43)*. Patients are taught to monitor their blood pressure, weight and drain volumes to identify if and when any change in their normal dialysis prescription may be needed. UF using glucose occurs quickly and early in the PD exchange, which can present a problem for the volume overloaded patient who requires a higher UF during the long overnight CAPD dwell or the daytime CCPD dwell *(43)*. In these circumstances, icodextrin can be used as an alternative dialysate. The recommended fluid allowance for patients receiving PD should be individualized for each patient with the goal of minimizing the use of hypertonic exchanges. The typical daily fluid allowance should not exceed 2 L; however, patients with a high urine output may need to have additional fluid *(22)*. Fluid allowance may be less depending on cardiac status, blood pressure control, rate of UF and amount of remaining residual renal function.

3.3.3. Nocturnal Hemodialysis

Sodium and fluid restriction in the NHD patient is dependent on fluid balance and blood pressure control. The current Dietary Reference

Intake (DRI) for sodium is 2400 mg/day and is a good starting point for the stable patient, but no formal guidelines for this cohort have been published. The hypotensive patient may require a more liberal sodium prescription. Fluid intake should be individualized based on interdialytic weight gain. A fluid restriction may not be required as long as the patient does not gain more than they can safely remove while maintaining hemodynamic stability. This may vary slightly from patient to patient based on the dialysis prescription, but it is approximately 2–4 L/night *(44)*.

3.4. Potassium

3.4.1. Hemodialysis

As the GFR falls, the kidneys lose their ability to filter potassium, and fecal potassium excretion increases *(20,67)*. Potassium removal during HD averages between 70 and 150 mEq per treatment *(68)*. This will vary depending on the dialyzer clearance and the potassium concentration of the dialysate, that is, the higher the dialysate concentration, the less potassium is cleared. Most patients receiving maintenance HD are placed on a dialysate bath of 2–3 mEq/L, with the standard being 2 mEq/L *(20,68)*. Use of low potassium dialysate (0–1 mEq/L) is rarely used in the outpatient dialysis setting because of the increased risk of cardiac arrest due to hypokalemia *(20)*. Patients with a low pre-dialysis serum potassium (<3.5 mEq/L) will generally require the upper range of the dialysate bath (3–4 mEq/L) especially if oral intake is poor. Hyperkalemia may be categorized as either mild (serum potassium 5.5–6.5 mEq/L) or moderate (>6.5 mEq/L) and may result in cardiac arrhythmias and cardiac arrest *(20,69)*.

The recommended dietary potassium restriction is 2–3 g/day or 40 mg/kg edema free body weight and should be individualized based on serum lab values *(20,33)*. Nutritional counseling regarding food sources of potassium and patient education about the complications of hyperkalemia are important to help the patient avoid elevated potassium levels during the interdialytic period. The primary sources of potassium are fruits, vegetables and dairy along with nuts, seeds, nut butters and dried beans and peas. Patients should also avoid salt substitutes, which contain potassium chloride, and check with their doctor or dietitian before using any herbal products or dietary supplements. For patients who are chronically non-adherent and whose dialysate bath cannot be lowered, a short-term dose of an oral sodium polystyrene sulfonate resin such as kayexalate may be given *(20,33)*. While the primary cause of hyperkalemia may be dietary intake, non-dietary factors such as medications, hyperglycemia, metabolic acidosis

Table 3
Dietary and Non Dietary Causes of Hyperkalemia in Hemodialysis Patients

Dietary causes
Use of salt substitutes
Use of herbal or over-the-counter supplements
Excessive consumption of high potassium foods
Excessive consumption of liquid nutritional supplements
Non-dietary causes
Severe, chronic constipation
Loss of remaining residual renal function
High dialysate potassium concentration
Frequent use of chewing tobacco
Metabolic acidosis: each 0.1 decrease in pH may increase potassium by 0.6–1 mEq/L
Inadequate dialysis
Blood transfusions
Hemolysis of blood sample due to error in blood draw or specimen handling
Hyperglycemia: potassium shifts from cell into serum
Conditions causing release of potassium through tissue destruction such as catabolism, starvation, infection, burns, surgical stress and chemotherapy
Drug interactions: beta blocking agents, spironolactone, angiotensin-converting enzyme inhibitors, cyclosporine and digoxin
Comorbid conditions such as Addison's disease and sickle cell anemia and hypoaldosteronism

Data from refs *(20,67,68)*.

and inadequate dialysis can also lead to elevated serum levels and should be investigated if dietary causes can be ruled out (see Table 3) *(67,68,70,71)*.

3.4.2. Peritoneal Dialysis

Hyperkalemia is less common in patients receiving PD due to the continuous nature of the dialysis, and some patients may not require a potassium restriction *(68,69)*. A 3–4 g/day dietary potassium restriction is recommended and should be adjusted based on laboratory values to maintain serum potassium within normal ranges *(39,41)*. It is advisable to have patients spread their high potassium food choices throughout the day. Patients with diabetes should be cautioned not to treat frequent low blood glucose levels with only high potassium fruit juices, such as orange juice, as this may cause hyperkalemia. Some patients may become hypokalemic due to nausea, vomiting, diarrhea or inadequate dietary intake and may require liberalization of the diet and change in the dialysate *(68,72)*.

3.4.3. Nocturnal Hemodialysis

A potassium restriction in NHD is normally not required unless mid week levels are high *(44,73)*. If a patient skips one night of treatment and has a high interdialytic potassium level, then that patient should be placed on a potassium restriction over their longest skip period *(44)*. Although no specific recommendations are available, clinical experience suggests that a 2.5–3 g potassium diet may be appropriate depending on the degree of hyperkalemia. Patients who are hypokalemic will need to increase their dietary potassium intake or be given a supplement if unable to maintain normal levels by diet alone *(44)*.

3.5. *Calcium/Phosphorus/Vitamin D*

3.5.1. Hemodialysis and Peritoneal Dialysis

Calcium and phosphorus balance in healthy individuals is maintained through interactions between the kidneys, parathyroid glands, bones and intestines. In CKD, the decline in GFR results in increased phosphorus retention and the decreased production of 1-25 dihydroxy-vitamin $D_3[1,25\text{-}(OH)_2D_3]$ or calcitriol, the active form of vitamin D. Decreases in calcitriol can result in reduced intestional calcium absorption, decreased mineral reabsorption/excretion by the kidneys, increased bone turnover and increased parathyroid hormone (PTH) production *(74)*. These metabolic changes, along with hyperphosphatemia, can lead to secondary hyperparathyroidism, renal osteodystrophy and elevated PTH levels. Active vitamin D can be given orally or intravenously during dialysis to correct vitamin D deficiency and suppress PTH production and secretion; however, calcitriol supplementation can also increase intestinal absorption of calcium and phosphorus and increase calcium mobilization from the bone leading to further mineral imbalance *(74)*. Newer analogs are available to suppress PTH with less impact on calcium and phosphorus. Regardless of therapy, all patients should be closely monitored to keep calcium, phosphorus, calcium × phosphorus (Ca × P) product and PTH levels within recommended guidelines.

Both excessive calcium load and hyperphosphatemia are associated with bone disease, vascular and soft tissue calcification and increased cardiovascular mortality because they contribute to an elevated Ca × P product *(75–77)*. Research indicates that a Ca × P product >55 is related to an increase in mortality in CKD patients *(78)*. Calcium load is affected by the amount of dialysate calcium, dietary calcium intake and use of calcium-based medications, specifically phosphorus binders, while serum phosphorus is controlled by adequate dialysate, diet restrictions and use of phosphorus binders *(20,79)*. Dietary calcium intake in dialysis patients is generally about 500 mg/day when high phosphorus foods are limited, the patient is compliant with restrictions and no calcium fortified foods are used. The increasing amount of calcium-fortified foods in the supermarket means that patients must be counseled to carefully read food labels to avoid these items. KDOQI recommends no more than 800–1000 mg phosphorus/day; however, in some cases this may be difficult to achieve as many high phosphorus foods are also high in protein. The phosphorus restriction needs to be adjusted to dietary protein requirements to prevent protein malnutrition. For patients with higher protein needs, calculating phosphorus based on protein requirements (using 10–12 mg phosphorus/gram protein) should provide a reasonable phosphorus restriction *(80)*. For PD patients, it may be more practical to use ≤17 mg phosphorus/kg ideal or adjusted ideal body weight *(59)*. Phosphorus is not currently required to be listed on food labels, and the use of phosphate-based food additives has contributed to hidden sources of phosphorus outside the traditionally known high phosphorus foods. The inorganic phosphates used as food additives are also 100% absorbable as compared to the 50–60% absorption rate from naturally occurring phosphorus *(81)*. Refer to Chapter 14 for a review of bone disease management in CKD.

3.5.2. Nocturnal Hemodialysis

With NHD, weekly phosphate removal is twice that of conventional HD, so phosphate levels are generally normal or low and calcium-phosphorus balance is normal *(73,82,83)*. Phosphorus restriction is not needed in the majority of patients, but it is dependent on the patients' appetite, oral intake and serum phosphorus levels. A large percentage of patients may be able to decrease or discontinue use of phosphate binders. Phosphorus levels should be monitored and the diet individualized. In patients who are hypophosphatemic and unable to increase dietary phosphorus intake, phosphate may be added to the bicarbonate bath *(82)*.

3.6. Lipids

3.6.1. Hemodialysis

There is a high prevalence of lipid abnormalities in the dialysis population, which is a contributing risk factor to cardiovascular disease (CVD). The mortality rate from CVD in patients undergoing maintenance HD is almost 50% *(20)*. HD patients generally have normal or high total cholesterol, low density lipoproteins (LDL) and triglycerides (TG) and normal or low high density lipoproteins (HDL) *(20,84)*. According to the Dialysis Morbidity and Mortality Study, only 20% of patients receiving maintenance HD meet the recommended normal lipid parameters outlined by the National Cholesterol Education Program (NCEP) Adult Treatment Panel (ATP III) *(85)*. The KDOQI recommended therapeutic lifestyle changes (TLC) are covered in detail in Chapter 7. Briefly, it recommends (i) a reduction in saturated fat to <7% of total calories; (ii) a reduction in total fat to 25–35% of total calories with monounsaturated fat providing up to 20% of calories and polyunsaturated fat up to 10%; (iii) total dietary cholesterol <200 mg/day; (iv) increased dietary fiber; and (v) modifications in calories to attain or maintain a desired weight along with exercise and smoking cessation *(85)*.

Before beginning any dietary modifications, the dietitian should assess the patient for any signs of PEM, as addition of a fat-restricted diet can compromise caloric intake which can further compromise nutritional status *(86)*. Patients also encounter difficulty trying to comply with the fat recommendations in addition to a complex renal diet. Dietary modification may be undertaken if the patient is well-nourished; however, pharmacological intervention with lipid lowering medication may be the only intervention used if the patient is unable or unwilling to further modify their diet. Encouraging general recommendations for modifying saturated fat along with smoking cessation and promoting exercise should be suggested.

3.6.2. Peritoneal Dialysis

Patients receiving PD often have elevated LDL, total cholesterol and TG levels along with abnormalities in serum apoproteins thought to be related to the glucose uptake from the dialysate and increased protein losses during dialysis *(22,43,87)*. There is little data as to the effectiveness of diet modification on lipid levels in PD with the exception of maintaining good glycemic control in patients with diabetes *(86,87)*. Strict fat-restricted diets may also compromise protein and calorie intake, and lipid-lowering medications are generally the first line of treatment along with encouraging general TLC such as smoking cessation and exercise. Maintaining adequate protein intake is the primary goal. Minimizing saturated fat and sugar intake should be recommended when possible, but strict dyslipidemia diets are not recommended for patients receiving PD.

3.6.3. Nocturnal Hemodialysis

While there are no specific recommendations for NHD, it would be prudent to encourage patients to follow general therapeutic lifestyle recommendations suggested by the KDOQI guidelines for dyslipidemias *(85)* due to the high risk of CVD in the dialysis population.

4. VITAMINS, MINERALS AND TRACE ELEMENTS

Vitamins, minerals and trace elements are micronutrients required by the body to help with normal metabolism, energy production, cell function and growth and recently have been shown to help lower the risk for CVD and cancer in the general population *(88,89)*. The Food and Nutrition Board of the Institute of Medicine has developed DRIs for most of the micronutrients for the general healthy population. The DRIs for the patient receiving maintenance dialysis have not been established, and the recommendations are the same for HD and PD. Kidney failure can cause impaired or excessive excretion of micronutrients due to the loss of glomerular filtration or impaired tubular function leading to either a deficiency or toxicity *(88)*.

The current recommendations for vitamins and minerals in NHD are the same as those for conventional HD. There is some concern that there may be an increased loss of water-soluble vitamins because patients are dialyzing twice as many days, but to date there has been no conclusive research in this area *(20,44,73)*.

4.1. Water-Soluble Vitamins

There are numerous causes of water-soluble vitamin deficiency in dialysis patients and these include patient anorexia, alterations in metabolism caused by renal failure, the dialysis process, drugs which may affect absorption and the renal diet restrictions *(20,22,90,91)*. Water-soluble vitamins are small, nonprotein-bound substances which are removed by dialysis and may be lost at a rate greater than normal urinary excretion *(92)*. Dialysis membrane pore size, surface area and increased flow rates can adversely affect water-soluble vitamin retention *(88,93)*. Some drugs such as immunosuppresants, anti-convulsants and chemotherapy drugs used to treat comorbid conditions can also interfere with vitamin absorption *(94)*.

Little research has been done to determine the exact requirements for biotin, riboflavin, pantothenic acid, niacin, cobalamin, and thiamine in kidney failure and serum levels are usually normal *(90,88)*. Thiamine deficiency has been reported in dialysis patients and several symptoms of thiamine deficiency, such as CHF with fluid overload (wet beriberi), lactic acidosis and unexplained encephalopathy, can mimic uremic complications making an early diagnosis difficult *(90,88,95)*. Biotin levels have been shown to be normal, but supplementation may help with dialysis-related intractable hiccups *(96,97)*.

The water-soluble vitamins most likely to be deficient are pyridoxine (B-6), folic acid, and vitamin C *(88,92,93)*. Adequate stores of B-6 are necessary for erythropoietin (EPO) to be effective in promoting red blood cell (RBC) formation, and deficiency symptoms include peripheral neuropathy and burning *(88,94)*. Levels of B-6 are low to normal in patients receiving conventional dialysis with higher losses for patients on high flux/high efficiency dialyzers or those receiving EPO therapy *(33,88,93)*. B-6, along with folic acid and vitamin B-12, is a co-factor in homocysteine metabolism, and low levels may result in hyperhomocysteinemia, which can contribute to increased cardiovascular risk in patients receiving maintenance dialysis *(91,98, 99,100)*. Consequently, more than the DRI for pyridoxine is required with the recommended amount being 10 mg/day *(88)*. Although the risk for pyridoxine toxicity is low, an upper limit has been established at 100 mg/day. Doses higher than this can cause a sensory neuropathy *(90,88,92)*.

Adequate serum levels and body stores of folic acid are important for RBC *(88,101)*. Levels are generally normal, but losses may be higher in hemodialysis especially with high flux dialysis. Folic acid supplementation may be indicated for the management of hyperhomocysteinemia as it enhances homocysteine removal *(100,101)*.

Hyperhomocysteinemia can damage vascular endothelial cells, cause lesion formation and promote platelet aggregation leading to CVD *(100)*. High dose folic acid therapy (5–15 mg daily) has been shown to reduce homocysteine levels by 25–30%, but does not normalize the levels. Further research on CVD outcomes are needed before conclusive recommendations on routine supplementation can be made *(88,101)*.

Vitamin B-12 plays a role in folic acid metabolism and the formation of RBC. Levels may be normal as vitamin B-12 may not be removed as much as other water-soluble vitamins *(20,94)*. There have been reports of B-12 deficiency in patients on high flux dialyzers and EPO, and high dose folic acid supplementation may increase requirements

(92,94,91). Vitamin B-12 levels may decrease as the length of time on dialysis increases *(20)*. Because vitamin B-12 relies on intrinsic factor for absorption, it may also be deficient in patients with malabsorption, partial or total gastrectomy, and intestinal resection *(20)*. A water-soluble renal vitamin supplement containing thiamine, niacin, biotin, riboflavin, pantothenic acid and cobalamin is generally recommended at levels to meet the DRIs for the general healthy population to help maintain normal levels (see Table 2).

Vitamin C has antioxidant properties, it regulates iron distribution and storage, and it may help to promote intestinal iron absorption *(33,90,88)*. Vitamin C levels can be low if not supplemented, as it is removed during dialysis, and the patient's dietary intake of vitamin C may be marginal *(88,92)*. The recommended daily dose of vitamin C is 60–100 mg. Higher doses can lead to oxalosis resulting in increased oxalate deposition in soft tissue, which can increase the risk of kidney stones, myocardial infarction, bone disease and shunt failure *(88,92,94)*. Whether or not vitamin C supplementation can help reduce inflammation and oxidative stress is still unclear and research is ongoing *(88)*. Vitamin C has been shown to reduce leg cramps when given with vitamin E; however, the long-term safety has not been established *(33,102)*.

4.2. Fat-Soluble Vitamins

Vitamin A is not removed by dialysis and can accumulate in kidney failure. Vitamin A levels increase with duration of time on dialysis but not frequency, and levels are generally two to five times higher in dialysis patients than in the general population *(20,88,103)*. Elevated vitamin A levels are thought to be due to the lack of removal of retinol binding protein (RBP) by the kidney. Toxicity occurs when the amount of retinol exceeds the binding capability of RBP. Symptoms include hypercalcemia, anemia, hypertriglyceridemia and increased alkaline phosphate levels. These symptoms may mimic uremia, and a diagnosis of toxicity cannot be made without assessing serum vitamin levels *(20, 88,103)*. Vitamin A supplementation is not recommended, and intake from food and/or supplements should not exceed the DRI. Patients with malabsorption syndromes may require supplemental vitamin A, but retinol levels should be checked before initiating therapy and then monitored regularly to avoid toxicity *(88,92,94)*.

Vitamin E is not removed by dialysis, and levels have been reported to be low, normal or high *(20,90,92)*. Low serum levels of vitamin E are thought to be associated with the development of atherosclerosis in the dialysis population, but research in this area and into the role of vitamin E in decreasing oxidative stress is ongoing *(88,92)*. Vitamin E (400 mcg), given with 250 mg vitamin C, has been shown to reduce leg cramps in patients receiving maintenance HD *(102)*. Vitamin E can cause an increased risk of deep vein thrombosis and a vitamin K-like responsive hemorrhagic condition in patients taking an anti-coagulant *(20,90,88)*. Recent research indicates that supplementation ≥400 IU may increase all-cause mortality in the general population, so amounts greater that the DRI is not recommended *(104)*. Vitamin K is not deficient in patients receiving maintenance dialysis; therefore, routine supplementation is not recommended *(20,90,94)*. Patients at possible risk for vitamin K deficiency are those receiving long-term antibiotic therapy, those eating poorly over an extended period of time or those patients on unsupplemented total parenteral nutrition *(20,90,91)*. Excessive vitamin K can interfere with anti-coagulant therapy therefore pts receiving vitamin K supplements should be closely monitored *(20,92,94)*. Adequate levels of Vitamin K may play a role in bone health by decreasing the frequency of bone fractures *(105)*.

4.3. Minerals and Trace Elements

Minerals and trace elements are mainly supplied by diet; however, serum levels can also be affected by environmental exposure, length of dialysis, concentrations of the dialysate, poor nutrition, impaired absorption or age *(90,88,91)*. Many minerals and trace elements are protein bound so uremia itself may alter levels; however, losses during dialysis are probably minimal *(90,88)*. Levels of some minerals and trace elements will be affected by the concentration gradient between the dialysate fluid and the serum.

Zinc deficiency can lead to taste and smell dysfunction, impaired wound healing, decreased resistance to infection and sexual dysfunction *(90,88,106)*. The prevalence of zinc deficiency in patients receiving maintenance dialysis is not known but can occur, and toxicity is rare *(20,88,106)*. Zinc is protein bound so levels may be falsely low when albumin levels are low. Calcium and iron supplements as well as fiber and alcohol intake can interfere with zinc absorption *(90,88,91)*. These factors decrease the reliability of using serum zinc alone as a diagnostic tool for zinc deficiency or for monitoring patients receiving zinc supplements *(88,91)*. Assessment of patient response to supplementation should use a combination of laboratory levels and changes in clinical symptoms. Chronic uremia can impair taste acuity; however, controversy exists as to whether zinc supplementation will improve taste perception *(90,107,108)*. Short-term zinc supplementation may be beneficial for wound healing; however, optimal dose levels have not been determined

(91). Therefore, until more definitive outcomes are determined, the recommended amount of zinc should not exceed the DRI of 15 mg/day.

Selenium levels have been found to be low in patients receiving maintenance dialysis and may be associated with low protein intakes *(90,88,109)*. Selenium supplementation may help improve immune function by decreasing oxidative stress *(90)*. Toxicity is rare but until more clinical evidence is established, regular supplementation is not recommended.

Copper levels have been shown to be normal, and deficiency is rarely seen unless the patient is receiving long-term parenteral nutrition with inadequate supplementation *(90,88)*. Serum copper levels can be affected by excessive intake, or high zinc intakes can interfere with copper absorption *(90,91)*. Copper supplementation is not recommended for the dialysis patient.

Aluminum does not appear to have any essential function in the human body, but it is important because of its potential toxicity in patients receiving maintenance dialysis *(88)*. Age, PTH, citrate, vitamin D and fluorine may increase aluminum absorption in the gut, and the length of time on dialysis may increase aluminum levels in bone *(20)*. Aluminum toxicity in dialysis patients is due to increased uptake and storage and is associated with dialysis encephalopathy syndrome, refractory anemia, and a reduction in bone formation leading to aluminum-induced adynamic bone disease (ABD) or osteomalacia *(20,90,88)*. A primary source of aluminum is aluminum-containing antacids, and patients should avoid long-term use of these medications. However, as stated in the KDOQI Guidelines for Bone Disease, aluminum-based binders may be used on a short-term basis for patients with chronic hyperphosphatemia (refer to Chapter 14).

Magnesium levels in maintenance dialysis patients are generally normal to mildly elevated due to a decrease in gastrointestinal absorption and the fact that high magnesium-containing foods, such as green leafy vegetables and legumes, are generally restricted *(20,33)*. Hypermagnesemia occurs primarily from excessive intake from drinking water, over-the-counter medications such as antacids or laxatives, alcoholism, or some phosphorus binders. Magnesium can be removed using a lower magnesium dialysate (0.75–1.5 mEq/L) *(20)*. Hypermagnesemia can cause hypertension, weakness and arrhythmias. Long-term hypermagnesemia may cause ABD by suppressing PTH secretion *(33,110)*. Deficiency can result in muscle weakness, seizures and arrhythmias and may interfere with the release of PTH leading to hypocalcemia *(20,110)*. Dietary restrictions are not instituted unless a patient develops hypermagnesemia and non-dietary sources are ruled out. Supplementation is not routinely prescribed unless the patient develops a magnesium deficiency.

Iron is essential in many metabolic processes including oxygen transport by hemoglobin and myoglobin. Iron is stored in the liver as ferritin and transported to cells as transferrin. Iron deficiency is common in patients receiving maintenance dialysis because of frequent blood sampling, dialysis-associated losses, decreased availability of ferritin, losses during surgery and gastrointestinal losses *(90,88)*. EPO therapy increases RBC production thereby increasing iron utilization *(90,111)*. Monitoring iron balance using transferrin saturation and ferritin is important to correct deficiency and prevent iron overload. Patients with CKD are unable to obtain adequate iron from diet alone, and oral iron supplementation may not be effective as absorption can be decreased by the presence of an inflammatory state, iron stores, age, sex, timing of supplement administration and simultaneous use of iron inhibiting medications such as calcium *(111–113)*.

Intravenous (IV) iron is usually administered during HD and when oral therapy in PD has failed. IV iron therapy has less gastrointestinal side effects and is more efficient. Iron management in PD is similar to recommendations for HD except that patients are brought in-center for an IV iron infusion *(41,88)*. See Chapter 13 for an in-depth review of anemia management.

5. SUMMARY

Diet is an essential component in the treatment of the maintenance dialysis patient. The diet for stage 5 CKD presents many challenges for the patient receiving maintenance dialysis that include lifestyle changes in food choices and preparation, adjusting to taking new medications, such as phosphorus binders with meals, and possibly having to combine the renal diet with other dietary modifications such as diabetes. The diet must be individualized for each patient to help promote adherence and maintain an optimal intake while balancing protein, sodium, potassium, phosphorus and fluid requirements. There are a variety of educational materials available to help educate both the patient and the professional. The National Kidney Foundation and the American Dietetic Association provide both patient and professional educational material in the area of kidney disease that can be accessed online or by contacting the organizations. Understanding the different types of RRT and their nutritional implications are important in maintaining nutritional status and improving outcomes. Nutrition therapy should involve routine nutrition assessment, ongoing monitoring of biochemical parameters and individualized patient nutrition education.

6. CASE STUDY

CT is a 34-year-old male, self-employed graphic designer who presents for training on CAPD after transferring from HD at another unit following failure of his 13-year-old kidney transplant. He has a history of esophageal reflux and hypertension. CT lives alone and his reported appetite is fair. He notes an 8 lb weight loss prior to starting HD and now complains of occasional indigestion stating food occasionally feels like it is lodged in his throat producing a feeling of choking. Height: 6′2″, weight 182 lb (82.7 kg), estimated dry weight 182 lb (82.7 kg), usual weight 190 lb (86.4 kg). Medications: prenatal vitamin, Tums as needed, 10 mg prednisone, 2 (400 mg) sevelamer (Renagel) as phosphorus binder three times a day with meals. Initial labs: Ca+ 9 mg/dL, P+ 6.3 mg/dL, Na− 141 mEq/L, K+ 4.7 mEq/L, Chol 102 mg/dL, BUN 39 mg/dL, Cr 7.8 mg/dL, Alb 3.4 g/dL, CO_2 22 mEq/L, Kt/V 2, urine output = 1500 mL. CAPD prescription: 7.5 L of 1.5% dextrose plus 2.5 L of 2.5% dextrose. Twenty-four hour diet recall indicates patient consumes an estimated 1800–2000 kcal and 70–80 g protein over three meals with an occasional snack. CT does not use the salt shaker at the table and has had difficulty decreasing his dairy intake. Estimated sodium intake is 2000–2500 mg, potassium 3000 mg and phosphorus 1800 mg.

After 3 months on PD, patient complains of increased dysphagia and decreased oral intake. There has been no change in the CAPD prescription. Weight 180 lb (81.8 kg). Lab: Ca+ 8.6 mg/dL, P+ 3 mg/dL, Na− 142 mEq/L, K+ 2.8 mEq/L, BUN 28 mg/dL, Alb 3.1 g/dL, Total Protein 6.1 g/dL, Chol 110 mg/dL, Kt/V 2.1, urine output 1400 mL. The patient is referred for a barium swallow that indicates a severe hiatal hernia and Schatzki's ring. CT had an esophageal dilation and dysphagia resolved. After 1 year on PD, CT's UF declines leading to poor clearance and fluid retention. He complains of early satiety, nausea, severe anorexia and develops 3+ edema to the knee. Use of diuretics and 4.25% dextrose is unsuccessful. A repeat Kt/V has dropped to 1.4. CT is switched to conventional HD on a 3.5 mEq calcium and 2 mEq potassium bath. After 1 week of HD, his appetite begins to improve and his labs are Ca+ 9.5 mg/dL, P+ 6 mg/dL, Na− 128 mEq/L, K+ 5.8 mEq/L, BUN 44 mg/dL, Cr 8 mg/dL, Alb 3 g/dL, CO_2 24 mEqQ/L, Chol 150 mg/dL, urine output 500 mL, 1+ edema, weight 184 lb (83.6 kg), estimated dry weight 182 lb (82.7 kg).

6.1. Case Questions and Answers

1. What are CT's estimated calorie and protein needs for CAPD? Using the simple formula for estimating glucose, how many calories is CT getting from his dialysate?
2. What are the diet recommendations for CT for CAPD based on his weight and initial labs?
3. Are CT's medications at the start of CAPD appropriate? Any recommendations for changes?
4. At 3 months, what diet and or medication changes are indicated for CT?
5. When CT changes his treatment modality to HD, what dietary modifications are necessary based on weight and laboratory data?

REFERENCES

1. US Renal Data System: USRDS 2005 Annual Data Report. www.usrds.org. Accessed 1/26/2006.
2. Gilbertson DT, Liu J, Xue JF, Louis TA, et al. Projecting the number of patients with end-stage renal disease in the United States to the year 2015. J Am Soc Nephrol 2005; 10:3736–3741.
3. Pupim, LB, Evanson JA, Hakim RM, Ikizler TA. The extent of uremic malnutrition at the time of initiation of maintenance dialysis is associated with the subsequent hospitalization. J Ren Nutr 2003; 13:259–266.
4. Hakim RM, Levin N. Malnutrition in hemodialysis patients. Am J Kidney Dis 1993; 21:125–137.
5. Zeir M. Risk of mortality in patients with end-state renal disease: the role of malnutrition and possible therapeutic implications. Horm Res 2002; 56:30–34.
6. Pupim LB, Caglar K, Hakim RM, Shyr Y, et al. Uremic malnutrition is a predictor of death independent of inflammatory status. Kidney Int 2004; 66:2054–2060.
7. Kopple JD. Effect of nutrition on morbidity and mortality in maintenance dialysis patients. Am J Kidney Dis 1994; 24:1002–1009.
8. Boxall MC, Goodship THJ. Nutritional requirements in hemodialysis. In: Mitch WE, Klahr S, ed. Handbook of Nutrition and the Kidney, 4th ed. Philadelphia: Lippincott Williams & Wilkins, 2005, pp. 218–227.
9. Dombros NV, Digenis GE, Oreopoulos DG. Nutritional markers as predictors of survival in patients on CAPD. Perit Dial Int 1995; 15:S10–S19.
10. National Kidney Foundation. Clinical practice guidelines for nutrition in chronic renal failure. Am J Kidney Dis 2000; 35:S17–S103.
11. Stenvinkel P, Heimburger O, Lindholm B, Kaysen GA, et al. Are there two types of malnutrition in chronic renal failure? Evidence for relationships between malnutrition, inflammation and atherosclerosis (MIA syndrome). Nephrol Dial Transplant 2000; 15:953–960.
12. Kalantar-Zadeh K, Block G, McAllister J, Humphreys MH, et al. Appetite and inflammation, nutrition, anemia and clinical outcomes in hemodialysis patients. Am J Clin Nutr 2004; 80:299–307.

13. Bossola M, Muscaritoli M, Tazza, L, Panocchia N, et al. Variables associated with reduced dietary intake in hemodialysis patients. J Ren Nutr 2005; 15:244–252.
14. Mehrotra R, Kopple JD. Nutritional management of maintenance dialysis patients: why aren't we doing better? Annu Rev Nutr 2001; 21:343–379.
15. Bergstrom J. Appetite in CAPD patients. Perit Dial Int 1995; 16:8–21.
16. Kopple JD, Kalantar-Zadeh K, Mehrotra R. Risks of chronic metabolic acidosis in patients with chronic kidney disease. Kidney Int 2005; 67:S21–S27.
17. Papadoyannakis NJ, Stefanidis CJ, McGeown M. The effect of the correction of metabolic acidosis on nitrogen and potassium balance of patients with chronic renalfailure. Am J Clin Nutr 1984; 40:623–627.
18. Walls J. Effect of correction of acidosis on nutritional status in dialysis patients. Miner Electrolyte Metab 1997; 23:234–236.
19. Kalantar-Zadeh K, Mehrotra R, Fouque T, Kopple JD, Metabolic acidosis and malnutrition inflammation complex syndrome in chronic renal failure. Semin Dial 2004; 17:455–465.
20. Kalantar-Zadeh K, Kopple JD. Nutritional management of patients undergoing maintenance hemodialysis. In: Kopple JD, Massry SG, ed. Kopple and Massry's Nutrional Management of Renal Disease, 2nd edn. Philadelphia: Lippincott Williams & Wilkins, 2004, pp. 433–458.
21. Bossola M, Muscaritoli M, Tazza L, Giungi S, et al. Malnutrition in hemodialysis patients: what therapy? Am J Kidney Dis 2005; 46:371–386.
22. Ikizler TA. Nutrition and peritoneal dialysis. In: Handbook of Nutrition and the Kidney, 5th edn. Philiadelphia: Lippincott, Williams & Wilkins, 2005, pp. 228–244.
23. Gutierrez A, Bergstrom J, Alvestrand A. Protein catabolism in sham-hemodialysis: the effect of different membranes. Clin Nephrol 1992; 38:20–29.
24. Ikizler TA, Flakoll PJ, Parker RA, Hakim RM. Amino acid and albumin losses during hemodialysis. Kidney Int 1994; 46:830–837.
25. Kaplan AA, Halley SE, Lapkin RA, Graeber CW. Dialysate protein losses with bleach processed polysuphone dialyzers. Kidney Int 1995; 47:573–578.
26. Ikizler TA, Pupim LA, Brouillette JR, Levenhagen DK, et al. Hemodialysis stimulates muscle and whole body protein loss and alters substrate oxidation. Am J Physiol Endocrinol Metab 2002; 282:E107–E116.
27. Pupim LB, Flakoll PJ, Ikizler TA. Protein homeostasis in chronic hemodialysis patients. Curr Opin Clin Nutr Metab Care 2004; 7:89095.
28. Bergstrom J. Why are dialysis patients malnourished? Am J Kidney Dis 1995; 26:229–241.
29. Jager KJ, Merkus MP, Husiman RM, Boeschoten EW, et al. Nutritional status over time in hemodialysis and peritoneal dialysis. J Am Soc Nephrol 2001; 12:1272–1279.
30. Rocco MV, Dwyer JT, Larive B, Greene T, et al. The effect of dialysis dose and membrane flux on nutritrional parameters in hemodialysis patients: results of the HEMO study. Kidney Int 2004; 65:2321–2334.
31. McCusker FX, Teehan BP, Thorpe KE, Keshaviah PR, et al. How much peritoneal dialysis is required for the maintenance of a good nutritional state? Canada-USA(CANUSA) peritoneal dialysis study group. Kidney Int Suppl 1996; 56: S56–S61.

32. National Kidney Foundation clinical practice guidelines for peritoneal dialysis adequacy. Am J Kidney Dis 2006; 48:S98–S129.
33. Biesecker R, Stuart N. Nutrition management of the adult hemodialysis patient. In: A Clinical Guide to Nutrition Care in Kidney Disease. 1st edn. Chicago: American Dietetic Asssociation; 2004, pp. 43–55.
34. Rocco MV, Paranandi L, Burrowes JD, Cockram DB, et al. Nutritional status in the HEMO study cohort at baseline. Am J Kidney Dis 2002; 39:245–256.
35. Burrowes JD, Larive B, Cockram DB, Dwyer J, et al. Effects of dietary intake, appetite and eating habits on dialysis and non-dialysis treatment days in hemodialysis patients: cross-sectional results from the HEMO study. J Ren Nutr 2003; 13:191–198.
36. Laville M, Fouque D. Nutritional aspects of hemodialysis. Kidney Int 2000; 58:133–139.
37. Grodstein GP, Blumenkrantz MJ, Kopple JD, Moran JK, et al. Glucose absorption during continuous ambulatory peritoneal dialysis. Kidney Int 1981; 19:564–567.
38. Bodnar DM, Busch S, Fuchs J, Piedmonte M, et al. Estimating glucose absorption in peritoneal equilibration tests. Adv Perit Dial 1993; 9:114–118.
39. McCann L. Pocket Guide to Nutrition Assessment of the Patient with Chronic Kidney Disease, 3rd edn. New York: National Kidney Foundation, 2002.
40. Burkart J. Metabolic consequence of peritoneal dialysis. Semin Dial 2004; 17:498–504.
41. McCann L. Nutrition management of the adult peritoneal dialysis patient. In: A Clinical Guide to Nutrition Care in Kidney Disease, 1st edn. Chicago: American Dietetic Association, 2004, pp. 57–60.
42. Hamburger RJ, Iknaus MA. Icodextrin fulfills unmet clinical need of PD patients: improved ultrafiltration. Dial Transplant 2003; 32:675–684.
43. Teitelbaum I, Burkart J. Peritoneal dialysis. Am J Kidney Dis 2003; 42: 1082–1096.
44. McPhatter LL. Nocturnal home hemodialysis. In: A Clinical Guide to Nutrition Care in Kidney Disease, 1st edn. Chicago: American Dietetic Association, 2004, pp. 233–235.
45. Cianciaruso B, Brunori G, Traverso G, Paanarello G, et al. Nutritional status in the elderly patient with uraemia. Nephrol Dial Transplant 1995; 10:65–68.
46. Chauveau P, Combe C, Laville M, Fouque D, et al. Factors influencing survival in hemo-dialysis patients aged older than 75 years: 2.5 year outcome study. Am J Kidney Dis 2001; 37:997–1003.
47. Wolfson M. Nutrition in elderly dialysis patients. Semin Dial 2002; 15:113–115.
48. Burrowes JD, Dalton S, Backstrand J, Levin NW. Patients receiving maintenance hemo-dialysis with low vs high levels of nutritional risk have decreased morbidity. J Am Diet Assoc 2005; 105:563–571.
49. Heimburger O, Stenvinkel P, Lindholm B. Nutritional effects and nutritional management of chronic peritoneal dialysis. In: Kopple JD, Massry SG, ed. Kopple and Massry's Nutritional Management of Renal Disease, 2nd edn. Philadelphia: Lippincott Williams & Wilkins, 2004, pp. 477–510.
50. Lindholm B. Bergstrom J. Protein and amino acid metabolism in patients undergoing continuous ambulatory peritoneal dialysis (CAPD). Clin Nephrol 1988; 30:S59–S63.

51. Blumenkrantz MJ, Kopple JD, Moran JK, Coburn JW. Metabolic balance studies and dietary protein requirements in patients undergoing continuous ambulatory peritoneal dialysis. Kidney Int 1982; 21:849–861.
52. Kopple JD, Blumenkrantz MJ. Nutritional requirements for patients undergoing continuous ambulatory peritoneal dialysis. Kidney Int 1983; 16:S295–S302.
53. Hagen JM, Boyle CA, Hamburger VE, Sandroni CC, et al. Treatment of malnutrition with 1.1% amino acid peritoneal dialysis solution: results of a multicenter outpatient study. Am J Kidney Dis 1998; 32:761–769.
54. Lindholm B, Bergstrom J. Nutritional aspects on peritoneal dialysis. Kidney Int 1992; 42:S165–S171.
55. Tjiong HL, van den Berg JW, Wattimena JL, Rietveld T, et al. Dialysate as food: combined amino acid and glucose dialysate improves protein anabolism in renal failure patients on automated peritoneal dialysis. J Am Soc Nephrol 2005; 16:1486–1493.
56. Ahmad S. Dietary sodium restriction for hypertension in dialysis patients. Semin Dial 2004; 17:284–287.
57. Wilson, J, Shah T, Nissenson AR. Role of sodium and volume in the pathogeneisis of hypertension in hemodialysis. Semin Dial 2004; 17:260–264.
58. Goldstein-Fuchs, J. Nutrition intervention for chronic renal diseases. In: Handbook of Nutrition and the Kidney, 5th edn. Philadelphia: Lippincott, Williams & Wilkins, 2005, pp. 267–301.
59. Wiggins KL. Guidelines for the nutrition care of renal patients, 3rd edn. Chicago: American Dietetic Association, 2002, pp. 19–35.
60. Testa A, Plou A. Clinical determinants of interdialytic weight gain. J Ren Nutr 2001; 11:155–160.
61. Moncrief JW. Continuous ambulatory peritoneal dialysis. Contrib Nephrol 1979; 17:139–145.
62. Nolph KD, Sorkin MI, Moore H. Autoregulation of sodium and potassium removal during continuous ambulatory peritoneal dialysis. Trans Am Soc Artif Intern Organs 1980; 26:334–338.
63. Wang X, Axelesson J, Lindholm B, Wang T. Volume status and blood pressure in continuous ambulatory peritoneal dialysis patients. Blood Purif 2005; 23: 373–378.
64. Lameire N, Van Biesen W. Hypervolemia in peritoneal dialysis patients. J Nephrol 2004; 17:58–66.
65. Krediet RT, van Westrhenen R, Zweers MM, Struijk DG. Clinical advantages of new peritoneal dialysis solutions. Nephrol Dial Transplant 2002; 17:16–18.
66. Davies SJ, Phillips L, Naish PF, Russel GI. Peritoneal glucose exposure and changes in membrane transport with time on peritoneal dialysis. J Am Soc Nephrol 2001; 12:1046–1051.
67. Beto J, Bansai VK. Hyperkalemia: evalutating dietary and non-dietary etiology. J Ren Nutr 1992; 2:28–29.
68. Musso CG. Potassium metabolism in patients with chronic kidney disease. Part II: patients on dialysis (stage 5). Int Urol Neprhol 2004; 36:469–472.
69. Beto J. Which diet for renal failure: making sense of the options. J Am Diet Assoc 1995; 95:898–903.
70. Constantino J, Roberts C. Life-threatening hyperkalemia from chewing tobacco in a hemodialysis patient. J Ren Nutr 1997; 7:106–108.

71. Ifudu O, Reydel K, Vlacich V, Delosreyes G, et al. Dietary potassium is not the sole determinant of serum potassium concentration in hemodialysis patients. Dial Transplant 2004; 33:684–688.
72. Rostand SG. Profound hypokalemia in continuous ambulatory peritoneal dialysis. Arch Intern Med 1983; 143:377–378.
73. Pierratos A, Ouwendyk M, Francoeur R, Vas S, et al. Nocturnal hemodialysis: three year experience. J Am Soc Nephrol 1998; 9:859–868.
74. Goodman WG. Calcium, phosphorus and vitamin D. In: Handbook of Nutrition & the Kidney, 5th edn. Philadelphia: Lippincott, Williams & Wilkins 2004, pp. 47–69.
75. Qunibi WY. Consequences of hyperphosphatemia in patients with end-stage renal disease (ESRD). Kidney Int 2004; 66:S8–S12.
76. Coladonato JA. Control of hyperphosphatemia among patients with ESRD. J Am Soc Nephrol 2005; 16:S107–S114.
77. Rodrigues-Benot A, Martin-Malo, A, Alvarez-Lara A, Rodriguez M, et al. Mild hyper-phosphatemia and mortality in hemodialysis patients. Am J Kidney Dis 2005; 46:68–77.
78. Block GA, Klassen PS, Lazarus M, Ofsthun N, et al. Mineral metabolism, mortality and morbidity in maintenance hemodialysis. J Am Soc Nephrol 2004; 15:2208–2218.
79. McCann L. Total calcium load in dialysis patients: an issue of concern for dietitians. Dial Transplant 2004; 33:282–289.
80. National Kidney Foundation. Clinical practice guidelines: bone metabolism and disease in chronic renal failure. Am J Kidney Dis 2003; 42:S12–S143.
81. Murphy-Gutekunst L. Hidden phosphorus in popular beverages: Part I. J Ren Nutr 2005; 15:e1–e6.
82. Lockridge RS, Spencer M, Craft V, Pipkin M, et al. Nightly home hemodialysis: five and one- half years of experience in Lynchburg, Virginia. Hemodial Int 2004; 8:61–69.
83. Schulman G. Nutrition in daily dialysis. Am J Kidney Dis 2003; 41:S112–S115.
84. Pagenkemper JJ. Dyslipidemias in chronic kidney disease. In: A Clinical Guide to Nutrition Care in Kidney Disease, 1st edn. Chicago: American Dietetic Association, 2004, pp. 107–119
85. National Kidney Foundation. KDOQI clinical practice guidelines for managing dyslipidemiaa in chronic kidney disease patients. Am J Kidney Dis 2003; 41: S1–S91.
86. Saltissi D, Morgan C, Knight B, Chang W, et al. Effect of lipid-lowering dietary recommendations on the nutritional intake and lipid profiles of chronic peritoneal dialysis patients and hemodialysis patients. Am J Kidney Dis 2001; 37: 1209–1215.
87. Fried L, Hutchison A, Stegmayr B, Prichard S, et al. Recommendations for the treatment of disorders in patients on peritoneal dialysis. Perit Dial Int 1999; 19:7–21.
88. Masud T. Trace elements and vitamins in renal disease. In: Handbook of Nutrition & the Kidney, 5th edn. Philadelphia: Lippincott, Williams & Wilkins, 2004, pp. 196–217.
89. Fletcher RH, Fairfield KM. Vitamins for chronic disease prevention in adults. JAMA 2002; 287:3127–3129.

90. Kalantar-Zadeh K, Kopple JD. Trace elements and vitamins in maintenance dialysis patients. Adv Ren Replace Ther 2003; 10:170–182.
91. Wiggins KL. Renal Care: Resources and Practical Applications. Chicago: American Dietetic Association, 2004, pp. 39–60.
92. Makoff R, Gonick H. Renal failure and the concomitant derangement of micronutrient metabolism. Nutr Clin Pract 1999; 14:238–246.
93. Kasama R, Koch T, Canals-Navas C, Pitone JM. Vitamin B-6 and hemodialysis: the impact of high-flux/high efficiency dialysis and review of the literature. Am J Kidney Dis 1996; 27:680–686.
94. Rocco MV, Makoff R. Appropriate vitamin therapy for dialysis patients. Semin Dial 1997; 10:272–277.
95. Hung SC, Hung SH, Tarng DC, Yang WC, et al. Thiamine deficiency and unexplained encephalopathy in hemodialysis and peritoneal dialysis patients. Am J Kidney Dis 2001; 38:941–947.
96. Jung U, Helbich-Endermann M, Bitsch R, Schenider BR, et al. Are patients with chronic renal failure (CRF) deficient in biotin and is regular biotin supplementation required? Z Ernahrungswiss 1998; 37:363–367.
97. Jones WO, Nidus BD. Biotin and hiccups in chronic dialysis patients. J Ren Nutr 1991; 1:80–83.
98. Shemin D, Bostom AG, Selhub J. Treatment of hyperhomocysteinemia in end-stage renal disease. Am J Kidney Dis 2001; 38:S91–S94.
99. Obeid R, Kuhlmann MK, Kohler H. Herrmann W. Repsonse of homocysteine cystathionine, and methylamalonic acid to vitamin treatment in dialysis patients. Clin Chem 2005; 51:196–201.
100. Clement L. Homocysteine: the newest uremic toxin? Renal Nutr Forum 2003; 23:1–4.
101. Schaefer RM, Tschner M, Kosch M. Folate metabolism in renal failure. Nephrol Dial Transplant 2002; 17:24–27.
102. Khajehdehi P, Mojerlou M, Behzadi S, Rais-Jalali GA. A randomized, double-blind, placebo-controlled trial of supplementary vitamins E, C and their combination for treatment of haemodialysis cramps. Nephrol Dial Transplant 2001; 16:1448–1451.
103. Muth I. Implications of hypervitaminosis A in chronic renal failure. J Ren Nutr 1991; 1:2–8.
104. Miller ER, Pastor-Barriuso R, Dalal D, Riemersma RA, et al. Meta-analysis: High dose vitamin E supplementation may increase all-cause mortality. Ann Intern Med 2005; 142:37–46.
105. Kohlmeier M, Saupe J, Shearer MJ, et al. Bone health of adult hemodialysis patients is related to vitamin K status. Kidney Int 1997; 51:1218–1221.
106. Erten Y, Kayata M, Sezer S, Ozdemir FN. Zinc deficiency: Prevalence and causes in hemodialysis patients and effect on cellular immune response. Transplant Proc 1998; 30:850–851.
107. Matson A, Wright M, Oliver A, Woodrow G, et al. Zinc supplementation at conventional doses does not improve the disturbance of taste perception in hemodialysis patients. J Ren Nutr 2003; 13:224–228.
108. van der Eijk I, Allman Farinelli MA. Taste testing in renal patients. J Ren Nutr 1997; 7:3–9.

109. Smith AM, Temple K. Selenium metabolism and renal disease. J Ren Nutr 1997; 7:69–72.
110. Ng AHM, Hercz G, Kandel R, Grynpas MD. Association between fluoride, magnesium, aluminum and bone quality in renal osteodsytrophy. Bone 2003; 34:216–224.
111. Skikine BS, Ahluwalia B, Fergusson B, Chonko A, et al. Effects of erythrpoieten therapy on iron absorption in chronic renal failure. J Lab Clin Med 2000; 135: 452–458.
112. Cook JD, Dassenko SA, Whittaker P. Calcium supplementation: effect on iron absorption. Am J Clin Nutr 1991; 53:106–111.
113. Hallberg L, Hulten L, Gramatkovski E. Iron absorption from the whole diet in men: how effective is the regulation of iron absorption? Am J Clin Nutr 1997; 66:347–356.

10 Transplantation

Pamela S. Kent

LEARNING OBJECTIVES

1. Discuss nutrient recommendations for adult kidney transplantation during acute and chronic post transplant phases.
2. Address post-transplant complications and their diagnosis and management.
3. Identify side effects of immunosuppressive therapy and recommend nutritional interventions.

Summary

Kidney transplantation is the preferred method of treatment for patients with stage 5 chronic kidney disease (CKD), and it is the most common solid organ transplant. Nutritional care of the kidney transplant recipient is a dynamic process. It involves the integration of clinical guidelines related to the care of CKD. A complete evaluation of the patient's nutritional status before transplantation is imperative to identify deficits and correct abnormalities. Continual reassessment of the nutrition goals and efficacy of medical nutrition therapy (MNT) will enable the adjustment of nutrition priorities during different phases of kidney transplant care. The registered dietitian is a valued member of the transplant team and, by providing timely MNT, he/she can enhance the transplant recipient's ability to prevent or minimize the nutrition-related complications.

Key Words: Kidney transplantation; nutrition; diabetes; cardiovascular disease; immunosuppressants; medical nutrition therapy.

From: *Nutrition and Health: Nutrition in Kidney Disease*
Edited by: L. D. Byham-Gray, J. D. Burrowes, and G. M. Chertow

1. INTRODUCTION

Kidney transplantation has become a viable option for patients with chronic kidney disease (CKD). Although the number of deceased donor transplants has been rising in recent years, there has been an increase in living-related and non-related organ donation. In 2005, there were 9357 deceased donor transplants and 6563 living donor transplants; however, there are currently 66,402 individuals on the United Network for Organ Sharing (UNOS) waiting list *(1)*. Insufficient organ donation accounts for the discrepancy between the number of recipients and candidates. Advances in surgical technique and immunosuppressive agents have led to significant improvements in morbidity and mortality.

Adequate nutrition is essential for the well-being of kidney transplant patients. To minimize nutrition depletion and optimize nutritional status, a complete and thorough evaluation by a registered dietitian (RD) should be performed. The evaluation should integrate the patient's complex medical condition related to CKD and the impact of ongoing therapeutic interventions on the patient's nutritional status. Providing adequate nutrition and reducing the long-term side effects are essential for graft survival in kidney transplant recipients.

There are three treatment phases to consider when providing care to kidney transplant recipients *(2)*. In the pre-transplant phase, the goal is to meet current education and nutrition needs, optimize the patient's nutritional status, and assist the patient in achieving desirable body weight based on the transplant facility guidelines. In the acute post-transplant phase (up to 8 weeks after transplantation), the goal is to support the increased metabolic demands of surgery and high-dose immunosuppressive therapy.

During the chronic post-transplant phase, the goal is to manage long-term complications related to immunosuppressive therapy, especially in individuals genetically predisposed to diabetes and cardiovascular disease *(3,4)*.

2. THE PRE-TRANSPLANT PHASE

Many of the chronic diseases that have caused kidney failure can negatively affect nutritional status and lead to malnutrition and vitamin and mineral deficiencies. A complete evaluation of the patient's nutritional status by a RD is imperative to optimize nutritional status, identify deficits, and, when possible, to correct these before kidney transplant surgery. It is important to remember that the candidates for kidney transplantation have been subjected to the effects of a chronic disease, and that not all deficiencies identified can be corrected without organ replacement.

When assessing the nutritional status of patients undergoing kidney transplantation, a combination of objective and subjective parameters should be used. The pre-transplant evaluation should include a medical and dietary history, medication review, anthropometric data, biochemical indices of nutritional status, evaluation of gastrointestinal (GI) issues, and dietary supplement intake including vitamin, mineral, and herbal formulations. The dietitian can then assess the nutritional needs and develop a nutrition care plan. The nutritional recommendations will be individualized based on the type of renal replacement therapy.

The impact of nutrition on kidney transplant outcomes has not been well defined in the literature. The effect of obesity on kidney transplant outcomes continues to be controversial. Obesity seems to influence delayed graft function (DGF), graft, and patient survival. A body mass index (BMI) $>35\,kg/m^2$ is significant for greater post-transplant complications, especially new onset transplant diabetes mellitus (DM), wound complications, and post-transplant weight gain *(5,6)*. After kidney transplantation, weight typically increases due to improved appetite and reversal of the uremic state. Attempts to identify individuals at risk for transplant weight gain should be made in the pre-transplant evaluation, when a comprehensive approach to prevention can be implemented.

Due to the limited availability of organs, the selection of transplant candidates who are likely to have a positive outcome remains an important issue. Both deceased donor transplant and living donor transplant recipients should meet specific weight criteria prior to transplantation. These criteria vary by medical facility, but the scientific literature indicates that the patient should have a BMI greater than 22, but less than 35 *(5,6)*.

Weight reduction may be required if the prospective candidate fails to achieve the target weight based on the criterion set by the transplant center. Successful weight loss can be achieved by targeting the complex behavioral, nutritional, and medical aspects of obesity. Future clinical trials should evaluate if weight modification can favorably impact outcomes in underweight and obese kidney transplant patients and if pharmacologic agents or surgical treatment for obesity are potential options.

3. THE ACUTE POST-TRANSPLANT PERIOD

3.1. *Nutritional Requirements*

The nutrition recommendations for adult kidney transplantation are summarized in Table 1. Nutrient requirements are increased during the acute post-transplant phase due to increased metabolic demands from surgery and high-dose immunosuppressive therapy.

3.1.1. Protein

The increase of nitrogen losses in the immediate postoperative period may be due to surgical stress, administration of large doses of corticosteroids, muscle catabolism, and pre-existing malnutrition *(7)*. Daily protein recommendations for patients with functioning grafts or those requiring temporary dialytic support are estimated at 1.3–2 g/kg of dry body weight or adjusted weight *(8)*.

3.1.2. Energy

Caloric requirements during the post-transplant recovery phase are higher than long-term caloric needs. Energy needs may be increased due to fever, infection, surgical stress, or high-dose corticosteroid therapy. Caloric requirements can be estimated by using 30–35 kcal/kg dry body weight or weight adjusted for obesity *(9)*. This recommendation seems adequate for maintaining or achieving neutral nitrogen balance. Estimation of energy needs can also be calculated using the Harris–Benedict equation to determine basal energy requirements, multiplied by a stress factor of 1.3–1.5 *(10)*.

3.1.3. Carbohydrate

Glucose intolerance is common before and after transplantation. New onset diabetes after transplantation (NODAT) can result from immunosuppressive treatment, surgical stress, genetic predisposition, obesity, increased age, and infection *(11)*. Hyperglycemia may be short-lived during this acute period and can be managed with exogenous insulin or oral hypoglycemic agents, as well as a carbohydrate-controlled diet. Complex carbohydrates may provide up to 70% of the estimated energy requirements *(10)*.

3.1.4. Fat

During the acute postoperative period, diet modification for fat is not warranted. The amount of fat is only limited by the appropriate energy level and often provides up to 30% of the total energy, with

Table 1
Nutrient Recommendations for Adult Kidney Transplantation during Acute and Chronic Post Transplant Phases

Nutrient	*Acute Phase*	*Chronic-post transplant phase*	*Comments*
Protein	1.3–2 g/kg dry body weight or adjusted body weight	0.8–1 g/kg dry body weight or adjusted body weight	• Protein catabolic rate is increased due to surgical stress and high-dose corticosteroids • Adequate amounts of protein are required for wound healing and to prevent infection • Additional losses are possible due to surgical drains, wounds, and indication for dialysis
Calories	130–150% of calculated basal energy expenditure	130–150% of calculated basal energy expenditure	• The upper range of calories is recommended for underweight patients; the lower range is recommended for overweight patients
	30–35 kcal/kg dry body weight or adjusted body weight	Maintain a healthy body weight	• Caloric needs may further increase in the presence of fever, infections, surgical stress, and high-dose corticosteroid therapy
Carbohydrate	50–70% of non protein calories	45–50% total calories. Emphasize complex carbohydrates	• Serum glucose levels may be increased because of medications (corticosteroids, cyclosporine, and tacrolimus), metabolic stress, or infection • Treat hyperglycemia with insulin dosed on a sliding scale; add other hyperglycemic agents as needed. • Initiate CHO-controlled diet

Fat	30–50% of non-protein calories, 10% saturated fat	<30% total calories, 7–10% saturated fat	• Dyslipidemia is of considerable concern but is not aggressively treated until the chronic post transplant period
Sodium	2–4 g/day; unrestricted if HTN/edema absent	Individualize	• Sodium should only be restricted in the acute postoperative period in the presence of poor allograft function or post-transplant HTN
Fluid	Individualize. May need to be limited if patient requires dialysis or if edema is present	Individualize	• Fluids should only be restricted in the acute postoperative period in the presence of poor allograft function or post-transplant HTN
Potassium	2–4 g/day	Individualize	• Serum levels may increase with administration of tacrolimus, cyclosporine, or potassium-sparing diuretics, renal insufficiency, or metabolic acidosis • Serum levels may decrease with administration of potassium-wasting diuretics or amphotericin
Phosphorus	Individualize	Individualize	• Serum levels may increase with renal insufficiency • Serum levels may decrease with administration of corticosteroids
Magnesium	Individualize	Individualize	• Serum levels may increase with renal insufficiency • Serum levels may decrease with administration of cyclosporine, tacrolimus, diuretics, and diarrhea

CHO, carbohydrate; HTN, hypertension.

intake of saturated fat limited to 10% of energy *(10)*. Dyslipidemia is usually addressed during the chronic post-transplant period. Nutrition education in the acute period should introduce the importance of a "heart-healthy" lifestyle.

3.1.5. Sodium

Sodium intake should be limited in the acute postoperative phase in the presence of poor allograft function or post-transplant hypertension. Some immunosuppressive medications can cause the development of hypertension and fluid retention, which may necessitate a 2–4 g/day sodium restriction *(12)*.

3.1.6. Potassium

Dietary potassium restriction of 2–4 g is indicated if serum potassium levels are elevated. Poor graft function, as well as impaired potassium excretion associated with cyclosporine immunosuppression and potassium-sparing diuretics, may contribute to hyperkalemia *(13)*. Hypokalemia may also occur in kidney transplant recipients due to potassium-wasting diuretics.

3.1.7. Vitamins, Minerals, and Trace Elements

Vitamin and mineral supplementation in the acute post-transplant phase should be individualized, depending on preexisting medical conditions and dietary intake.

Information about the effect of kidney transplantation on trace elements is scarce. Because patients receiving maintenance dialysis may be have abnormal zinc metabolism, it may beneficial to monitor these levels *(14,15)*.

Hypomagnesemia has been reported in kidney transplant recipients treated with cyclosporine A (CsA), and serum magnesium levels should be monitored regularly and supplemented if indicated *(16)*.

3.1.8. Herbals

Herbal products and botanicals are very popular in the United States *(17,18)*. Some herbal preparations like Ginseng are promoted to enhance the immune system which theoretically may increase the risk of organ rejection. Others, like St. John's Wort, can cause drug–drug interactions, requiring higher doses of immunosuppression to maintain trough levels *(19)*. The use of herbals and botanicals in the transplant population is contraindicated (refer to Chapter 22).

4. COMMON PROBLEMS POST-TRANSPLANT IN THE ACUTE CARE SETTING

4.1. Hyperglycemia

The prevalence of DM is greater among patients with solid organ transplants than in the general population. Insulin resistance is a common post-surgical complication due immunosuppressive therapy with corticosteroids, tacrolimus, and cyclosporine *(20)*. Prior personal or family history of DM and obesity predisposes a patient to insulin resistance, especially during the period of high-dose corticosteroids used during the initial postoperative days or during treatment for acute rejection *(61)*. Oral hypoglycemic agents or insulin therapy combined with nutrition intervention may be necessary for glycemic control.

4.2. Gastrointestinal Issues

There is an increased incidence of GI complications in kidney transplant recipients due to infections, mucosal injury, and ulceration, which can manifest anywhere in the GI tract from the mouth to the anus. Oral lesions may be caused by viral and fungal infections or can be a side effect from cyclosporine or sirolimus. The most common esophageal disorder is fungal esophagitis, which is caused by candida *(21)*. Transplant programs typically prescribe prophylactic antifungal agents to prevent fungal infections. Other gastroduodenal disorders are caused by cytomegalovirus and herpes simplex infection *(22,23)*. Bacterial infections of the GI tract are also diagnosed in the transplant recipient and include *Clostridium difficile* colitis. Symptoms include diarrhea and abdominal tenderness and usually respond to appropriate medical treatment *(21)*. Diarrhea is a frequent disorder, which may be caused by pathogen microorganisms or by immunosuppressive agents.

Several factors can be responsible for ulcer formation after transplant and include the administration of corticosteroids and a prior history of peptic ulcer disease. Although the incidence is low, the stress of surgery, nonsteroidal anti-inflammatory drugs, corticosteroids, or immunosuppression may impact the formation of ulcers after renal transplantation *(24)*. One study documented a 70% incidence of *Helicobacter pylori* and a 65% incidence of gastritis *(25)*. Transplant candidates need to be evaluated for *H. pylori* if there is no other obvious indication for peptic ulcer disease.

Some transplant programs prescribe prophylactic histamine-2 receptor blockers.

4.3. Hypophosphatemia

Abnormalities of serum phosphate levels occur even with stable, functioning allografts. Hypophosphatemia can occur when the glomerular filtration rate normalizes in transplant recipients with preexisting hyperparathyroidism *(26)*. Other causes of hypophosphatemia include the use of phosphate binders or calcium supplements with meals, or an inadequate phosphorus intake *(27)*. If phosphorus levels remain suboptimal after correcting for obvious causes, supplementation may be indicated. Some phosphorus supplements such as Neutraphos-K contain appreciable amounts of potassium and require close monitoring of serum potassium levels *(27)*.

4.4. Hyperkalemia

Hyperkalemia can develop because of poor graft function, cyclosporine therapy, and cell lysis related to the catabolic effect of both surgery and corticosteroids. Nutritional interventions should include restriction of potassium intake and the provision of adequate calories and protein to minimize catabolism *(28)*. An increased dietary phosphorus intake, which may be prescribed to treat hypophosphatemia, can also contribute to hyperkalemia.

4.5. Inadequate Intake

Successful kidney transplantation corrects anorexia and uremia typically observed in the patient with CKD. Kidney transplant recipients normally consume an oral diet. There are a few patients whose post-surgical experience is more complicated due to a temporary loss or a slow return of appetite. In these cases, nutritional intervention to facilitate surgical recovery is the priority and may include nutritional supplementation and/or specialized nutrition support.

Nutritional support is rare after kidney transplantation. If a patient is unable to meet his/her metabolic demands orally and has a functional GI tract, standard high-nitrogen tube feedings should be initiated. Nutrient-dense enteral formulas may be indicated in the event of poor allograft function with volume overload. If parenteral nutrition is required, the formula should be tailored to consider allograft function, urine output, electrolytes, and type of renal replacement therapy (refer to Chapter 11).

5. THE CHRONIC POST-TRANSPLANT PHASE

The nutrition goals of the chronic post-transplant phase are to provide adequate nutrition, prevent infection, and manage long-term nutritional complications. During the chronic post-transplant phase, overnutrition may lead to transplant complications, including obesity, hyperlipidemia, DM, and hypertension. Providing adequate nutrition and reducing the long-term side effects are essential for allograft survival among kidney transplant recipients.

5.1. Nutritional Requirements

Kidney transplant recipients without complicating factors can enjoy a dietary regimen consistent with guidelines for the healthy population, addressing specific issues as needed. In general, the diet is moderate in sodium and based on the patient's blood pressure and presence of edema. Sodium intake can range from 2 to 4 g or a no-added-salt diet depending on the medical condition of the patient. The dietary recommendations for fat intake should be less than 30% of total calories. It is prudent to have a protein intake of 0.8–1 g/kg of dry body weight or adjusted body weight. Energy requirements can be calculated using the Harris–Benedict equation to determine basal energy requirements, multiplied by a stress factor of 1.3 and modified based on weight management.

5.1.1. Calcium

Preexisting secondary hyperparathyroidism after kidney transplantation has been cited in the literature *(29–31)*. Post-transplant hyperparathyroidism is present in 43% of long-term renal transplant patients with normal serum creatinine *(32)*. The 1,25-dihydroxy vitamin D levels can be suboptimal in transplant candidates and remain abnormal for at least 6 months following transplantation *(33)*.

Long-term corticosteroid therapy with resultant inhibition of bone formation and the stimulation of bone resorption can also cause osteoporosis *(31)*. Corticosteroids decrease the intestinal absorption of elemental calcium by approximately 42%. The amount of calcium excreted in the urine is also increased with corticosteroid therapy *(34)*. Some transplant centers individualize calcium and vitamin D supplementation to maintain bone mineral metabolism.

5.1.2. Vitamins and Minerals

Typically, vitamin supplementation is not required after kidney transplantation if patients regularly consume a balanced and adequate diet. However, some transplant programs recommend a general multi-vitamin with supplemental minerals.

6. PHARMACOLOGY UPDATE

Advances in immunosuppressive therapies have greatly improved the success of kidney transplantation. Immunosuppression is used to prevent rejection and maintain long-term graft survival. The typical immunosuppressive regimen consists of a calcineurin inhibitor (cyclosporine and tacrolimus), an anti-proliferative agent (azathioprine and mycophenolate mofetil), and corticosteroids (prednisone) *(39)*. In multi-drug therapy, each drug mediates the immunocompetence cascade at a different point. The mechanism of immunosuppression is to inhibit the adaptive immune response while allowing nonspecific immune functions to remain intact *(35)*. Immunosuppressive agents also have non-immunologic side effects. The common use of multi-therapy regimens is an attempt to use lower doses of individual agents to minimize associated side effects. Unfortunately, this therapy has yet to be perfected and standard immunosuppressive regimens are being challenged as novel immunosuppressive regimens are being investigated (Table 2).

6.1. Cyclosporine A (Sandimmune/Neoral)

CsA (Novartis Pharmaceuticals Corporation, St. Louis, MO 63166-6556) is a calcineurin inhibitor that was considered the gold standard for maintenance immunosuppression *(36)*. CsA selectively inhibits adaptive immune responses, but has several side effects, which may include gingival hyperplasia, GI disturbances, hyperglycemia, hyperkalemia, hypophosphatemia, hypomagnesemia, hepatotoxicity, and nephrotoxicity. This drug is absorbed in the upper small intestine and can be affected by food, drug–drug interactions, bile flow, and the lipoprotein and hematocrit status *(37)*. Due to the nephrotoxic side effects, CsA peak and trough levels and kidney function are closely monitored. Neoral (Novartis Pharmaceuticals Corporation, 63166-6556) is a microemulsion preparation of CsA and has better absorption because it is not dependent on bile. There are usually fewer side effects with Neoral.

6.2. Tacrolimus (Prograf/FK506)

Tacrolimus (Fujisawa Healthcare, Inc., Chantilly, VA) is also a calcineurin inhibitor but is 10–100 times more potent than CsA *(38)*. Whereas both tacrolimus and CsA inhibit interleukin-2 (IL-2) synthesis and release, each displays a different mechanism at the cellular level.

Table 2
Nutritional Side Effects of Immunosuppressive Medications

Medication	*Complication*	*Suggested interventions*
Anti-Lymphocyte globulin[a] (Thymoglobulin)	• Fever and chills • Increased risk of infection, profound leucopenia, thrombocytopenia	• Provide nutrient-dense foods • Ensure patient is receiving adequate protein
Azathioprine (Imuran)[b]	• Nausea/vomiting • Diarrhea • Mucositis • Macrocytic anemia • Pancreatitis	• Try antiemetic medications if vomiting does not subside • Review medications and substitute for potential medications that may be causing diarrhea, make sure patient is receiving adequate fluid to replace losses • Provide foods that will not irritate the throat • Make sure patient is receiving adequate folate intake • Initiate parenteral nutrition if pancreatitis is severe
Basiliximab (Simulect)[c]	• None reported	
Corticosteroids, (prednisone[d], prednisolone[d], Solu-Medrol[e]	• Hyperglycemia • Sodium retention • Ulcers • Osteoporosis • Hyperphagia • Impaired wound healing and increased risk of infection	• Monitor blood sugar and need for hypoglycemic agents and CHO-controlled diet • Limit high sodium foods • Ensure adequate intake of calcium, vitamin D, vitamin A, vitamin C, and zinc • Behavior modification to prevent overeating

	• Hypertension • Pancreatitis (rare)	
Cyclosporine (Neoral, Sandimmune)[c]	• Hyperkalemia • Hypomagnesemia • Hypertension • Hyperglycemia • Dyslipidemia • Gingival hyperplasia • GI disturbances • Hypophosphatemia • Hepatoxicity • Nephrotoxicity	• Restrict high potassium foods • Limit high sodium foods • Monitor blood sugar and need for hypoglycemic agents and CHO-controlled diet • Limit fat to <30% of total calories during the long-term phase
Daclizumab (Zenapax)[f]	• None reported	
Muromonab—CD3 (OKT3)[g]	• Nausea, vomiting • Diarrhea • Anorexia • Fever, chills, myalgias	• Try antiemetic medications • Review medications and substitute for those that may cause diarrhea; make sure the patient receives adequate fluid to replace losses • Offer frequent meals of nutrient-dense foods
Mycophenolate mofetil (CellCept)[f]	• Diarrhea • Nausea • Vomiting	• Review medications and substitute for those that may cause diarrhea; make sure the patient receives adequate fluid to replace losses • Hyperlipidemia • GI disorders (constipation, diarrhea, nausea, vomiting, dyspepsia)

(Continued)

Table 2
(Continued)

Medication	*Complication*	*Suggested interventions*
Sirolimus (Rapamycin)[h]	• Hypokalemia • Increased liver function tests • Delayed wound healing	• Limit fat intake to <30% of total calories as fat during long-term phase, maintain a healthy weight • Monitor for adequate nutrient intake
Tacrolimus (Prograf, FK506)[i]	• Nausea, vomiting • Hyperkalemia • Hyperglycemia • Abdominal distress • Diarrhea • Constipation • Hypophosphatemia • Hypomagnesemia	• Try antiemetic medications • Limit high potassium foods • Monitor blood sugar and need for hypoglycemic agents and CHO-controlled diet • Monitor oral intake; consider alternate methods of nutrition support if intake suboptimal

CHO, carbohydrate; GI, gastrointestinal.
[a] Genzyme, Cambridge, MA 02142.
[b] Faro Pharmaceuticals, San Diego, CA, 92121.
[c] Novartis Pharmaceuticals, St.Louis, MO 63166-6556.
[d] Watson Laboratories, Inc., Corona CA 92880.
[e] Pharmacia and UpJohn, Peapack, NJ 07977.
[f] Roche, Centerville, VA 20120.
[g] Ortho Biotech, Bridgewater, NJ, 08807-0914.
[h] Wyeth Pharmaceuticals, Philadelphia, PA 19101.
[i] Fujisawa Healthcare, Inc., Chantilly, VA 20153-1644.

The ingestion of food with tacrolimus affects the rate and extent of the absorption of the drug. Side effects include insulin resistance, hyperkalemia or hypokalemia, hypophosphatemia, and hypomagnesemia, as well as GI distress including anorexia, nausea, vomiting, and diarrhea or constipation.

6.3. Mycophenolate Mofetil (CellCept/RS-61443)

Mycophenolate mofetil (Roche, Centerville, VA, 20120) is mainly used as an adjunctive agent in multi-therapy protocols with CsA, corticosteroids, or tacrolimus. The primary effect on the immune system is to inhibit T-cell proliferation *(35)*. There are several side effects of this drug, but the most common involves GI distress with diarrhea, nausea, or vomiting.

6.4. Sirolimus (Rapamycin/Rapamune)

Sirolimus (Rapamune, Wyeth Pharmaceuticals, Philadelphia, PA 19101) is a macrolide antibiotic that is structurally similar to tacrolimus and inhibits the proliferation of immune cells. Potential side effects include dyslipidemias (hypertriglyceridemia and hypercholesterolemia), increased liver enzymes, delayed wound healing, and hypokalemia. When sirolimus is used in combination therapy with CsA and corticosteroids, the dyslipidemias can be further exacerbated *(40)*.

6.5. Azathioprine (Imuran)

Azathioprine (Faro Pharmaceuticals, San Diego, CA 92121) is a nonspecific immunosuppressant whose mode of action is to inhibit the proliferation of immunocompetent cells. It is typically used in multi-drug therapy and may be administered intravenously or orally. Common GI side effects include diarrhea and cholestasis *(20)*.

6.6. Corticosteroids (Prednisone, prednisolone, methylprednisone, Solu-Medrol, and Solu-Cortef)

The most commonly prescribed corticosteroids used in transplant programs include prednisone and methylprednisolone, which have anti-inflammatory properties and inhibit the production of lymphokines. This class of immunosuppressants can be administered in either high oral or parenteral doses for acute rejection or as oral pulse doses, which are then tapered to maintenance levels or in some cases discontinued *(41,42)*. Associated side effects are believed to be dose-dependent and may include impaired wound healing, avascular necrosis of long bones, upper GI ulceration, protein catabolism, hypertension, steroid-induced

DM, cataract formation, and stimulation of appetite with resultant weight gain, among others *(43)*.

6.7. Other Agents

Monoclonal antibodies (muromonab-CD3, Orthoclone OKT3, Ortho Biotech, Bridgewater, NJ 08807-0914) and antithymocyte globulin (Genzyme, Cambridge, MA 02142), are examples of immunosuppressants used either perioperatively for induction therapy or for acute rejection episodes. These agents can cause GI distress as well as flu-like symptoms, and appropriate adjustments in nutritional therapy should be made *(35)*. Daclizumab (Zenapax), Roche, is a genetically engineered monoclonal antibody that is an IL-2 receptor antagonist. This medication can be combined with traditional immunosuppressive therapy. There are no known side effects associated with this drug *(44)*. Another IL-2 antagonist without known side effects is Basiliximab (Simulect), Roche. It is used in combination therapy with CsA and corticosteroids. This medication is administered preoperatively and can be administered on the fourth postoperative day *(45)*.

Balanced control of the immune system with immunosuppressive therapy is the cornerstone of graft survival. Dietitians should provide important nutrition education to assist transplant recipients in preventing and controlling side effects from their immunosuppressive therapy.

7. LONG-TERM CARE CHALLENGES

7.1. Cardiovascular Disease

Cardiovascular disease is the leading cause of death after kidney transplantation, with an incidence considerably higher than that in the general population due to the high prevalence and accumulation of classical risk factors before and after transplantation. The treatment of risk factors must be introduced early in the course of kidney failure and continued following transplantation. The cardiovascular risk factors that accompany post-renal transplantation include an atherogenic lipid profile, hypertension, DM, chronic prothrombotic state, and obesity *(46,47)*.

Dyslipidemias following kidney transplantation are common. Approximately 70% of kidney transplant recipients will experience serum lipid abnormalities during the post-transplant course *(46)*. Contributing factors for dyslipidemias include immunosuppressive therapy (CsA, corticosteroids, tacrolimus, sirolimus, etc.), graft dysfunction, obesity, diuretic or antihypertensive drug therapy, age, gender, DM, and proteinuria.

Although most risk factors cannot be corrected, diet and obesity are two factors that may be modified. Therapeutic lifestyle changes should be considered as a first step in the treatment of obesity and lipid disorders and involves reducing intake of saturated fat and cholesterol, increasing physical activity, and maintaining a healthy weight *(47)*. Lipid profiles can improve with dietary intervention, but often patients still require lipid-lowering medications to reach target lipid levels as suggested by the Adult Treatment Panel guidelines *(47)*. The HMG-CoA reductase inhibitors are prescribed for treatment of post-transplant hyperlipidemia in combination with therapeutic lifestyle changes *(48)*.

Hyperhomocysteinemia has been reported in kidney transplant recipients *(49–54)*. The mechanism for elevated homocysteine levels in transplant recipients is unknown, but factors such as inadequate folate status or renal dysfunction have been identified. It has also been reported that CsA can interfere with the folate-assisted remethylation of homocysteine *(55)*. Folate supplementation can be an effective intervention in reducing homocysteine levels in renal transplant recipients *(54)*.

7.2. *Obesity*

The negative impact of obesity on kidney transplant outcomes is well established. Weight gain following kidney transplantation can occur in both obese and non-obese patients, which adversely effects blood pressure, glucose metabolism, and lipids, thereby contributing to atherosclerosis *(56)*. The cause of post-transplantation weight gain is multifactorial and includes hyperphagia from steroid administration, elimination of the cachectic effects of dialysis, lack of physical activity, genetic predisposition, age, gender, and race *(57,58)*.

Weight management is a difficult challenge. Lifestyle changes involving diet, behavior modification, and physical activity are the cornerstone of successful weight control. In one study, the effect of early intensive dietary intervention on recently transplanted kidney patients showed reduced weight gain *(59)*. Future studies should investigate whether pre-transplant and post-transplant weight reduction can favorably impact outcomes.

Few studies have evaluated the impact of exercise following kidney transplantation *(60)*. Physical activity offers significant health benefits, including the reduction of cardiovascular risk factors, improvement in diabetes management and bone health, and weight management. Patients are typically cleared to resume exercise 6 weeks after transplant.

7.3. New Onset Diabetes After Transplantation

NODAT is associated with DGF, graft failure, and infection *(62,63)*. The incidence of post-transplant DM reported in the literature ranges from 3 to 20% *(62)*. The etiology of NODAT is multifactorial and includes genetic predisposition, diabetogenic effects of immunosuppressant, age, gender, post-transplant weight gain, and ethnicity *(63)*.

NODAT is diagnosed when the serum glucose is higher than 200 mg/dl or fasting plasma glucose is greater than or equal to 126 mg/ml *(63)*. Nutritional management of NODAT typically includes a carbohydrate-controlled diet, exercise, and weight management. Hypoglycemic agents may also be prescribed.

7.4. Hypertension

Hypertension occurs in 90% of kidney transplant recipients *(64,65)*. The etiology of hypertension is multifactorial and includes impaired renal function, use of calcineurin inhibitors, uncontrolled renin secretion from the native organ, stenosing lesions of the transplant artery, genetic predisposition, and obesity *(65)*. Mild hypertension can accelerate the loss of allograft function and impact the survival of the transplant recipient *(65)*. Antihypertensive treatment may include sodium modification, weight management, diuretics, and calcium channel blockers.

8. GENERAL CONSIDERATIONS

8.1. Interactions

Grapefruit juice and grapefruit products can affect absorption of certain medications and should not be consumed by patients taking calcium channel blockers, some immunosuppressants, antilipemics, and estrogen products. Grapefruit products inhibit the cytochrome P450 isoenzyme CYP3A4 mechanism in the GI tract. The pharmacokinetics of cyclosporine demonstrates intraindividual variability when this medication was administered with grapefruit *(66)*. The duration of the effect of grapefruit juice can last 24 hours, and with repeated intake, it can have a cumulative effect *(67)*.

8.2. Food Safety

Foodborne infections are common and sometimes can be life-threatening *(68)*. There are currently 250 known foodborne diseases, but information is limited on the incidence of infection from food sources in the immunosuppressed patient *(68–71)*. Transplant recipients infected with foodborne microorganisms exhibit symptoms ranging from mild intestinal distress to severe dehydration, which can jeopardize immunosuppression. Given the increased prevalence of foodborne outbreaks, nutrition education post-transplant should include guidelines on food safety.

9. CONCLUSION

The nutritional interventions associated with kidney transplantation vary according to the phase of care as well as the nutritional status of the transplant recipient. The RD provides nutritional assessments and interventions throughout the transplant process for recovery and maintenance of transplant function. The RD should also monitor the patient for general health issues and comorbidities, such as DM, obesity, and lipid disorders, and provide MNT. The RD is a valued member of the transplant team and, by providing timely MNT, he/she can enhance the transplant recipient's ability to prevent or minimize the nutrition-related complications. The entire team should be encouraged to participate in educational activities to increase successful patient outcomes.

10. CASE STUDY

B.G. is a 65-year-old BM with CKD stage 5 due to type 2 DM. The patient was diagnosed with diabetes for 25 years and HTN for 12 years. The patient had been receiving hemodialysis for 3 years. The transplant selection committee placed qualifying criteria of a 20-pound weight loss prior to surgery. He achieved this goal weighing in at 246 pounds, and is now 6 weeks post-deceased donor kidney transplant surgery with a healing incision.

The patient regularly checks his blood glucose levels three times a day pre-meal to assess need for dosing of exogenous insulin, per his history of blood sugars logged in his glucometer. Both the patient and his wife attended DM education classes approximately 10 years ago. Prior to transplant, his HgbA1C (glycated hemoglobin) was 6.6% with use of thiazolidinedione and sulfonylurea. He has been taking Lipitor for 6 years, at which time he also quit smoking.

His retinopathy has been treated with laser surgery, and vision is adequate but marginal. Digestion has shown variability between constipation and/or diarrhea, but more recently consistent diarrhea. His ability to exercise is compromised by neuropathy, but he does periodically walk at a recreation center.

Anthropometrics	*Labs (6 weeks s/p transplant)*	*Medications*
Height :5′ 11″ without shoes	Glu 250 mg/dl (non-fasting)	Prednisone 20 mg qd
Frame: large	HgbA1c 8.1%	Cellcept 250 mg, 4 bid
Weight: 260 lb, edema 2+	Creat 1.8 mg/dl	Prograf, 1 mg bid
	BUN 25 mg/dl	Oscal 500 Plus D, tid
	K+ 6.1 mEq/L	MVI supplement
	CO_2 20 mEq/L	Toprol XL
	Phos 2.0 mg/dl	Neurontin, 100 mg tid
	Ca 8.5 mg/dl	Nexium, 40 mg qd
	Mg 1.8 mg/dl	Neutra-Phos bid
	Alb 3.4 g/l (no proteinuria)	MagTab SR
	Chol 273 mg/dl	Lantus, 20 units q PM
	HDL 36 mg/dl	Regular insulin
	TG 352 mg/dl	
	LDL 128 mg/dl	

He has a supportive wife who works part-time in a billing office, and two adult children who have families out of town. His wife prepares meals, avoids added salt and high sodium seasonings. She also assists patient with some self-care issues regarding his DM. The church community has shown support during their more stressful situations and sometimes provides meals to the family.

10.1. Case Study Questions

1. What are this patient's protein and calorie needs?
2. Are there suggestions which may help his digestive concerns?
3. What types of exercise might be appropriate for this gentleman?
4. How could calcium/phosphorus balance be improved?
5. What glycemic control issues could be addressed?
6. Is the patient on an appropriate multivitamin/mineral (MVI) product?
7. Could patient's edema be related to diet?
8. What are some factors that could be affecting patient's serum potassium?
9. How would you deal with patient's hyperlipidemia?

REFERENCES

1. Organ Procurement and Transplantation Network. View data reports. Available at: http://www.optn.org. Accessed April 14, 2006.

2. Kumar MR, Coulston AM. Nutritional management of the cardiac transplant patient. J Am Diet Assoc 1983;83:463–465.
3. Aguilar-Salinas CA, Diaz-Polanco A, Quintana E, Macias N, Arellano A, Ramirez E, Ordonez ML, Velasquez-Alva C, Gomez Perez FJ, Alberu J, Correa-Rotter R. Genetic factors in the pathogenesis of hyperlipidemia posttransplantation. Am J Kidney Dis 2002;40:169–177.
4. Baltar J, Ortega T, Ortego F, Laures A, Rebollo P, Gomez E, Alvarez-Grande J. Posttransplantation diabetes mellitus: prevalence and risk factors. Transplant Proc 2005;37:3817–3818.
5. Martins C, Pecoits-Filho R, Riella MC. Nutrition for post-renal transplant recipients. Transplant Proc 2004; Jul–Aug, 36(6):1650–1654.
6. Massarweh NN, Clayton JL, Mangum CA, Florman SS, Slakey DP. High body mass index and short and long-term renal allograft survival in adults. Am J Transplant 2006; 6(2):357–363.
7. Seagraves A, Moore EE, Moore FA, Weil R, III. Net protein catabolic rate after kidney transplantation: impact of corticosteroid immunosuppression. JPEN J Parenter Enteral Nutr 1986;10:453–455.
8. Jequler E. Energy, obesity and body weight standards. Am J Clin Nutr 1987;45:1035.
9. Glynn C, Greene G, Winkler M, Albina J. Predictive versus measured energy expenditure using limits-of-agreement analysis in hospitalized, obese patients. JPEN J Parenter Enteral Nutr 1999;23:147–154.
10. Hasse JM, Weseman RA. Solid organ transplantation: In: Gottschlich MM, Fuhrman MP, Hammond KA, Holcombe BJ, Seidner DL, eds. Nutrition Support Dietetics Core Curriculum. Dubuque, Iowa: ASPEN, 2001, p. 605.
11. Blue LS. Nutrition considerations in kidney transplantation. Top Clin Nutr 1992;7:7–23.
12. Blue LS. Adult kidney transplantation. In: Hasse JM, Blue LS, eds. Comprehensive Guide to Transplant Nutrition. Chicago, IL: American Dietetic Association, 2002, p. 49.
13. Rosenberg ME. Nutrition and transplantation. Kidney 1986;18:19.
14. Mahajan SK, Prasad AS, Rabbani P, Briggs WA, McDonald FD. Zinc deficiency: a reversible complication of uremia. Am J Clin Nutr 1982;36:1177–1183.
15. Sandstead HH. Trace elements in uremia and hemodialysis. Am J Clin Nutr 1980;33:1501–1508.
16. Thakur V, Kumar R, Dhawan IK. Correlation between serum magnesium and blood cyclosporine A concentrations in renal transplant recipients. Ann Clin Biochem 2002;39:70–72.
17. Bell. AD. Herbal medicine and the transplant patient. Nephrol Nurs J 2002;29:269–274.
18. Dietary Supplement Health and Education Act of 1994. US Food and Drug Administration, Center for Food Safety and Applied Nutrition. Available at: http://vm.cfsan.fda.gov/~dms/dietsupp.html. Accessed April 12, 2006.
19. Ruschitzka F, Meier PJ, Turina M. Acute heart transplant rejection due to St. John's wort. Lancet 2000;355:548–549.
20. DiCecco SR, Francisco-Ziller N, Moore D. Overview and immunosuppression. In: Hasse JM, Blue LS, eds. Comprehensive Guide to Transplant Nutrition. Chicago, IL: American Dietetic Association, 2002, pp. 1–30

21. Ponticello C, Passenni P. Gastrointestinal complications in renal transplant recipients. Transpl Int 2005;18(6):643–650.
22. Bardaxoglou E, Maddern G, Ruso L, Siriser F, Campion JP, Le Pogamp P, Catheline JM, Launois B. GI surgical emergencies following kidney transplantation. Transpl Int 1993;6:148–152.
23. Mosimann F, Cuenoud PF, Steinhauslin F, Wauters JP. Herpes simplex esophagitis after renal transplantation. Transpl Int 1994;7:79–82.
24. Troppman C, Papalois BE, Chiou A, Benedetti E, Dunn DL, Matas AJ, Najarian JS, Gruessner RW. Incidence, complications, treatment, and outcome of ulcers of the upper GI tract after renal transplantation during the CsA era. J Am Coll Surg 1995;80:433–443.
25. Ozgur O, Boyacioglu S, Ozdogan M, Gur G, Telatar H, Haberal M. Helicobacter pylori infection in haemodialysis patients and renal transplant recipients. Nephrol Dial Transplant 1997;12:289–291
26. Torres A, Lorenzo V, Salido E. Calcium metabolism and skeletal problems after transplantation. J Am Soc Nephrol 2002;13:551–558.
27. Drug Facts and Comparisons. St Louis, MO: Wolters Kluwer, 2000.
28. Beto J, Bansal VK. Hyperkalemia: evaluating dietary and nondietary etiology. J Ren Nutr 1992;1;28–29.
29. Rubello D, Giannini S, D'Angelo A, Nobile M, Carraio G, Rigotti P, Marchini F, Zaninotto M, Dalle Carbonare L, Sartori L, Nibale O, Carpi A. Secondary hyperparathyroidism is associated with vitamin D receptor polymorphism and bone density after renal transplantation. Biomed Pharmacother 2005 Aug;59(7): 402–407.
30. Palmer SC, Strippoli GF, McGregor DO. Interventions for preventing bone disease in kidney transplant recipients: a systematic review of randomized controlled trials. Am J Kidney Dis 2005 Apr;45(4):638–649.
31. Torres A, Lorenzo V, Salido E. Calcium metabolism and skeletal problems after transplantation. J Am Soc Nephrol 2002;13:551–558.
32. Messa P, Sindici C, Cannella G, Miotti V, Risaliti A, Gropuzzo M, Di Loreto PL, Bresadola F, Mioni G. Persistent secondary hyperparathyroidism after renal transplantation. Kidney Int 1998;54:1704–1713.
33. Querings K, Girndt M, Geisel J, Georg T, Tilgen W, Reichrath J. 25-hydroxyvitamin D deficiency in renal transplant recipients. J Clin Endocrinol Metab 2006;91(2):526–529.
34. Haef J, Jakobsen V, Tuedegoard E, Kanstrupl L, Fogh-Anderson N. Dietary habits of nutritional status of renal transplant patients. J Ren Nutr 2004;14(1):20–25.
35. Fisher JS, Woodle ES, Thistlethwaite JR. Kidney transplantation: graft monitoring and immunosuppression. World J Surg 2002;26:185–193.
36. Morris PJ. Cyclosporine. In: Morris PJ, ed. Kidney Transplantation: Principles and Practice. Philadelphia, PA: WB Saunders Co, 1994, pp. 179–201.
37. Bennett W, Burdmann E, Andoh T, Houghton D, Lindsley J, Elzinga L. Nephrotoxicity of immunosuppressive drugs. Nephrol Dial Transplant 1994;9:141–145.
38. Kahan B. Cyclosporine. N Engl J Med 1989;321:1725–1738.
39. Gelone DK, Lake KD. Transplantation pharmacotherapy. In: Cupples SA, Ohler L, eds. Solid Organ Transplantation: A Handbook for Primary Health Care Providers. New York, NY: Springer Publishing, 2002, pp. 114–116.

40. Gonzalez-Posada JM, Rodriquez AP, Tamajon P, Hdez-Marrero D, Maceira B. Immunosuppression with sirolimus, cyclosporine and prednisone in renal transplantation. Transplant Proc 2002;34:99.
41. Starzl T, Marchiono T, Waddell W. The reversal of rejection in human renal homografts with subsequent development of homograft tolerance. Surg Gynecol Obstet 1963;117:385–395.
42. Gelone DK, Lake KD. Transplantation pharmacotherapy. In: Cupples SA, Ohler L, eds. Solid Organ Transplantation: A Handbook for Primary Health Care Providers. New York, NY: Springer Publishing, 2002, pp. 120–121.
43. McCann L. Pocket Guide to Nutrition Assessment of the Patient with Chronic Kidney Disease, 3rd edn. Redwood City, CA: Satellite Healthcare, 2002.
44. Vincenti F, Nashan B, Light S. Double Therapy and the Triple Therapy Study Groups. Daclizumab: outcome of phase III trials and mechanisms of action. Transplant Proc 1998;30:2155–2158.
45. Kahan B, Rajagopalan P, Hall M. Reduction of the occurrence of acute cellular rejection among renal allografts recipients treated with basiliximab, a chimeric anti-interleukin-2 receptor monoclonal antibody. Transplantation 1999;67: 276–284.
46. Kasiske, B, Cosio FG, Beto J, Bolton K, Chavers BM, Grimm R, Levin A, Masri B, Parekh R, Wanner C, Wheeler DC, Wilson PWF. Clinical practice guidelines for managing dyslipidemias in kidney transplant patients: a report from the Managing Dyslipidemias in Chronic Kidney Disease Work Group of the National Kidney Foundation Kidney Disease Outcomes Quality Initiative. Am J Transplant 2004;4(Suppl 7):13–53.
47. Expert Panel on Detection, Evaluation, and Treatment of High Blood Cholesterol in Adults. Executive summary of the Third Report of the National Cholesterol Education Program (NCEP) Expert Panel on Detection, Evaluation, and Treatment of High Blood Cholesterol in Adults (Adult Treatment Panel III). 2004. Available at: http://www.nhlbi.nih.gov/guidelines/cholesterol/. Accessed by April 14 2006.
48. Martinez-Castelao, A, Grinyo JM, Gil-Vernet S, Seron D, Castineiras MJ, Ramos R, Alsina J. Lipid-lowering long-term effects of six different statins in hypercholesterolemic renal transplant patients under cyclosporine immunosuppression. Transplant Proc 2002;34:398–400.
49. Savaj S, Ghorbani G, Ghoda AJ. Hyperhomocysteinemia in renal transplant recipients. Transplant Proc 2002;33:2701.
50. Mor E, Helfmann L, Lustig S, Bar-Nathan N, Yussim A, Sela BA. Homocysteine levels among transplant recipients: effect of immunosuppressive protocols. Transplant Proc 2001;33:2945–2946.
51. Zantvoort FA, Waldeman J, Colic D, Lison AE. Incidence and treatment of hyperhomocysteinemia after renal transplantation. Transplant Proc 2001;33: 3679–3680.
52. Merouani A, Rozen R, Clermont MJ, Genest J. Renal function, homocysteine and other plasma thiol concentrations during the postrenal transplant period. Transplant Proc 2002;34:1159–1160.
53. Friedman AN, Rosenberg IH, Selhub J, Levey AS, Bostom AG. Hyperhomocysteinemia in renal transplant recipients. Am J Transplant 2002;2: 308–313.

54. Savaj S, Rezakhani S, Porooshani F, Ghods AJ. Effect of folic acid therapy on serum homocysteine level in renal transplant recipients. Transplant Proc 2002;34:2419.
55. Arnadottir M, Hultberg B, Vladov V, Nilsson-Ehle P, Thysell H. Hyperhomocysteinemia in cyclosporine-treated renal transplant recipients. Transplantation 1996;61:509–512.
56. Hricik DE. Weight gain after kidney transplantation. Am J Kidney Dis 2001;38:409–410.
57. Hasse JM. Nutritional issues in adult transplantation. In: Cupples SA, Ohler L, eds. Solid Organ Transplantation: A Handbook of Primary Health Care Providers. New York, NY: Springer Publishing, 2002, pp. 64–87.
58. Clunk J, Lin C, Curtis J. Variables affecting weight gain in renal transplant recipients. Am J Kidney Dis 2001;73:53–55.
59. Patel MG. The effect of dietary intervention on weight gains after renal transplantation. J Ren Nutr 1998;8:137–141.
60. Bernadi A, Biasia F, Pati T, Piva M, Scaramuzzo P, Stoppa F, Bucciante G. Factors affecting nutritional status, response to exercise and progression of chronic rejection in kidney transplant recipients. J Ren Nutr 2005;15(1):54–57.
61. Sezar S, Bilgic A, Uvar M, Arat Z, Ozdemir FN, Haberal M. Risk factors for development of posttransplant diabetes mellitus in renal transplant recipients. Transplant Proc 2006;38(2):529–532.
62. Markell MS. Post-transplant diabetes: incidence, relationship to choice of immunosuppressive drugs, and treatment protocol. Adv Ren Replace Ther 2001;8:64–69.
63. Davidson J, Wilkinson A, Danta J, et al. New onset diabetes after transplantation: International Consensus Guidelines. Transplantation 2003;75(10):SS3–24.
64. Schwenger V, Zeier M, Ritz E. Hypertension after renal transplantation. Ann Transplant 2001;6:25–30.
65. Zeier M, Dikow R, Ritz E. Blood pressure after renal transplantation. Ann Transplant 2001;6:21–24.
66. Hollander A, van Rooij J, Lentjes E, Arbouw F, van Bree J, Schoemaker R, van Es L, van der Woude F, Cohen A. The effect of grapefruit juice on cyclosporine and prednisone metabolism in transplant patients. Clin Pharmacol Ther 1995;57:318–324.
67. Bailey DG, Malcolm J, Arnold O, Spence JD. The effect of grapefruit juice on CsA and prednisone metabolism in transplant patients. Br J Clin Pharmacol 1998;46:101–110.
68. National Institute of Allergy and Infectious Diseases. National Institutes of Health. Fact Sheet: Foodborne Diseases. February 2005. Available at: http://www.niaid.nih.gov/factsheets/foodbornedis.htm. Accessed April 14, 2006.
69. Daniels N, Shafie A. A review of pathogenic vibrio infections for clinicians. J Infect Med 2000;17:665–685.
70. US Food and Drug Administration, Center for Food Safety and Applied Nutrition. Foodborne Pathogenic Microorganisms and Natural Toxins Handbook, Bag Bug Book. (2000) Available at: http://www.cfsan.fda.gov/~mow/intro.html. Accessed April 14, 2006.
71. US Environmental Protection Agency. Safe drinking water – guidance for people with severely weakened immune systems. June 1999. Available at: http://www.epa.gov/OGWDW/crypto.html. Accessed April 14, 2006.

2 Management

11 Protein-Energy Malnutrition

Kamyar Kalantar-Zadeh

LEARNING OBJECTIVES

1. To recognize the prevalence of protein-energy malnutrition, its potential causes and its association with inflammation, anorexia, anemia and other untoward conditions in chronic kidney disease stage.
2. To describe the epidemiologic association between the markers of protein-energy malnutrition, such as hypoalbuminemia, and cardiovascular disease and poor survival in dialysis patients.
3. To identify the tools and measures used to assess the protein-energy malnutrition in stage 5 chronic kidney disease and their limitations.
4. To recognize current and emerging treatment modalities for protein-energy malnutrition in dialysis patients.

Summary

Protein-energy malnutrition is common in stage 5 chronic kidney disease (CKD) and usually associated with anorexia, chronic inflammation, kidney disease wasting (KDW), refractory anemia and poor outcome including high risk of cardiovascular disease and death. Uremia, endocrine and gastrointestinal disorders, and oxidative and carbonyl stress may engender this type of malnutrition. The dialysis treatment and techniques may also contribute to the protein-energy malnutrition. The term Malnutrition-Inflammation-Cachexia Syndrome (MICS) underscores the close link between the protein-energy malnutrition and inflammation and their concomitant contribution to the KDW. The MICS is associated with poor short-term survival and may result in paradoxical

From: *Nutrition and Health: Nutrition in Kidney Disease*
Edited by: L. D. Byham-Gray, J. D. Burrowes, and G. M. Chertow

associations between some cardiovascular risks, and such relationships are continuing to be studied in the emerging theory known as the obesity paradox or "reverse epidemiology." Four major nutritional evaluation categories in dialysis patients include the assessment of nutritional intake, body composition and laboratory markers and the nutritional scoring systems. Dietary supplements and nutritional and pharmacological interventions that can stimulate appetite, mitigate inflammation and enhance protein and energy intake may improve survival in patients with stage 5, CKD although scientific evidence for or against such effects are currently inadequate.

Key Words: Protein-energy malnutrition, anorexia, Malnutrition-Inflammation-Cachexia Syndrome, kidney disease wasting, cardiovascular disease, reverse epidemiology.

1. INTRODUCTION

Among individuals with stage 5 chronic kidney disease (CKD) who undergo maintenance dialysis treatment to survive, currently one out of every five people dies each year in the United States. This unacceptably high mortality rate has not changed substantially in recent years despite many advances in dialysis techniques and patient care *(1)*. Maintenance dialysis patients also have a high hospitalization rate and a low self-reported quality of life *(2)*. Cardiovascular diseases comprise the bulk of morbidity and mortality in dialysis patients *(3)*. The dialysis-dependent stage 5 CKD population grows constantly and fast, reaching over half a million by 2010 in the United States *(4)*, and continues to consume a disproportionately large component of the Medicare budget; hence, identifying factors that lead to poor dialysis outcome and their successful management is of outmost importance *(5)*. It was once believed that the traditional cardiovascular risk factors and/or conditions related to dialysis treatment and technique are the main causes of poor clinical outcome; however, recent randomized controlled trials have failed to show an improvement of mortality by lowering serum cholesterol (the 4D Trial) *(6)* or increasing dialysis dose [the HEMO *(7)* and ADEMEX *(8)* studies]. Evidence suggests that conditions other than the traditional cardiovascular risk factors must be related to the enormous cardiovascular epidemic and high death rate in this population. Among the potential candidates for the poor clinical outcomes in maintenance dialysis patients, the protein-energy malnutrition continues to be at the top of the list. Epidemiologic studies have repeatedly and consistently shown a strong association between survival and measures of nutritional status in maintenance dialysis patients *(9,10)*.

2. PROTEIN-ENERGY MALNUTRITION: DEFINITION AND ETIOLOGY

A workable definition of protein-energy malnutrition in CKD was recently presented as "Protein-energy malnutrition is the state of decreased body pools of protein with or without fat depletion or a state of diminished functional capacity, which is caused at least partly by inadequate nutrient intake relative to nutrient demand and/or which is improved by nutritional repletion in patients with CKD" *(11)*. Hence, the protein-energy malnutrition is engendered when the body's need for protein or energy fuels or both cannot be satisfied by the current dietary intake. Various studies using different criteria have been used to establish the presence of protein-energy malnutrition in the dialysis population. Its reported prevalence varies between 18 and 75% among dialysis patients according to the type of dialysis modality, nutritional assessment tools and the origin of patient population *(12,13)*. Although per definition the protein-energy malnutrition should not involve micronutrients that are indeed believed to be adequate or even abundantly retained in the setting of renal insufficiency, many protein-energy malnourished dialysis patients may also have a relative deficiency in vitamins and trace elements *(14)* (see Chapters 4 and 9).

The etiology of protein-energy malnutrition in dialysis patients is not clear, but some probable causes are listed in Table 1 and shown in Fig. 1. Uremia, endocrine and gastrointestinal disorders, volume overload, and oxidative and carbonyl stress may lead to the CKD-associated malnutrition. The origin of protein-energy malnutrition appears to precede dialysis treatment, because it is observed progressively as the glomerular filtration rate falls below 60 ml/min/1.73 m^2, that is, the stage 3 CKD and greater *(15)*. Diminished appetite (anorexia) is a main cause of protein-energy malnutrition and may be related to elevated circulating levels of cytokines or may be engendered via signaling through the central melanocortin system *(16)*. Dialysis patients with a poor appetite have higher levels of inflammatory markers, including C-reactive protein (CRP), and almost two-fold increased risk of death *(17)*. Moreover, dietary restrictions imposed by nephrologists and/or dietitians to prevent hyperphosphatema or hyperkalemia may lead to low protein intake.

Dialysis treatment and techniques may also contribute to engendering or worsening protein-energy malnutrition. Nutrient loss may happen through hemodialysis membrane or via peritoneal membrane, although its contribution to protein-energy malnutrition may not be substantial. High prevalence of comorbid conditions and metabolic

Table 1
Potential Contributors of the Protein-Energy Malnutrition in Stage 5 Chronic Kidney Disease and Dialysis Patients

A. Inadequate nutrient intake
  1. Anorexia, secondary to:
    a. Uremic toxicity
    b. Impaired gastric emptying (e.g., in diabetes mellitus)
    c. Inflammation with or without comorbid conditions[a]
    d. Emotional and/or psychological disorders
  2. Dietary restrictions
    a. Prescribed restrictions: Low potassium, low phosphate regimens
    b. Social constraints: Poverty, inadequate dietary support
    c. Physical incapacity: Inability to acquire or prepare food or to eat
B. Nutrient losses during dialysis
  1. Loss through hemodialysis membrane into hemodialysate
  2. Adherence to hemodialysis membrane or tubing
  3. Loss into peritoneal dialysate
C. Hypercatabolism due to comorbid illnesses
  1. Cardiovascular diseases[a]
  2. Diabetic complications
  3. Infection and/or sepsis[a]
  4. Other comorbid conditions[a]
D. Hypercatabolism associated with dialysis treatment
  1. Negative protein balance
  2. Negative energy balance
E. Endocrine disorders of uremia:
  1. Resistance to insulin
  2. Resistance to growth hormone and/or IGF-1
  3. Increased serum level of or sensitivity to glucagons
  4. Hyperparathyroidism
  5. Other endocrine disorders
F. Acidemia with metabolic acidosis
G. Concurrent nutrient loss with frequent blood losses

IGF-1 insulin-like growth factor 1.

[a] Me given factor may also be associated with inflammation. Adapted from ref *(11)*

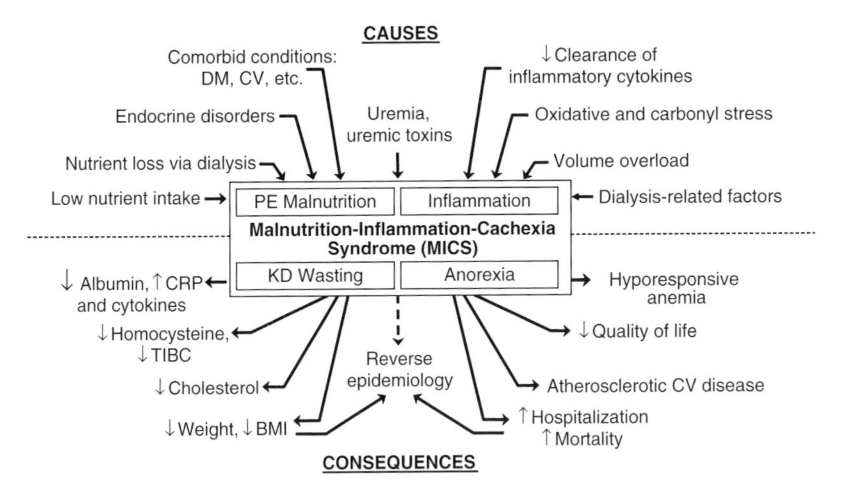

Fig. 1. Schematic representation of the causes and consequences of malnutrition-inflammation-cachexia (or complex) syndrome (MICS). BMI, body mass index; CRP, C-reactive protein; CV, cardiovascular; DM, diabetes mellitus; KD, kidney disease; PE, protein energy; TIBC, total iron binding capacity (also known as transferrin).

disorders including insulin resistance and acidosis may also lead to hypercatabolism and/or wasting. A higher than normal resting energy expenditure is also reported in dialysis patients independent of comorbidity *(18)* (Table 1).

3. ASSESSMENT OF PROTEIN-ENERGY MALNUTRITION

Methods and tools to assess protein-energy malnutrition in CKD patients are classically divided into four major categories: (i) assessment of appetite and dietary intake, (ii) biochemical measures, (iii) body composition and (iv) scoring systems (Table 2). A normal appetite is essential to maintain adequate food intake and to avoid undernourishment. Even though a diminished appetite (anorexia) is one of the early signs of uremia progression in CKD and is implicated as one of the underlying etiologies of protein-energy malnutrition in dialysis patients, *(17)* its uniform assessment may not be reliable because of its inherent subjectivity. It has been argued that inflammation is a cause of diminished appetite in dialysis patients *(17)*. If this hypothesis should be true, then the inflammation may be causally linked to the protein-energy malnutrition in CKD patients.

Table 2

Proposed Classification of the Assessment Tools for Evaluation of Protein-Energy Malnutrition in Maintenance Dialysis Patients (adapted from ref. *(11)*).

A. Nutritional intake
   1. Appetite assessment
   2. Direct: Diet recalls and diaries, food frequency questionnaires
   3. Indirect, based on urea nitrogen appearance: nPNA (nPCR)

B. Body composition
   1. Weight-based measures: BMI, weight for height, edema-free fat-free weight
   2. Skin and muscle anthropometry via caliper: skinfolds, extremity muscle mass
   3. Total body elements: Total body potassium, total body nitrogen
   4. Energy-beam based methods: DEXA, BIA, NIR
   5. Other methods: Underwater weighing

C. Laboratory values
   1. Visceral proteins (negative acute phase reactants): Albumin, prealbumin, transferrin
   2. Somatic proteins and nitrogen surrogates: Creatinine, SUN
   3. Lipids: Cholesterol, triglycerides, other lipids and lipoproteins
   4. Growth factors: IGF-1, leptin
   5. Peripheral blood cell count: Lymphocyte count

D. Scoring systems
   1. Conventional SGA and its modifications [e.g., DMS *(23)*, MIS *(24)*, CANUSA-version*(33)*]
   2. Other scores: HD-PNI*(34)*), others [e.g., Wolfson*(35)*, Merkus *(36)*, Merckman *(37)*]

BIA, bioelectrical impedance analysis; BMI, body mass index; CANUSA, Canada-USA study-based modification of the SGA *(33)*; CRP, C-reactive protein; DEXA, dual energy X-ray absorptiometry; DMS, dialysis malnutrition score *(23)*; HD-PNI, hemodialysis prognostic nutritional index *(34)*; IGF-1, insulin-like growth factor 1; IL, interleukin (e.g, IL-1 and IL-6); MIS, malnutrition inflammation score *(24)*; NIR, near infra-red interactance; nPCR, normalized protein catabolic rate; nPNA, normalized protein nitrogen appearance; SAA, serum amyloid A; SGA, subjective global assessment of nutritional status; SUN, serum urea nitrogen; TNF-α, tumor necrosis factor α.

A traditional nutritional evaluation is dietary assessment, because both the quality and quantity of the ingested nutrients can be assessed with a high degree of reproducibility. However, dietary assessment methods including 24-h recall, 3-day food diary with interview and food frequency questionnaires are difficult to accurately implement or interpret in dialysis patients. A more routinely used and readily available method is the calculation of the weight-normalized protein equivalent of total nitrogen appearance (nPNA), also known as the normalized protein catabolic rate, which is derived from the rate of urea generation between the two subsequent dialysis treatment sessions. This urea kinetic estimate of the protein intake has significant survival predictability *(10)*. However, the nPNA assessment can be reliable only if there is no appreciable residual renal function and if the patient is in no negative or positive nitrogen balance.

Anthropometry and body composition measures are among traditional indicators of nutritional status in dialysis patients. Weight for height and body mass index (BMI = weight/height^2) can be conveniently calculated and are also known to predict outcomes in dialysis patients *(19,20)*. However, the reliability of these measures to represent the true body composition is questionable, especially because a high BMI can occur with both high total body fat and very high muscle mass. Caliper anthropometry including mid-arm muscle mass and skinfold thickness has a poor reproducibility. More reliable methods such as underwater weighing and total nitrogen or potassium measurements are costly and rare if ever used in dialysis patients, although they are considered gold standards. Energy-beam methods may provide more pragmatic alternatives. Portable devices such as those based on the bioelectrical impedance analysis or near infra-red interactance technology are evaluator-friendly and patient-friendly *(21)*, whereas dual energy X-ray absorptiometry is a more elaborate and costly method that requires both resources and expertise.

Serum concentrations of albumin, prealbumin (transthyretin), transferrin (total iron binding capacity or TIBC), cholesterol, urea nitrogen and creatinine can be evaluated as markers of nutritional status and outcome predictors in dialysis patients. However, these laboratory values may significantly be confounded by such non-nutrition factors as inflammation, oxidative stress, iron stores, liver disease or residual renal function. Serum albumin is one of the most sensitive mortality predictors (Fig. 2) *(9)*; a fall in serum albumin concentration by as low as 0.6 g/dL from baseline over a 6-month interval is associated with the doubling of the death risk in hemodialysis patients *(9)*.

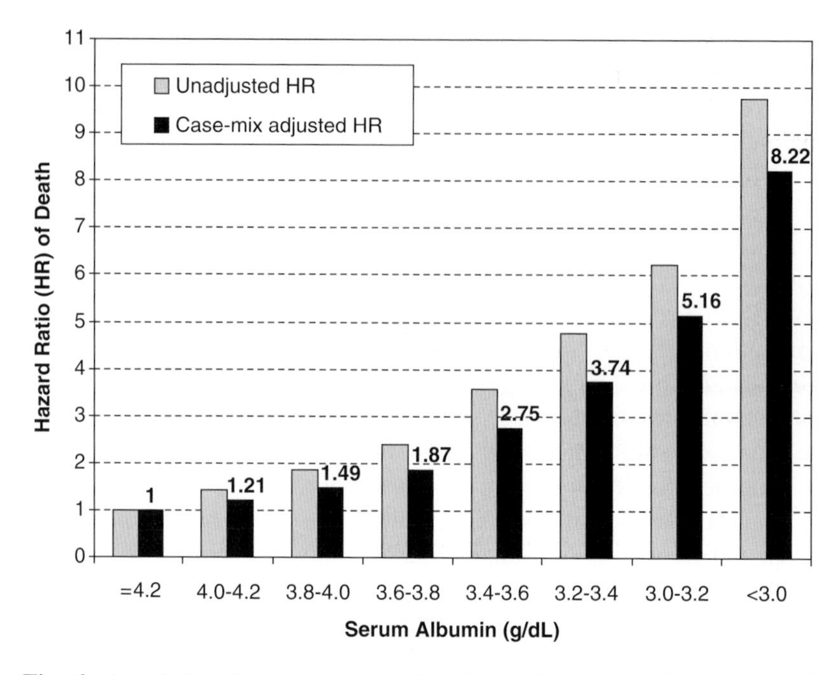

Fig. 2. Association between serum albumin levels (averaged over 3-month intervals) and subsequent 2-year death risk in 58,058 maintenance hemodialysis patients, 2001–2003 (based on data published data in ref. *(9)*).

In the past few years, several scoring systems have been introduced or developed to assess the overall nutritional aspects of CKD patients. The Subjective Global Assessment (SGA) is probably the most known scoring tool, which has also been recommended by the Kidney Disease Outcomes Quality Initiative (KDOQI) Nutrition guidelines for the periodic assessment of dialysis patients *(22)*. Among the limitations of the SGA are the inherently "subjective" characteristics of the assessment components and the semi-quantitative scoring. Fully quantitative versions of the SGA that have been developed for dialysis patients including the Dialysis Malnutrition Score (DMS) *(23)* and the Malnutrition-Inflammation Score (MIS) *(24)*. The reproducibility and objectivity of the DMS and MIS may be superior to the conventional SGA *(22)*.

4. INFLAMMATION AND KIDNEY DISEASE WASTING

Inflammation is defined as a protective response elicited by injury or destruction of tissues, which serves to destroy, dilute or sequester both the injurious agent and the injured tissue *(11)*. This important defense mechanism, which is inherently "acute" and should happen on an "as-needed" basis, may become harmful to the organism if it becomes "chronic." Evidence suggests that dialysis patients with protein-energy malnutrition are more likely to have abnormally high circulatory levels of inflammatory markers and pro-inflammatory cytokines such as CRP and interleukin-6 (IL-6), both known to be strong predictors of

Table 3
Possible Causes of Inflammation in Chronic Kidney Disease Patients

A. Causes of inflammation due to CKD or decreased GFR
   1. Decreased clearance of pro-inflammatory cytokines
   2. Volume overload with or without concomitant endotoxinemia[a]
   3. Oxidative stress (e.g., oxygen radicals)[a]
   4. Carbonyl stress (e.g., pentosidine and advanced glycation end-products)
   5. Decreased levels of antioxidants (e.g., vitamin E, vitamin C, carotenoids, selenium, glutathione)[a]
   6. Deteriorating protein-energy nutritional state and food intake[a]

B. Coexistence of comorbid conditions
   1. Inflammatory diseases with kidney involvement (e.g., SLE or HIV disease)
   2. Increased prevalence of comorbid conditions (CVD, DM, advanced age, etc)[a]
   3. Remnant allograft from a previous solid organ transplantation[a]

C. Additional inflammatory factors related to dialysis treatment
   I. Hemodialysis:
      1. Exposure to dialysis tubing
      2. Dialysis membranes with decreased biocompatiblility
      3. Impurities in dialysis water and/or dialysate
      4. Back-filtration or back-diffusion of contaminants
      5. Foreign bodies (such as PTFE) in dialysis access grafts
      6. Intravenous catheter
   II. Peritoneal dialysis:
      1. Episodes of overt or latent peritonitis[a]
      2. PD-catheter as a foreign body and its related infections
      3. Constant exposure to PD solution

CVD, cardiovascular disease; DM, diabetes mellitus. GFR, glomerular filtration rate; HIV, human immune deficiency virus; SLE, systemic lupus erythematosus PD.

[a] Me given factor may also be associated with protein-energy malnutrition. Adapted from ref. *(11)*.

poor outcome *(25)*. It is not clear why chronic inflammation occurs commonly in CKD patients; however, some potential causes have been listed in Table 3.

Inflammatory markers are associated with anorexia in dialysis patients *(17)*. Chronic inflammation may also lead to increased rate of protein depletion in skeletal muscle and other tissues, muscle and fat wasting, hypoalbuminemia, and hypercatabolism, leading to the so-called kidney disease wasting (KDW). Because both the protein-energy malnutrition and inflammation are usually concurrent, act on the same direction on laboratory markers and body proteins, and are both associated with the KDW and atherosclerotic cardiovascular disease in dialysis patients, a so-called Malnutrition-Inflammation Complex (or cachexia) Syndrome (MICS) has been defined to underscore the close link between these two conditions (Fig. 1) *(11)*. However, there is currently no conclusive consensus with regard to the nature or direction of the association between the protein-energy malnutrition and inflammation and their pathophysiologic link to the KDW and survival *(13)*.

5. CONSEQUENCES OF PROTEIN-ENERGY MALNUTRITION

In addition to anorexia, hypoalbuminemia and the KDW, malnutrition may have other clinically important consequences in dialysis patients.

5.1. Refractory Anemia

Anemia appears to be more common in those dialysis patients who also suffer from protein-energy malnutrition and/or inflammation *(26)*. A blunted response to erythropoiesis stimulating agents (ESA) is usually associated with increased levels of pro-inflammatory cytokines such as IL-6 *(26)*. An inverse association between such markers of nutritional state as serum prealbumin, transferrin and total cholesterol concentration and blood lymphocyte count and the required ESA dose has also been reported *(26)*. In a meta-analysis, L-carnitine administration that is used to improve nutritional state was associated with improved hemoglobin and a decreased ESA dose and in anemic dialysis patients *(27)*. Moreover, anabolic steroids have also been used successfully to simultaneously improve both nutritional state and anemia in dialysis patients.

5.2. Atherosclerotic Cardiovascular Disease

Dialysis patients with coronary heart disease often have hypoalbuminemia and elevated levels of inflammatory markers *(9)*. Epidemiologic evidence suggests that inflammation may be linked to cardiovascular disease in malnourished or cachectic dialysis patients. Emerging data even in the general population imply that such indicators of inflammation as an increased serum CRP level are stronger predictors of cardiovascular events than low density lipoprotein hyper-cholesterolemia *(28)*. The association between elements of MICS and atherosclerosis has been underscored by some investigators who have chosen the term "Malnutrition-Inflammation-Atherosclerosis" syndrome for this entity. Chronic inflammation may be the missing link that causally ties protein-energy malnutrition to poor outcome and high death rate in these individuals.

5.3. Reverse Epidemiology

The emergence of the reverse epidemiology hypothesis may have a bearing on the management of dialysis patients. In highly industrialized, affluent countries, protein-energy malnutrition is an uncommon cause of poor outcome in the general population, whereas over-nutrition is associated with a greater risk of cardiovascular disease and shortened survival. In contrast, in maintenance dialysis patients, "under-nutrition" is one of the most common risk factors for adverse cardiovascular events. Similarly, certain markers that predict a low likelihood of cardiovascular events and an improved survival in the general population, such as decreased BMI or lower serum cholesterol levels, are risk factors for increased cardiovascular morbidity and death in dialysis patients. Hence, obesity, hypercholesterolemia and hypertension appear paradoxically to be protective features that are associated with a greater survival among dialysis patients *(29)*. The association between under-nutrition and adverse cardiovascular outcome in dialysis patients, which stands in contrast to that seen in the general population, has been referred to as "reverse epidemiology" *(29)*. Possible causes of these paradoxical findings include survival bias and time discrepancy between competitive risk factors, that is, under-nutrition, which is the short-term killer, versus over-nutrition, which is the long-term killer. It is possible that new standards or goals for such traditional risk factors as body mass, serum cholesterol and blood pressure should be considered for these individuals, especially if they suffer from protein-energy malnutrition.

6. MANAGEMENT OF PROTEIN-ENERGY MALNUTRITION

Because protein-energy malnutrition and inflammation are powerful predictors of death risk in dialysis patients, it is possible that nutritional and anti-inflammatory interventions improve poor outcome in dialysis patients. Ample evidence suggests that maintaining an adequate nutritional intake in patients with a number of acute or chronic catabolic illnesses may improve their nutritional status irrespective of its etiology. However, evidence as to whether nutritional treatment may improve morbidity and mortality in dialysis patients is quite limited. There are no large-scale, randomized prospective interventional studies that have examined these questions. However, secondary data analyses have indicated that a high protein intake, for example, between 1.2 and 1.4 g/kg body weight per day, is associated with the best survival in dialysis patients *(10)*. According to the KDOQI guidelines on nutrition, dialysis patients can receive at least 35 kcal/kg body weight per day of energy and 1.2 g/kg body weight per day of protein *(22)*.

Table 4 summarizes selected nutritional interventions that have been tried or recommended in dialysis patients. Enhancing food intake by either dietary counseling or positive reinforcement may be helpful, especially if renal dietitians take a proactive role to this end. Many nephrologists and dietitians advocate oral supplementations as an adjunct therapy. However, it is important to appreciate that simultaneously imposed dietary restrictions to control potassium, phosphorus and/or calcium intake or to manage diabetes mellitus or dyslipidemia may interfere with or even contradict the foregoing efforts and lead to confusion for both patients and health care providers.

Tube feeding and parenteral interventions may reinforce protein and energy intake even among anorectic patients. A metabolic study demonstrated that intradialytic parenteral nutrition promoted a large increase in whole-body protein synthesis and a significant decrease in proteolysis in non-inflamed but malnourished hemodialysis patients *(30)*. Hormonal or pharmacological interventions may be associated with many side effects that mitigate the enthusiasm of using them, although emerging data suggests dietary supplements with anti-inflammatory interventions *(31)*, especially if associated with simultaneous appetite stimulating properties (such as megesterol acetate *(32)* or pentoxiphylline), may improve nutritional status and outcomes in dialysis patients. A number of other techniques have been employed or recommended for the prevention or treatment of protein-energy malnutrition before the onset of dialysis therapy, maintenance of an adequate

Table 4
Overview of Nutritional/Anti-Inflammatory Interventions for Dialysis Patients with Protein-Energy Malnutrition (adapted from *(38)*)

A. Oral interventions
  1. Increasing food intake
  2. Oral supplements
B. Enteral interventions
  1. Tube feeding
C. Parenteral interventions
  1. Intradialytic parenteral nutrition
  2. Other parenteral interventions
D. Hormonal interventions
  1. Androgens
  2. Growth factors/hormones
E. Non-hormonal medications
  1. Anti-inflammatory agents (e.g., borage oil, pentoxiphylline)
  2. Anti-oxidants (e.g. vitamin E, N-acetylcysteine)
  3. Appetite stimulators (e.g., megesterol)
  4. Carnitine
  5. Others (e.g. fish oil)
F. Dietary counseling
  1. In-center supervision/counseling
G. Dialysis treatment related
  1. Dialysis dose and frequency
  2. Membrane compatibility

dose of dialysis, avoidance of acidemia, and aggressive treatment of superimposed catabolic illness.

7. SUMMARY/CONCLUSIONS

In stage 5 CKD patients who undergo maintenance dialysis treatment, protein-energy malnutrition, with or without concurrent inflammation, is undoubtedly one of the most challenging and unresolved issues of contemporary medicine and nephrology. Protein-energy malnutrition and its concomitant wasting syndrome are unacceptably common and associated with inflammation and hypoalbuminemia. Malnutrition may be a major cause of poor outcome and high death rate in the fast growing dialysis patient population The impact of malnutrition on deteriorating the short-term survival in dialysis patients is that it overwhelms and even reverses the conventional associations between such traditional cardiovascular risk factors as obesity or hypercholesterolemia and survival. There is currently no uniform assessment tool for the detection or accurate grading of the protein-energy malnutrition in CKD, although emerging scoring systems appear promising. There is a paucity of information concerning the effect of nutritional therapy, appetite-stimulating agents or anti-inflammatory modalities on morbidity and mortality in dialysis patients. Simultaneous dietary restrictions to avoid hyperkalemia or hyperphosphatemia in CKD patients may handicap efforts to prevent or treat malnutrition in this patient population. Randomized clinical trials are needed to compare the effect of the nutritional support and anti-inflammatory agents, with or without appetite stimulating effects, in individuals with CKD who suffer from protein-energy malnutrition. The most optimal nutritional intervention that can overcome the high death rate linked to the enormous cardiovascular epidemic and poor outcome in CKD patients is yet to be determined.

ACKNOWLEDGMENT

Supported by a grant from the National Institute of Diabetes Digestive and Kidney Diseases of the National Institutes of Health (DK16612).

REFERENCES

1. United States Renal Data System: Excerpts from the USRDS 2004 Annual Data Report. Am J Kidney Dis 45 (suppl 1):S1–S280, 2005.
2. Kalantar-Zadeh K, Kopple JD, Block G, Humphreys MH: Association among SF36 quality of life measures and nutrition, hospitalization, and mortality in hemodialysis. J Am Soc Nephrol 12:2797–2806, 2001.
3. Foley RN, Parfrey PS, Sarnak MJ: Epidemiology of cardiovascular disease in chronic renal disease. J Am Soc Nephrol 9:S16–23, 1998.
4. United States Renal Data System: US Department of Public Health and Human Services, Public Health Service, National Institutes of Health, Bethesda, 2001.
5. Morbidity and mortality of dialysis. NIH Consens Statement 11:1–33, 1993.
6. Wanner C, Krane V, Marz W, Olschewski M, Mann JF, Ruf G, Ritz E: Atorvastatin in patients with type 2 diabetes mellitus undergoing hemodialysis. N Engl J Med 353:238–248, 2005.
7. Eknoyan G, Beck GJ, Cheung AK, Daugirdas JT, Greene T, Kusek JW, Allon M, Bailey J, Delmez JA, Depner TA, Dwyer JT, Levey AS, Levin NW, Milford E, Ornt DB, Rocco MV, Schulman G, Schwab SJ, Teehan BP, Toto R: Effect of dialysis dose and membrane flux in maintenance hemodialysis. N Engl J Med 347:2010–2019, 2002.
8. Paniagua R, Amato D, Vonesh E, Correa-Rotter R, Ramos A, Moran J, Mujais S: Effects of increased peritoneal clearances on mortality rates in peritoneal

dialysis: ADEMEX, a prospective, randomized, controlled trial. J Am Soc Nephrol 13:1307–1320, 2002.

9. Kalantar-Zadeh K, Kilpatrick RD, Kuwae N, McAllister CJ, Alcorn Jr H, Kopple JD, Greenland S: Revisiting mortality predictability of serum albumin in the dialysis population: time dependency, longitudinal changes and population-attributable fraction. Nephrol Dial Transplant 20:1880–1889, 2005.
10. Shinaberger CS, Kilpatrick RD, Regidor DL, McAllister CJ, Greenland S, Kopple JD, Kalantar-Zadeh K: Longitudinal associations between dietary protein intake and survival in hemodialysis patients. Am J Kidney Dis 48:37–49, 2006.
11. Kalantar-Zadeh K, Ikizler TA, Block G, Avram MM, Kopple JD: Malnutrition-inflammation complex syndrome in dialysis patients: causes and consequences. Am J Kidney Dis 42:864–881, 2003.
12. Mehrotra R, Kopple JD: Nutritional management of maintenance dialysis patients: why aren't we doing better? Annu Rev Nutr 21:343–379, 2001.
13. Kalantar-Zadeh K, Kopple JD: Relative contributions of nutrition and inflammation to clinical outcome in dialysis patients. Am J Kidney Dis 38:1343–1350, 2001.
14. Kalantar-Zadeh K, Kopple JD, Deepak S, Block D, Block G: Food intake characteristics of hemodialysis patients as obtained by food frequency questionnaire. J Ren Nutr 12:17–31, 2002.
15. Kopple JD, Greene T, Chumlea WC, Hollinger D, Maroni BJ, Merrill D, Scherch LK, Schulman G, Wang SR, Zimmer GS: Relationship between nutritional status and the glomerular filtration rate: Results from the MDRD study. Kidney Int 57:1688–1703, 2000.
16. Cheung W, Yu PX, Little BM, Cone RD, Marks DL, Mak RH: Role of leptin and melanocortin signaling in uremia-associated cachexia. J Clin Invest 115: 1659–1665, 2005.
17. Kalantar-Zadeh K, Block G, McAllister CJ, Humphreys MH, Kopple JD: Appetite and inflammation, nutrition, anemia and clinical outcome in hemodialysis patients. Am J Clin Nutr 80:299–307, 2004.
18. Ikizler TA, Wingard RL, Sun M, Harvell J, Parker RA, Hakim RM: Increased energy expenditure in hemodialysis patients. J Am Soc Nephrol 7:2646–2653, 1996.
19. Kopple JD, Zhu X, Lew NL, Lowrie EG: Body weight-for-height relationships predict mortality in maintenance hemodialysis patients. Kidney Int 56:1136–1148, 1999.
20. Kalantar-Zadeh K, Kopple JD, Kilpatrick RD, McAllister CJ, Shinaberger CS, Gjertson DW, Greenland S: Association of morbid obesity and weight change over time with cardiovascular survival in hemodialysis population. Am J Kidney Dis 46:489–500, 2005.
21. Kalantar-Zadeh K, Kuwae N, Wu DY, Shantouf RS, Fouque D, Anker SD, Block G, Kopple JD: Associations of body fat and its changes over time with quality of life and prospective mortality in hemodialysis patients. Am J Clin Nutr 83:202–210, 2006.
22. Steiber AL, Kalantar-Zadeh K, Secker D, McCarthy M, Sehgal A, McCann L: Subjective Global Assessment in chronic kidney disease: A review. J Ren Nutr 14:191–200, 2004.
23. Kalantar-Zadeh K, Kleiner M, Dunne E, Lee GH, Luft FC: A modified quantitative subjective global assessment of nutrition for dialysis patients. Nephrol Dial Transplant 14:1732–1738, 1999.

24. Kalantar-Zadeh K, Kopple JD, Block G, Humphreys MH: A malnutrition-inflammation score is correlated with morbidity and mortality in maintenance hemodialysis patients. Am J Kidney Dis 38:1251–1263, 2001.
25. Kalantar-Zadeh K, Stenvinkel P, Pillon L, Kopple JD: Inflammation and nutrition in renal insufficiency. Adv Ren Replace Ther 10:155–169, 2003.
26. Kalantar-Zadeh K, McAllister CJ, Lehn RS, Lee GH, Nissenson AR, Kopple JD: Effect of malnutrition-inflammation complex syndrome on EPO hyporesponsiveness in maintenance hemodialysis patients. Am J Kidney Dis 42:761–773, 2003.
27. Hurot JM, Cucherat M, Haugh M, Fouque D: Effects of L-carnitine supplementation in maintenance hemodialysis patients: a systematic review. J Am Soc Nephrol 13:708–714, 2002.
28. Ridker PM, Rifai N, Rose L, Buring JE, Cook NR: Comparison of C-reactive protein and low-density lipoprotein cholesterol levels in the prediction of first cardiovascular events. N Engl J Med 347:1557–1565, 2002.
29. Kalantar-Zadeh K, Block G, Humphreys MH, Kopple JD: Reverse epidemiology of cardiovascular risk factors in maintenance dialysis patients. Kidney Int 63: 793–808, 2003.
30. Pupim LB, Flakoll PJ, Brouillette JR, Levenhagen DK, Hakim RM, Ikizler TA: Intradialytic parenteral nutrition improves protein and energy homeostasis in chronic hemodialysis patients. J Clin Invest 110:483–492, 2002.
31. Kalantar-Zadeh K, Braglia A, Chow J, Kwon O, Kuwae N, Colman S, Cockram DB, Kopple JD: An anti-inflammatory and antioxidant nutritional supplement for hypoalbuminemic hemodialysis patients: a pilot/feasibility study. J Ren Nutr 15:318–331, 2005.
32. Rammohan M, Kalantar-Zadeh K, Liang A, Ghossein C: Megestrol acetate in a moderate dose for the treatment of malnutrition-inflammation complex in maintenance dialysis patients. J Ren Nutr 15:345–355, 2005.
33. Anonymous. Adequacy of dialysis and nutrition in continuous peritoneal dialysis: association with clinical outcomes. Canada-USA (CANUSA) Peritoneal Dialysis Study Group. J Am Soc Nephrol 7:198–207, 1996.
34. Beto JA, Bansal VK, Hart J, McCarthy M, Roberts D: Hemodialysis prognostic nutrition index as a predictor for morbidity and mortality in hemodialysis patients and its correlation to adequacy of dialysis. Council on Renal Nutrition National Research Question Collaborative Study Group. J Ren Nutr 9:2–8, 1999.
35. Wolfson M, Strong CJ, Minturn D, Gray DK, Kopple JD: Nutritional status and lymphocyte function in maintenance hemodialysis patients. Am J Clin Nutr 39:547–555, 1984.
36. Merkus MP, Jager KJ, Dekker FW, de Haan RJ, Boeschoten EW, Krediet RT: Predictors of poor outcome in chronic dialysis patients: The Netherlands Cooperative Study on the Adequacy of Dialysis. The NECOSAD Study Group. Am J Kidney Dis 35:69–79, 2000.
37. Marckmann P: Nutritional status and mortality of patients in regular dialysis therapy. J Intern Med 226:429–432, 1989.
38. Kalantar-Zadeh K, Stenvinkel P, Bross R, Khawar OS, Rammohan M, Colman S, Benner D: Kidney insufficiency and nutrient-based modulation of inflammation. Curr Opin Clin Nutr Metab Care 8:388–396, 2005.

12 Nutrition Support

Marcia Kalista-Richards
and Robert N. Pursell

LEARNING OBJECTIVES

1. Discuss the challenges faced by the clinician in providing nutrition support to the patient with stage 5 chronic kidney disease.
2. Identify nutrition support modalities that may be used for the patient with stage 5 chronic kidney disease.
3. Identify considerations in the provision of nutrition support for the patient with stage 5 chronic kidney disease.

Summary

The use of regular foods is always the preferred method to provide nourishment; however, at times specialized nutrition support is indicated. A review of the different nutrition support modalities including enteral and parenteral nutrition, parenteral nutrition via hemodialysis, known as Intradialytic Parenteral Nutrition (IDPN) and via the peritoneal dialysis, known as Intraperitoneal Nutrition (IPN) is covered. Considerations in the provision of such modalities are discussed.

Key Words: Nutrition support for stage 5 chronic kidney disease; parenteral nutrition; enteral nutrition; intradialytic parenteral nutrition; intraperitoneal parenteral nutrition.

1. INTRODUCTION

This chapter addresses nutrition support of adults with stage 5 chronic kidney disease (CKD). Information is presented on enteral

From: *Nutrition and Health: Nutrition in Kidney Disease*
Edited by: L. D. Byham-Gray, J. D. Burrowes, and G. M. Chertow

nutrition support including tube feeding administration, complications and management. When the gastrointestinal (GI) tract cannot be used, nutrients can be provided using parenteral nutrition (PN). Information is included on the fundamental components (protein, carbohydrate, fat, and electrolytes) in addition to other additives such as vitamins and trace elements along with guidelines and considerations for PN use. PN support is appropriate for the patient with stage 5 CKD; however, it is necessary to be cautious with respect to the total volume infused in addition to monitoring various electrolyte and mineral concentrations such as potassium and phosphorus. In addition, individuals with stage 5 CKD requiring dialysis have two rather unique methods of nutrition support; either intradialytic parenteral nutrition (IDPN) and intraperitoneal nutrition (IPN), each being specific to the type of dialysis selected. The indications, administration, adverse effects and the clinical outcomes are briefly described for each respective method.

2. ENTERAL NUTRITION SUPPORT

2.1. Oral Intake and Supplements

Individuals with stage 5 CKD require adequate nutrients to meet the demands of dialysis and associated comorbid states, to treat or prevent nutritional depletion, and to treat the malnutrition-inflammation complex syndrome *(1)*. At the same time, it is necessary to balance the many dietary restrictions imposed by kidney failure in order to maintain acceptable laboratory levels and to achieve safe fluid balance between dialysis treatments. Many factors impact on an individual's ability to consume adequate oral nourishment such as taste changes, GI disorders and depression (Table 1).

The clinician needs to be skillful in optimizing a balance between dietary needs and restrictions through the individualization of the nutrient prescription for the person with stage 5 CKD. Goals of the nutrition treatment plan should be threefold: (i) to provide foods normally eaten and tolerated; (ii) to promote meal satisfaction; and (iii) to limit any unnecessary restrictions. When a suboptimal intake is experienced, nutrition supplements should be encouraged; however, they may actually replace a meal and interfere with achieving a balanced intake. The concentrated amount of nutrients in supplements may lead to a sense of fullness. It is vital to carefully assess what an individual is eating and why they are not eating as well as to reverse or treat any situation that is interfering with intakes *(2)*.

Medications such as antiarrythmics, antineoplastics, calcium salts, oral iron preparations and some vasodilators have been associated with

Table 1
Factors Associated with Inadequate Oral Intake for the Stage 5 Chronic Kidney Disease Individual

Factors
Changes in taste acuity
Dietary restrictions and combination diets related to comorbitities
Dietary counseling or comprehension of information
Dental status affecting chewing abilities; denture fit and availability
Gastrointestinal problems including anorexia, nausea, vomiting and diarrhea
Constipation secondary to limited fluid and fiber intake, medications such as iron, phosphorus binders
Nutrient deficiencies such as iron causing pica symptoms or zinc related to taste change
General fatigue or that related to anemia
Dialysis schedule and frequent hospitalizations
Social issues related to income, depression, food security, housing
Cultural food preferences, religious practices
Polypharmacology

anorexia *(3)*. When medications interfere with intake, a review of those being used as well as the meal schedule is warranted, with consideration for the use of alternative medications when feasible *(2)*. Appetite stimulants have been used to improve oral intake for individuals with stage 5 CKD. Megestrol acetate, cyproheptidine and dronabinol are examples of pharmacological agents available to increase appetite and promote anabolism *(4,5,6,7)*.

2.2. Tube Feeding

When oral intake, including nutritional supplements, is inadequate, tube feeding should be considered if medically appropriate *(2)*. An individual's clinical condition, the presence and severity of malnutrition, the degree of food inadequacy and time period (days to weeks) that oral intake is less than optimal, will influence when enteral tube feeding support should be initiated *(2)*. Guidelines from the American Society of Parenteral and Enteral Nutrition *(8)* suggest that nutrition support should be initiated in patients with inadequate oral intake for 7–14 days, or in whom inadequate intake is anticipated for that time period *(8)*. Tube feeding is the preferred method to provide an individual with nutrients, as this method has been associated with fewer infectious and metabolic complications in comparison to PN. The philosophy continues to hold true that when the gut works, use it. Tube feeding can be the sole source of nutrition when necessary or can be used to meet the balance of nutrients not consumed orally.

2.2.1. Tube Feeding Access and Administration

The type of access for tube feeding administration will be influenced by the need for short-term use (few weeks) versus long-term use (greater than 6–8 weeks), patient comfort, comorbid states and dialysis treatment modality, risk of aspiration, presence of gastroparesis, safety considerations in placement and maintenance, and ease of use for patient or caregiver *(9,10)*. Nasogastric and nasoenteric access is used short term and can be used for gastric or small bowel feedings. Gastrostomy or gastrojejunostomy feedings into the stomach or small bowel, respectively, are generally used for long-term feeding. There is less reported use of percutaneous gastrostomy (PEG) feedings in patients receiving peritoneal dialysis (PD). Fein *(11)* reported that patients who have a functional PEG can safely start PD, whereas inserting a PEG tube in a patient currently on PD has been associated with a high rate of peritonitis.

The tube size must allow for a smooth flow of the formula especially if a more viscous formula is used (e.g., a concentrated 2 kcal/mL or fiber-containing formula). A tube lumen size of number 8 or larger is usually suggested for formulas containing fiber or for other viscous formulas administered by infusion pump *(12)*. The use of a feeding pump is typical in an acute care setting in order to administer a constant rate of formula. Intermittent or bolus feedings are more likely to be used for individuals who are ambulatory, in rehabilitation or long-term care settings and those who receive dialysis in an outpatient facility. In the home setting, enteral feeding pumps are used when gravity drip is not tolerated due to reflux, aspiration, severe diarrhea, dumping syndrome, uncontrolled glucose levels, circulatory overload, or infusion via jejunostomy tube at rates less than 100 mL *(13)*.

2.2.2. Tube Feeding Formulas

Formulas used for tube feeding range from those which require complete digestion to predigested formulas developed for individuals with digestive disorders, such as malabsorption or short gut syndrome. These formulas provide 1–1.2 kcal/mL to more concentrated products of 1.5–2 kcal/mL. Products have also been developed to meet specialty needs of individuals with disease-specific disorders such as liver or kidney disease. As products evolve and outcome-based research advances, it is necessary for the clinician to appropriately match the needs of the patient with the best product available in a safe and cost effective manner.

Products developed for patients with stage 5 CKD on dialysis require digestion and are lower in sodium, potassium and phosphorus with a concentrated source of calories and protein. Vitamin and mineral contents are designed to match the needs of the dialysis patient with added amounts of folic acid and pyridoxine and limited amounts of vitamins C and A. The Dietary Reference Intakes for most nutrients are met using 1 L of formula. Enteral formulas contain a source of vitamin K in concentrations ranging from 50 to 80 μg/1000 kcal of formula *(14)*. As a result of the early case reports of warfarin resistance related to the use of enteral feeding products, knowledge of the vitamin K content is important for patients receiving warfarin. The dose may need to be adjusted and coagulation protocols monitored with tube feedings, any oral intake or transition to oral intake *(14)*. If an individual is not adequately nourished, they may develop vitamin K deficiency, which could also lead to problems associated with coagulation.

Renal specific products may need to be used cautiously or be replaced with non-renal specific products in cases where an individual has been undernourished and is at risk for refeeding syndrome. Those individuals with a history of poor intake, chronic alcoholics, malnourished patients, and particularly marasmic patients or obese patients with significant weight loss, are at risk for refeeding syndrome *(15,16)*.

This syndrome can be defined as severe electrolyte and fluid shifts associated with metabolic abnormalities in malnourished patients undergoing aggressive refeeding, whether enterally or parenterally *(17)*. Hypokalemia, hypophosphatemia, hypomagnesemia, abnormal glucose metabolism, fluid balance abnormalities and thiamine deficiency can occur. Prevention includes the slow administration of nutrients at a caloric level below maintenance needs, with careful monitoring of phosphorus, potassium and magnesium as it should be anticipated that these values drop *(18,15,17)*. A low serum potassium, magnesium or phosphorus is generally not expected in patients with CKD; therefore, it is of particular importance to be aware of the problems related to refeeding syndrome. The lower potassium, phosphorus and magnesium content of renal formulas may actually precipitate a serious decline in serum electrolytes once nutrition support is initiated. It is advantageous if the clinician is familiar with the individual's previous nutrition intake before the start of nutrition support in order to select the most appropriate product and to manage the patient wisely. It is at this time that a concentrated non-renal specific formula is used rather than a renal product that is low in potassium, magnesium and phosphorus. Once serum values are normalized and the process of refeeding has been implemented and tolerated, the need for a renal specific formula can be reassessed with selection of the most appropriate enteral product to meet the patient's needs. Other non-renal concentrated formulas that provide 1.5–2 kcal/mL may also be used depending on the individual's state of hydration, dialysis schedule, and laboratory results. Modular carbohydrate, protein or fat sources are available and can be used to tailor individual needs.

Products that are concentrated (2 kcal/mL) and low in electrolytes and protein (~30 g of protein/L) have been marketed towards the individual with advanced CKD not on dialysis. As these products are very low in protein, the addition of a modular protein supplement with careful monitoring of overall nutrient intake may be necessary. Adherence to a low protein diet carries the risk of nutrient deficits and malnutrition (Table 2).

Over the years, closed enteral systems have become more widely used in the hospitalized setting. These products contain sterile tube feeding, are ready to hang for up to 24–72 hours, have been associated with less contamination and require less nursing intervention time *(19)*. The closed system solutions generally are 1–1.5 L of volume and are concentrated formulas (i.e., 2 kcal/mL). Disadvantages of the closed system include that it does not allow for the addition of additives, and it has the potential for greater waste if the tube feeding orders change or if the feeding is stopped prematurely or abruptly.

2.2.3. Tube Feeding Management

The management of GI complications including diarrhea, malabsorption, nausea and vomiting, along with mechanical complications such as tube occlusions, are similar for the patient with stage 5 CKD as well as other individuals who require enteral support. The patient requiring dialysis needs further monitoring of fluid status, electrolyte management, constipation and gastroparesis. The amount of free water flushes with feedings or the administration of medications needs to be strictly limited especially on non-dialysis days for the hemodialysis (HD) patient. The risk for constipation is greater for the dialysis patient as a result of both medications and limited fluid allowances. If a fiber-containing product is used, it is essential to achieve a balance between providing adequate fluid intake to promote stool output with the need for restricting fluids.

Tube occlusions are a potential problem particularly for someone using intermittent or cyclic feedings where multiple water flushes

Table 2
Renal-Specific, Calorically Dense and Modular Products

Formula Manufacturer	*Energy*	*Protein Content*	*Fat Content*	*Carbohydrate (Kcal)*	*Sodium (g)*	*Potassium (g)*	*Phosphorus (g)*	*Osmolality (M/Osm/kg)*	*Volume to Meet DRI*	*Water (mg)*
Nepro with Carb Steady (Abbott Nutrition)	1800	81	96	166.8	1060	1060	700	585	948	725
Novosource Renal (Novartis)	2000	74	100	200	1600	1100	650	960	1000	709
Nutrirenal (Nestle)	2000	70	104	204	740	1256	650	650	750	–
Novosource 2.0 (Novartis)	2000	90	88	220	800	1520	1100	790	1000	948
TwoCal HN (Abbott Nutrition)	2000	83.5	90.5	218.5	1450	2440	1050	725	948	700
Isosource 1.5 (Novartis)	1500	68	65	250	1290	2250	1070	650	933	778
Oxepa (Abbott Nutrition)	1500	62.7	93.8	105.3	1310	1960	1060	535	946	785
Suplena with Carb Steady (Abbott Nutrition)	1795	45	96	205	785	1120	–	600	948	735
Renalcal (Nestle)	2000	34.4	82.4	290.4	N/A	N/A	600	600	1000	700

(Continued)

Table 2
(Continued)

Formula Manufacturer	*Energy*	*Protein Content*	*Fat Content*	*Carbohydrate (Kcal)*	*Sodium (g)*	*Potassium (g)*	*Phosphorus (g)*	*Osmolality (M/Osm/kg)*	*Volume to Meet DRI*	*Water (mg)*
Ensure Plus (Abbott Nutrition)	1500	54.9	53.3	200	1012	1857	680	1185	760	770
Boost Plus	1520	59	58	200	720	1610	1310	720	946	780
Modular Formulas										
Beneprotein (Novartis) 7 g/1 scoop	25	6	0	0	15	15	20	–	–	–
ProMod (Abbott Nutrition) 6.6 g/1 scoop	28	5	0.60	0.67	25	45	35	–	–	–
Polycose (Abbott Nutrition) 100 mL	200	0	0	50	70	6	3	70 gram water	–	–

Source: Product information from respective companies were used.

are usually required between feedings and/or with administration of medications. In general, feeding tubes should be flushed routinely with about 20–30 mL of warm water every 4 hours during continuous feeding and before and after each intermittent or bolus feedings and medication administration *(8,18)*. Flushes contribute extra water and can be a concern for the renal patient. Calorically dense and fiber-containing products are more viscous, and formulas that are administered at very low rates can potentially aggravate the problem, as the formula may not flow easily through the tube *(20)*. In addition to water, carbonated beverages such as cola and cranberry juice have been used to flush feeding tubes; however, these products are not recommended *(20)* and they can negatively impact on serum potassium and phosphorus levels. Cranberry juice has also been reported to interfere with warfarin metabolism and therefore may predispose the patient to coagulation problems *(21)*. Liquid irrigants, enzyme solutions and mechanical devices have also been used to unclog a tube. A "prudent approach" used at one facility in home tube feeding patients included use of activated Viokase solution for 30 min/week in the nasoenteric and jejunal feeding, which successfully resolved sluggish infusions and prevented tube occlusions *(20)*.

Tube feeding rates are often reduced on dialysis days because of the dialysis schedule. Therefore, an increased amount of feeding may need to be given on non-dialysis days if tolerated, in order to meet the patient's nutritional needs, to prevent weight loss and a compromised nutrition status. An individual's compliance to the dialysis schedule will affect the ability to provide adequate nutrition. If dialysis is completed as scheduled, an individual is more likely to achieve fluid and electrolyte balance with clearance of uremic toxins, whereas under-dialysis results in fluid and electrolyte imbalances, uremic symptoms of nausea and elevated blood urea nitrogen (BUN) levels. The individual receiving PD may have fullness or distention problems as a result of the dialysate and the tube feeding could negatively affect the ability to provide adequate nutrition.

Fluid restriction is determined by whether the patient has residual renal output, the ability of the dialysis treatment to remove fluid accumulated between treatments, and other comorbid factors ranging from congestive heart failure to the presence of an ileostomy. Strict fluid restrictions would require the use of 2 kcal/mL formula; serum potassium and phosphorus values would dictate which enteral product would best meet the needs of the patient. Acute or chronic diarrhea or vomiting can decrease serum potassium levels, which need to be factored into the decision regarding product selection.

Gastroparesis is common in individuals with diabetes and delayed gastric emptying and is associated with dialytic procedures, elevated BUN values and hyperglycemia thereby necessitating close monitoring of the tolerance to enteral tube feedings *(22)*. GI tolerance may also be related to whether a patient is fed into the stomach versus post-pylorically and if intermittent versus continuous feedings are used. Tube feeding volume should be assessed and whether promotility agents such as metoclopramide or erythromycin are being used to promote gastric emptying *(18)*. The patient's position during dialysis should be considered to determine whether the tube feeding can be given. For example, if a patient needs to be in a supine position because of hypotensive episodes during dialysis, the feeding needs to be held in order to prevent the risk of aspiration. A semi-recumbent position (>30 degree elevation) is recommended to prevent aspiration for individuals requiring tube feeding *(18)*.

Patients with stage 5 CKD require medications with noted side effects such as constipation, diarrhea, nausea or vomiting *(23)*. Tube feedings have been associated with similar side effects; therefore, it is important to identify the etiology of the symptoms and to treat the problem based on the actual cause rather than an assumption. Prokinetic agents and sorbitol containing medications, as well as *Clostridium difficile* colitis, are also common causes of diarrhea *(8)*.

Metabolic complications can occur in individuals on tube feedings; the health care team must assess various chemistries independently to determine potential reasons for increased or decreased values (Table 3). Baseline metabolic and nutrition assessment parameters should be obtained with the initiation of enteral feedings, and follow-up parameters should be based on the individual's needs and clinical condition.

2.2.4. Guidelines for Administration

Initiation of enteral feedings is generally 25–50 mL/h, unless trophic feeding of 10–20 mL/h are used. Advancement of tube feeding rates or volume to the goal rate should be based on patient tolerance and status. It is important to be attentive to the caloric density of the product selected. When a 2 kcal/mL formula is used, nutrient needs can be met at a rate of 40–50 mL/h. For example, if there is a 2400 kcal goal, the volume given will differ significantly based on the caloric density of the product. A 2 kcal/mL formula at 50 mL/h equals 1200 mL or 2400 kcal and 85 g of protein, whereas a standard 1.2 kcal/mL formula at 80 mL/h equals 1920 mL or 2304 kcal and 106 g of protein. This is a difference of 720 mL (Table 4).

Table 3
Metabolic Complications of Tube Feeding

Complication	*Related Cause*	*Treatment*
Tube Feeding Syndrome (Azotemia, hypernatremia) Dehydration	High protein feedings with high renal solute load and inadequate water	Adequate fluid especially in older adult Avoid excess protein
(Refeeding Syndrome) Hypokalemia, hypomagnesemia Hypophosphatemia	Overzealous initiation of feedings	Conservative infusion of nutrition; Low carbohydrate infusion Start at approximately 20 kcal/kg body weight; Replete thiamine
Electrolyte Imbalance	Insulin therapy or alcoholism (hypokalemia, hypophosphatemia)	
Platin chemotherapy agents e.g. (Cisplatin, Carboplatin) (hypomagnesemia)	Diarrhea, Diuretics (hypokalemia, hypomagnesemia)	Therapeutic replacement
	Renal failure (hyperkalemia, hypermagnesemia hyperphosphatemia)	Use of enteral formulas with lower amounts

(Continued)

Table 3
(Continued)

Complication	*Related Cause*	*Treatment*
Hyperglycemia	Overfeeding, excess glucose infusion, diabetes mellitus, steroids, metabolic stress, sepsis	Calorie and Carbohydrate control Match nutrition to needs, prevent overfeeding. Calculate all sources of nutrition/dextrose. Insulin, oral hypoglycemic agent management. Fiber containing tube feeding may help. Assess use of and dose of steroids. Treat stress, sepsis.
Fluid Management Dehydration	Inadequate fluid intake, excess losses (diarrhea, ostomy, fistula, fever, diuresis, wounds); high protein feedings without adequate fluid	Meet fluid needs, replace losses Adequate water flushes via tube
Overhydration	Excess fluid, cardiac, renal or liver disease, excess water flushes or formula;	Meet fluid needs without excesses Consider all sources of fluid

Source: See ref. *(15)*.

Table 4
Comparison of Renal versus Non Renal Products

Formula	*Rate*	*Volume*	*Kcal*	*Protein*	*Fat mL*	*Carbohydrate*	*Sodium (gm)*	*Potassium (mg)*	*Phosphorus (mg)*	*Free Water (mL)*
Product #1 Renal High Protein	50 mL/h	1200	2160	97	115	200	1272	1272	840	870
Product #2 Renal Low Protein	50 mL/h	1200	2154	54	115	246	942	1344	840	882
Standard Product #1 1.2 kcal/mL	80 mL/h	1920	2304	107	75	302	2572	3475	2304	1574
Concentrated Product #2 1.5 kcal/mL	60 mL/h	1440	2160	90	71	293	2016	2592	1440	1094

An individual who is 54 years old; height 5'8", weight 154 pounds/70 kg; Estimated needs: 35 kcal/kg = 2450 kcal Protein needs: 1.2 gm/kg = 84 g.

Source: Product information book Abbott Nutrition.

3. PARENTERAL NUTRITION SUPPORT FOR THE ADULT WITH STAGE 5 CKD

3.1. Use of Parenteral Nutrition

PN is the administration of nutrients using an intravenous (IV) method intended for individuals who do not have a functional GI tract or in situations where administration of nutrients using the GI tract was not tolerated or could not be safely used. PN is also indicated for patients who are at nutrition risk, which can be defined as the loss of at least 10% of pre-illness weight and not taking food by mouth for 5–7 days *(8)*. PN is not beneficial unless it is used for at least 7–14 days *(8)*. This form of nutrition allows for the administration of amino acids (AA), dextrose, intravenous fat emulsions (IVFE), vitamins, minerals, electrolytes and fluids. Contraindications for the use of PN include a functional GI tract, if the risks outweigh the benefits, or if the individual is in a terminal or untreatable state.

PN can be administered using various access methods. Central administration of IV nutrition uses a large vein such as the internal jugular. The accesses typically used are either a percutaneous catheter or tunneled catheter, implanted port, or peripherally inserted central catheter. Because a large vein is used, osmolarity limits of substrates are not a concern as with peripheral administration of nutrition.

Peripheral parenteral nutrition (PPN) uses a peripheral vein in an upper extremity via a conventional needle access; however, it is recommended that substrates be limited to 900 mOsm/L in order to avoid sclerosis of the vein *(24,25)*. This generally limits the solution to an 8.5% AA and 20% dextrose or a final concentration of 4.25% AA and 10% dextrose. IVFE are isotonic, therefore, they can be administered peripherally and provide a concentrated source of calories. The site of the needle access may need to be changed frequently as a result of vein irritation and instability; however, the addition of heparin and hydrocortisone to a peripheral solution has been noted to improve tolerance *(24,25)*. PPN requires a larger volume of solution such as 3 L to meet the nutritional requirements, which is an amount unrealistic for the individual with stage 5 CKD who has fluid restrictions. Thus, central access is the preferred and necessary route for PN in this population.

PN solutions may be provided as a “2-in-1” solution which the carbohydrate, protein, vitamins, electrolytes, minerals, additives and sterile water are combined in one solution. IVFE can be administered separately from the “2-in-1” solution. When IVFE are added to the PN solution, it forms a total nutrient admixture (TNA) or a

"3-in-1" solution. Most individuals receiving home PN utilize the TNA solution. A TNA solution allows for a more concentrated solution and use of 30% IVFE for individuals with fluid limits. A 30% IVFE is intended for use only in a TNA, and it is not intended for direct infusion *(26)*. A 30% emulsion has a lower-phospholipids-to-triglyceride-ratio that may provide an advantage over 10% and 20% emulsions; however, further research is needed *(26)*.

3.2. Nutrient Substrates

3.2.1. Protein

Protein is provided in the PN formula to provide a source of nitrogen. Crystalline AA are the source of protein, with commercial solutions being available in concentrations of 3.5–20%. The concentration used will be based on an individual's protein requirements, fluid allowance and availability in the formulary. The adult patient with stage 5 CKD who has an increased protein requirement and often limited fluid allowance would generally use more concentrated AA solutions such as 10, 15, or 20%. Standard AA solutions include a balance of essential amino acids (EAA) and nonessential amino acids (NEAA). Renal failure formulas are modified in AA content and contain EAA or a combination of EAA and NEAA. The administration of only EAA for longer than 2–3 weeks may result in the development of hyperammonemia and metabolic encephalopathy *(8)*. For individuals with renal failure, arginine, ornithine and citrulline should be supplied to enable detoxification of ammonia via the Kreb's urea cycle *(8)*.

3.2.2. Carbohydrate

Carbohydrates provide a source of calories and are supplied as an anhydrous dextrose monohydrate in sterile water. It is available in concentrations ranging from 5 to 70% and provides 3.4 kcal/g of dextrose. The amount of dextrose provided is intended to support the energy needs of the individual with careful consideration of achieving tight glycemic control and minimizing the carbohydrate load, particularly for the individual requiring ventilator support. Although exact carbohydrate requirements have not been clearly identified, a minimum of 100 g/day is suggested, with general recommendations to provide 60% of total kcal as carbohydrates, not to exceed 5 mg/kg/min or 7 g/kg/day *(8,16,22,24)*. For the individual with CKD, fluid constraints combined with energy needs and glycemic control impact the selection of the percentage of dextrose and volume used. A more concentrated solution of 40–70% dextrose is often used.

3.2.3. Fat

Lipids provide a source of essential fatty acids and a concentrated source of calories. Long-chain fatty acid emulsions are the only source of IVFE currently available and may be made from either soybean oil or a combination of safflower and soybean oils. IVFE are available as either 10, 20, or 30% solutions; the 30% solution is reserved for TNA. IVFE contain egg phospholipids as an emulsifier with glycerol to adjust the osmolarity. IVFE are contraindicated in individuals who have an egg allergy because of the egg phospholipids and require evaluation for individuals with a soy allergy. IVFE contribute to the phosphorus and vitamin K intake.

General recommendations are to provide 20–35% of total calories as fat or 1 g/kg/day in critical illness *(8,16)*. Lipids should not exceed more than 60% of total calories, or supply more than 2.5 g/kg/day *(16,27)*. Lipid requirements can be met by providing 2–4% of calories as linoleic acid, 0.25–0.5% of total calories as α-linolenic acid or about 10% of total calories as a commercial lipid emulsion *(16,24)*. As newer commercial lipid emulsions are introduced, the minimum amount given is based on the linoleic acid content rather than the percentage of total calories *(24)*. Other lipid sources such as propofol contribute fat calories and should be calculated in the individual's overall caloric and lipid intake. Propofol provides 1.1 kcal/mL and 0.1 g fat/mL.

Serum triglyceride, alanine aminotransferase, aspartate aminotransferase and alkaline phosphatase should be monitored before the initiation of PN and routinely thereafter following administration of IVFE *(22)*. An increase in these parameters suggest impaired hepatic clearance of the lipid load *(22)*. A serum triglyceride level greater than 250 mg/dL after 6 hours of infusion suggests poor clearance of lipids *(27)*. Another suggestion is to maintain triglyceride values <400 mg/dL while IVFE are infusing *(24)*.

IVFE of 20–30% concentrations are usually recommended to conserve fluids, and it should be infused over a 12 hours period *(24)*. IVFE products need to be used within 12 hours of opening when given as a separate infusion because of the concern for microbial contamination *(16,28)*. The IVFE infusion rate should not exceed 0.125 g/kg/h *(16)*. Omega-6 fatty acid intake requires monitoring as high doses are associated with hypoxemia, bacteremia and suppression of immune function *(22)*. The use of IVFE as part of long-term therapy may include hepatomegaly, jaundice due to central lobular cholestasis, splenomegaly, thrombocytopenia, leucopenia, transient increases in liver function tests, overloading syndrome and deposition of brown pigment (fat pigment) in the reticuloendothelial tissue of the liver of which this cause is unknown *(29)*. IVFE containers should never be used if visual changes or precipitates are seen, nor should any partly used container be stored for later use *(28,29)*.

3.3. *Parenteral Additives*

Electrolytes, vitamins and minerals may be added to the solution to provide complete nutrition. PN should not be used as a vehicle for the administration of medications *(16,30)*. Although insulin is routinely added to PN, it has been associated with harmful effects; therefore, when insulin is added, adherence to defined protocols is paramount *(16)*. Medication stability and compatibility with TNA or 2 in 1 solutions need to be checked as data are limited *(30)*.

3.3.1. Fluid and Electrolytes

Impaired kidney function impacts on the ability of the kidneys to maintain normal fluid and electrolyte balance. The individual with stage 5 CKD requiring HD needs to limit fluid intake, as urinary output generally declines in proportion to the decreasing glomerular filtration rate. Daily fluid intake usually corresponds to fluid output. If urine output is greater than 1 L/day, 2 L of fluid intake/day is recommended to keep intradialytic weight gain to less than 5% of the individual's dry weight *(31)*. If urine output is less than 1 L/day, fluid intake of 1–1.5 L/day is recommended *(31)*. Over time, patients with stage 5 CKD become oliguric (i.e., urine output declines to $\leq$400 mL/day) or anuric (urine output $<$ 100 mL/day), which necessitates tighter fluid management. Volume is also limited in the individual with heart failure. The fluid recommendation for PD is 1–2 L/day and needs to be individualized to patient tolerance with minimal use of hypertonic solutions to maintain fluid balance.

Electrolyte requirements have been established for individuals with normal kidney function along with suggested values for individuals with kidney failure *(8)*. An individual with stage 5 CKD on PN will require an assessment of body weight, nutrition status, serum electrolyte values, residual kidney function, comorbid diseases and medications. Advanced renal failure results in more severe alterations of fluid and electrolyte balance. Serum potassium, magnesium and phosphorus levels may increase with declining kidney function because of impaired excretion. Management of electrolytes requires careful monitoring of laboratory values to minimize the potential risk of complications and to meet the patient's needs.

Sodium and potassium may be available as chloride, acetate, phosphate or lactate. Magnesium is most often available as the sulfate salt. Calcium gluconate is used most frequently as it is the most soluble *(24)*. Parenteral AA solutions include small amounts of electrolytes and acetate need to be calculated into the total solution. Acetate, a bicarbonate precursor, is used for PN rather than bicarbonate because of concerns of pH changes and the risk of insoluble precipitation with calcium and magnesium *(24)*. Acetate can be metabolized to bicarbonate by the liver. Individuals with acidosis may be treated with the addition of acetate; however, the amount will be dependent on sodium and potassium additives. Bicarbonate may also be used to treat acidosis but would require access using a separate line. The amount of calcium and phosphorus requires careful monitoring as excess amounts added to the PN solution may result in insoluble precipitation causing crystal deposition that may lead to death *(24)*.

3.3.2. Vitamins and Trace Minerals

The American Medical Association Nutrition Advisory Committee has made recommendations for the inclusion of multivitamins (MVI) and trace elements for individuals who require PN *(8)*. Recommendations for individuals with renal failure are published by the American Society of Parenteral and Enteral Nutrition *(8)*. Ambler and Kopple *(32)* recommend 10 μg chromium, 1 mg copper, 75 μg iodine, 500 μg manganese, 60 μg selenium and 5 mg zinc once per week for the individual with stage 5 CKD who receives PN for greater than 2 weeks.

Commercially available parenteral MVI preparations for adults contain 12 of the known vitamins; vitamin K has been added to create MVI 13. The current MVI 13 contains 150 μg vitamin K, whereas MVI 12 does not contain any Vitamin K. IVFE also contains vitamin K related to the oils, which lipid emulsions are manufactured from (soybean or a combination of soybean and safflower). Soybean oil contains between 150 and 300 μg vitamin K/100 g of oil, whereas safflower oil contains 6–12 μg of vitamin K/100 g of oil *(14)*. It is expected that those IVFE that contain more soybean oil would be higher in vitamin K. Recommendations for vitamin K are 1 mg daily *(8)*, or 2–4 mg vitamin K/week *(33)*, which can be met using the IV MVI preparation and IVFE. All sources of vitamin K (IVFE, MVI, intramuscular injections or oral or tube feeding products if used) should be assessed. Consideration should be given to adjusting dosage of anticoagulation in this setting.

Parenteral MVI contains 1 mg vitamin A, whereas oral vitamins used for dialysis patients do not contain vitamin A. Total vitamin A intake, whether oral or parenteral should be monitored, particularly in individuals who require long-term PN support as toxicity may cccur in presence of renal failure.

The Food and Nutrition Board of the National Academy of Sciences recommended that vitamin C in parenteral preparations be increased from 100 to 200 mg daily *(34)*. Doses of IV vitamin C at this level can increase urinary oxalate excretion that could potentially elevate the risk of calcium oxalate crystallization in individuals susceptible to nephrolithiasis *(34,35)*. Long-term use of PN in individuals receiving home support, in those with risk factors for nephrolithiasis (e.g., significant GI disease), in those with reduced kidney function, and in those with other dietary sources of vitamin C, require careful monitoring *(34)*.

The use of IV iron in PN remains controversial, and supplementation has been associated with increased proliferation of microorganisms *(36)*; therefore, iron is recommended via the enteral route when possible *(16)*. Iron cannot be added to a TNA solution because of its incompatibility with lipids. When iron dextran is added to non-IVFE-containing PN solutions, compatability limits should be monitored *(16)*.

The selection and amount of trace elements administered to patients with CKD varies considerably. Trace elements may be either withheld, given several times per week, administered in a half dose, or specific trace elements are ordered individually. It is often difficult to predict the exact amount required for this population as many of the patients have comorbid conditions and unique nutritional demands. A combination of clinical judgment, assessment of symptoms and evaluation of laboratory data should be considered when deciding to administer trace elements and amounts. It is necessary to consider the route of excretion and losses associated with renal replacement therapy (RRT). Additionally, the clinician should evaluate for potential risk for losses (malabsorption, fistula or stool) versus risk for toxicity secondary to impaired renal or biliary excretion. For example, selenium may be a concern for anuric patients as selenium homeostasis is maintained through urinary excretion. Furthermore, selenium deficiency has been reported in critically ill patients and those receiving long-term PN *(36)*.

Individuals with hepatobiliary disease have impaired excretion therefore should receive reduced amount of manganese and copper *(16)*. These trace elements should be avoided in individuals with chronic hyperbilirubinemia.

There appears to be a protective effect of the GI tract as a barrier for aluminum absorption *(37)*. However, because aluminum is cleared primarily via the kidney, individuals with impaired renal function are at risk for toxicity *(37)*. Most published reports of toxicity relate to the use of PN and/or the use of aluminum hydroxide phosphate binders *(37,38)*. Therefore, individuals with renal dysfunction and long-term PN require monitoring due to the combined risk factors *(38)*. Changes in bone formation and parathyroid hormone (PTH) secretion along with anemia and central nervous system changes have been associated with aluminum toxicity *(16,21)*. As of July 2004, the Food and Drug Administration (FDA) mandated that manufacturers of PN products measure the aluminum content and disclose it on the label *(16)*. The maximum limit of aluminum is 25 μg/L for large volume parenterals (i.e., AA solutions, concentrated dextrose solutions, IVFE and sterile water for injection) *(16)*. The maximum amount of aluminum identified by the FDA that can be safely tolerated is 5 μg/kg/day *(21)*. Continued research, reporting of outcomes and observations which helps strengthen evidence-based guidelines are needed in this area.

3.3.3. Other Additives

Choline can be made from the AA methionine. Choline is used to make the phospholipid lecithin and the neurotransmitter acetylcholine. Further research is needed to determine the need for restriction or supplementation of choline in the stage 5 CKD population, particularly when requiring long-term PN where choline deficiency has been associated with hepatic steatosis *(39)*.

Other additives to the PN solution may include heparin to reduce complications of catheter occlusion related to fibrin formation around the catheter tip *(24)*, with a general recommendation of 1 U/mL volume for central access *(40)*. Heparin should be avoided in patients with heparin antibodies and patients who have had a bone marrow transplant *(40)*.

Regular insulin can be added using 0.2–0.3 units per gram of carbohydrate per 24 hours to control glucose levels in patients with type 1 diabetes *(40)*. For patients with type 2 diabetes or those with iatrogenic hyperglycemia insulin can be administered as two-thirds of the previous 24 hours insulin coverage *(40)*. H2-receptor blockers including famotidine, ranitidine or cimetidine can be added with dose adjustments according to the stage of kidney disease.

3.4. Initiation and Monitoring of Parenteral Nutrition Support

PN orders should consider the patient's fluid, electrolyte and acid base balance status, glycemic control and access availability *(41,42)*. When initiating PN, it is recommended to use a volume of 1 L or to limit total calories until an assessment of patient's tolerance and laboratory parameters are made. The total volume can be delivered safely with limiting the dextrose calories; however, the clinician needs to assess all other sources of fluid and combine or adjust sources based on status. When the total fluid volume needs to be reduced, total infusion of PN may be limited to 1–1.5 L/day. Meeting of RRT goals and monitoring overall volume status of patient will determine fluid allowances when starting PN and identifying the safe total volume allowance.

Correction of electrolyte disturbances is recommended prior to starting PN. Refeeding syndrome can be avoided by limiting the amount of carbohydrate in feedings *(8)*. Consider refeeding abnormalities that can occur and what correction factors have been implemented using RRT. For example, if an individual is at risk for refeeding syndrome (which can lower serum potassium), the same individual may have hyperkalemia because of kidney failure, therefore, the dialysate solution can be adjusted to a low potassium bath. Caution should be taken that the potassium does not fall to an undesirable level with the initiation of PN. Thus, it is important to assess the whole picture.

Current literature supports normoglycemia to reduce the morbidity and mortality of critically ill patients *(41)*. Hyperglycemia has been associated with decreased immune function and increased risk of infectious complications; therefore, efforts to monitor and achieve blood glucose control are recommended. Conservative initiation of PN and glycemic control takes precedence over aggressive feedings. Dextrose may be limited to 100–150 g/day for the first day, particularly in individuals with diabetes who are at risk for refeeding syndrome, or with steroid use *(24)*. Advancement to goal rates should be made as glucose levels permit.

Protein can be administered at full dose. IVFE may be started at full dose provided serum triglyceride levels are within normal range *(24)*, or administered at 0.5 gm/kg/day and advanced to meet the individual's needs while checking plasma triglyceride levels *(42)*. Hypertriglyceridemia should be avoided in order to prevent the potential development of pancreatitis *(8)*.

Acid base abnormalities may be a result of the individual's underlying condition, although nutrition support can also influence the values as well. Manipulation of the acetate and chloride content of the PN may aid in the correction of such abnormalities. Serum electrolytes should be monitored closely.

3.5. Discontinuing Parenteral Nutrition

Hypoglycemia can occur if the solution is discontinued abruptly; however, this is an uncommon event. Prevention tactics include tapering the PN solution, hanging 10% dextrose at the same rate as the PN, monitoring serum glucose levels when insulin is being given, and assessing other sources of nutrition (IV, tube feedings or oral intake).

3.6. Monitoring Clinical and Laboratory Parameters

General guidelines for monitoring clinical and laboratory parameters in patients receiving PN include obtaining a baseline comprehensive metabolic panel, daily weights, intakes and output, daily laboratory values until stable then weekly thereafter. Serum glucose three times per day until stable. Serum triglyceride levels should be checked prior to the infusion of lipids. Liver enzymes, bilirubin and a complete blood count should be checked when PN is initiated, then two to three times per week until stable, then weekly *(27)*. Vitamin D, calcium and phosphorus, aluminum and PTH should be checked monthly for stage 5 CKD patients requiring dialysis and receiving PN, along with monthly assessment of transport proteins *(27)*. Frequency of monitoring will be based on acuity and results. Less frequent monitoring can be used as patient stabilizes (Table 5).

4. INTRADIALYTIC PARENTERAL NUTRITION

IDPN consists of the infusion of nutrients through the venous drip chamber during the HD treatment. The nutrients can consist of AA (EAA, NEAA or both), glucose and IVFE (Refer to Table 6 for sample IDPN formulas). The use of IDPN was first reported by Heidland and Kult *(43)* in 1975 when they infused EAA plus histidine during the later part of a HD treatment. There is concern that AA loss during the HD treatment may contribute to the overall poor nutritional status of patients on chronic HD. In fact, Giordano et al. *(44)* showed an average loss of 1.5–2 g of free AA per hour during HD. Furthermore, Wolfson et al. *(45)* demonstrated that with standard dialyzers, approximately

Table 5
Monitoring Guidelines for Parenteral Nutrition

Parameter	*at Baseline*	*if Critically Ill*	*if Stable*	*using IDPN*	*using IPN*
Weight	Yes	Daily	2–3×/Week	Dialysis Days	Daily
Intake/output	Yes	Daily	Daily unless fluid can be assessed by weight and physical exam	Daily	Daily
Subjective global assessment	As able	Daily	Monthly then every 6 months	Monthly then every 6 months	Monthly then every 6 months
Capillary glucose	3×/day	3×/day until less than 150 mg/dL	3×/day until less than 150 mg/dL	Line draw before, during, after hemodialysis	Daily
Electrolytes, BUN/creatinine	Yes	Daily	1–2×/week, then monthly if long term	Every dialysis, until stable, then monthly	Monthly
Calcium, phosphorus magnesium	Yes	Daily until stable, then 2–3×/week	Weekly, then monthly if long term	Each dialysis, then monthly when stable	Monthly
Liver function studies	Yes	2–3×/week until stable	Weekly, then monthly when stable	Monthly	Monthly
Triglycerides	Yes	Weekly	Monthly	6 hours after treatment, then monthly	Monthly
Transport proteins	Weekly	Weekly	Monthly	Monthly	Monthly

Adapted from refs. *(72,73)*. IDPN, intradialytic parenteral nutrition; IPN, introperitoneal nutrition.

Table 6
Different Formulations for Intradialytic Parenteral Nutrition

Type	*Composition*	*kcal*	*Non-protein Kcal*	*Volume*
250 mL D50W 250 mL 10% AA	125 g dextrose 25 g protein	525	425	500 mL
250 mL D50W 500 mL 8.5% AA 250 mL 20% lipids	125 g dextrose 43 g protein 50 g fat	1097	925	1000 mL
250 mL D50W 250 mL 10% AA 250 mL 20% lipids	125 g dextrose 25 g protein 50 g of fat	1025	925	750 mL
500 mL D50W 550 mL 10% AA	250 g dextrose 55 g protein	1070	850	1050 mL
250 mL D50W 250 mL 10% AA	125 g dextrose 55 g protein	815	595	500 mL
250 mL D70W 500 mL 15% AA 250 mL 20% lipid	175 g dextrose 75 g protein 50 g of fat	1395	1095	1000 mL

AA, amino acid.
Source: Product information.

8 g of AA was lost into the dialysate during a routine HD treatment. On the other hand, the same group was able to demonstrate that when an infusion of AA plus glucose was given during HD, 79% of the AA was retained. Therefore, AA and glucose supplementation during HD may be used to avoid net AA loss and potentially provide a nutritional supplement. Some research has been reported on the effectiveness of IDPN, but larger clinical trials are needed to determine the impact of IDPN on patient outcomes *(43,46–57)*.

4.1. Indications and Qualifications of Therapy

More specific guidelines for initiating IDPN in chronic HD include the following: a weight of less than 90% of the ideal body weight, a weight loss of more than 10% over 6 months, triceps skinfold thickness less than 6 mm in males or less than 12 mm in females, serum albumin less than 3.4 g/L, or a protein catabolic rate of less than 0.8 g/kg/day *(58,59)*.

High cost and reimbursement issues limit the use of IDPN in the HD population. In fact, Foulks *(60)* indicated that the cost of IDPN may be as high as $30,000 per year, per patient. In order to justify IDPN in a patient, a chronic GI disorder, such as severe diabetic gastroparesis, must be present, which would prohibit the use of the enteral route for feeding. An intact GI tract would otherwise allow the use of oral feeding with supplements or even tube feedings to meet the nutritional needs. Furthermore, it is understood that IDPN can only meet part of the nutritional needs of a given patient and that a minimal oral intake is necessary which may need to make up as much as 50% of the nutritional requirements. Lastly, IDPN is not a substitute for PN. If the nutritional needs exceed what IDPN supplemented with oral intake can provide, PN should be utilized.

4.2. Monitoring

Hyperglycemia may occur as a result of the high dextrose load, and blood sugars should be monitored before, during and after HD. If insulin is required, it can be given during the treatment with the IDPN. Rebound hypoglycemia may also occur after dialysis, and it has been recommended that a snack consisting primarily of carbohydrates be given towards the end of dialysis. Hypokalemia, hypomagnesemia and hypophosphatemia have been reported and may result from high insulin levels, or the development of an anabolic process that would increase the intracellular concentrations of these various electrolytes. If electrolytes are added to the IDPN, the concentrations should be monitored accordingly. When IVFE are utilized, the patient should be observed for signs of lipid intolerance during the first administration. Protocols have been suggested for lipid monitoring *(61)*. Generally, IVFE infusions should not exceed 2.5 g/kg/day.

The additional volume of IDPN which can vary between 500 and 1000 mL does not seem to be a problem as it can generally be removed with ultrafiltration. Lastly, pain or cramping in the access arm has infrequently been reported during the infusion. The cause has not been defined; however, it may be related to the hypertonicity of the infused solution. A number of adverse effects can occur with IDPN. For the most part, these are infrequent and mild in severity and can sometimes be eliminated with various adjustments in the treatment.

5. INTRAPERITONEAL NUTRITION

PD consists of a solute and fluid exchange between the peritoneal capillary blood and the dialysate solution within the peritoneal cavity across the peritoneal membrane. The standard peritoneal fluid contains a concentration ranging from 1.5 to 4.25% glucose serving as the osmotic agent. IPN consists of the infusion of AA and/or glucose into the peritoneal cavity. This process involves the replacement of one or up to two exchanges of dextrose-based dialysate with 1–2% AA solution. The net absorption of AA has been shown to have a positive impact on the protein-calorie malnutrition found in the peritoneal dialysis population. The potential for the peritoneum to absorb fluid and nutrition was suggested in 1918 by Blackfan and Maycy *(62)* when normal saline was infused into the peritoneal cavity of dehydrated children. This in fact preceded the first report of PD being used to treat renal failure by Ganter in 1923 *(63)*. In 1979, a report by Oreopoulous et al. *(64)* described the use of AA instead of glucose as both an osmotic agent and a nutritional source.

The use of AA-based dialysis solutions must address the degree of net AA absorption and the potential for ultrafiltration. Williams et al. *(65)* reported that the absorption of AA is between 70 and 90% after a 6 hours dwell. This may vary with the patient's membrane characteristics and dwell time. Therefore a 2-L bag of 1% AA solution will provide approximately 14 g of AA. Furthermore, it has been shown by Oren et al. *(66)* that there may be general improvement in the plasma AA profile. Generally, one 2-L bag of AA-based solution is substituted for a dextrose-containing bag in the evening exchange. Therefore, it has been demonstrated that there is a net AA reabsorption that could potentially have a positive impact on the nutritional status in patients on PD *(67–70)*

5.1. Indications

According to the National Kidney Foundation Kidney Disease Outcome Quality Initiative recommendations *(71)*, patients may benefit from AA-based solutions if they satisfy the following criteria: (i) evidence of protein malnutrition and inadequate dietary protein intake; (ii) inability to administer or tolerate adequate oral protein nutrition, including food supplements, or enteral tube feeding; (iii) the combination of some oral or enteral intake, which when combined with AA-based solutions will meet the individual's nutritional goals. Furthermore, in patients who have difficulty controlling hyperglycemia, hypercholesterolemia or hypertriglyceridemia that is related to excessive carbohydrate absorption from peritoneal dialysate, AA-based solutions may reduce serum glucose and lipid levels. It should be understood that AA-based solutions should not take the place of total PN when a patient's daily protein requirements are such that PN would be the treatment of choice.

5.2. Monitoring

Side effects of AA dialysate can occur and, for the most part, are mild in nature. Azotemia may worsen and can be associated with GI symptoms such as nausea and vomiting. One should monitor the BUN, and symptoms generally resolve with decreased AA load. Serum bicarbonate level should be monitored. If there is evidence of metabolic acidosis, it is usually mild and decreases with an adjustment of the AA load or the addition of an alkalizing agent.

6. CONCLUSION

Patients with stage 5 CKD are at nutritional risk and require a thorough nutrition assessment to formulate appropriate nutrition prescriptions. The decision to initiate nutrition support and what type of support should be based on the patient's overall clinical condition, nutritional status and laboratory results. Individuals requiring nutrition support must be monitored closely to avoid potential complications.

Periodic reassessment of the patient with stage 5 CKD is important to assure nutrient adequacy, the need for formula changes based on the individual's needs, weight status, and goals, along with overall tolerance. An interdisciplinary team approach should be implemented when providing patient care and nutrition support. It is vital to look at the whole picture, and evaluate the patient's clinical status, mode of dialysis and medications in order to provide the most comprehensive care and assure the safety of the patient.

7. CASE STUDY

An 81-year-old male on HD for several years with multiple medical problems developed a state of severe malnutrition after being hospitalized for respiratory failure and pneumonia. Patient's albumin decreased to 2.4 gm/dL and lost approximately 7 kg of body weight. Patient was started on IDPN twice a week as an outpatient. The formulation consisted of 500 mL of 50% dextrose in addition to 50 g of protein. Serum albumin increased from 2.4 to 2.9 g/L in 4 months. Renal dietary restrictions were relaxed and he was encouraged to eat whatever he wished. Subsequently, his appetite actually improved and he gained 3 kg of lean body weight. His activity level also increased and he began to travel with his family and attend social functions.

7.1. Questions

1. What frequently precipitates severe malnutrition in a previously stable CKD 5 patient?
2. How often is IDPN administered?
3. How is nutritional improvement recognized?
4. Can one's appetite improve?
5. What are the potential benefits of this form of nutritional support?

REFERENCES

1. Colman S, Bross R, Benner D, et al. The nutritional and inflammatory evaluation in dialysis patients (NIED) study: Overview of the NIED study and the role of dietitians. J Ren Nutr 2005;15:231–243.
2. National Kidney Foundation. Kidney disease outcomes quality initiative clinical practice guidelines for nutrition in chronic renal failure. Am J Kidney Dis 2000:35(suppl 2):S1–S140.
3. Smith B, Garney P. Medications commonly prescribed in chronic kidney disease. In: Byham-Gray L, Wiesen K, eds. A Clinical Guide to Nutrition Care in Kidney Disease, Chicago: American Dietetic Association, 2004:175–194.
4. Gullickson C. The use of appetite stimulants and anabolic agents in hemodialysis patients. Renal Nutrition Forum 2005;23:1–6.
5. Burrowes JD, Bluestone PA, Wang J, et al. The effects of moderate doses of megestrol acetate on nutritional status and body composition in a hemodialysis patient. J Ren Nutr 1999;9:89–94.
6. Boccanfuso JA, Hutton M, McAllister B. The effects of megestrol acetate on nutritional parameters in a dialysis population. J Ren Nutr 2000;10:36–43.
7. Rammohan M, Kalantar-Zadeh K, Liang A, et al. Megestrol acetate in a moderate dose for the treatment of malnutrition-inflammation complex in maintenance dialysis patients. J Ren Nutr 2005;15:345–355.
8. A.S.P.E.N. Board of Directors and the Clinical Guidelines Task Force. Guidelines for the use of parenteral and enteral nutrition in adult and pediatric patients. JPEN J Parenter Enteral Nutr 2002;26:(suppl 1):1–138SA.
9. DeChicco RS, Matarese LE. Determining the nutrition support regimen. In: Matarese LE, Gottschlich MM, eds. Contemporary Nutrition Support Practice: A Clinical Guide, 2nd edn, Philadelphia: Saunders, 2003:181–187.
10. Neven, A. Enteral access device and care. In: Kalista-Richards M, Marian M, eds. In: Sharpening Your Skills as a Nutrition Support Dietitian. The American Dietetic Association Dietitians in Nutrition Support, Chicago: 2003:109–120.
11. Fein, PA. Safety of PEG tubes in peritoneal dialysis patients. Semin Dial 2002;15:213–214.
12. Lysen LK. Enteral equipment. In: Matarese LE, Gottschlich MM, eds. Contemporary Nutrition Support Practice: A Clinical Guide, 2nd edn, Philadelphia: Saunders, 2003:201–214.
13. Winkler M. Discharge planning and reimbursement. In Kalista-Richards M, Marian M, eds. In: Sharpening Your Skills as a Nutrition Support Dietitian. Chicago: The American Dietetic Association Dietitians in Nutrition Support, 2003:173–186.

14. Singh H, Duerksen DR. Vitamin K and nutrition support. Nutr Clin Pract 2003;18:359–365.
15. Beyer PL. Complications of enteral nutrition. In: Matarese LE, Gottschlich MM, eds. Contemporary Nutrition Support Practice: A Clinical Guide, 2nd edn, Philadelphia: Saunders, 2003:215–226.
16. Task Force for the Revision of Safe Practices for parenteral Nutrition. Safe practices for parenteral nutrition. JPEN J Parenter Enteral Nutr 2004;28:S39–S70.
17. Crook MA, Hally V, Panteli JV. The importance of the refeeding syndrome. Nutrition 2001:17:632–637.
18. Lord L, Harrington M. Enteral nutrition implementation and management. In: Merritt, R, ed. A.S.P.E.N. Nutrition Support Practice Manual, 2nd edn, Silver Spring: A.S.P.E.N., 2005:76–89.
19. Vanek WV. Closed versus open enteral delivery systems: A quality improvement study. Nutr Clin Pract 2000;15:234–243.
20. Lord LM. Restoring and maintaining patency of enteral feeding tubes. Nutr Clin Pract 2003;18:422–426.
21. Drug nutrient interactions. Hayes E. In: Merritt, R, ed. A.S.P.E.N. Nutrition Support Practice Manual, 2nd edn. Silver Spring: A.S.P.E.N., 2005:118–136.
22. Goldstein-Fuchs DJ, McQuiston B. Renal failure. In: Matarese LE, Gottschlich MM, eds. Contemporary Nutrition Support Practice: A Clinical Guide, 2nd edn, Philadelphia: Saunders, 2003:460–483.
23. Smith B, Garney P. Medications commonly prescribed in chronic kidney disease. In: A Byham-Gray L, Wiesen K, eds. A Clinical Guide to Nutrition Care in Kidney Disease, 1st edn, Chicago, IL: The American Dietetic Association 2004:175–194.
24. Skipper, A. Parenteral nutrition. In: Matarese LE, Gottschlich MM, eds. Contemporary Nutrition Support Practice: A Clinical Guide, 2nd edn, Philadelphia: Saunders, 2003:227–241.
25. Krzywda EA, Edmiston CE. Parenteral nutrition access and infusion equipment. In: Merritt, R, ed. A.S.P.E.N. Nutrition Support Practice Manual, 2nd edn, Silver Spring: A.S.P.E.N., 2006:90–96.
26. Product Information. Clintec Nutrtition Division. Baxter Healthcare Corporation, One Baxter Parkway, Deerfield, IL 60015-4633.
27. Fuhrman, MP. Parenteral nutrition in kidney disease. In: Byham-Gray, Wiesen K, eds. A Clinical Guide to Nutrition in Kidney Disease, Chicago, IL: The American Dietetic Association, 2004:159–174.
28. Sacks GS, Mayhew S, Johnson D. Parenteral nutrition implementation and management. In: Merritt, R, ed. A.S.P.E.N. Nutrition Support Practice Manual, 2nd edn, Silver Spring: A.S.P.E.N., 2006:108–117.
29. Liposyn Product Information. Hospira, Inc. Medical Communications, Dept 098 Q Bldg H-2, 275 N. Field Drive, Lake Forest IL 60045.
30. Kumpf VJ, Mirtallo JM, Petersen C. Parenteral nutrition formulations: Preparation and ordering. In: Merritt R, ed. A.S.P.E.N. Nutrition Support Practice Manual, 2nd edn, Silver Spring: A.S.P.E.N., 2006:96–107.
31. Kopple JD. National Kidney Foundation KDOQI clinical practice guidelines for nutrition in chronic renal failure. Am J Kidney Dis 2001;37(suppl):S66–S70.
32. Amber C, Kopple JD. Nutrition support for patients with renal failure. In: Klein S, ed. The A.S.P.E.N. Nutrition Support Practice Manual. Silver Spring: American Society for Parenteral and Enteral Nutrition, 1998:16-1–16-12.

33. National Advisory Group on Standards and Practice Guidelines for Parenteral Nutrition. Safe practices for parenteral nutrition formulation. JPEN J Parenter Enteral Nutr 1998:22:49–66.
34. Pena de la Vega L, Lieske J, Milliner D, et al. Urinary oxalate excretion increases in home parenteral nutrition patients on a higher intravenous ascorbic acid dose. JPEN J Parent Enteral Nutr 2004;28:435–441.
35. Council on Renal Nutrition of the National Kidney Foundation. Nutrient calculations. In: McCann L, edn, Pocket Guide to Nutrition Assessment of the Patient with Chronic Kidney Disease, 3rd edn, New York: National Kidney Foundation, 2002:3-1–3-18.
36. Fuhrman MP. Complication management in parenteral nutrition. In: Matarese LE, Gottschlich MM, eds. Contemporary Nutrition Support Practice A Clinical Guide, 2nd edn, Philadelphia: Saunders, 2003:242–262.
37. Fuhrman MP, Parker M. Micronutrient assessment. Support Line 2004;26:17–24.
38. Klien GL. Nutritional aspects of aluminum toxicity. Nutr Res Rev 1990;3: 117–141.
39. Buchman, A. Choline deficiency during parenteral nutrition in humans. Nutr Clin Pract 2003;18:353–358.
40. Mataresse LM. Writing parenteral nutrition orders. In: Kalista-Richards M, Marion, M, eds. Sharpening Your Skills as a Nutrition Support Dietitian. Chicago: The American Dietetic Association Dietitians in Nutrition Support, 2003:89–98.
41. van den Berghe G, Wouters PJ, Bouillon R, et al. Outcome benefit of intensive insulin therapy in the critically ill: Insulin dose versus glycemic control. Crit Care Med 2003;31:359–366.
42. Barco K. Total parenteral nutrition for adults with renal failure in acute care. Renal Nutrition Forum 2003;22:1–7,11–12.
43. Heidland A, Kult J. Long term effect of essential amino acid supplementation in patients on regular dialysis treatment. Clin Nephrol 1975;3:234–239.
44. Giordano C, Pascale C, Christofaro D, et al. Protein malnutrition in the treatment of chronic uremia. In: Livingstone BG, ed. Nutrition and Renal Disease, Edinburgh, Livingstone, London, 1968:23.
45. Wolfson M, Jones M, Kopple J. Amino acid losses during hemodialysis with infusion of amino acids and glucose. Kidney Int 1982;21:500–506.
46. Thunberg B, Jaih D, Patterson P, et al. Nutritional measurements and urea kinetics to guide intradialytic hyperalimentation. Proc Clin Dial Transplant Forum 1980;10:22–28.
47. Powers D, Jackson A, Piraino A, et al. Prolonged intradialysis hyperalimentation in chronic hemodialysis patients with amino acid solution renamin formulated for renal failure. In: Kinney JM, Borum PR, eds. Prospectives In Clinical Nutrition, Baltimore, Munich: Urban & Schwarzenberg, 1989:191–205.
48. Czeklaski S, Hozejowski R, Malnutrition Working Group. Intradialytic amino acid supplementation in hemodialysis patients with malnutrition: The results of a multicenter study. J Ren Nutr 2004;14:82–88.
49. Guarniere B, Faccini L, Lipartiti T, et al. Simple methods for nutritional assessment in hemodialyzed patients. Am J Clin Nutr 1980;33:1598–1607.
50. Hecking E, Port F, Brehm H, et al. A controlled study on the value of amino acid supplementation with essential amino acids and ketoanalogs in chronic hemodialysis. Proc Dial Transplant Forum 1977:7:157–161.

51. Chertow G, Ling J, Lew J, et al. The association of intradialytic parenteral nutrition administration with survival in hemodialysis patients. Am J Kidney Dis 1994;24:912–920.
52. Olshan R, Bruce J, Schwartz A, et al. Intradialytic parenteral nutrition administration during outpatient hemodialysis. Dial Transplant 1987;16:495–496.
53. Madigan K, Yingling D. Effectiveness of intradialytic parenteral nutrition in diabetic patients with end stage renal disease. J Am Diet Assoc 1990;6:861–863.
54. Foulks C. The effect of intradialytic parenteral nutrition on hospitalization rate and mortality in malnourished hemodialysis patients. J Ren Nutr 1994;4:5–10.
55. Blondin J. Nutritional status: A continuous quality improvement approach. Am J Kidney Dis 1999;33:198–202.
56. Capelli J, Kushner H, Camiscioli T, et al. The effect of intradialytic parenteral nutrition on mortality rates in end stage renal disease care. Am J Kidney Dis 1994;23:808–816.
57. Ruggian J, Fishbane S, Frei G, et al. The effect of intradialytic parenteral nutrition on hypoalbuminemic hemodialysis patients {abstract}. J Am Soc Nephrol 1994;5:502.
58. The American Dietetic Association. A Clinical Guide to Nutritional Care in End Stage Renal Disease, 2nd edn, Chicago: American Dietetic Association, 1994.
59. Kopple J, Foulks C, Piriano B, et al. National kidney foundation position paper on proposed healthcare financing administration guidelines for reimbursement of enteral and parenteral nutrition. J Ren Nutr 1996;6:45–47.
60. Foulks C. An evidence of base evaluation of intradialytic parenteral nutrition. Am J Kidney Dis 1999;33:186–192.
61. Goldstein D, Strom J. Intradialytic parenteral nutrition: Evolution and current concepts. J Ren Nutr 1999;1:9–22.
62. Blackfan K, Maycy K. Intraperitoneal injection of saline solution. The American Journal of Diseases in Children 1918;15:19.
63. Ganter G. Ueber Beseitigung Giftiger Staffe Aus Dem Blute Dirch Dialyse. Nunch Med Wochschr 1923;70:1478.
64. Oreopoulous D, Clayton S, Dombros N, et al. Amino acids as an Osmotic Agent (Instead of Glucose) in Continuous Ambulatory Peritoneal Dialysis. Continuous Ambulatory Peritoneal Dialysis Proceedings of an International Symposium, Paris, 1979,335.
65. Williams P, Marliss E, Anderson G, et al. Amino acid absorption following intraperitoneal administration in CAPD patients. Perit Dial Bull 1982;2:124.
66. Oren W, Wu G, Anderson G, et al. The effective use of amino acid dialysate over four weeks in CAPD patients. Perit Dial Bull 1983;3:66.
67. Chertow G, Lazarus M, Lyden M, et al. Laboratory surrogates of nutritional status after administration of intraperitoneal amino acid based solutions in ambulatory peritoneal dialysis patients. J Ren Nutr 1995;5:116.
68. Kopple J, Bernard D, Messana J, et al. Treatment of malnourished peritoneal dialysis patients with amino acid-based dialysate. Kidney Int 1995;47:1148.
69. Young G, Dibble J, Hobson S, et al. The use of amino acid-based CAPD fluid over twelve weeks. Nephrol Dial Transplant 1989;4:285.
70. Dibble J, Young G, Hobson S, et al. Amino acid-based CAPD fluid over twelve weeks: the effects on carbohydrate and lipid metabolism. Perit Dial Int 1990;10:71.

71. National Kidney Foundation. Kidney disease outcomes quality initiative clinical practice guidelines on nutrition and chronic renal failure. Am J Kidney Dis 2001;37(suppl):S49.
72. Fuhrman MP. Parenteral nutrition in kidney disease. In: Byham-Gray L, Wiesen K, eds. A Clinical Guide to Nutrition Care in Kidney Disease, Chicago: American Dietetic Association, 2004:159–174.
73. Sacks GS, Mayhew S, Johnson D. Parenteral nutrition implementation and management. In Merritt R, ed. A.S.P.E.N. Nutrition Support Practice Manual. 2nd edn, Silver Spring: A.S.P.E.N, 2005:108–117.

13 Anemia Management

Arthur Tsai and Jeffrey S. Berns

LEARNING OBJECTIVES

1. Identify the epidemiology and principal causes of anemia in patients with chronic kidney disease.
2. Describe the physiology of red blood cell production and how erythropoietin deficiency, iron deficiency, and other factors contribute to anemia of chronic kidney disease.
3. Discuss how recombinant human erythropoietin and other erythropoietic stimulating agents as well as oral and intravenous iron are used to treat anemia in patients with chronic kidney disease.
4. Discuss the recent Kidney Disease Outcomes Quality Initiative Guidelines for the diagnosis and treatment of anemia in patients with chronic kidney disease.

Summary

Anemia is a significant cause of morbidity and mortality in patients with chronic kidney disease and end-stage renal disease. The anemia in this setting results primarily from inadequate erythrocyte production by the bone marrow due to a deficiency of erythropoietin. Other factors may contribute to the development of anemia, including most notably iron deficiency, inflammation, and malnutrition. Treatment options, including recombinant human erythropoietin and darbepoetin alfa along with appropriate oral and intravenous iron therapy, are effective in correcting anemia in these patients and result in improved quality of life and clinical outcomes.

Key Words: Anemia; erythropoietin; kidney disease; dialysis; iron; epoetin; ascorbate; L-carnitine; darbepoetin.

From: *Nutrition and Health: Nutrition in Kidney Disease*
Edited by: L. D. Byham-Gray, J. D. Burrowes, and G. M. Chertow

1. INTRODUCTION

Hemoglobin (Hb) is an iron-containing protein found in erythrocytes. Oxygen that diffuses from the lungs into the bloodstream is taken up by Hb in circulating erythrocytes, which deliver it to body tissues for use in metabolic processes. Thus, the Hb concentration in the blood is a good measure of the body's overall oxygen carrying capacity. Normal mean Hb levels are 15.5 g/dL in men and postmenopausal women, and 14 g/dL in premenopausal women *(1)*. While there is not one universally accepted definition, the World Health Organization has defined anemia as an Hb level of less than 13 g/dL in men and postmenopausal women and less than 12 g/dL in premenopausal women *(2)*. Another measure of red cell volume, the hematocrit (Hct), is a less accurate test, so Hb measurement is preferred.

Anemia arises as an early complication of chronic kidney disease (CKD), often as the glomerular filtration rate (GFR) falls below about 60 ml/min/1.73 m^2. The prevalence and severity of anemia in CKD patients increase with worsening kidney function. Anemia is present in as many as 90% of patients with stages 4 and 5 CKD (GFR $<$30 ml/min/1.73 m^2) and is almost universally present when patients reach the need for dialysis *(3)*. The anemia of CKD is normocytic and normochromic with a low corrected reticulocyte count. Although bone marrow examination is not routinely performed in clinical practice to evaluate anemia in patients with CKD, its appearance is generally normal.

Clinical practice guidelines and recommendations for anemia in patients with CKD from the National Kidney Foundation's (NKF) Kidney Disease Outcomes Quality Initiative (KDOQI) have recently been revised *(4,5)*. It is now recommended that Hb testing should be carried out at least annually in all patients with CKD, regardless of stage or etiology, with a diagnosis of anemia being made and further evaluation undertaken when the Hb is less than 13.5 g/dL in adult males and 12 g/dL in adult females. This evaluation should include a complete blood count, serum red blood cell indices [mean corpuscular volume (MCV), mean corpuscular hemoglobin concentration (MCHC)], reticulocyte count, iron parameters such as percentage transferrin saturation (TSAT; serum iron/total iron binding capacity $\times$ 100), Hb content of reticulocytes (CHr), and serum ferritin, and a test for stool occult blood.

2. ERYTHROPOIETIN AND ANEMIA IN PATIENTS WITH KIDNEY DISEASE

Anemia in patients with CKD is multifactorial, but it is caused primarily by insufficient production of erythrocytes by the bone marrow due to a deficiency of erythropoietin *(6)*. Erythropoietin is a glycoprotein made primarily by peritubular interstitial cells in the kidneys, and to a much lesser extent by the liver. When erythropoietin-producing renal cells sense decreased oxygen delivery as a result of anemia or hypoxia, they upregulate their production of erythropoietin several 100-fold *(7)*. Transcription of the erythropoietin gene is regulated by a variety of factors, the most important of which is the transcription factor hypoxia inducible factor (HIF). HIF is stabilized in the presence of hypoxia and enhances transcription of the erythropoietin gene. Erythropoietin binds to receptors on the surface of erythropoietic precursor cells in the bone marrow. The successful maturation of these cells into circulating red blood cells requires the presence of erythropoietin. This activity increases Hb levels, improves oxygen delivery to the erythropoietin-producing cells in the kidneys, and decreases the stimulus for making erythropoietin in a classic feedback loop system. The kidneys of patients with CKD make erythropoietin, but in an amount that is inadequate to maintain normal levels of erythrocyte production *(6)*.

Erythropoietin was first purified from over 2500 L of human urine in 1977. The gene for erythropoietin was cloned in 1985, and in 1989, based on clinical trials in patients on hemodialysis, the U.S. Food and Drug Administration (FDA) approved recombinant human erythropoietin (epoetin) for use in patients on dialysis *(8,9)*. FDA approval of epoetin for other indications followed shortly thereafter, including for patients with CKD not on dialysis. The success of replacement therapy with epoetin, as well as other erythropoietic stimulating agents (ESAs), has served to confirm that erythropoietin deficiency is the primary etiology of anemia in patients with CKD. Specific details regarding ESA use are discussed below.

3. OTHER FACTORS CONTRIBUTING TO ANEMIA OF CKD

While erythropoietin deficiency is clearly the single most common underlying factor in the anemia of CKD, other potential factors exist, including decreased red blood cell life span, inhibition of the responsiveness of the erythroid bone marrow to the effects of erythropoietin, gastrointestinal tract and surgical blood loss, nutritional deficiency (i.e., folic acid or vitamin B12 deficiency), infectious and inflammatory disorders, hemolysis, severe hyperparathyroidism, malignancy, and infiltrative processes of the bone marrow (Table 1). A causative role for uremic inhibitors of erythropoiesis has been suggested, but remains speculative. Besides erythropoietin deficiency, iron deficiency is the next most common factor contributing to the anemia of CKD.

Table 1
Causes of Anemia in Patients with CKD and ESRD

Erythropoietin deficiency
Uremic inhibition of erythropoiesis
Inflammation
Infection
Malnutrition-Inflammation Complex
Iron deficiency
Folic acid deficiency
Vitamin B12 deficiency
Hyperparathyroidism
Primary bone marrow disorders
HIV infection
Hemolysis
Malignancy

CKD, chronic kidney disease; ESRD, end-stage renal disease; HIV, human immunodeficiency virus

3.1. Iron Deficiency

Iron is an essential component of Hb and is necessary for normal production of red blood cells. Iron is stored as part of the storage protein ferritin in the liver, spleen, and bone marrow, until it is released into the blood for red blood cell synthesis. The primary iron source for erythropoiesis is circulating iron that is bound to the transport protein transferrin, which delivers iron to the bone marrow where it is incorporated into Hb. At the end of their lifespan, erythrocytes are removed from the circulation by processing cells called macrophages and broken down to recover their iron content. The iron is recycled into the bloodstream to be bound by transferrin, ready to be used again by the bone marrow for red blood cell production. Without an adequate supply of iron, red blood cell production declines even if sufficient erythropoietin is present. Iron deficiency is a major correctable cause of anemia in CKD and end-stage renal disease (ESRD) patients *(10,11)*.

Two blood tests are most commonly used clinically to assess iron status *(4,12)*. The serum ferritin level is used as an indirect marker of total body iron storage. It is also an "acute-phase reactant" and is increased in the presence of inflammatory processes independent of iron stores. For this reason, it is not a very sensitive marker of iron deficency. The serum TSAT reflects the amount of circulating iron that is available for delivery to the bone marrow for erythropoiesis.

This test also has significant limitations, for it has high variability and only modest specificity for iron deficiency. Another test, the reticulocyte hemoglobin content (CHr) may be a better marker of iron deficiency than TSAT or serum ferritin but is not yet widely used *(13)*.

Absolute iron deficiency is a state where circulating iron and the body's total iron stores are depleted. This is often defined in patients with CKD by a TSAT less than 20% (or CHr less than 29 pg/mL) and a serum ferritin of less than 100 ng/mL *(4)*. Functional iron deficiency is a condition in which iron stores are adequate, but the need for iron for erythropoiesis is greater than the amount that can be mobilized from the body's iron stores and delivered to the bone marrow. It is typically characterized by a TSAT less than 20% and serum ferritin in the range of 500–800 ng/mL or higher. Initiation of ESA treatment may accelerate bone marrow utilization of iron, resulting in functional iron deficiency. Unfortunately, acute and chronic inflammatory conditions may also be associated with similar iron indices, thus confounding the identification of iron deficiency as the cause for the anemia.

3.2. Inflammation

CKD appears to be a state of chronic inflammation, although why this is so is not yet clear. Both acute and chronic inflammation are associated with production by immune cells of proteins called cytokines, including tumor necrosis factor-alpha (TNF-alpha) and interleukins (ILs), and activation of the acute-phase response associated with elevation of C-reactive protein (CRP) levels and reduction in serum albumin level *(14,15)*. Inflammation contributes to anemia by inhibiting the bone marrow's responsiveness to endogenous erythropoietin and ESA therapy *(16)*. Pro-inflammatory cytokines like TNF-alpha inhibit erythropoietin production and suppress development of red blood cell precursors in the bone marrow. Inflammation also causes cytokine-mediated dysregulation of iron balance in the body, reduces oral iron absorption, and inhibits release of iron from its storage sites in the liver, spleen, and bone marrow *(17)*.

An emerging understanding of the importance of cytokine-mediated disruption of normal iron metabolism during chronic inflammation has recently begun to take shape. Hepcidin is a small circulating regulatory peptide that appears to act as a regulator of iron homeostasis and iron availability *(18)*. In states of chronic inflammation where the generation and release of pro-inflammatory cytokines is increased, hepcidin production has been found to be increased as well. Increased circulating hepcidin inhibits macrophage iron release, reduces gastrointestinal tract absorption of iron, and increases sequestration of iron into iron stores.

Thus, hepcidin directly causes less iron to be available to the bone marrow for use in erythropoiesis.

Inflammation is also believed to contribute to protein calorie malnutrition via cytokine effects in mediating anorexia and increased protein catabolism. The term Malnutrition-Inflammation Complex Syndrome (MICS) has been used to encompass both the malnutrition and inflammation components, which appear to perpetuate one another and are often found concurrently in CKD and end-stage renal disease (ESRD) patients with anemia *(19)*. The MICS has been associated with markers of inflammation, such as low serum transferrin, high CRP and IL-6, and decreased cholesterol and pre-albumin levels (reflecting malnutrition), as well as a lower subjective global assessment of nutrition scores. MICS and its attendant inflammatory state have been found to correlate with anemia and hyporesponsiveness to erythropoietin therapy in these patient populations.

A recent theory proposes that leptin, a hormone produced by the body's fat cells that functions to maintain body fat content, is protective against refractory anemia in patients with CKD and ESRD. Malnutrition leads to decreased fat cells and therefore decreased leptin production and signaling. Serum leptin levels correlate inversely with epoetin requirements in chronic hemodialysis patients and CKD patients not yet on dialysis *(20,21)*.

3.3. Infection

The presence of impaired kidney function is associated with an increased risk of infection because of a chronically immunodeficient state mediated by circulating "uremic toxins," the presence of percutaneous dialysis catheters or indwelling synthetic arteriovenous bridge grafts, and the presence of diabetes mellitus and other conditions that may increase infection risk. In patients with CKD and ESRD, infection impairs the responsiveness to native erythropoietin as well as ESA therapy by mediating increased cytokine release, similar to the process in non-infectious, inflammatory states. With recurrent bouts of infection or persistent infection, chronic inflammation develops in which elevated levels of circulating cytokines inhibit erythropoietin production and suppress bone marrow responsiveness to erythropoietin.

3.4. Folic Acid and Vitamin B12 Deficiency

Folic acid and vitamin B12 are important for the proper formation and maturation of erythrocytes. Deficiency of either folic acid or vitamin B12 can result in the development of macrocytic anemia, a form of anemia in which abnormally large red blood cells are formed. The most common method for detecting folate and vitamin B12 deficiency is the measurement of blood levels, although assessment of red cell folate is more accurate but not as widely available.

Hemodialysis can result in the loss of folic acid into the dialysate *(22,23)*. Vitamin B12 is protein bound and is thus removed to a much smaller extent by dialysis. Despite its clearance by dialysis, folic acid deficiency is uncommon in chronic hemodialysis patients as long as dietary intake is adequate. Folic acid and vitamin B12 deficiency are only very rarely the cause of anemia or ESA hyporesponsiveness in CKD and ESRD patients. This is probably in part because certain foods are fortified with folic acid and because daily oral multivitamins and oral folate supplementation are prescribed for many dialysis patients.

3.5. Treatment

3.5.1. ESA Therapy with Epoetin and Darbepoetin

The ability to treat anemia in dialysis patients and subsequently CKD patients not on dialysis was dramatically improved when the first recombinant human erythropoietin (epoetin alfa) was approved for use in 1989 as Epogen (Amgen, Inc.) based on initial clinical studies in hemodialysis patients *(8,9)*. Epoetin alfa is produced by recombinant DNA technology and has an amino acid structure identical to the native human hormone *(24)*. Other forms of recombinant human erythropoietin (epoetin) are now manufactured and marketed under a variety of different names in the United States and abroad. In 2001, darbepoetin alfa was approved for use in dialysis and non-dialysis CKD, along with other indications. Darbepoetin alfa is a novel ESA with a structure that differs in five amino acids compared to the native hormone. It has five N-linked carbohydrate chains while epoetin has three chains. This results in a two-fold to three-fold longer pharmacologic half-life compared to epoetin alfa, which allows for less frequent dosing *(25)*. Epoetins and darbepoetin alfa interact with the same receptor on erythroid progenitor cells to stimulate erythropoiesis as does the native hormone. They have similar clinical efficacy and side effect profiles.

Currently available ESAs can be given intravenously or by subcutaneous injection. While epoetins are more effective when given by subcutaneous injection, this is not the case for darbepoetin alfa, for which the intravenous and subcutaneous routes of administration appear to be of similar efficacy *(26)*. Patients with CKD and those on PD typically receive these agents subcutaneously, while HD patients can receive subcutaneous or intravenous therapy. In the United States, many HD patients receive epoetin intravenously, mainly for reasons of convenience and patient preference, and this route is now specifically recommended in the FDA-approved package insert because of risk of pure red cell aplasia (PRCA) with subcutaneous ESA therapy. Outside the United States, the majority of HD patients receive subcutaneous epoetin treatment due primarily to the lower overall cost of this route, although intravenous injection has become more common as the result of an awareness of ESA-related PRCA. Typical dosing strategies for these agents are summarized in Table 2.

Current KDOQI guidelines recommend that in patients with CKD receiving ESA therapy, the Hb target should generally be in the range of 11.0 to 12.0 g/dl and that the Hb target should not be greater than 13.0 g/dl *(4)*. Treatment of anemia in patients with CKD to these Hb levels, particularly compared to the Hb levels of 6-8 g/dL that were typically seen in HD patients prior to the availability of these therapies, has been shown in prospective clinical studies to result in marked improvement in quality of life (QOL), exercise capacity, markers of cardiovascular disease, and neurocognitive function. An association between higher Hb levels and lower hospitalization rates, hospital length of stay, and mortality has also been observed in retrospective cohort studies *(27–30)*. Evidence of this from large, prospective, randomized clinical trials is limited, however. Above Hb levels of 13 g/dL, there has been concern about adverse effects, including HD access thrombosis and cardiovascular morbidity and mortality, primarily as the result of a large clinical trial in hemodialysis patients with underlying cardiac disease *(31)*. There is little evidence

Table 2
Initial Dosing for ESA Therapy

Generic name	*Brand name*	*Initial dose for CKD and PD*	*Initial dose for HD*
Epoetin alfa	Epogen, Procrit	4000–10,000 U SC every 1–2 weeks	50–100 U/kg IV or SC at each HD treatment
Darbepoetin alfa	Aranesp	40–100 mcg SC every 2 weeks to monthly	40–100 mcg IV or SC every 1–2 weeks

CKD, chronic kidney disease; ESA, erthyropoietic stimulating agents; HD, hemodialysis; IV, intravenously; PD, peritoneal dialysis; SC, subcutaneously.

Epogen labeled for use in end-stage renal disease; Procrit labeled for use in CKD.

that targeting or maintaining Hb levels above 13 g/dL confer significant benefits other than in some QOL assessments *(5,27)*. In addition, results from the Correction of Hemoglobin and Outcomes in Renal Insufficiency (CHOIR), which compared target Hb levels of 13.5 g/dL and 11.3 g/dL in epoetin-treated CKD patients, showed no improvement in mortality or cardiovascular outcomes in the higher Hb target group, and the study was terminated early because of a statistically significant higher incidence of a composite end-point of mortality, stroke, heart attack, and hospitalization for congestive heart failure in the higher Hb group *(32)*.

Side effects of epoetin and darbepoetin alfa are uncommon. Worsening hypertension occurs in some dialysis patients. A rapid increase in red cell mass with altered vascular hemodynamics and effects of nitrous oxide, endothelin and other vasoactive peptides may be involved, but the exact mechanism remains uncertain *(33)*. Other side effects include flu-like symptoms, as well as HD vascular access thrombosis with higher Hb levels. In 2001, a rather sudden and marked increase in cases of PRCA was reported in patients with CKD receiving subcutaneous injections of primarily one specific epoetin preparation (Eprex, Ortho-Biotech) that was not available in the United States. PRCA is characterized by severe anemia that, in these patients, was associated with inhibition of red cell production due to development of anti-erythropoietin antibodies *(34)*. An extensive analysis has led to the conclusion that the vast majority of these cases were associated with one specific form of epoetin alfa, probably due to a change in the pharmacologic preservative and certain components of the syringe in which the agent was delivered *(35)*. With changes in route of administration and the use of other agents, the incidence of PRCA has fortunately dramatically decreased, although some cases continue to be reported with various ESAs.

3.5.2. Iron Therapy

An adequate supply of iron is necessary for normal erythropoiesis as well as optimal response to ESA therapy. Hemodialysis patients, who may lose 2 g of iron or more per year from the hemodialysis procedure along with other sources (surgery, gastrointestinal losses, etc.), are particularly prone to iron deficiency *(10)*. Most patients on ESA therapy will require iron supplementation, which may be given in oral or intravenous form. Oral iron is easy to administer, although it must be given 1 hour before or more than 2 hours after meals for optimal absorption, and it may be poorly tolerated due to gastrointestinal side effects and compliance is often poor. Aluminum- and calcium-containing phosphate binders interfere with absorption of dietary iron. Oral iron has been found to be ineffective for replacing and maintaining adequate stores in hemodialysis patients *(36,37)*.

Because of the poor tolerability and efficacy of oral iron, and the ease with which hemodialysis patients in particular can receive intravenous medications, intravenous iron has become the preferred mode of iron repletion for these patients *(38)*. Hemodialysis patients who regularly receive maintenance intravenous iron require less epoetin and sustain better Hb levels than patients who receive no regular iron supplementation or oral iron *(39)*. There has also been increasing use of intravenous iron therapy in CKD and peritoneal dialysis patients *(40)*.

Intravenous iron formulations consist of large molecular weight complexes of iron and a carbohydrate moiety. Several intravenous iron formulations are currently available in the United States: iron dextran, sodium ferric gluconate in sucrose complex, and iron sucrose (see Table 3). Rare anaphylactic reactions to iron dextran, some

Table 3
Intravenous Iron Formulations

Generic name	*Brand name*	*Initial loading dose for treating iron deficiency in CKD, PD, HD*[c]	*Maintenance dosing in HD patients*
Iron dextran[a]	InFed, DexFerrum	HD:100 mg × 10 doses; PD and CKD: 500 mg IV × 2 doses	50–100 mg weekly
Iron sucrose[b]	Venofer	HD: 100 mg weekly × 10 doses; PD and CKD: 300 mg twice then 400 mg once	25–100 mg weekly to monthly
Sodium ferric gluconate in sucrose[b]	Ferrlecit	HD: 125 mg × 8 doses; PD and CKD: 125 mg × 8 doses or 250 mg × 4 doses	32.5–125 mg weekly to monthly

CKD, chronic kidney disease; HD, hemodialysis; PD, peritoneal dialysis.

[a] Before the full dose of iron dextran is given, a 25 mg test dose should be administered.

[b] Iron sucrose and sodium ferric gluconate do not require administration of a test dose.

[c] Iron loading dose of 1000 mg, repeated as necessary.

resulting in death, have been attributed to the dextran component of the iron complex *(41,42)*. These may occur with any exposure, even if no reaction resulted from an initial test dose. Other common side effects associated with intravenous iron administration include muscle cramping, nausea, vomiting, hypertension, injection site discomfort, diarrhea, and hypotension with rapid administration. Ferric gluconate and iron sucrose can be used safely in patients who have had anaphylactic reactions to iron dextran *(43,44)*.

KDOQI guidelines recommend iron treatment as needed to maintain TSAT greater than 20% and serum ferritin above 100 ng/mL in CKD and peritoneal dialysis patients and above 200 mg/dL in hemodialysis patients *(4,5)*. A CHr greater than 29 pg/cell can be used instead of the TSAT of 20% if available. The KDOQI guidelines note that routine administration of IV iron is not recommended if serum ferritin is greater than 500 ng/mL. At least 200 mg of elemental iron per day given in three divided doses is recommended for adults receiving oral iron. For ferrous sulfate, the least expensive formulation available, this requires three 325-mg tablets per day (Table 4). In hemodialysis patients, intravenous iron can be given as an initial large dose to correct iron deficiency, typically using 1000 mg over 8–10 consecutive treatments, repeated as necessary. Maintenance iron supplementation with 25–125 mg weekly to replace chronic ongoing blood losses is often prescribed, adjusted according to need with the intent of maintaining KDOQI recommended levels of TSAT and serum ferritin. Patients on peritoneal dialysis or with CKD not on dialysis in whom regular

Table 4
Commonly Used Oral Iron Formulations for Adults

Generic name	*Formulation*	*Elemental iron content*
Iron polysaccharide	150-mg capsules	150 mg
Ferrous sulfate	160-mg tablet	50 mg
	200-mg, 300-mg 324-mg, 325-mg tablets	60–65 mg (various formulations)
Ferrous gluconate	240-mg tablet	27 mg
	246-mg tablet	28 mg
	300-mg tablet	34 mg
	325-mg tablet	36 mg
Heme iron polypeptide	tablets	12 mg

intravenous access is not available can receive intermittent dosing as needed via a peripheral intravenous catheter *(40,45,46)*.

3.5.3. Folic Acid and Vitamin B12

In CKD and hemodialysis patients, serum folic acid and vitamin B12 levels should be monitored at the time of initial assessment of anemia and periodically thereafter, as well as if a patient appears to have developed hyporesponsiveness to ESA therapy without other explanation. In particular, this testing should be performed in the presence of an elevated erythrocyte mean corpuscular volume (MCV) indicating macrocytic anemia. If there is documented deficiency of folic acid and/or vitamin B12, then the appropriate supplementation should be administered. KDOQI guidelines do not recommend routine supplementation of folate and vitamin B12 in CKD patients who do not have documented deficiency *(5)*.

3.5.4. Pharmacologic Adjuvants to ESA Therapy

Several pharmacologic agents have been assessed for possible effects of enhancing erythropoiesis directly or by increasing the effect of epoetin therapy, particularly in patients with what is often called epoetin- or ESA-hyporesponsiveness, which has been defined as failure to maintain Hgb above 11 g/dL despite epoetin doses greater than 500 IU/kg/week (or comparable darbepoetin dose) *(5)*. None of these agents has been demonstrated to have proven efficacy or long-term safety.

Vitamin C, also known as ascorbic acid, is able to mobilize the release of iron from ferritin stored in the liver, spleen, and bone marrow tissues and enhance the incorporation of iron into heme to form Hb *(47)*. Some small studies suggested that intravenous vitamin C increased Hb levels and decreased epoetin requirements in anemic hemodialysis patients, particularly those who have functional iron deficiency *(48,49)*. However, KDOQI guidelines found that there is insufficient evidence of safety and efficacy to recommend the use of vitamin C as an ESA adjuvant *(5)*. A concern with prolonged vitamin C use in dialysis patients is secondary oxalosis, in which elevated plasma oxalate levels lead to calcium oxalate deposition in the heart muscle, blood vessels, and other tissues *(50)*.

L-carnitine is important in the transport of long chain fatty acids across the inner mitochondrial membrane for oxidation to provide energy in the form of ATP. It also acts to remove intracellular acyl-CoA molecules, which can be toxic to cells and which accumulate in ESRD *(51)*. The presence of a carnitine deficiency syndrome in hemodialysis patients has been proposed, with muscle weakness, intradialytic hypotension, cardiac dysfunction, and epoetin-hyporesponsive anemia *(52)*. Although an NKF Carnitine Consensus Conference recommended use of IV L-carnitine in some anemic hemodialysis patients *(53)* and a recent meta-analysis concluded that L-carnitine was beneficial in some dialysis patients *(54)*, the most recent KDOQI guidelines concluded that there was insufficient evidence of efficacy to recommend L-carnitine as an adjuvant to ESA therapy *(5)*.

Prior to the availability of epoetin, intramuscular androgens such as nandrolone decanoate were used to stimulate erythropoiesis *(48)*. Use of androgens dramatically decreased after the widespread adoption of epoetin therapy. Side effects of androgens, including abnormal liver function tests, dyslipidemia, hirsutism, acne, priapism, dysmenorrhea, and masculinization limit their use. The KDOQI Guidelines state that androgens should not be used as adjuvant therapy due to the lack of evidence for efficacy as well as safety concerns *(5)*.

4. CONSEQUENCES OF ANEMIA IN CKD AND ESRD AND EFFECTS OF TREATMENT

Lower Hb levels have been found to be associated with increased risk of death from non-cardiac and cardiac causes, increased frequency and duration of hospitalizations, and increased severity of cardiac dysfunction in CKD and ESRD patients *(28,55–58)*. Anemia of CKD is also associated with reduced QOL with increased fatigue, depression, shortness of breath, and exercise intolerance. These QOL indicators have been shown in prospective clinical studies to improve substantially with ESA treatment to raise Hb levels *(31,59,60)*. There is less evidence from prospective studies that using ESAs to achieve higher Hb levels reduces overall morbidity, hospitalization, or mortality *(31)* although a substantial body of literature from observational studies supports such an association *(28,58)*.

4.1. Mortality

Several large, observational studies have shown that anemia of CKD correlates with increased mortality in patients with CKD and ESRD. In one such study of 44,550 hemodialysis patients who had been followed for a minimum of 6 months, patients with mean serum Hb values of less than 9 g/dL had more than two-fold increased risk mortality compared to patients with mean serum Hb values between 11 g/dL and 12 g/dL *(61)*. Other studies have reported similar findings *(28,58,62,63)*. However, these retrospective studies cannot determine whether the anemia of CKD is a direct cause of increased risk of death in these patients.

Rather, anemia may simply be a marker for more chronically ill patients who tend to have a greater risk for death over time.

Until recently, there has been only one prospective, randomized controlled trial that has been large enough to assess mortality with CKD or dialysis patients. In this study, hemodialysis patients with prior cardiac disease were randomized to target Hct levels of 30% (low) or 42% (normal) using epoetin alfa *(31)*. After nearly two and one-half years, overall mortality in the normal Hct group was higher than in the lower Hct group, and the study was stopped for safety reasons, even though this did not reach the level of statistical significance. As mentioned above, results from the CHOIR study, which compared target Hb levels of 13.5 and 11.3 g/Dl in epoetin-treated CKD patients, showed no improvement in mortality or cardiovascular outcomes in the higher Hb target group and found a statistically significant higher incidence of a composite end-point of mortality, stroke, heart attack, and hospitalization for congestive heart failure in the higher Hb group *(32)*. Thus, although it is commonly believed that anemia of CKD may be a direct cause of increased risk for death, stronger evidence of a causal link between effects of ESA treatment and reduced mortality, including the impact of specific Hb levels, is still needed, particularly for Hb levels approaching the normal range.

4.2. Cardiac Disease

Anemia is associated with increased risk of cardiovascular morbidity and mortality in CKD and ESRD patients and is an independent risk factor for the development of left ventricular hypertrophy (LVH) and left ventricular dilatation in these patients. LVH is independently associated with a significantly increased risk of death in both non-renal and renal patient populations *(55,64)*. Ischemic heart disease is also more common in patients with CKD and ESRD than in age- and sex-matched patients without kidney disease *(56)*. While uncontrolled observations and small prospective studies have suggested that there is some improvement in LVH with epoetin therapy in CKD and hemodialysis patients when Hb levels are raised from very severely anemic levels to Hb levels consistent with mild anemia *(65,66)*, more recent larger randomized controlled trials have found that treating patients to raise the Hb to more normal levels does not further reduce development of LVH or LV dilatation compared to more moderate anemia *(67,68)*. Another study, the Cardiovascular Risk Reduction by Early Anemia Treatment with epoetin beta trial, which compared target Hb groups of 13–15 g/dL and 10.5–11.5 g/dL in epoetin-treated patients with CKD found no benefit of the higher Hb levels in terms of overall cardiovascular events or on prevention or progression of LVH *(69)*.

4.3. Quality of Life

Anemia of CKD is associated with symptoms that are characteristic of chronically impaired oxygen delivery to body tissues, such as fatigue, impaired thinking, sleep disturbance, depression, shortness of breath, and exercise intolerance. These symptoms adversely impact the CKD patient's QOL and functional status. The anemia-associated symptoms have been shown to improve with treatment to raise serum Hb *(59,60,70,71)*. Neurophysiologic measures including cognitive test scores and sleep disturbances also improve with higher Hb levels with epoetin therapy *(72)*.

5. CONCLUSIONS

Anemia of CKD is multifactorial, but is caused primarily by insufficient production of erythrocytes by the bone marrow due to a deficiency of erythropoietin. Other contributing factors include decreased red blood cell life span, inhibition of the responsiveness of bone marrow to the effects of erythropoietin, gastrointestinal or surgical blood loss, nutritional deficiency, and infectious and inflammatory disorders. Besides erythropoietin deficiency, iron deficiency is the next most common factor contributing to the anemia of CKD. Currently available ESAs include epoetins and darbepoetin alfa, which can be given intravenously or by subcutaneous injection. In the setting of iron deficiency, intravenous iron has become the preferred mode of iron repletion for CKD and dialysis patients because of the poor tolerability and efficacy of oral iron, and the ease with which hemodialysis patients in particular can receive intravenous medications. Several pharmacologic agents have been assessed for possible effects of enhancing erythropoiesis or increasing the effect of epoetin therapy, but none of these agents has shown proven efficacy or long-term safety.

Lower Hb levels have been found to be associated with increased risk of death from non-cardiac and cardiac causes, increased frequency and duration of hospitalizations, increased severity of cardiac dysfunction, and reduced QOL in CKD and ESRD patients. Recently revised KDOQI guidelines should be helpful to clinicians caring for patients with all stages of CKD, including those on peritoneal and hemodialysis, as a guide to the diagnosis, evaluation, and treatment of the anemia of CKD.

6. CASE STUDY

J.W. is 68 years old with type 2 diabetes mellitus of 20-year duration, long-standing hypertension, and progressive CKD (estimated GFR 23 ml/min/1.73 m^2) who has developed increasing fatigue, reduced exercise tolerance, and has lost interest in his hobbies. He no longer has the energy to take his grandson fishing. His appetite is good, and he has no nausea. He does not have any edema. He is not short of breath at rest, and does not have orthopnea, but he gets easily "winded" when climbing the short flight of stairs to his front door. He lives with his wife, who confirms that he has been eating well, but reports that he just "does not have any energy anymore." His blood pressure and blood glucose have generally been under good control. A recent serum HbA1C was 6.2%.

His medications include glipizide 5 mg twice daily, furosemide 40 mg daily, a multivitamin, metoprolol 50 mg twice daily, losartan 100 mg daily. He has used sildenafil citrate (Viagra®) in the past but has not asked for a refill of this in several months. He has no known drug allergies. He does not smoke or drink alcohol.

On physical examination, he is moderately obese. His blood pressure is 128/72 mmHg and pulse is 72/min. The lungs are clear to auscultation, heart sounds are normal, and he has no edema. The remainder of the exam is also normal. Laboratory studies show a serum creatinine of 3.4 mg/dL and Hb 8.7 g/dL.

6.1. Key Questions

1. What studies should be done to evaluate the anemia in this patient?
2. Initial labs return with a TSAT of 13%, ferritin of 48 ng/mL; stool occult blood is negative, vitamin B12 and folate levels are normal, the MCV is low at 76 fl, and reticulocyte count is low at 1.1%. At this point, what would be the next steps in his management?
3. A colonoscopy revealed a small benign polyp. Upper gastrointestinal tract endoscopy was normal. The patient is treated with 1000 mg of intravenous iron. Repeat lab studies show a serum Hb of 9.5 g/dL, TSAT of 28%, and ferritin of 210 ng/mL. What is the next step in his anemia management?

REFERENCES

1. Lee GR, Bithell TC, Foerster J, et al. Appendix A, in Clinical Hematology (9th edn), Philadelphia, Lea and Febiger, 1993, p. 2303.
2. World Health Organization. Nutritional Anaemias: Report of a WHO Scientific Group., in, Geneva, Switzerland, World Health Organization, 1968.
3. Kazmi WH, Kausz AT, Khan S, et al. Anemia: an early complication of chronic renal insufficiency. Am J Kidney Dis 38:803–812, 2001.

4. National Kidney Foundation. KDOQ1 clinical practice guidelines and clinical practice recommendations for anemia in chronic kidney disease: 2007 update of hemoglobin target. Am J Kidney Dis 50:474–530, 2007.
5. Anonymous. KDOQI clinical practice guidelines and clinical practice recommendations for anemia in chronic kidney disease. Am J Kidney Dis 47:S11–S145, 2006.
6. Eschbach JW. The anemia of chronic renal failure: pathophysiology and the effects of recombinant erythropoietin. Kidney Int 35:134–148, 1989.
7. Cheung JY, Miller BA. Molecular mechanisms of erythropoietin signaling. Nephron 87:215–222, 2001.
8. Eschbach JW, Abdulhadi MH, Browne JK, et al. Recombinant human erythropoietin in anemic patients with end-stage renal disease. Results of a phase III multicenter clinical trial. Ann Intern Med 111:992–1000, 1989.
9. Eschbach JW, Egrie JC, Downing MR, et al. Correction of the anemia of end-stage renal disease with recombinant human erythropoietin. Results of a combined phase I and II clinical trial. N Engl J Med 316:73–78, 1987.
10. Eschbach JW, Cook JD, Scribner BH, et al. Iron balance in hemodialysis patients. Ann Intern Med 87:710–713, 1977.
11. Fishbane S, Maesaka JK. Iron management in end-stage renal disease. Am J Kidney Dis 29:319–333, 1997.
12. Fishbane S, Kowalski EA, Imbriano LJ, et al. The evaluation of iron status in hemodialysis patients. J Am Soc Nephrol 7:2654–2657, 1996.
13. Fishbane S, Galgano C, Langley RC, Jr.,et al. Reticulocyte hemoglobin content in the evaluation of iron status of hemodialysis patients. Kidney Int 52:217–222, 1997.
14. Barany P, Divino Filho JC, Bergstrom J. High C-reactive protein is a strong predictor of resistance to erythropoietin in hemodialysis patients. Am J Kidney Dis 29:565–568, 1997.
15. Goicoechea M, Martin J, de Sequera P, et al. Role of cytokines in the response to erythropoietin in hemodialysis patients. Kidney Int 54:1337–1343, 1998.
16. Gunnell J, Yeun JY, Depner TA, et al. Acute-phase response predicts erythropoietin resistance in hemodialysis and peritoneal dialysis patients. Am J Kidney Dis 33:63–72, 1999.
17. Feelders RA, Vreugdenhil G, Eggermont AM, et al. Regulation of iron metabolism in the acute-phase response: interferon gamma and tumour necrosis factor alpha induce hypoferraemia, ferritin production and a decrease in circulating transferrin receptors in cancer patients. Eur J Clin Invest 28:520–527, 1998.
18. Roy CN, Andrews NC. Anemia of inflammation: the hepcidin link. Curr Opin Hematol 12:107–111, 2005.
19. Kalantar-Zadeh K, McAllister CJ, Lehn RS, et al. Effect of malnutrition-inflammation complex syndrome on EPO hyporesponsiveness in maintenance hemodialysis patients. Am J Kidney Dis 42:761–773, 2003.
20. Axelsson J, Qureshi AR, Heimburger O, et al. Body fat mass and serum leptin levels influence epoetin sensitivity in patients with ESRD. Am J Kidney Dis 46:628–634, 2005.
21. Yilmaz A, Kayardi M, Icagasioglu S, et al. Relationship between serum leptin levels and body composition and markers of malnutrition in nondiabetic patients on peritoneal dialysis or hemodialysis. J Chin Med Assoc 68:566–570, 2005.

22. Teschner M, Kosch M, Schaefer RM. Folate metabolism in renal failure. Nephrol Dial Transplant 17 Suppl 5:24–27, 2002.
23. Wolfson M. Use of water-soluble vitamins n patients with chronic renal failure. Semin Dial 1:28–32, 1988.
24. Egrie J. The cloning and production of recombinant human erythropoietin. Pharmacotherapy 10:3S–8S, 1990.
25. Macdougall IC. Darbepoetin alfa: A new therapeutic agent for renal anemia. Kidney Int Suppl May (80):55–61, 2002.
26. Macdougall IC, Matcham J, Gray SJ. Correction of anaemia with darbepoetin alfa in patients with chronic kidney disease receiving dialysis. Nephrol Dial Transplant 18:576–581, 2003.
27. Berns JS. Should the target hemoglobin for patients with chronic kidney disease treated with erythropoietic replacement therapy be changed? Semin Dial 18:22–29, 2005.
28. Collins AJ, Li S, St Peter W, et al. Death, hospitalization, and economic associations among incident hemodialysis patients with hematocrit values of 36 to 39%. J Am Soc Nephrol 12:2465–2473, 2001.
29. Jones M, Ibels L, Schenkel B, et al. Impact of epoetin alfa on clinical end points in patients with chronic renal failure: A meta-analysis. Kidney Int 65: 757–767, 2004.
30. Roger SD, McMahon LP, Clarkson A, et al. Effects of early and late intervention with epoetin alpha on left ventricular mass among patients with chronic kidney disease (stage 3 or 4): Results of a randomized clinical trial. J Am Soc Nephrol 15:148–156, 2004.
31. Besarab A, Bolton WK, Browne JK, et al. The effects of normal as compared with low hematocrit values in patients with cardiac disease who are receiving hemodialysis and epoetin. N Engl J Med 339:584–590, 1998.
32. Singh AK, Szczech L, Tang KL, Barnhart H, Sapp S, Wolfson M, Reddan D; for the CHOIR Investigators. Correction of anemia with epoetin alfa in chronic kidney disease. N Engl J Med 355:2085–98, 2006.
33. Vaziri ND. Mechanism of erythropoietin-induced hypertension. Am J Kidney Dis 33:821–828, 1999.
34. Casadevall N, Nataf J, Viron B, et al. Pure red-cell aplasia and antierythropoietin antibodies in patients treated with recombinant erythropoietin. N Engl J Med 346:469–475, 2002.
35. Boven K, Stryker S, Knight J, et al. The increased incidence of pure red cell aplasia with an Eprex formulation in uncoated rubber stopper syringes. Kidney Int 67:2346–2353, 2005.
36. Macdougall IC, Tucker B, Thompson J, et al. A randomized controlled study of iron supplementation in patients treated with erythropoietin. Kidney Int 50: 1694–1699, 1996.
37. Markowitz GS, Kahn GA, Feingold RE, et al. An evaluation of the effectiveness of oral iron therapy in hemodialysis patients receiving recombinant human erythropoietin. Clin Nephrol 48:34–40, 1997.
38. Besarab A, Amin N, Ahsan M, et al. Optimization of epoetin therapy with intravenous iron therapy in hemodialysis patients. J Am Soc Nephrol 11: 530–538, 2000.

39. Besarab A, Kaiser JW, Frinak S. A study of parenteral iron regimens in hemodialysis patients. Am J Kidney Dis 34:21–28, 1999.
40. Macdougall IC, Roche A. Administration of intravenous iron sucrose as a 2-minute push to CKD patients: A prospective evaluation of 2,297 injections. Am J Kidney Dis 46:283–289, 2005.
41. Fishbane S, Ungureanu VD, Maesaka JK, et al. The safety of intravenous iron dextran in hemodialysis patients. Am J Kidney Dis 28:529–534, 1996.
42. Fletes R, Lazarus JM, Gage J, et al. Suspected iron dextran-related adverse drug events in hemodialysis patients. Am J Kidney Dis 37:743–749, 2001.
43. Michael B, Coyne DW, Fishbane S, et al. Sodium ferric gluconate complex in hemodialysis patients: Adverse reactions compared to placebo and iron dextran. Kidney Int 61:1830–1839, 2002.
44. Charytan C, Levin N, Al-Saloum M, et al. Efficacy and safety of iron sucrose for iron deficiency in patients with dialysis-associated anemia: North American clinical trial. Am J Kidney Dis 37:300–307, 2001.
45. Van Wyck DB, Roppolo M, Martinez CO, et al. A randomized, controlled trial comparing IV iron sucrose to oral iron in anemic patients with nondialysis-dependent CKD. Kidney Int 68:2846–2856, 2005.
46. Panesar A, Agarwal R. Safety and efficacy of sodium ferric gluconate complex in patients with chronic kidney disease. Am J Kidney Dis 40:924–931, 2002.
47. Bridges KR, Hoffman KE. The effects of ascorbic acid on the intracellular metabolism of iron and ferritin. J Biol Chem 261:14273–14277, 1986.
48. Berns JS, Mosenkis A. Pharmacologic adjuvants to epoetin in the treatment of anemia in patients on hemodialysis. Hemodial Int 9:7–22, 2005.
49. Tarng DC, Wei YH, Huang TP, et al. Intravenous ascorbic acid as an adjuvant therapy for recombinant erythropoietin in hemodialysis patients with hyperferritinemia. Kidney Int 55:2477–2486, 1999.
50. Canavese C, Petrarulo M, Massarenti P, et al. Long-term, low-dose, intravenous vitamin C leads to plasma calcium oxalate supersaturation in hemodialysis patients. Am J Kidney Dis 45:540–549, 2005.
51. Hoppel C. The role of carnitine in normal and altered fatty acid metabolism. Am J Kidney Dis 41:S4–S12, 2003.
52. Evans A. Dialysis-related carnitine disorder and levocarnitine pharmacology. Am J Kidney Dis 41:S13–S26, 2003.
53. Eknoyan G, Latos DL, Lindberg J. Practice recommendations for the use of L-carnitine in dialysis-related carnitine disorder. National Kidney Foundation Carnitine Consensus Conference. Am J Kidney Dis 41:868–876, 2003.
54. Hurot JM, Cucherat M, Haugh M, et al. Effects of L-carnitine supplementation in maintenance hemodialysis patients: A systematic review. J Am Soc Nephrol 13:708–714, 2002.
55. Levin A, Singer J, Thompson CR, et al. Prevalent left ventricular hypertrophy in the predialysis population: Identifying opportunities for intervention. Am J Kidney Dis 27:347–354, 1996.
56. Jurkovitz CT, Abramson JL, Vaccarino LV, et al. Association of high serum creatinine and anemia increases the risk of coronary events: Results from the prospective community-based atherosclerosis risk in communities (ARIC) study. J Am Soc Nephrol 14:2919–2925, 2003.

57. McClellan WM, Flanders WD, Langston RD, et al. Anemia and renal insufficiency are independent risk factors for death among patients with congestive heart failure admitted to community hospitals: A population-based study. J Am Soc Nephrol 13:1928–1936, 2002.
58. Ma JZ, Ebben J, Xia H, et al. Hematocrit level and associated mortality in hemodialysis patients. J Am Soc Nephrol 10:610–619, 1999.
59. Moreno F, Sanz-Guajardo D, Lopez-Gomez JM, et al. Increasing the hematocrit has a beneficial effect on quality of life and is safe in selected hemodialysis patients. Spanish Cooperative Renal Patients Quality of Life Study Group of the Spanish Society of Nephrology. J Am Soc Nephrol 11:335–342, 2000.
60. McMahon LP, Dawborn JK. Subjective quality of life assessment in hemodialysis patients at different levels of hemoglobin following use of recombinant human erythropoietin. Am J Nephrol 12:162–169, 1992.
61. Ofsthun N, Labrecque J, Lacson E, et al. The effects of higher hemoglobin levels on mortality and hospitalization in hemodialysis patients. Kidney Int 63:1908–1914, 2003.
62. Xia H, Ebben J, Ma JZ, et al. Hematocrit levels and hospitalization risks in hemodialysis patients. J Am Soc Nephrol 10:1309–1316, 1999.
63. Robinson BM, Joffe MM, Berns JS, et al. Anemia and mortality in hemodialysis patients: Accounting for morbidity and treatment variables updated over time. Kidney Int 68:2323–2330, 2005.
64. Parfrey PS, Foley RN. The clinical epidemiology of cardiac disease in chronic renal failure. J Am Soc Nephrol 10:1606–1615, 1999.
65. Silberberg J, Racine N, Barre P, et al. Regression of left ventricular hypertrophy in dialysis patients following correction of anemia with recombinant human erythropoietin. Can J Cardiol 6:1–4, 1990.
66. Foley RN, Parfrey PS, Morgan J, et al. Effect of hemoglobin levels in hemodialysis patients with asymptomatic cardiomyopathy. Kidney Int 58: 1325–1335, 2000.
67. Parfrey PS, Foley RN, Wittreich BH, et al. Double-blind comparison of full and partial anemia correction in incident hemodialysis patients without symptomatic heart disease. J Am Soc Nephrol 16:2180–2189, 2005.
68. Levin A, Djurdjev O, Thompson C, et al. Canadian randomized trial of hemoglobin maintenance to prevent or delay left ventricular mass growth in patients with CKD. Am J Kidney Dis 46:799–811, 2005.
69. Drueke TB, Locatelli F, Clyne N, Eckardt EU, Macdougall IC, Tsakiris D, Burger HU, Scherhag A; for the CREATE Investigators. Normalization of hemoglobin level in patients with chronic kidney disease and anemia. N Engl J Med 16:2071–84, 2006.
70. Benz RL, Pressman MR, Hovick ET, et al. A preliminary study of the effects of correction of anemia with recombinant human erythropoietin therapy on sleep, sleep disorders, and daytime sleepiness in hemodialysis patients (The SLEEPO study). Am J Kidney Dis 34:1089–1095, 1999.
71. Moreno F, Aracil FJ, Perez R, et al. Controlled study on the improvement of quality of life in elderly hemodialysis patients after correcting end-stage renal disease-related anemia with erythropoietin. Am J Kidney Dis 27:548–556, 1996.
72. Pickett JL, Theberge DC, Brown WS, et al. Normalizing hematocrit in dialysis patients improves brain function. Am J Kidney Dis 33:1122–1130, 1999.

14 Bone and Mineral Metabolism and Disease

Linda McCann

LEARNING OBJECTIVES

1. To identify the development of bone and mineral abnormalities in chronic kidney disease and their classification based on signs and symptoms, biochemical parameters, and pathological findings.
2. To describe the impact of bone and mineral abnormalities in chronic kidney disease.
3. To discuss current therapies and treatment recommendations for bone and mineral abnormalities in chronic kidney disease.

Summary

Bone and mineral abnormalities begin early in the course of chronic kidney disease (CKD) and continue as kidney function diminishes. These abnormalities represent complex pathologic processes that can significantly impact morbidity, mortality, and quality of life for those individuals who have CKD. Additionally, the focus on bone and mineral abnormalities has changed as new assessment techniques and treatment options have evolved. Recognizing the magnitude of these bone and mineral abnormalities and the diversity in practice for treating them, the National Kidney Foundation (NKF) Kidney Disease Outcomes Quality Initiative (KDOQI) developed and published evidence-based practice guidelines for CKD bone metabolism and disease in 2003. These guidelines address methods for identifying and classifying renal bone disease, recommend treatment options, and propose research initiatives. Early and continuous evaluation and treatment of bone and mineral abnormalities have the potential to improve outcomes of those with CKD.

From: *Nutrition and Health: Nutrition in Kidney Disease*
Edited by: L. D. Byham-Gray, J. D. Burrowes, and G. M. Chertow

Key Words: Uremic or renal osteodystrophy; adynamic bone disease; active vitamin D; hyperparathyroidism; KDOQI; KDIGO; practice guidelines; calcimimetics; parathyroid hormone; RO; CKD-MBD; phosphate; calcium.

1. INTRODUCTION

Bone and mineral abnormalities are common in those who have chronic kidney disease (CKD). Some of these abnormalities are a result of the disease process and others may, at least in part, be produced by therapies used to treat kidney disease and the associated comorbidities. Abnormalities range from adynamic bone state with low bone turnover to severe hyperparathyroid bone disease with accelerated bone turnover *(1)*. Regardless of the manifestation or the causes, there is significant evidence that these bone and mineral abnormalities have the potential for increasing morbidity and mortality in the CKD population *(1–3)*.

In the years since dialysis became routinely available, the profile of bone and mineral abnormalities has changed, molded by the dialysis process and various therapies. The focus on hyperparathyroid bone disease and osteomalacia has expanded to include concern for low turnover bone abnormalities, mixed bone disease, and soft tissue mineralization. It is now recognized that bone and mineral abnormalities in CKD can impact many areas of the body. Bone disease, in those individuals on dialysis therapy, is typically asymptomatic until late and even then symptoms may be nonspecific and unobtrusive *(4,5)*. Previously common symptoms such as pruritus, bone pain, fractures, deformities, and muscle weakness *(6)* have been overshadowed by metastatic and extraskeletal calcifications which have the potential to adversely impact many areas of the body. Extraskeletal calcification can be localized in arteries, eyes, visceral organs, around joints, and in the skin *(1)*. Manifestation of abnormal bone and mineral metabolism is summarized in Table 1 *(3–6)*. It is important to understand the widespread incidence and impact of bone and mineral abnormalities in CKD and to use caution in applying therapies, always weighing the risks and benefits.

The importance of this problem was recognized by the NKF KDOQI when it facilitated the development of evidence-based practice guidelines for bone metabolism and disease which were published in 2003 *(1)*. The international group, Kidney Disease: Improving Global Outcomes (KDIGO), was established to develop international guidelines for various issues that impact outcomes in CKD. Recognizing that the science holds true for all populations, publication of KDIGO global bone and mineral guidelines is slated for 2008. These guidelines

Table 1
Manifestations of Abnormal Bone and Mineral Metabolism in Chronic Kidney Disease (CKD)

Manifestation	*Characteristics/description*
Altered vitamin D metabolism	Deficiency in calcitriol Defective intestinal absorption of calcium Hypocalcemia Stimulation of PTH synthesis
Abnormal handling of, calcium phosphorus, and magnesium by the kidneys	Hyperphosphatemia Hypocalcemia
Secondary hyperparathyroidism	Decreased skeletal response to PTH Altered degradation of PTH Abnormal regulation of calcium-dependent PTH secretion Impaired skeletal response to PTH Increased parathyroid gland chief cell proliferation Bone disease
Metastatic and extra-skeletal calcifications	Calcification of coronary arteries and cardiac valves, potential skin ulceration, or soft tissue necrosis
Fractures	Incidence is increased in CKD
Bone pain	Uncommon, due to decrease in prevalence of aluminum bone disease reported; expressed as a general ache
Pruritis	Associated with high PTH levels, hypercalcemia, high CaP, and metastatic calcification
Dialysis-related amyloidosis	Disabling arthropathy after long vintage dialysis
Proximal myopathy and muscle weakness	Usually limited to proximal muscles, caused by SHPT, phosphorus depletion, aluminum toxicity, or low vitamin D levels

CaP, calcium times phosphorus product; PTH, parathyroid hormone; SHPT, secondary hyperparathyroidism.

Adapted from refs. *(1,3,4,6)*.

Table 2
Features and Treatment of CKD-MBD

Type	*Features*	*Common lab profile*	*Usual treatment*
HPT–mild	↑osteoblasts and osteoclasts; ↑bone turnover	High PTH Low to normal Ca^{++} Normal to high P	Control P Vitamin D analog Maintain Ca^{++}mid normal
HPT–severe	↑osteoblasts and osteoclasts; excessive bone turnover	Very high PTH Normal to high Ca^{++} High P	Control $Ca^{++}$and P with binders, dialysis, diet Calcimimetics
Adynamic	↑osteoblasts and osteoclasts; ↓bone turnover; potential for extraskeletal calcification	Low PTH High Ca^{++} Normal to high P	Minimize Ca^{++}load/serum level Stimulate PTH Liberalize P level/decrease binders
Osteomalacia	↓bone turnover; abnormal mineralization; associated with aluminum toxicity	Low to normal PTH Normal to high Ca^{++} Low to normal P	Eliminate/avoid aluminum Treat aluminum overload, if present Minimize abnormal biomarkers
Mixed	↑bone turnover; abnormal mineralization	Normal to high PTH Normal to high Ca^{++} Normal to high P	Minimize abnormal biomarkers Titrate therapies to avoid cyclic bone turnover
Extraskeletal calcification	Calcification of coronary/ aortic/other vessels; potential CUA	Low or high PTH Normal to high Ca^{++} Normal to high P	Control Ca^{++} (binders, diet, dialysate) Avoid excess calcium load Control P (binders, diet, dialysis)

CUA, calcific uremic asteriolopathy; P, phosphours; PTH, parathyroid hormone; MBD, Mineral and Bone Disease.

are expected to promote consistent, high quality patient care, facilitate comparison of research data from different countries, identify and promote vital research, and ultimately improve patient outcomes.

The KDIGO group has proposed a simple, clinically relevant differentiation of terms to help enhance communication, facilitate clinical decision making, and promote the development of global evidence-based clinical practice guidelines for bone and mineral abnormalities in CKD *(7)*. Renal osteodystrophy (RO), which has been inconsistently used to describe renal bone and mineral disorders, should be used exclusively to define alterations in bone pathology associated with CKD, assessed by histomorphometry with the results based on a classification system that includes parameters of bone turnover, mineralization, and volume. The term Chronic Kidney Disease Mineral and Bone Disorder (CKD-MBD) is used to describe the clinical syndrome that develops as a systemic disorder of mineral and bone metabolism due to CKD which is manifested by one or a combination of (i) abnormalities of calcium, phosphorus, parathyroid hormone (PTH), and vitamin D metabolism; (ii) abnormalities of bone turnover, mineralization, volume, and strength; and (iii) vascular and/or other soft tissue calcification *(7)*. These terms are used as described in this chapter. Features and treatments of CKD-MBD are described in Table 2.

2. PATHOGENESIS OF BONE AND MINERAL ABNORMALITIES IN CKD

The cascade of events that begins the course of bone and mineral abnormalities begins early in CKD. The kidneys help maintain the balance of calcium and phosphorus in the body by regulating the net excretion of these minerals in the urine. Balance is also maintained by changes in calcium and phosphorus absorption in the intestines and through the exchange of those ions between bone and the extra cellular fluid. Bone level calcium and phosphorus stores help support metabolic and homeostatic requirements. PTH, calcitriol, and phosphatonins coordinate the responses of the kidneys, intestines, and bones. Progressive loss of kidney function causes disturbances in mineral metabolism and bone integrity *(1,4,6)*.

Two general characteristics define the state of the bone—high turnover and low turnover. Secondary hyperparathyroidism (SHPT), considered high turnover disease, is characterized by abnormal and increased bone remodeling, including osteitis fibrosa and mixed bone disorders. Low turnover bone states are characterized by decreased bone mineralization and formation, including osteomalacia and adynamic bone disorder (ABD) *(1,4)*.

Once the glomerular filtration rate (GFR) drops below 60 ml/min/1.73 m^2 (stage 3 CKD), PTH levels begin to rise in the blood *(1)*, setting in motion the development of SHPT *(6,8)*. This rise in PTH appears to be in response to several factors, including vitamin D deficiency, increasing phosphate retention, and a skeletal resistance to PTH, all of which lead to hypocalcemia and further stimulation of PTH. With progressive loss of kidney function, there seems to be a decrease in the number of vitamin D receptors (VDRs) and calcium sensing receptors (CaRs) in the parathyroid glands, making them resistant to the actions of vitamin D and calcium. Dietary phosphorus modification may modulate PTH levels even if the serum phosphorus levels are not elevated above the normal range *(4,5)*. Hyperphosphatemia occurs later in the progression of CKD, usually when GFR drops to about 20–30 ml/min/1.73 m^2 and significantly influences the function and growth of parathyroid glands *(1,5)*.

By the time dialysis is required, most CKD patients have some degree of SHPT which is characterized by hypersecretion of PTH and eventually hyperplasia of the parathyroid glands. In the historic trade-off hypothesis, (Fig. 1) *(1,4,9–10)*, it is suggested that as GFR declines, phosphorus excretion also declines, serum phosphorus levels

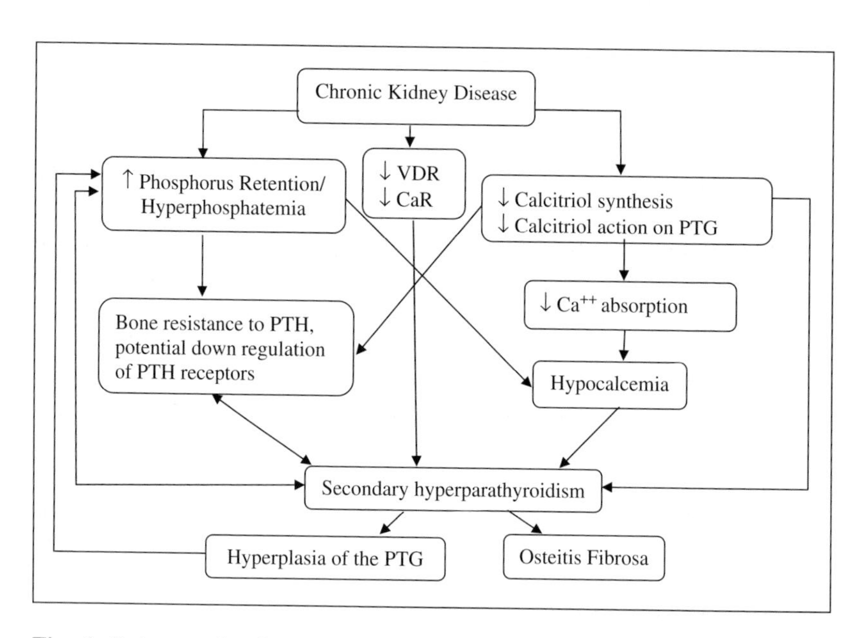

Fig. 1. Pathogenesis of secondary hyperparathyroidism. CaR, calcium sensing receptor; PTH, parathyroid hormone; VDR, vitamin D receptor; PTG, parathyroid gland; Ca^{++}, calcium.

increase, serum calcium decreases, and production of calcitriol is inadequate to meet physiological needs. Reduced calcitriol levels hinder the absorption of calcium from the intestines. These factors lead to hypocalcemia, the primary stimulus for increased production and secretion of PTH. Increased PTH levels stimulate phosphorus excretion and calcitriol production to correct the hypocalcemia. In the later stages of CKD, increased PTH production and secretion can no longer counterbalance the abnormal serum levels of calcium and phosphorus *(1,4,5)*. In addition, the kidneys lose their ability to adequately degrade and clear PTH from the body. Increased PTH production and secretion along with decreased PTH degradation lead to SHPT *(4,7–10)*.

Phosphorus retention is also a key element in the development of SHPT. Hyperphosphatemia helps regulate the production of calcitriol by reducing the activity of the enzyme that activates 25(OH) vitamin D *(10)*. Hyperphosphatemia also has an effect on PTH gene expression and indirectly increases PTH production *(11)*. Normally, higher PTH levels decrease phosphorus reabsorption to restore serum phosphorus levels to normal. In the late stages of CKD, this compensatory mechanism is inadequate to maintain the serum phosphorus levels *(11)*. Additionally, with significant SHPT, phosphorus is released directly from the bone into the blood contributing to hyperphosphatemia. Chronically elevated phosphorus levels are associated with parathyroid gland size and are central to parathyroid gland hyperplasia and continued high PTH levels *(1,4,10,11)*.

In CKD, production of calcitriol by the kidney is reduced. Low levels of calcitriol contribute to SHPT both directly and indirectly *(9,11,12)*. Calcitriol exerts a direct negative feedback control on the parathyroid gland, inhibiting preproparathyroid hormone production and gene transcription *(1,4,5)*. Indirectly, low calcitriol levels hinder the absorption of calcium from the intestine and mobilization of calcium from the bone. These actions suggest that calcium, rather than vitamin D, predominantly regulates PTH *(9)*.

Hypocalcemia results from increased calcium–phosphate complexes and from a decrease in absorption of dietary calcium from the intestine *(12,13)*. In addition, the ability of the bone to release calcium into the blood is hindered. The CaR provides a regulatory mechanism involving release of PTH to maintain calcium homeostasis *(10)*. Even slight physiologic, normal range changes in serum calcium seem to modulate the development of SHPT.

Skeletal resistance to the calcemic action of PTH is also a factor in the development of SHPT. As CKD progresses, increasingly higher levels of PTH are needed to induce PTH effects and maintain normal bone remodeling activity *(4,14,15)*. Skeletal resistance is thought to be multifactorial, perhaps from altered regulation of PTH receptors in the bone that makes them less sensitive to PTH as well as from phosphorus retention and calcitriol deficiency *(6,16)*.

As hypocalcemia, hyperphosphatemia, and calcitriol deficiency continue, the parathyroid gland continually increases production of PTH leading initially to parathyroid cell hypertrophy. With chronic stimulation, the parathyroid cells proliferate and diffuse hyperplasia develops. This proliferation of parathyroid cells makes it difficult to modulate PTH levels. Nodular hyperplasia is characterized by cells that have fewer CaRs and VDRs and are significantly resistant to vitamin D therapy *(1,5,17,18)*.

Osteitis fibrosa is caused by SHPT and historically has been the most common form of bone disease in CKD *(9,19)*. It is characterized by marrow fibrosis and increased bone turnover due to both bone resorption and bone formation. The bone resorption is caused by an increase in the number and activity of osteoclasts, and bone formation is due to increased osteoblasts and osteoid deposition *(1,4,6)*.

While mixed bone disease has features of both high and low turnover abnormalities, it is generally classified as a high turnover disease. Mixed bone disease has been associated with aluminum accumulation, hypocalcemia, and hypophosphatemia *(4,9)*.

3. LOW TURNOVER BONE STATES

Osteomalacia due to aluminum overload was common in the 1970s and early 1980s secondary to aluminum levels in the dialysate and the use of aluminum hydroxide phosphate binders. With the change in dialysate standards and limitations of aluminum ingestion, aluminum-related osteomalacia is uncommon, but toxicity can still occur *(4,9)*. The KDOQI guidelines recommend that aluminum hydroxide be used only in extreme hyperphosphatemia and for a short duration. Other sources of aluminum should be identified and eliminated if the patient has elevated serum aluminum (>60 mcg/L), a positive desferroxamine (DFO) challenge, or clinical symptoms that are associated with aluminum toxicity *(4)*. There is also a potential for developing osteomalcia related to vitamin D deficiency, metabolic acidosis, hypophosphatemia, and deficiencies in the trace elements, fluoride and strontium *(4)*. Osteomalacia is characterized by a decreased bone formation rate, widened osteoid seams, as well as decreased formation and resorption surfaces *(4,9)*.

ABD is characterized by a lack of new bone formation, low cellular activity, low numbers of osteoblasts, and normal or reduced osteoclasts.

Increased bone matrix is the primary defect. Mineralization is usually decreased, without excess osteoid deposition or abnormal thickness. The reduction in osteoblasts and limited bone formation may be a result of a relative deficiency in PTH. Other systemic PTH inhibitory factors may also play a role in ABD *(9,19,20)*.

Several subgroups of CKD stage 5 dialysis (CKD stage 5D) patients may be more likely to have ABD. These include peritoneal dialysis patients, elderly patients, and those with diabetes *(19,21)*. Plasma levels of PTH are generally higher than normal in CKD, even when associated with ABD. In uremia, a relative reduction in PTH is able to induce a low turnover bone state even at laboratory normal PTH levels *(22,23)*. There are also racial differences in response to PTH. Blacks tend to have reduced skeletal sensitivity to PTH and less likelihood of developing overt bone disease despite higher plasma levels of PTH *(24)*.

In addition to bone abnormalities, extraskeletal calcification is a significant finding in CKD-MBD. Arterial calcification is found early in CKD and progresses over time. Early research suggests that coronary calcification is more likely to occur in individuals who have higher serum phosphorus, higher calcium times phosphorus product (CaP) levels, and a higher daily calcium load *(25)*. Guérin et al. found that the severity of calcification is correlated with age, dialysis vintage, fibrinogen levels, and the prescribed dose of calcium-based phosphate binders. They showed that vascular calcification is associated with increased stiffness of the large capacity arteries such as the carotid artery and the aorta *(26)*. Other research shows that the presence and severity of arterial calcifications predict cardiovascular and all cause mortality *(27)*.

Calcific uremic arteriolopathy (CUA), previously termed calciphylaxis, is associated with painful skin lesions, subcutaneous nodules, tissue ischemia, and necrosis of the skin or subcutaneous tissue of the extremities. CUA symptoms have been linked to high levels of PTH and phosphorus. The risk of CUA is six times higher in females than males *(28)*, and the presence of obesity seems to be a predisposing factor *(29)*. Although CUA is less common today, it can be a life-threatening complication of SHPT.

Another abnormality in CKD is atypical accumulation of β_2-microglobulin (β_2MA), a polypeptide that is involved with lymphocyte-mediated immune response. Accumulation is progressive due to decreased catabolism and excretion by the kidneys. Symptoms seldom occur until the patient has been on dialysis therapy for a long time, that is, 5–15 years. The most common first symptom is carpal tunnel syndrome. Kidney transplantation is currently the only therapy that stops the progression of β_2MA. Current treatment focuses the use of high-flux, biocompatible dialyzer membranes to enhance clearance of β_2MA during dialysis and on easing joint pain and inflammation *(30,31,40)*.

Osteoporosis is a skeletal disorder commonly found in the elderly. Because a large percentage of those receiving dialysis are over the age of 65, osteoporosis may occur as an adjunct problem to CKD-MBD. Traditional therapies used to treat osteoporosis in otherwise healthy adults are thought to be inappropriate for dialysis patients, especially without benefit of a bone biopsy *(32)*. Diagnosis of osteoporosis in CKD is more complicated as CKD-MBD may have similar manifestations *(33)*.

4. IDENTIFICATION OF BONE AND MINERAL DISORDERS—RO OR CKD-MBD?

When utilizing the KDIGO definitions of RO and CKD-MBD, it becomes obvious that much of the current research is really describing CKD-MBD *(4)*. Standardization of terminology will help facilitate research comparisons and the developed practices that have the potential to improve patient outcomes.

5. BONE BIOPSY

Bone biopsy is the most accurate diagnostic tool for determining bone lesions in CKD. All other assessment parameters should be compared to bone biopsy as the gold standard for assessing bone metabolism. Historically, bone biopsy has been viewed as significantly invasive and potentially painful. Additionally appropriate sample processing techniques, expert interpretation, and standardized reporting terminology have been lacking. As bone biopsy becomes more common, these limitations may diminish. Biopsy with tetracycline labeling allows the classification of bone pathology based on static and dynamic parameters that diagnose RO *(4,34)*. Routine bone biopsy is not recommended for CKD stage 5D patients unless the clinical picture or symptoms are inconsistent with the biochemical profile *(1)*.

6. RADIOGRAPHY, PULSE PRESSURE, AND ELECTRON BEAM COMPUTED TOMOGRAPHY

While X-rays provide limited information on CKD stage 5D-specific bone abnormalities, they do help in the assessment and identification of extraskeletal calcification and osteoporosis. Lateral abdominal

X-rays are a simple low cost way to detect vascular calcification *(4)*. Pulse pressure (the difference between systolic and diastolic blood pressure) has been shown to predict arterial stiffness, and cardiac calcification contributes to arterial stiffness. Increased pulse pressure in dialysis patients is associated with increased mortality risk *(35,36)*. However, the true value of this simple calculation in predicting vascular calcification is uncertain. Another method of identifying soft tissue calcification is electron beam computed tomography (EBCT); however, EBCT is not routinely available. EBCT studies have shown that dialysis patients have coronary artery calcium scores that are several fold higher than individuals without CKD *(37)*.

7. BONE MINERAL DENSITY

Bone mineral density (BMD) is most commonly measured by dual X-ray absorbtiometry. This procedure measures the mineral content and the density of the bone, but does not provide information about bone turnover, does not correlate well with bone histology, or identify the type of lesion in CKD stage 5D *(33)*. The KDOQI bone workgroup did not find evidence that BMD was helpful in guiding therapy for CKD-MBD, except for the evaluation of osteoporosis and in patients who have fractures *(1)*.

8. BIOCHEMICAL MARKERS OF BONE AND MINERAL METABOLISM IN CKD

With some limitations, a number of biochemical parameters can assist in the diagnosis and management of CKD-MBD. Much of the current research is focused on correlating specific biochemical parameters to bone biopsy, thus enhancing their clinical value. Concomitant, serial monitoring of several biomarkers is useful. These include total serum calcium, corrected calcium, serum phosphorus, alkaline phosphatase (AP), and plasma PTH *(1,5,34)*.

Ionized calcium is the fraction of blood calcium that is critical to physiological processes. It is more difficult to accurately measure than total calcium, but is the most indicative of an individual's true calcium status. Routine monitoring of total calcium over ionized calcium is recommended because it is usually more reproducible, less costly, and adequate, especially in the presence of normal plasma proteins *(6)*.

Total calcium may under-represent ionized calcium in protein compromised patients. Because a significant portion of serum calcium is bound to protein, predominately albumin, it has been suggested that correcting total calcium for low albumin may more accurately estimate ionized calcium. A myriad of formulas have been proposed to make this correction. After evaluating the evidence, the KDOQI bone guidelines provide two acceptable formulas for adjusting calcium *(38,39)*. However, these two formulas may not yield similar results (Fig. 2).

These formulas may also correlate differently to ionized calcium because of the methods utilized for measuring either albumin or calcium. Correction formulas are thought to inaccurately predict calcium status in up to 30% of patients when compared to actual ionized calcium levels *(39)*. Adjusting total calcium downward in patients with normal albumin levels greater than 4.0 g/dL is questionable at best *(4)*.

High serum calcium is associated with bone abnormalities as well as morbidity and mortality in CKD patients on dialysis *(1,2)*. KDOQI recommends maintaining serum total calcium at the lower end of normal range for the specific laboratory (8.4–9.5 mg/dL). In addition, the total calcium ingestion per day, including dietary calcium and other sources such as phosphate binders, should not exceed 2000 mg/day. This amount is slightly less than the upper tolerable limit, but above what is recommended for the general population *(1)*. An excessive calcium load should be avoided, especially if serum levels are above the mid-range of normal. Extraskeletal calcification is multifactorial, but calcium load likely plays some role *(1)*.

Albumin 30 g/L (3.0 g/dL)
Corrected calcium (mg/dL) = Total calcium (mg/dL) + 0.0704 x [34 - serum albumin (g/L)] (20)
Example: Corrected calcium = 8.9 mg/dL + 0.0704 x (34 – 30)
Corrected calcium = 8.9 + (0.0704 x 4)
Corrected calcium = 8.9 + .2816
Corrected calcium = 9.2 mg/dL

Albumin 3.0 g/dL (30 g/L)
Corrected calcium (mg/dL) = Total calcium (mg/dL) + 0.8 × [4-serum albumin (g/dL)] (21)
Corrected calcium = 8.9 mg/dL + 0.8 × (4 – 3.0)
Corrected calcium = 8.9 mg/dL + (0.8 x 1.0)
Corrected calcium = 8.9 + 0.8
Corrected calcium = 9.7 mg/dL

Fig. 2. KDOQI Formulas for corrected calcium.

Phosphorus, one of the most common chemical elements in the body, is involved with a wide variety of metabolic and enzymatic processes. It circulates and is measured as phosphate ions in the serum, but is usually reported as elemental phosphorus concentrations *(4)*. Concentration of phosphorus in the serum varies significantly depending on the time of day and recent dietary phosphorus intake which may be significantly increased by many phosphate-laden food additives. This may help explain the significantly high and variable levels seen in individual CKD patients. Fasting phosphorus measures are ideal, but unlikely in chronic dialysis patients. Falsely high levels may be due to breakdown of blood cells if specimens are not being processed correctly *(4)*. Chronic hyperphosphatemia is associated with bone and mineral abnormalities, worsening SHPT, as well as morbidity and mortality *(2,3)*. Hyperphosphatemia is also aggravated by severe SHPT, where bone phosphorus is released directly into the blood and is unavailable to phosphate binders *(6)*. KDOQI recommends maintaining serum phosphorus between 3.5 and 5.5 mg/dL in dialysis patients. Methods for control of serum phosphorus include dietary modification, adequate dialysis, and the use of phosphate binders *(1,4)*.

Calcium and phosphorus are evaluated jointly. The CaP is a calculated biochemical indicator that has predictive power for abnormal mineral metabolism, morbidity, and mortality *(2,3)*. It is possible to have a target range CaP, even if one or both of the components are out of range. Thus, solitary monitoring of CaP should not be substituted for independent evaluation of serum calcium and phosphorus. KDOQI recommends maintaining CaP at $\leq 55\ mg^2/dL^2$ in those patients who are on dialysis *(1)*.

PTH is the most important biomarker for the evaluation of CKD-MBD. It plays a critical role in the regulation of mineral and bone homeostasis, and its secretion is regulated by serum ionized calcium. Levels of phosphorus and vitamin D also affect synthesis and secretion of PTH. While PTH targets may be extrapolated from earlier assays and methods, PTH is routinely used to monitor and treat CKD-MBD *(1)*. Intact PTH (iPTH) levels >400–500 pg/mL have significant predictive power for high turnover bone state *(1,40)*. Low turnover bone state is likely when iPTH levels are <100 pg/mL *(1,41)*. However, when the iPTH is between 100 and 500 pg/mL, a definitive diagnosis requires a bone biopsy *(1,40)*. It must be remembered that PTH has a very short half life and values may fluctuate dramatically from measurement to measurement. KDOQI recommends that trends, rather than single measurements, be used to determine therapy.

Biologically active PTH (1–84) polypeptide is synthesized and secreted by the parathyroid cells. However, along with 1–84 PTH, varied PTH fragments are released. As previously discussed, many factors modulate PTH gene expression, PTH production, PTH secretion, and parathyroid cell proliferation *(42)*. Calcium is the primary determinant of minute to minute PTH secretion, whereas calcium, phosphorus, and vitamin D levels regulate PTH production and cell proliferation. The 1–84 molecule is degraded into smaller fragments within minutes of release *(4,42)*. This degradation is modulated by parathyroid cells and the serum calcium concentration. Thus, when serum calcium is low, PTH degradation decreases and when calcium is high, greater PTH breakdown occurs. PTH fragments have varying half lives *(42)*. They also have diverse biologic activity on PTH receptors. Elimination of PTH fragments is primarily through glomerular filtration and tubular degradation, therefore they accumulate in CKD. These variable circumstances have generated questions regarding the predictive value and interpretation of plasma PTH *(1,42)*.

Recent advances in PTH measurement suggest that iPTH assays capture both 1–84 and other fragments. The metabolic significance of PTH fragments is not fully understood. The action of the largest known PTH fragment (7–84) may oppose the action of the 1–84 molecule and contribute to the PTH resistance seen in CKD stage 5D. Newer generation PTH assays measure only the 1–84 molecule. These are referred to as whole PTH or bioactive PTH, depending on the manufacturer *(43,44)*. Because these newer assays ignore PTH fragments, concentrations are approximately 50% lower. In general, results from these two generations of assays are highly correlated across a wide spectrum of PTH concentrations, but individual patient correlations may vary. While some researchers advocate using the ratio between 1–84 PTH and 1–84 PTH plus fragments to evaluate CKD-MBD, the value of this practice is still under investigation. Further research is needed to fully elucidate the opposing action of PTH fragments and to correlate second generation assay results to bone histology *(45)*.

The normal range of iPTH, 10–65 pg/mL (1.1–7.15 pmol/L), reflects normal bone turnover in those without CKD. In CKD, with progressive skeletal resistance to PTH, normal bone turnover more closely correlates with plasma iPTH levels of 2.5–4 times the normal upper limit *(4)*. Thus, a normal PTH values are not normal in CKD patients on dialysis.

AP can also add information about the state of bone turnover. AP is an isoenzyme that is produced primarily in the liver and by osteoblasts in the bone. Other sites of AP production are the intestines, placenta, and kidneys, although these sources contribute negligible amounts under normal conditions. With normal liver function, AP is a useful indicator of bone cell activity. Most research indicates that in CKD, elevated serum AP levels are due to bone AP and correlate with other markers of high turnover bone disease *(46)*.

Bone-specific AP (BSAP) is the fraction of AP that is generated by the osteoblasts. BSAP correlates well with iPTH and other indices of SHPT. However, the KDOQI bone workgroup did not find evidence that BSAP adds significant information above the total AP measurement in CKD patients when liver function is normal *(1)*.

Osteocalcin or bone GLA-protein is a noncollagenous protein that is produced by the osteoblasts and is found within the bone matrix. While the physiologic role of osteocalcin is not clear, it seems to impact bone formation. The value of measuring osteocalcin is questionable as it is excreted by the kidneys and fragments accumulate in CKD. Correlation of bone GLA-protein with other established biomarkers of CKD-MBD is mixed. Assays that detect only intact osteocalcin hold promise especially for identification of adynamic bone in CKD stage 5D *(47)*.

Primary causes of aluminum toxicity have been eliminated, thus serum aluminum is measured less frequently than in the past. Measures of serum aluminum reflect recent aluminum exposure and the potential for accumulation. They may identify hidden sources of aluminum to which a patient is exposed. Serum levels of aluminum should be $<10\,\mu g/L$ *(1)*. In CKD patients on dialysis, serum aluminum levels of >60 mcg/L have notable specificity, sensitivity, and predictive value for the diagnosis of aluminum-related bone disease. KDOQI recommends a DFO test be performed in any symptomatic patient with a serum aluminum level between 60 and $200\,\beta g/L$ or one being considered for a parathyroidectomy (PTX) *(1)*. The guidelines give specific recommendations for the performance of the DFO test and interpretation of results.

9. TREATMENT OF CKD-MBD

Treatment goals for CKD-MBD are recommended within the KDOQI clinical practice guidelines for bone metabolism and disease *(1)*. The primary goal is to promote standardized, best practice management of bone and mineral abnormalities as a way to improve patient outcomes. Evidence-based practice guidelines are meant to help clinicians in decision making, but not to dictate practice or set absolute standards. Where evidence is missing, experts provide a consensus of opinion for practice recommendations. It is expected that clinicians will apply the guidelines and recommendations based on evaluation of an individual patient's status and needs *(4,6)*.

The KDOQI practice guidelines for CKD-MBD provide aggressive biochemical targets as summarized in Table 3. The Dialysis Outcomes and Practice Patterns Study (DOPPS) study, although done prior to publication of the guidelines, indicated that very few dialysis patients consistently meet similar targets *(3)*. KDOQI dietary modification recommendations are summarized in Table 4. Thus, the guidelines set a higher standard of care for which to strive *(6)*.

Table 3
KDOQI Recommended Laboratory Measures, Target Levels, and Frequencies for CKD Stage 5D

Biomarker	*Frequency of measure*	*Target levels*
Calcium	At least monthly	8.4–9.5 mg/dL, lower end of normal
Phosphorus	At least monthly	3.5–5.5 mg/dL
CaP	At least monthly	$\leq 55\ mg^2/dL^2$
iPTH	At least quarterly	150–300 pg/mL

CaP, calcium times phosphorous product; CKD stage 5D, chronic kidney disease stage 5 dialysis; PTH, intact parathyroid hormone; KDOQI, Kidney Disease Outcomes Quality Initiative.

Table 4
KDOQI-Recommended Dietary Modifications Related to CKD-MBD

Calcium	*Phosphorus*
Limit total elemental calcium to <2000 mg/day with no more than 1500 mg from phosphate binders.	Limit dietary phosphorus (800 and 1000 mg/day) recognizing that protein requirements may prevent individual patients from meeting this limitation. Nutrition status must be preserved; additional phosphate binders may be necessary to maintain target serum phosphorus levels when higher protein goals are prescribed.

CKD-MBD, Chronic Kidney Disease Mineral and Bone Disorder; KDOQI, Kidney Disease Outcomes Quality Initiative.

Treatment approaches are intended to normalize blood levels of phosphorus and calcium and to optimize PTH levels for normal blood levels bone turnover, while minimizing extraskeletal complications such as calcification. Controlling these biochemical markers is important to prevent or reduce progression of bone disease. The following treatment options are available to help achieve KDOQI recommended goals *(1,4,48)*.

- Limit dietary phosphate intake, recognizing that processed foods have additives that contribute to the phosphate load. Limiting phosphate intake must be balanced with ensuring adequate protein intake.
- Optimize renal replacement therapy (adequate dialysis).
- Prescribe and titrate phosphate binders as needed to control serum phosphorus.
- Limit calcium based binders to an elemental calcium load of <1500 mg/day and avoid them in patients who are hypercalcemic, hypoparathyroid, or who have indications of soft tissue calcification.
- Evaluate calcium intake, and limit total elemental calcium to <2000 mg/day. Monitor the patient for signs of hypocalcemia and supplement calcium if symptoms are present.
- Prescribe and titrate vitamin D or its analogs to control PTH.
- Prescribe and titrate calcimimetics to control PTH, which can be done without increasing CaP.
- Treat carefully to achieve PTH levels that promote normal bone turnover, rather than suppressing PTH at a level that might be consistent with adynamic bone state.
- Educate the patient on all aspects of bone and mineral management including the rationale for various treatments and the consequences of nonadherence.
- Consider PTX, if all other treatments fail or surgery is deemed most appropriate for the patient.
- Above all consider the impact of various therapies and weigh the risk to benefit profile.

Phosphate removal on dialysis depends on the serum phosphorus concentration at the beginning of treatment, the ultrafiltration rate, the dialyzer capabilities, as well as the frequency and duration of dialysis. The majority of phosphorus clearance takes place early in the treatment and is followed by a slow equilibrium from the intracellular compartment. While phosphorus removal continues throughout the treatment, it is removed more slowly as serum concentrations decline. Patients who are unable to control serum phosphorus with a conventional dialysis schedule may benefit from one or more extra days of dialysis *(49)*.

A typical 4-h dialysis treatment, three times per week, does not maintain a net zero balance between phosphorus intake and clearance. Reports vary as to how much phosphorus is removed by high flux dialysis with each treatment, but it is estimated to be about 800 mg (range 400–1100 mg) *(4,48,49)*. Even when the patient limits dietary phosphate to the recommended level of 800–1000 mg/day, there is still a significant positive balance of phosphorus which generally requires the use of phosphate binders. In severe SHPT, hyperphosphatemia may be aggravated by excessive bone turnover with the release of phosphorus directly into the blood. Thus, hyperphosphatemia may not always be due to dietary excess or binder nonadherence.

The most appropriate phosphate binders are patient-specific and readily available. They must be well tolerated, without creating other problems or adding an excessive pill burden *(1)*. Appropriate prescription of phosphate binders requires an assessment of dietary phosphate intake and titration of the dose to control serum phosphorus in the target range. Prior to each modification of the binder prescription, it is critical to assess patient adherence to that binder prescription. The binder dose should be titrated to the phosphate content of the meal and include binders for additional sources of phosphate such as snacks. Commonly available phosphate-binding compounds are summarized in Table 5.

The move toward more frequent dialysis holds great promise for ameliorating CKD-MBD *(4)*. Patients who dialyze more frequently than three times per week seldom have trouble controlling phosphorus and some require supplementation. While phosphorus clearance is better with more frequent dialysis, phosphate binder doses may not change as many of these patients eat more heartily.

The KDOQI target range for serum calcium was based primarily on a consensus of expert opinion, but since publication of the recommendations has been supported by retrospective analysis and observational studies. Maintaining serum calcium in the target range involves regulating the calcium in the dialysate, as well as monitoring and modifying the calcium load from the diet and medications such as calcium-based binders or antacids. Many foods are fortified with calcium and can significantly add to the dietary calcium load. It is important that patients be made aware of calcium sources and recommendations for the daily elemental calcium load.

KDOQI recommends the routine use of dialysate with a 2.5 mEq/L calcium concentration to minimize the movement of calcium from blood to dialysate or dialysate to blood *(1)*. With 3.5 mEq/L calcium dialysate concentrations, most patients have a positive flux

Table 5
Available Phosphate Binder—Compounds

Binder source	*Rx*	*Available forms*	*Content (mineral/ metal/ element)*	*Potential advantages*	*Potential disadvantages*
Aluminum hydroxide	No	Liquid, tablet, capsule	Aluminum (Varies from 100 mg to >200 mg)	Effective, high phosphate binding capacity; various forms	Aluminum toxicity; altered bone mineral integrity, dementia
Calcium acetate	Yes	Gelcap, tablet	25% elemental Ca^{++} (169 mg elemental Ca^{++}/667 mg cap)	Effective, potentially better binding capability than $CaCO_3$ with less Ca^{++} absorption	Adds to Ca^{++} load with potential for hypercalcemia, soft tissue calcification, and PTH suppression; Prescription, more costly than $CaCO_3$; GI symptoms
Calcium carbonate	No	Liquid, tablet, chewable, capsule, gum	Contains 40% elemental Ca^{++} (200 mg elemental Ca^{++}/500 mg $CaCO_3$)	Effective, inexpensive, readily available	Additional Ca^{++} load with potential hypercalcemia, soft tissue calcification, PTH suppression; GI symptoms
Calcium citrate	No	Tablet, capsule, liquid	Contains 22% elemental Ca^{++}	Not recommended	Citrate enhances aluminum absorption,

(*Continued*)

Table 5
(Continued)

Binder source	*Rx*	*Available forms*	*Content (mineral/ metal/ element)*	*Potential advantages*	*Potential disadvantages*
Magnesium carbonate/ Calcium carbonate	No	Tablet	Contains approximately 28% Mg^{++}/ total Mg carbonate and 25% elemental Ca^{++}/total $CaCO_3$	Effective; potential for lower calcium load than calcium-based binders	GI symptoms, potential for hypermagnesemia
Sevelamer HCL	Yes	Caplet	None	Effective; no calcium/metal; not absorbed; shown to reduce coronary/aortic calcification when compared to calcium-based binders; reduces plasma lipid concentration	Cost; potential for decreased bicarbonate levels; may require calcium supplement if symptoms of hypocalcemia are present; GI symptoms
Lanthanum carbonate	Yes	Wafer, chewable	Contains 250 or 500 mg elemental lanthanum	Effective, chewable wafers	Potential for accumulation of lanthanum, although effect is not fully elucidated; GI symptoms

PTH, parathyroid hormone; Rx, prescription; GI, gastrointestinal
Newer phosphate binders which address some of the limitations of current binders are under investigation.

of calcium. Conversely, lower concentrations of dialysate calcium promote negative calcium flux. As with phosphorus, the movement of calcium depends on serum levels, dialysate levels, and the duration of exposure to dialysate *(50)*. Serum calcium is impacted by calcium load, presence of active vitamin D, and circulating PTH. Severe SHPT promotes release of calcium from the bone, adding another source for hypercalemia. Serum calcium may also be elevated when PTH levels are low and blood calcium is not being incorporated into the bone. A retrospective analyses of a large dialysis data base validates the KDOQI recommendation for serum calcium, showing that calcium levels above the mid-normal range increase mortality, whereas lower calcium levels are not associated with increased mortality risk *(2)*.

There are two categories of pharmacologic agents that are routinely used to treat SHPT, vitamin D and its analogs, and calcimimetics. A description of these agents and their distinctive actions is given in Table 6.

Vitamin D products are commonly used to control PTH levels in CKD-MBD. Vitamin D, vitamin D analogs, and vitamin D derivatives are described in Table 7. Each of these vitamin D products has specific structure and actions. Vitamin D doses are typically based on the elevation of plasma PTH. Calcitriol is the naturally occurring form of vitamin D hormone. It is used to treat hypocalcemia and SHPT and is available in oral and IV forms. It has been associated

Table 6
Actions of Pharmacologic Agents Used to Treat SHPT

Vitamin D/analogs	*Calcimimetic*
Acts on the genomic receptor	Acts on the cell surface receptor
Slow onset and recovery (days to weeks)	Rapid onset (minutes) and recovery (hours to days)
Inhibits PTH synthesis	Decreases serum levels of $Ca^{++}$and P
Potential for increasing serum $Ca^{++}$and P	Inhibits PTH synthesis and secretion
Little or no impact on hyperplasic/ nodular glands	Inhibits gland hyperplasia
	Can work even when glands are nodular

P, phosphorus; PTH, Parathyroid hormone; SHPT, secondary hyperparathyroidism; Ca^{++}, calcium.

Table 7
Vitamin D Products (US)

Product	*Brand Name*	*Route of Administration*	*Target CKD Population*
Calcitriol 1α,25$(OH)D_3$	Calcitriol generic	Oral, IV	Stages 3–5
First-generation D_3 molecule;	Rocaltrol® (Roche)	Oral	Stage 5D
chemically different from D_2 analogs	Calcijex® (Abbott)	IV	Stage 5D
Doxercalciferol 1α $(OH)D_2$	Hectorol® Capsules	Oral	Stages 3–5
Prohormone, activated by liver	Hectorol®	IV	Stage 5D
Synthetic D2	(Genzyme)		
Paricalcitol 1α $(OH)_2$ 19-nor-D_2	Zemplar® Capsules	Oral	Stages 3,4, 5D
Synthetic D2	Zemplar®	IV	Stage 5D
	(Abbott)		

CKD, Chronic kidney disease; IV, intravenous, 5D, Stage 5, Chronic kidney disease.

with hypercalcemia, partly due to increased gastrointestinal calcium absorption.

Paricalcitol [19-nor-1,25$(OH)_2$] is a sterol derived from vitamin D_2. It is missing a carbon-19 methylene group that is present in all natural vitamin D metabolites. Paricalcitol has lower calcemic effect through VDR selectivity at the tissue level of bone and intestine, while having greater activity in the parathyroid tissue. Clinical trials have demonstrated the effectiveness of paricalcitol at controlling PTH *(51, 52)*. Additionally, a recent uncontrolled retrospective study suggests that paricalcitol provides a survival advantage over calcitriol *(53)*.

Doxercalciferol [1α-hydroxyvitamin D_2] is a prohormone that requires hepatic conversion to its active form, 1,25$(OH)_2D_2$. Doxercalciferol, IV or oral forms, has been shown to be clinically effective in controlling PTH in CKD patients *(54)*.

Dihydrotachysterol$_2$ is one of the first vitamin D derivatives used to treat CKD-MBD, but there are few convincing reports on its efficacy and safety. It is available for oral administration in tablet and liquid forms *(55)*. There are several other vitamin D products that are used outside the United States, which will not be discussed here but can be reviewed in the literature *(4,55)*.

Deficiencies of nutritional vitamin D are recognized in the general public and in CKD. Supplementation of ergocalciferol or cholecalciferol to replete 25(OH) in those on dialysis is thought to be beneficial and is becoming more common. However, the exact doses, frequency of administration, and target blood levels have not been fully determined.

9.1. Calcimimetics

Calcimimetics are a class of compounds that act on the CaRs in the parathyroid cells. They lower the threshold for receptor activation by extracellular calcium ions and suppress PTH secretion. Unlike vitamin D products, calcimimetics reduce plasma PTH with either no change or a decrease in serum calcium and phosphorus levels. Cinacalcet HCL (Sensipar™) has been shown to be effective in lowering PTH, even in patients who have been unresponsive to vitamin D due to gland hyperplasia. The action of cinacalcet is rapid with peak reduction of PTH levels within 2–6 hours after administration. The drug is taken orally and is generally well tolerated *(56–58)*.

With the potential decrease in serum calcium, it is important for patients to have normal serum calcium levels before starting calcimimetic therapy. Patient response to cinacalcet is varied and biochemical markers should be measured routinely; calcium and phosphorus should be checked within 1 week and PTH within 1–4 weeks after initiation of therapy. The initial dose is 30 mg/day with titration every 4 weeks up to a maximum dose of 180 mg/day. Cinacalcet can be used in conjunction with phosphate binders and vitamin D products to maximize treatment of CKD-MBD. Long-term studies are underway to determine the full benefits of calcimimetics at the level of the bone *(56,58)*.

Patient education is a vital part of the long-term management of CKD-MBD. Adherence to treatment regimens depends partially on the patient's understanding of their importance. While the ultimate choice of following the advice of the healthcare team resides with the patient, the clinical team must provide information at the level the patient can understand, utilizing a common message and varied teaching techniques.

PTX, subtotal or total, is an option for those patients who do not respond to medical or pharmacologic management of SHPT. The KDOQI bone guidelines give specific recommendations for consideration of a PTX and address the post-surgical management of the patient *(1)*. With the sudden reduction in plasma PTH after surgery, flux of calcium and phosphorus into the bone can be remarkable. This condition is referred to as "hungry bone syndrome" which results in significant hypocalcemia that requires close monitoring and often IV calcium administration *(59)*.

10. TREATMENT OPTIONS FOR ADYNAMIC BONE

Patients with adynamic bone have an increased risk for fracture *(2)*, hypercalcemia, and extraskeletal calcifications. In dialysis patients with biopsy-documented adynamic bone or iPTH levels below 100 pg/mL, bone turnover can be stimulated by allowing the iPTH to rise. This may be accomplished by decreasing the total calcium load, lowering serum calcium, and/or by reducing or discontinuing agents that suppress PTH synthesis and secretion *(1,21,22)*.

11. SUMMARY

Whether treating biopsy documented RO or CKD-MBD, identified by abnormal biomarkers, bone and mineral abnormalities, or extraskeletal calcifications, the issues are complex. There have been significant advances in the understanding of bone and mineral abnormalities in CKD patients, including their significant potential for increasing morbidity and mortality. Additionally, techniques for monitoring and treating these abnormalities continue to evolve. Improving patient outcomes requires early identification of abnormalities and appropriate utilization of all the therapeutic options—nutritional, pharmacologic, and dialytic to minimize the progression and complications of RO or CKD-MBD. Furthermore, successful management of bone and mineral abnormalities in CKD requires the full commitment and participation from the entire healthcare team, most importantly the patient.

12. CASE PRESENTATION

DC is a 54-year-old Asian American male who has been on hemodialysis for over 20 years. He is a very intelligent, educated man who has been reasonably compliant with medical advice. His long history parallels the changes in bone and mineral management for dialysis patients. He has taken various binders (aluminum hydroxide, calcium acetate, magnesium/calcium carbonate, sevelamer HCL) and all available forms of vitamin D (Calcijex®-calcitriol, Zemplar®-paricalcitol, Hectorol®-doxercalciferol, generic calcitriol). He has been successfully treated for aluminum overload with DFO after a long duration of therapy with aluminum-based binders. As of April 2004, his PTH remained significantly elevated in spite of aggressive vitamin D and binder therapy. Additionally, he was hypercalcemic and hyperphosphatemic. His bone-related medications included 10 pills® per meal and 15 mcg Hectorol® per treatment. His biochemical parameters were calcium 10.7 mg/dL, phosphorus 6.5 mg/dL, and biPTH 604 pg/mL (approximately 1200 pg/mL iPTH).

1. What are the possible causes of hypercalcemia, hyperphosphatemia?
2. What therapy options would you consider for this patient based on his current status?

12.1. Follow-Up

This patient was originally scheduled for a PTX in April, 2004. However, cinacalcet HCL (Sensipar®) was initiated at 30 mg/day. After 1 month of therapy, he had a significant reduction in serum calcium (from 10.5 to 9.7 mg/dL), phosphorus (from 6.5 to 4.2 mg/dL), and biPTH (from 604 to 354 pg/mL). The patient is currently on 120 mg/day Sensipar®, 4 mcg Hectorol®, and 10 Renagel® per meal. His iPTH level is 363 pg/mL (estimated biPTH 180 pg/mL), calcium is 9.6 mg/dL, and phosphorus is 4.6 mg/dL. The patient continues to work full time and has routinely declined more frequent treatment.

REFERENCES

1. National Kidney Foundation. KDOQI clinical practice guidelines for bone metabolism and disease in chronic kidney disease. Am J Kidney Dis 2003; 42: 1–201.
2. Block GA, Klassen PS, Lazarus JM, Ofsthun N, Lowrie E, Chertow GM. Mineral metabolism, mortality, and morbidity in maintenance hemodialysis. J Am Soc Nephrol 2004; 15:2208–2218.
3. Young EW, Akiba T, Albert JM, McCarthy JT, Kerr PG, Mendelssohn DC, Jadoul M. Magnitude and impact of abnormal mineral metabolism in hemodialysis patients on Dialysis Outcomes and Practice Patterns Study (DOPPS). Am J Kidney Dis 2004; 44:S34–S38.
4. Olgaard K, Ed. Clinical Guide to Bone and Mineral Metabolism in CKD. National Kidney Foundation, New York, NY, 2006.
5. Llach F, Yudd M. Pathogenic, clinical and therapeutic aspects of secondary hyperparathyroidism in chronic renal failure. Am J Kidney Dis 1998; 32 Suppl 2:S3–S12.
6. Malluche HH, Monier-Faugere MC. Hyperphosphatemia: Pharmacologic intervention yesterday, today, and tomorrow. Clin Nephrol 2000; 54(4):309–317.
7. Moe S, Drueke T, Cunningham J, Goodman W, Martin K, Olgaard K, Ott S, Sprague S, Lameire N, Eknoyan G. Definition, evaluation, and classification of renal osteodystrophy: A position statement from Kidney Disease: Improving Global Outcomes (KDIGO). Am J Kidney Dis 2006; 69(11):1945–1953.
8. Slatopolsky E, Delmez JA. Pathogenesis of secondary hyperparathyroidism. Am J Kidney Dis 1994; 23(2):229–236.
9. Sutton RA, Cameron EC. Renal osteodystrophy: Pathophysiology. Semin Nephrol 1992; 12(2):91–100.
10. Silver J, Levi R. Cellular and molecular mechanisms of secondary hyperparathyroidism. Clin Nephrol 2005; 63(2):119–126.
11. Slatopolsky E, Finch J, Denda M, Ritter C, Zhong M, Dusso A, MacDonald PN, Brown AJ. Phosphorus restriction prevents parathyroid gland growth. High phosphorus directly stimulates PTH secretion in vitro. J Clin Invest 1996; 97(11):2534–2540.
12. Slatopolsky E, Brown A, Dusso A. Calcium, phosphorus, and vitamin D disorders in uremia. Contrib Nephrol 2005; 149:261–271.
13. Coburn JW, Koeppel MH, Brickman AS, Massry SG. Study of intestinal absorption of calcium in patients with renal failure. Kidney Int 1973; 3:264–272.
14. Rodriguez M, Martin-Malo A, Martinez ME, Torres A, Felsenfeld AJ, Llach F. Calcemic response to parathyroid hormone in renal failure: role of phosphorus and its effect on calcitriol. Kidney Int 1991; 40(6):1055–1062.
15. Massry SG, Stein R, Garty J, Arueff AI, Coburn JW, Norman AW, Friedler RM. Skeletal resistance to the calcemic action of parathryroid hormone in uremia: Role of 1,25 $(OH)_2$ D_3. Kidney Int 1976; 9:467–474.
16. Hoyland JA, Picton ML. Cellular mechanisms of renal osteodystrophy. Kidney Int Suppl 1999; 73:S8–S13.
17. Silver J, Kilav R, Naveh-Many T. Mechanisms of secondary hyperparathyroidism. Am J Physiol Renal Physiol 2002; 283(3):F367–F376.
18. Felsenfeld AJ. Considerations for the treatment of secondary hyperparathyroidism in renal failure. J Am Soc Nephrol 1997; 8(6):993–1004.

19. Couttenye MM, D'Haese PC, Verschoren WJ, Behets GJ, Schrooten I, De Broe ME. Low bone turnover in patients with renal failure. Kidney Int Suppl 1999; 73:S70–S76.
20. Salusky IB, Goodman WG. Adynamic renal osteodystrophy: Is there a problem? J Am Soc Nephrol 2001; 12(9):1978–1985.
21. Coen G. Adynamic bone disease: An update and overview. J Nephrol 2005; 18(2):117–122.
22. Carmen Sanchez M, Auxiliadora Bajo M, Selgas R, Mate A, Millan I, Eugenia Martinez M, Lopez-Barea F. Parathormone secretion in peritoneal dialysis patients with adynamic bone disease. Am J Kidney Dis 2000; 36(5):953–961.
23. Galassi A, Spiegel DM, Bellasi A, Block GA, Raggi P. Accelerated vascular calcification and relative hypoparathryoidism in incident hemodialysis diabetic patients receiving calcium binders. Nephrol Dial Transplant 2006; 21(11): 3215–3222.
24. Gupta A, Kallenback LR, Zasuwa G, Devine GW. Race is a major determinant of secondary hyperparathyroidism in uremic patients. J Am Soc Nephrol 2000; 11(2):330–334.
25. Goodman WG, Goldin J, Kuizon BD, Yoon C, Gales B, Sider D, Wang Y, Chung J, Emerick A, Greaser L, Elashoff RM, Salusky IB. Coronary artery calcification in young adults with end stage renal disease who are undergoing dialysis. N Engl J Med 2000; 342(20):1478–1483.
26. Guérin AP, London GM, Marchais SJ, Metivier F. Arterial stiffening and vascular calcifications in end stage renal disease. Nephrol Dial Transplant 2000; 15(7): 1014–1021.
27. Blacher J, Guérin AP, Pannier B, Marchais SJ, London GM. Arterial calcification, arterial stiffness, and cardiovascular risk in end stage renal disease. Hypertension 2001; 38(4):938–942.
28. Mazhar AR, Johnson RJ, Gillen D, Stivelman JC, Ryan MJ, Davis CL, Stehman-Breen CO. Risk factors and mortality associated with calciphylaxis in end stage renal disease. Kidney Int 2001; 60(1):324–332.
29. Coates T, Kirkland GS, Dymock RB, Murphy BF, Brealey JK, Mathew TH, Disney AP. Cutaneous necrosis from calcific uremic arteriolopathy. Am J Kidney Dis 1998; 32(3):384–391.
30. Kuchle C, Fricke H, Held E. Schiffl H. High-flux hemodialysis postpones clinical manifestation of dialysis related anyloidosis. Am J Nephrol 1996; 16(6): 484–488.
31. Akalin T, Abayli E, Azak A, Yuksel C, Atilgan G, Dede F, Akalin T, Abayli E, Camlibel M. The effect of high-flux hemodialysis on dialysis-associated amyloidosis. Ren Fail 2005; 27(1):31–34.
32. Moe SM, Drüeke TB. A bridge to improving healthcare outcomes and quality of life. Am J Kidney Dis 2004; 43(3):552–557.
33. Miller PD. Treatment of osteoporosis in chronic kidney disease and end-stage renal disease. Curr Osteoporos Rep 2005; 3(1):5–12.
34. Martin KJ, Olgaard K, Coburn JW, Coen GM, Fukagawa M, Langman C, Malluche HH, McCarthy JT, Massry SG, Mehls O, Salusky IB, Silver JM, Smogorzewski MT, Slatopolsky EM, McCann L. Bone Turnover Work Group. Diagnosis, Assessment, and Treatment of Bone Turnover Work Group. Turnover Abnormalities in Renal Osteodystrophy. Am J Kidney Dis 2004; 43(3):558–565.

35. Safar ME, Blacher J, Pannier B, Guerin AP, Marchais SJ, Guyonvarc'h PM, London GM. Central pulse pressure and mortality in end-stage renal disease. Hypertension 2002; 39(3):735–738.
36. Miwa Y, Tsushima M, Arima H, Kawano Y, Sasaguri T. Pulse pressure as an independent predictor for the progression of aortic wall calcification in patients with well controlled hyperlipidemia. Hypertension 2004; 43(3):536–540.
37. Braun J, Oldendorf M, Moshage W, Heidler R, Zeitler E, Luft FC. Electron beam computed tomography in the evaluation of cardiac calcification in chronic dialysis patients. Am J Kidney Dis 1996; 27(3):394–401.
38. Clase CM, Norman GL, Beecroft ML, Churchill DN. Albumin corrected calcium and ionized calcium in stable haemodialysis patients. Nephrol Dial Transplant 2000; 15(11):1841–1846.
39. Ladenson JH, Lewis JW, Boyd JC. Failure of total calcium corrected for protein, albumin, and pH to correctly assess free calcium status. J Clin Endocrinol Metab 1978; 46(6):986–993.
40. Qi Q, Monier-Faugere MC, Geng Z, Malluche HH. Predictive value of serum parathyroid hormone levels for bone turnover in patients on chronic maintenance dialysis. Am J Kidney Dis 1995; 26(4):622–631.
41. Sherrard DJ. Aplastic bone: A non disease of medical progress. Adv Ren Replace Ther 1995; 2(1):20–23.
42. Martin KJ, Hruska KA, Lewis J, Anderson C, Slatopolsky E. The renal handling of parathyroid hormone. Role of peritubular uptake and glomerular filtration. J Clin Invest 1977; 60(4):808–814.
43. Reichel H, Esser A, Roth HJ, Schmidt-Gayk H. Influence of PTH assay methodology on differential diagnosis of renal bone disease. Nephrol Dial Transplant 2003; 18(4):759–768.
44. Goodman WG. The evolution of assays for parathyroid hormone. Semin Dial 2005; 18(4):296–301.
45. Martin KJ, Akhtar I, Gonzalez EA. Parathyroid hormone: New assays, new receptors. Semin Nephrol 2004; 24(1):3–9.
46. Canavese C, Barolo S, Gurioli L, Cadario A, Portigliatti M, Isaia G, Thea A, Marangella M, Bongiorno P, Cavagnino A, Peona C, Boero R, D'Amicone M, Cardelli R, Rossi P, Piccoli G. Correlations between bone histopathology and serum biochemistry in uremic patients on chronic hemodialysis. Int J Artif Organs 1998; 21(8):443–450.
47. Morishita T, Nomura M, Hanaoka M, Saruta T, Matsuo T, Tsukamoto Y. A new assay that detects only intact osteocalcin. Two step noninvasive diagnosis to predict adynamic bone disease in hemodialyzed patients. Nephrol Dial Transplant 2000; 15:659–667.
48. Goodman WG. Medical Management of secondary hyperparathyroidism in chronic renal failure. Nephrol Dial Transplant 2003; 18(3):iii2–iii8.
49. Gutzwiller JP, Schneditz D, Huber AR, Schindler C, Gutzwiller CE. Estimating phosphate removal in haemodialysis: An additional tool to quantify dialysis dose. Nephrol Dial Transplant 2002; 17:1037–1044.
50. Malberti F, Ravani P. The choice of dialysate calcium concentration in the management of patients on haemodialysis and haemodiafiltration. Nephrol Dial Transplant 2003; 18(Suppl 7): vii37–vii40.

51. Brown, AJ. Vitamin D Analogues. Am J Kidney Dis 1998; 32(4) Suppl 2:S25–S39.
52. Llach F, Keshav G, Goldblat MV, Lindberg JS, Sadler R, Delmez J, Arruda J, Lau A, Slatopolsky E. Suppression of parathyroid hormone secretion in hemodialysis patients by a novel vitamin D analog: 19-Nor-1,25-dihydroxyvitamin D_2. Am J Kidney Dis 1998; 32(4) Suppl 2: S48–S54.
53. Teng M, Wolf M, Lowrie E, Ofstun N, Lazarus JM, Thadhani R. Survival of patients undergoing hemodialysis with paricalcitol or calcitriol therapy. N Engl J Med 2003; 349:446–456.
54. Frazzo JM, Elangovan L, Maung HM, Chesney RW, Acchiardo SR, Bower JD, Kelly BJ, Rodriguez HJ, Norris KC, Robertson JA, Levine BS, Goodman WG, Gentile D, Mazess RB, Kylklo DM, Douglass LL, Bishop CW, Coburn JW. Intermittent doxercalciferol (1α hydroxyvitamin D2) therapy for secondary hyperparathyroidism. Am J Kidney Dis 2000; 36:550–561.
55. Morii H, Ishimura E, Inoue T, Tabata T, Morita A, Nishii Y, Fukushima M. History of vitamin D treatment of renal osteodystrophy. Am J Nephrol 1997; 17:382–386.
56. Block GA, Martin KJ, de Francisco AL, Turner SA, Avram MM, Suranyi MG, Hercz G, Cunningham J, Abu-Alfa AK, Messa P, Coyne DW, Locatelli F, Cohen RM, Evenepoel P, Moe SM, Founier A, Braun J, McCary LC, Zani VJ, Olson KA, Druke TB, Goodman WG. Cinacalcet for secondary hyperparathyroidism in patients receiving hemodialysis. N Engl J Med 2004; 350:1516–1525.
57. Nemeth EF, Heaton WH, Miller M, Fox J, Balandrin MF, Van Wagenen BC, Colloton M, Karbon W, Scherrer J, Shatzen E, Rishton G, Scully S, Qi M, Harris R, Lacey D, Martin D. Pharmacodynamics of type II calcimimetic compound cinacalcet HCL. J Pharmacol Exp Ther 2004; 308:627–635.
58. Goodman WG. Calcimimetics: a remedy for all problems of excess parathyroid hormone activity in chronic kidney disease? Curr Opin Nephrol Hypertens 2005; 14(4):355–360.
59. Richards ML, Wormuth J, Bingener J, Sirinek K. Parathyroidectomy in secondary hyperparathyroidism: Is there an optimal operative management? Surgery 2006; 139(2):174–180.

15 Chronic Diseases

Diabetes, Cardiovascular Disease, and Human Immunodeficiency Virus Infection

Sharon R. Schatz

LEARNING OBJECTIVES

1. To describe the systemic magnitude of diabetes mellitus beyond kidney disease and how this impacts comprehensive nutrition care.
2. To delineate the heart–kidney integral relationship regarding hemodynamic and regulatory function and effects on medical nutrition therapy.
3. To discuss need for a multidimensional approach to adequately address nutrition risk factors and metabolic complications associated with human immunodeficiency virus infection and stage 5 chronic kidney disease.
4. To describe the causal and non-causal risk factors which unite these three separate diseases with uremia and dialysis.

Summary

Although diabetes mellitus, cardiovascular disease, and human immunodeficiency virus infection are three separate entities, each has causal and non-causal risk factors that are common in the stage 5 chronic kidney disease population. The medical nutrition therapies are similar, which emphasize adequate protein and energy intakes, fluid control, and possibly carbohydrate and fat modifications. Each patient requires an individualized evaluation, taking into account the comorbid conditions, age, psychosocial situation, rehabilitation potential, and

From: *Nutrition and Health: Nutrition in Kidney Disease*
Edited by: L. D. Byham-Gray, J. D. Burrowes, and G. M. Chertow

type of renal replacement therapy. The overall impact of the disease states need to be considered beyond the restrictions imposed by kidney failure, if mortality rates are to decline while improving quality of life.

Key Words: Medical nutrition therapy; diabetes mellitus; cardiovascular disease; HIV infection; hemodialysis; peritoneal dialysis.

1. INTRODUCTION

Causal and non-causal risk factors are statistically related markers of the probability for morbid events. Causal risk factors signal the presence and severity of disease which often guide treatment, whereas non-causal risk factors are useful for prognostic purposes or for monitoring the evolution of a given disease *(1)*. These concepts become especially relevant when evaluating nutrition-related issues of dialysis in the presence of other chronic illnesses such as diabetes mellitus (DM), cardiovascular disease (CVD), and human immunodeficiency virus (HIV) infection. While these are three distinct disease states, each one has common causal and non-causal risk factors that are united by the presence of uremia and dialysis in the stage 5 chronic kidney disease (CKD) population. In addition, although they are dissimilar, there are common manifestations whose links ultimately guide medical nutrition therapy (MNT).

2. DIABETES

In 2005, the total prevalence of DM was 7% in the U.S. population; yet disproportionately, 54% of patients receiving maintenance dialysis had DM, with DM being the primary diagnosis for kidney failure in 44% of the dialysis population *(2,3)*. IgA nephropathy is the most common cause of non-diabetes-related kidney disease in people with type 2 diabetes (T2DM) *(4)*. Growing rates of DM-related stage 5 CKD in the United States and Canada reflect a higher incidence of T2DM and people living longer *(3,5–7)*. In addition, the prevalence of DM is greater in minority groups such as African Americans, Hispanics, and Native Americans, with black women aged 55–74 years accounting for the greatest number of newly diagnosed patients with stage 5 CKD secondary to DM *(3,6)*.

MNT is at the forefront of diabetes-related prevention strategies targeted by the U.S. Department of Health and Human Services, which focuses on metabolic control to prevent (secondary) or control (tertiary) complications of diabetes *(8)*. Understanding the systemic magnitude of DM beyond kidney disease is crucial to comprehensive nutrition care. Patients with stage 5 CKD with DM as the primary diagnosis have the lowest survival rate, poorest rehabilitation potential, and highest incidence of hospitalizations, mostly attributable to cardiovascular events *(5,7,9)*.

2.1. Renal Replacement Therapy in People with Diabetes

2.1.1. Hemodialysis

Hemodialysis (HD) is the most common form of renal replacement therapy (RRT) used to treat people with DM *(3,5)*. Poor glycemic control upon initiation of HD indicates worse survival *(6,10)*. An unfavorable 5-year survival rate is associated with DM and CVD, infection, advanced age, dyslipidemia, left ventricular hypertrophy (LVH), hypoalbuminemia, and more severe peripheral vascular disease (PVD) *(6)*.

The American College of Endocrinology emphasizes the need to understand the action and duration of insulin and the effect of food intake timing on insulin action, which have particular relevance in dialysis *(11)*. Insulin has a prolonged half-life that is related to a decreased catabolic rate in kidney disease, which may intensify its effect *(12)*. Treatment schedules may interfere with a patient's usual mealtime, and post-treatment fatigue may impede intake. These combined factors can precipitate hypoglycemia; so a patient may need to eat before coming to treatment and/or bring a snack to consume afterwards, with nutrition supplements offering a convenient option. The schedule could also affect insulin dosing, as a patient may alter injection times on treatment days, contributing to too much circulating insulin compounded by extended duration promoting hypoglycemia, or conversely, insufficient insulin coverage contributing to hyperglycemia. Thus, it is important to determine when the person administers insulin and eats each day and encourage regular habits to promote consistent blood glucose patterns. Medications may need adjustment with traditional split doses of insulin being converted to regimens with greater flexibility or sliding scales.

2.1.2. Peritoneal Dialysis

Survival with DM is better during the initial 2 years of peritoneal dialysis (PD) treatment, but the likelihood of losing residual renal function after 1 year on PD is doubled *(13–15)*. PD may thus be used before transitioning to HD *(15)*. Side effects due to intraperitoneal (IP) exposure to high glucose concentrations include an inflammatory state, hyperlipidemia, fibrosis, enhanced protein loss, generalized intra-abdominal fat accumulation, increased risk of CVD, weight gain, obesity, and acute hyperglycemia; and the underlying diabetic state may compound or exacerbate these problems *(16–18)*. Problems with ultrafiltration (UF) develop from peritoneal fibrosis, neoangiogenesis, and increased membrane permeability, with glucose either directly or indirectly contributing to peritoneal membrane alteration via glucose degradation product generation or advanced glycation end products formation *(16,19,20)*. Strategies to reduce glucose-related toxicity include dual chamber systems and amino acid-based fluids in addition to icodextran, which may provide better glycemic control, increase UF volume, improve blood pressure (BP) control, and be less hyperlipidemic *(16,19–21)*. However, icodextran (maltose based) solutions may interfere with glucose dehydrogenase-based glucometers giving false readings, so a compatible glucometer is needed *(22)*.

Glycemic control is complex with PD. Poor glycemic control is associated with fluid retention *(23)*. Conversely, sodium and fluid control is essential to glycemic control in PD to deter UF resulting in lower percent glucose dialysate dwells to lessen glucose absorption. There is a uremia-associated insulin resistant state, as insulin-stimulated glucose uptake by muscle is diminished and further aggravated by IP glucose absorption. In addition, β-cell function deteriorates with time, which further affects insulin requirements *(23)*. IP insulin administration affords better glycemic control and insulin sensitivity due to its more rapid absorption directly into the portal vein system with less endogenous hepatic glucose production. However, IP doses are 30% higher than the subcutaneous (SC) route due to insulin activity loss, as there is dilution with the PD infusion to delay the absorption of insulin as well as adsorption to the delivery system's plastic surface *(17,20,24)*. IP insulin may increase the risk of peritonitis due to contamination, but similar incidences of gram-positive infecting organisms does not support this theory *(20,25)*. Instilling insulin (mixed with saline) into an empty abdominal cavity may lower insulin requirements *(24)*. IP insulin may cause hepatic subcapsular steatosis that is associated with hypertriglyceridemia, obesity, fatty liver infiltration, and non-alcoholic steatohepatitis *(17,20,26)*.

Patients using PD may maintain weight with calorie absorption from the dialysate; however, gastrointestinal (GI) issues may negatively influence protein consumption. PD is associated with more gastroesophageal reflux disease (GERD) symptoms and eating dysfunction *(18)* Severe gastric emptying may be present even when patients are tested with an empty peritoneal cavity *(27)*. Susceptibility to peritonitis is higher due to gram-negative organisms, although those without peritonitis had higher albumin at the start of continuous ambulatory peritoneal dialysis therapy *(25)*.

2.2. Monitoring Glycemic Control

Adequate glycemic control reduces the risk of additional diabetes-related complications; however, present monitoring of glucose levels in dialysis patients may be inadequate *(22,28)*. The goals for monitoring glycemic control in people receiving maintenance dialysis are similar to those without RRT (refer to Chapter 6), but exact target levels for best outcomes have not been clearly established. The Kidney Disease Outcomes Quality Initiative (KDOQI) Guidelines for Diabetes recommend that the American Diabetes Association (ADA) guidelines for monitoring glycemic control in people with DM be followed *(29)*.

HbA1C reflects mean glycemia over the preceding 2–3 months *(30)*. HbA1C is the largest glycated hemoglobin fraction and the most consistent index of circulating glucose with formation determined by temperature, pH, hemoglobin and glucose concentrations, and exposure length to glucose. Uremic acidosis possibly increases the rate of formation of HbA1C and it may elevate the levels *(22)*. Concerns about accuracy and interpretation of the values in stage 5 CKD relate to decreased red blood cell survival, hemolysis, and iron deficiency decreasing values; while carbamylated hemoglobin formation due to urea dissociation may falsely increase values *(22,29,31)*. The accepted value for HbA1C is 7 in the non-dialysis population; however, a value greater than 8 cannot be deemed good control even if there is overestimation. Tracking trends in the values may be preferable to focusing on an absolute value. Nevertheless, HbA1C can provide valid, reliable results for most patients with stage 5 CKD with appropriate methodology and target is <7% irrespective of CKD *(29)*. On an individualized basis, a higher goal may be set if there is significant hypoglycemia *(22,29,32)*.

Another potential method used to monitor glucose is fructosamine, which measures glycemia over a shorter 2- to 3-week period. A high protein turnover rate such as in dialysis may affect use of fructosamine, while a high urate level may interfere with assay results *(22)*. Fructosamine is less reliable than HbA1C, especially with PD *(22,29,31)*.

Self-monitoring of blood glucose (SMBG) enables better blood glucose control by allowing patients to monitor influences of food, exercise, and medications and to make adjustments accordingly *(30,33)*. SMBG is especially indicated with insulin use to detect and deter asymptomatic hypoglycemia and hyperglycemia; it is usually recommended three to four times/day, with unknown optimal frequency when oral agents are prescribed *(30)*. A single random test is a poor guide to overall glycemic control; and checking fasting values alone may be insufficient, as postprandial glucose reduction may significantly impact HbA1C while lowering cardiovascular risk *(22,34,35)*. The ineffectiveness of SMBG to capture nighttime glucose control with PD signifies a potential role for continuous glucose monitoring *(22,36)*. Glycemic control is best evaluated by monitoring both SMBG and HbA1C *(30)*.

2.3. *Medical Nutrition Therapy*

The nutrition needs of people with DM who are receiving maintenance dialysis are similar to those without DM based on the type of RRT (HD or PD), with extra emphasis on achieving euglycemia, minimizing dyslipidemia, and attaining a total energy intake appropriate for weight management. The ADA nutrition recommendations and interventions for people with DM include the following *(8)*:

- Carbohydrate monitoring, which is essential for achieving glycemic control with doses of insulin and insulin secretagogues being matched to the carbohydrate content of the meals.
- Use of the glycemic index and load may provide modest additional benefits.
- Low carbohydrate diets which restrict total carbohydrate intake to less than 130 g/day are not recommended.
- Monitoring plasma triglycerides (TGs) with the use of high carbohydrate diets.
- Encourage fiber intake of 14 g/1000 calories.
- Limit saturated fat to less than 7% of total calories and dietary cholesterol to less than 200 mg/day and minimize use of *trans* fats.

Protein requirements may be increased due to infection, for wound healing, or for repletion of depleted visceral proteins. Two or more servings of fish per week (not commercially fried) are recommended to provide omega-3 fatty acids (ω-3 FAs) *(8)*. Other low fat protein sources such as egg whites, egg substitutes, and skinless poultry could be promoted as well as use of soy products with adjustment for phosphorus content. Energy goals are based on the individual's weight and metabolic factors to achieve and maintain a desirable body weight while promoting optimal nitrogen utilization. Advise use of mono- and poly insaturated fats that are low in potassium, sodium, and phosphorus.

Oral nutrition supplements (ONS) specially formulated for diabetes may not be necessary, as the ADA standards allow substitution of sucrose for other carbohydrates in the meal plan *(8)*. The choice of a ONS should be based on overall nutrient profile to meet the individual's metabolic needs and the palatability and affordability of the product.

The ADA promotes a diabetic diet that considers the personal and cultural preferences and lifestyle of the patient while respecting the individual's desires and willingness to change *(8)*. It is unwise to assume that a good understanding of diabetes management already exists when a patient with diabetes starts dialysis. The patient is often overwhelmed by multiple dietary restrictions and foods previously allowed may now be limited or avoided. There may also be difficulty coordinating the dietary recommendations into a cohesive meal plan. The ability to interchange carbohydrate units is complicated by the potassium and phosphorus content of the foods. Alternates to traditional bedtime snacks and treatment for hypoglycemia may be indicated, although long-standing habits are not easy to change. Renal diabetic exchange lists can classify food groups for meal patterns, but it is important that the patient understand and be able to integrate the rationale and basic concepts of nutrition goals for DM and dialysis to further dietary acceptance and foster compliance.

2.3.1. Malnutrition

Diabetes is the most significant predictor of loss of lean body mass (LBM) in dialysis patients, independent of inadequate dialysis dose, metabolic acidosis, and insufficient protein intake *(9)*. Patients receiving maintenance HD with DM have increased muscle breakdown that appears nonresponsive to insulin, which suggests existing protein homeostasis anomalies with either effects of insulin resistance (IR)/deficiency being additive to uremia or vice versa *(37)*. A higher incidence of protein malnutrition with dialysis and DM may not be a direct death risk *(38)*. However, the correlation with other death risk factors cannot be dismissed.

2.3.2. Fluid Control

Fluid control may be more difficult in people with DM because hyperglycemia may increase thirst and diabetic neuropathy may predispose symptoms of dry mouth, decreased salivary flow rates, and the effects of xerogenic drugs *(39–43)*. Fluid intake may be higher, if dental problems impair chewing or if gastroparesis is severe and solid food is not well tolerated.

2.3.3. Gastric Autonomic Neuropathy

Diabetic gastroparesis is prevalent in about 50% of people with DM, although the entire GI system can be affected by gastric autonomic neuropathy *(44–48)*. GERD is present in about 30% of people with DM; it is partially related to vagal neuropathy and can promote heartburn and dysphagia for solid foods *(44–46)*. Diabetic diarrhea (which is profuse, watery, and often nocturnal, alternating with constipation) is present in approximately 20% of people with DM and may be due to decreased GI motility, reduced fluid absorption, bacterial overgrowth, pancreatic insufficiency, coexistent celiac disease, and abnormal metabolism of bile acid salt *(44)*. Fecal incontinence is related to anal sphincter incompetence or diminished rectal sensation, whereas fecal impaction and bowel perforation can result when constipation, a dysfunction of intestinal neurons with diminished gastrocolic reflex, is severe *(44,46)*.

Diabetic gastroparesis, non-obstructive delayed gastric emptying with impaired gastric acid secretion and GI motility, is linked to disrupted vagus nerve function and neuron dysfunction *(44,46)*. Symptoms include early satiety, postprandial fullness, anorexia, nausea, bloating, belching, epigastric discomfort, abdominal pain, and emesis of previously consumed undigested food *(44–49)*. Food remaining in the stomach for a long period of time can lead to fermentation with bacterial overgrowth or hardening into solid masses (known as bezoars) causing nausea, vomiting, or obstruction, which blocks passage to the small intestines. Nutrient delivery to the small bowel is impeded, which affects medication absorption; and meal-related protein synthesis will probably be decreased *(37,44,45,50)*. Acute glycemic changes have reversible influences, as hyperglycemia slows and hypoglycemia accelerates gastric emptying. Poor glycemic control worsens early satiety, fullness, and nausea *(44,45,50,51)*.

Although normalization of blood glucose levels may improve gastric emptying, gastroparesis complicates the balance between insulin dose and food absorption. Erratic blood glucose levels often suggest gastroparesis prior to the diagnosis. Gastric emptying, the main determinant of postprandial glycemia, is also influenced by nutrient delivery into the small intestines. Factors that affect glucose appearance into portal circulation potentially include complex carbohydrate breakdown to glucose, mucosal absorption of glucose, small intestine motor patterns, and splanchic blood flow *(44,50,51)*. Frequent preprandial and postprandial blood glucose monitoring enables proper adjustments in insulin dose *(51)*. Less insulin is required with decreased gastric emptying to maintain euglycemia, and rapid acting insulin is probably contraindicated *(50,51)*. Timing of insulin doses may also need to be changed. Prolonged insulin action with stage 5 CKD further complicates this scenario.

Low fat, low fiber, small, frequent feedings are recommended for patients with diabetic gastroparesis *(44)*. Food should be chewed thoroughly *(48)*. Patients should be advised to eat high protein foods first, to limit "filling" or high volume foods, and to minimize fluids with food to combat early satiety. When symptoms are severe, only liquids may be tolerated *(47,49)*. Liquids could be taken an hour pre- or post-meal to reduce feelings of fullness and carbonated beverages should be avoided. When diarrhea is present, soluble fiber should be restricted and other approaches such as the use of a gluten-free or lactose-free diet can be implemented *(46)*.

2.4. Oxidative Stress

Oxidative stress (OS) in CKD is evidenced by an overabundance of lipid, carbohydrate, and protein oxidation products. It is aggravated by DM with an additive effect when combined with HD *(38,52,53)*. Adequate glucose and lipid control are essential for successful management. This is compounded by a potassium restriction, which limits intake of fruits and vegetables that contain natural phytochemicals and antioxidants *(52)*. Intake of fruits and vegetables is advocated, but the few clinical studies involving DM and functional foods (e.g. tea, cocoa, and coffee) are inconclusive; and supplementation with vitamins E and C and carotene is not advised *(8)*.

2.5. Physical Inactivity and Exercise

Exercise is contraindicated in people with DM who are receiving maintenance dialysis if serum glucose is greater than 300 mg/dl. Carbohydrate and insulin adjustments may be needed during exercise *(8,54)*. Low impact moderate exercise should be encouraged to improve insulin sensitivity, raise high density lipoprotein (HDL), and lower BP medication doses. Referral to a physical therapist to develop appropriate exercise programs can increase functional capacity and improve quality of life for patients with stage 5 CKD *(28,54)*.

3. CARDIOVASCULAR DISEASE

CVD remains the single largest cause of mortality in stage 5 CKD because of the increased incidence of coronary artery disease (CAD), congestive heart failure (CHF), myocardial infarction (MI), and LVH *(3,55–58)*. CKD appears even more strongly related to CVD in

African Americans *(59)*. The heart–kidney relationship is integral to hemodynamic and regulatory functions, as these two organs communicate at multiple levels through the sympathetic nervous system, the renin-angiotensin-aldosterone system, antidiuretic hormone, the endothelin, and natriuretic peptides. In addition, there are alterations in the neurohormonal system due to kidney failure, which affects the heart *(59)*.

In CKD, there are two subtypes of arterial vascular disease: atherosclerosis and arteriosclerosis. Clinical presentation includes ischemic heart disease (IHD) (namely angina), MI, PVD, endocarditis, cerebrovascular disease, heart failure, and sudden death, which is common in CKD *(55)*. Mortality increases with the number of traditional or non-traditional risk factors for atherosclerotic CVD (ASCVD) *(60)*. Traditional risk factors include advanced age, HTN, DM, hypercholesterolemia, low HDL, high low density lipoprotein (LDL), obesity, LVH, male gender, menopause, smoking, physical inactivity, and family history. Examples of nontraditional risk factors include vascular calcification, hyperphosphatemia, electrolyte imbalance, chronic inflammation, hyperhomocysteinemia, albuminuria, OS, and malnutrition. Traditional risk factors may have different qualitative and quantitative risk relationships with CVD in CKD; some of these, especially high cholesterol and HTN, may interact with non-traditional factors such as inflammation, malnutrition, and other comorbidities to alter the overall association with incident ASCVD *(28,57)*. DM, HTN, dyslipidemia, and older age have been found to be independent predictors of CVD in CKD *(28,58)*.

3.1. Congestive Heart Failure, Left Ventricular Hypertrophy, and Ischemic Heart Disease

CHF is more likely to develop with stage 5 CKD *(3,56)*. The three mechanical contributors to CHF are pressure overload due to HTN, volume overload, and cardiomyopathy *(58,59)*. LVH is a more important survival determinant than CAD; it is due to HTN, fluid overload, anemia, IHD, myocardial calcification, and systemic diseases *(55,59,61)*. IHD is often due to the combined effect of volume overload and LVH, with risk factors including HTN, smoking, DM, IR, lipid disorders, and vascular calcification *(55,61)*.

Heart failure is often difficult to diagnose as UF removes salt and retained water, leaving only low BP, fatigue, and anorexia as clues to its presence *(55)*. A patient could lose dry weight and replace it with fluid when volume overloaded, if target dry weight is kept constant *(61)*. Therefore, reassessment of dry weight is essential, especially with loss of muscle mass due to DM.

3.2. Dialysis-Related Dyslipidemia

Although hyperlipidemia is present in 40% of patients receiving HD and in 62% of those on PD, it is often under-diagnosed and under-treated in the United States in patients with stage 5 CKD *(28,62)*. It is likely that chronic inflammation, illness, and poor nutritional status confound the relationship of dyslipidemia and ASCVD *(63)*. Screening is recommended for secondary causes such as medications or underlying diseases of alcoholism, hypothyroidism, and liver disease *(21,48)*. The uremic dyslipidemic pattern is similar to metabolic syndrome, and relying only on the measurement of LDL may be insufficient because it does not account for atherogenicity of very low density lipoprotein (VLDL) and its remnant particles, intermediate density lipoproteins (IDL) *(21,64)*. Non-HDL cholesterol (total cholesterol- HDL) correlates highly with apoB-containing lipoproteins (ApoB), representing the sum of atherogenic LDL, IDL, and VLDL; and it may be more stable than other lipid parameters in HD *(64)*. Non-HDL cholesterol in patients receiving maintenance HD has been shown to be an independent predictor of ASCVD mortality *(21,64)*. A substantial number of HD patients have elevations in remnant ApoB in the absence of elevated TG levels, so treatment for non-HDL greater than 130 mg/dl may be warranted independent of TG levels, contrary to KDOQI guidelines *(21)*.

3.3. Insulin Resistance (IR)

IR is an independent predictor of cardiovascular mortality in a cohort of non-diabetic patients with stage 5 CKD. It plays an important role in clustering atherosclerotic risk factors such as HTN, dyslipidemia, and abnormal glucose metabolism *(65)*. An increase in ApoB may be stimulated by increased free fatty acid availability with IR *(21)*. IR is acknowledged as a non-traditional risk factor for CVD, but it is not commonly termed pre-diabetes in the CKD population. Impaired fasting glucose (fasting plasma glucose between 100 and 125 mg/dl) and impaired glucose tolerance (2-h plasma glucose between 140 and 199 mg/dl) have been officially termed "pre-diabetes" *(66)*. By acknowledging IR as pre-diabetes, early interventions such as therapeutic diet and lifestyle changes may minimize CVD risk as mild elevations in glucose may be seen prior to the development of abnormalities in the lipid profile. However, the question arises: what is the likelihood of the progression from pre-diabetes to DM in people with stage 5 CKD?

3.4. *Hyperhomocysteinemia*

Hyperhomocysteinemia (tHcy) has been diagnosed in about 85% of maintenance dialysis patients and it is correlated with increased mortality in this cohort. However, the causal relationship between CVD and tHcy is uncertain *(59,67)*. High doses of folic acid, up to 25 g, have been unable to normalize homocysteine levels in some patients, although smaller doses (1–5 g) have been effective in the pre-dialysis population *(59)*. The proposed mechanism for the role of tHcy in the progression of vascular disease includes endothelial dysfunction related to oxidative damage, oxidation of LDL, smooth muscle cell proliferation, and coagulation abnormalities *(67)*.

3.5. *Inflammation*

High levels of C-reactive protein (CRP) may indicate an inflammatory state caused by the dialysis procedure itself, vessel inflammation due to atherosclerosis, or OS; however, it is unknown if this acute phase response is merely an epiphenomenon reflecting established atherosclerotic disease or if acute phase reactants are involved in initiation, progression of atherosclerosis, or both *(68,69)*. Lowering CRP cannot be viewed as a primary strategy, as there is no evidence that this will decrease cardiovascular risk. However, weight reduction, glucose control, lipid control, and smoking cessation may be helpful *(1,59)*.

3.6. *Medical Nutrition Therapy*

MNT guidelines to minimize CVD risk are similar to diabetes: attain/maintain a desirable body weight, control fat intake, and improve glycemic control. Goals for dietary fat intake include 20–35% of total calories from fat, with saturated fat less than 7%, polyunsaturated fat up to 10% and monounsaturated fat up to 20%, total cholesterol less than 200 mg/day, and a minimal amount of *trans* fats *(8,48)*. The goals for fiber are 20–30 g/day, with 5–10 g of viscous or soluble fibers such as beta glucans, gums, mucilages (e.g., psyllium), pectins, and some hemicelluloses. Functional fiber products in powder or pill form could be incorporated into the diet with adjustment for potassium and phosphorus contents *(48,70)*. Viscous and soluble fiber intake lowers total and LDL cholesterol. However, all fibers may be beneficial in lowering BP and improving insulin response *(70)*. Laxatives may be needed if constipation develops due to a high fiber diet with a limited fluid intake *(48)*.

The Dietary Reference Intakes recommend 45–65% of total calories as carbohydrates *(8)*. Evidence for a significant association between total carbohydrate and CVD is lacking, but there is good evidence that diets high in glycemic load and refined sugars may raise serum TG levels, and moderate evidence that high intakes of sucrose or fructose (usually in the form of high fructose corn syrup) increase the risk of excessive energy intake *(71)*.

Hyperkalemia has a direct effect on cardiac function with potential for atrial fibrillation and even sudden death. Patient education regarding dietary potassium intake should encompass dietary control and other factors that may influence serum potassium levels. Patients need to know that unexpected occurrences such as natural climatic or environmental disaster or a clotted access can delay the HD treatment and ultimately lead to hyperkalemia.

3.6.1. Plant Sterols and Stanols

Plant sterols and stanols are plant compounds with similar chemical structure and biological functions as cholesterol that may also have anti-inflammatory and anti-oxidative properties *(72,73)*. Incorporating these products into the diet may help lower LDL and total cholesterol levels by competing with cholesterol for uptake into mixed micelles, resulting in the absorption of less dietary and biliary cholesterol *(8,48,73)*. Daily intake of 2–3 g of plant sterols or stanols reduces LDL by approximately 10% with an additive effect on statin therapy *(48,73)*. Esterification allows the incorporation of these products into a variety of foods, and soft gel capsules containing plant sterols are also available *(8,72)*.

3.6.2. Omega-3 Fatty Acids

Omega-3 fatty acids (ω-3 FAs) contain eicosapentanoic acid (EPA) and decahexanoic acid (DHA), which have been shown to reduce adverse CVD outcomes such as hypertriglyceridemia, atherosclerotic plaque formation, and decreased endothelial function. Sources of ω-3 FAs include cold-water fish such as salmon, mackerel, tuna, butterfish, whitefish, rainbow trout, halibut, and sablefish. Alpha linolenic acid (ALA) is another essential FA that is converted in the body to EPA and DHA; it is found in flaxseed, soybean, canola, and walnut oils. Evidence for reduced adverse CVD outcomes with use of ALA products is sparse and inconclusive *(8)*. The American Heart Association recommendations for ω-3 FAs are the following *(74)*:

- if no documented CHD, eat a variety of fish more than two times/week and include ALA-rich oils;
- if documented CHD, consume about 1 g of EPA + DHA/day, preferably from fatty fish (although fish oil supplements could be considered in consultation with the physician);
- if TGs are elevated, use 2–4 g of EPA + DHA/day provided as capsules only under a physician's care.

There is a risk of bleeding when high doses of ω-3 FAs are used; therefore, individuals receiving anticoagulation medications need to be monitored *(75)*. Side effects of fish oil supplements include "fish burps" (that may be lessened by freezing capsules), loose stools and potential vitamin A and mercury toxicity *(76)*.

3.6.3. Fluid Control

Regulation of sodium and fluid intakes are essential to control interdialytic weight gain and to deter volume overload, which adversely affects cardiomyopathy, LVH, CHF, IHD, and BP. Hemodynamic stability during the HD treatment is affected by removing large fluid volumes, and fluctuations in fluid and electrolytes contribute to CVD risk. Use of hypertonic saline during HD may cause post-treatment thirst, further compounding the problem. Increased fluid retention with PD necessitates using higher dialysate concentrations for UF, and the concomitant glucose absorption contributes to TG production and extra calories resulting in weight gain.

4. HUMAN IMMUNODEFICIENCY VIRUS

Acquired Immunodeficiency Syndrome (AIDS) was first reported in the United States in 1981, and, in 1983, researchers discovered the primary causal virus, HIV type 1 *(77,78)*. HIV progressively destroys the body's ability to fight infection, predisposing the individual to life-threatening illnesses. AIDS is the most advanced stage of HIV; it includes all HIV-infected people having CD4+ T cells less than 200 mm^3, which is 20% of the level present in healthy adults *(77)*.

4.1. HIV and Nephropathy

HIV in patients with stage 5 CKD is primarily due to HIV-associated nephropathy (HIVAN), although unrelated intrinsic diseases could occur. Patients receiving RRT could develop HIV from unprotected sexual contacts, needle sharing, blood transfusions, or organ transplants, although the exact cause may not be fully known *(79)*. The true incidence and prevalence of HIV in patients receiving maintenance dialysis may be underestimated as HIV screening is not routinely done *(3,80,81)*.

HIVAN is a collapsing form of focal and segmental glomerulosclerosis, which is usually accompanied by massive proteinuria, although peripheral edema and HTN are uncommon *(78,81,82)*. The risk factors for HIVAN include decreased CD4 cell count, high viral load, low lymphocyte count, and a family history of kidney disease; it is unclear if IV drug use is a risk or confounding factor *(81,82)*.

4.2. Renal Replacement Therapy

The type of RRT chosen should be based on the individual's lifestyle, preference, and availability of support *(79)*. Both HD and PD are effective modalities, with the type of treatment not predictive of survival *(79,83)*. PD is an acceptable treatment despite a higher incidence of peritonitis, although associated protein losses could worsen co-existing wasting syndrome *(79,81)*.

If HD is chosen, biocompatible membranes are recommended to lessen white blood cell activation and cytokine release. The Centers for Disease Control and Prevention does not prohibit the reuse of dialyzers in HIV *(80,84)*. Standard infection control policies are adequate with transmission risk $< 0.5\%$ after percutaneous or mucous membrane exposure *(80)*.

4.3. Highly Active Antiretroviral Therapy

Combination therapies using three or more anti-HIV medications, referred to as highly active antiretroviral therapy (HAART), are usually recommended *(85)*. Drug-related side effects include liver problems, DM, abnormal fat distribution, high cholesterol, decreased bone density, pancreatitis, nerve problems, GI intolerance, nausea, diarrhea, stomatitis, anorexia, emesis, and metallic and other taste perversions *(84–86)*. Food in the GI tract can influence drug absorption and complex medication-food schedules, and side effects can compromise treatment adherence and tolerability *(87)*.

4.4. Nutrition Risk and Wasting Syndrome

The Nutrition for Healthy Living (NFHL) Study designated weight loss as the strongest independent predictor of mortality in people with AIDS despite the availability of HAART *(88)*. Nutrition risk stratification encompasses many parameters: a comprehensive dietary assessment including a diet history and estimates of current intake, subjective global assessment, anthropometric measurements, biochemical markers, and an evaluation of comorbidities, medications, and psychosocial/economic issues *(87,89–91)*. GI disease and associated symptoms play a major role in nutrition risk (Table 1). Decreased dietary intake may be due to

Table 1
Gastrointestinal Manifestations with HIV

Dysphagia and odynophagia
- Stomatitis: aphthous and chemotherapy, induced
- Neoplasms: of the oral cavity, posterior pharynx or esophagus
- Esophagitis: related to candida, cytomegalovirus, herpes simplex virus, idiopathic ulceration, acid-reflux, Kaposi's sarcoma

Nausea and/or Vomiting: often an adverse effect of medication abdominal pain
- Stomach: gastritis, focal ulcer, outlet obstruction
- Small bowel: enteritis, obstruction, perforation
- Colon: colitis, obstruction, perforation, appendicitis
- Liver and/or spleen: infiltration, cholecystitis, papillary stenosis, cholangitis
- Pancreatitis
- Mesentery infiltration

Diarrhea:
- Infections: protozoa/helminths, bacterial, viral
- HAART
- Bowel: irritable bowel, inflammatory bowel disease, small bowel overgrowth (clinical syndrome of diarrhea and malabsorption of fat, vitamin B12, and carbohydrates)
- Inadvertent use of cathartics: megadoses of vitamin C, lactose intolerance, sorbitol-containing foods
- AIDS enteropathy

Malabsorption
- Steatorrhea
- Intestinal mucosal disease
- Small bowel overgrowth
- Achlorhydria
- Rapid transit secondary to infectious diarrhea
- Lactose intolerance
- Biliary obstruction secondary to tumor or infection
- Pancreatic insufficiency
- Mucosal atrophy secondary to protein-calorie malnutrition

Anorectal disease
- Bacteria
- Viruses
- Protozoa
- Neoplasms

Miscellaneous: condyloma acuminatum, idiopathic ulcers, perirectal abscess, fistula

Jaundice and hepatomegaly
Hepatitis B and C
GI bleeding
GI tumors
Kaposi's sarcoma,
Non-Hodgkin's lymphoma

AIDS, Acquired Immunodeficiency Syndrome, GI, gastrointestinal: HAART, highly active antiretroviral therapy; HIV, human immunodeficiency virus.
Compiled from refs *(88,91,92)*.

dysgeusia, oral pathology (poor dentition, thrush), depression, neurological impairment and changes in mental status, fear of worsening diarrhea or abdominal pain, and poor supportive care *(87,90,92)*. Anorexia may also be present, possibly related to HAART toxicity and less often from opportunistic conditions of the oral cavity, upper GI tract, endocrine system, or central nervous system, especially if the CD4 count is < 200 cells/mm^3 *(88)*. An inability to eat due to medication regimens or fatigue adds to nutrition risk, while chemical dependency and socioeconomic factors can limit access to proper food and adequate nutrition *(87,88)*. Uremia, the dialysis schedule, inadequate dialysis, and renal-related dietary restrictions may further affect dietary intake. The NFHL cohort revealed that increased use of dietary supplements such as herbal remedies, vitamins, and chemical supplements including protein drinks needs to be evaluated *(90)*. High nutrition risk in HIV patients is associated with dialysis, poorly controlled DM, and >10% unintentional weight loss over 4–6 months or > 5% within 4 weeks, while moderate nutrition risk is related to dyslipidemia, evidence of body fat redistribution, oral thrush, and chronic nausea, vomiting, and diarrhea *(87)*.

Monitoring changes in body weight should not be the sole method used to detect nutritional deficiencies, and use of BMI allows for comparisons to population norms only *(88,91)*. Traditional anthropometric measurements can be performed, although the reference tables are not validated for people with HIV-associated wasting. However, waist and hip circumferences have relevance with waist : hip ratios of > 0.85 in women and > 0.95 in men, indicating lipodystrophy *(90,91)*. A marked decrease in one body compartment may be counterbalanced by increases in another with little net change in total body weight *(88)*. Decreases in body cell mass (BCM) (all non-adipose cells + aqueous compartments of the fat cells) may be seen in proximal muscle wasting, weakness of deltoids/iliopsoas, and peripheral edema and may be reflected in laboratory values by increased blood urea nitrogen (BUN)/creatinine and decreased albumin *(88)*.

Wasting associated with HIV, particularly loss of metabolically active lean tissue, is associated with higher mortality rates, accelerated disease progression, loss of muscle protein mass, and impaired strength and functional status *(89)*. This may be due to inadequate intake, impaired deglutition, malabsorption, metabolic alterations, and endocrine factors such as adrenal insufficiency and hypogonadism, excessive cytokine production, and neoplasms such as lymphoma and Kaposi's sarcoma *(88–92)*. Rapid weight loss (greater than 4 kg in less than 4 months) has been associated with acute infection, and chronic, more gradual weight loss (greater than 4 kg in more than 4 months) often reflects GI complications *(91)*. There could also be non-classical wasting from lactic acidosis accompanied by rapid weight loss, abdominal pain, and fatigue *(89)*.

4.5. MNT and Other Interventions

Energy balance is particularly relevant in patients with HIV who are receiving maintenance dialysis, with the goal of achieving a steady state. Negative energy balance leading to weight loss may be primarily due to calorie deficits, as well as possible increases in resting energy expenditure (REE) linked to plasma viral load, secondary infections, HAART, drug stimulants such as cocaine, and constitutional symptoms such as fever and night sweats *(90,91)*. REE may be increased by as much as 10% in patients with HIV *(88)*. Elevated REE may be a cofactor in accelerated weight loss and evidence of failure to compensate for decreased energy intake. Total energy expenditure (TEE) may be decreased despite increased REE if physical activity is reduced in an effort to conserve energy with secondary opportunistic infections. It is postulated that the lethargy and fatigue accompanying malnutrition in AIDS may be compensatory in nature *(88,90,91)*. Conversely, a positive energy balance will promote weight gain given increased intake with decreased energy expenditure.

The actual energy needs of patients with HIV undergoing maintenance dialysis are unknown, and in weight-stable patients, TEE is not significantly elevated and may be comparable to the value in healthy subjects *(91)*. Activity level may be significantly increased with compulsive exercise or decreased, as previously discussed *(90)*.

Maintenance of intact visceral proteins might be more significant in this population given the propensity toward infection which could become worse during dialysis. Thus, energy requirements need to be individually factored for metabolic needs and activity. The KDOQI nutrition guidelines for energy and protein in patients receiving maintenance dialysis would be applicable with frequent adjustments after monitoring the patient's weight, protein status, and overall medical condition. An extra 5–10 kcal/kg weight may be required to achieve anabolic effects *(88)*. Evidence is lacking to support the recommendation of 1.5 g protein/kg weight/day, especially given the need to balance protein and calories for proper utilization *(90,91)*. Sodium, potassium, and fluid requirements would be determined by considering the individual's GI symptoms and metabolic profile.

A multidimensional approach to treating the HIV-infected patient is indicated, including nutrition counseling, appetite stimulants, anabolic hormones, and exercise *(89)*. Treatment of secondary infections and other complications is paramount, with aggressive control of underlying problems such as diarrhea and nausea and attention paid to meal timing with medications and symptoms *(90,91)*. Safety issues regarding food and water handling should be emphasized in nutrition education. ONS can be useful, with the type of supplement based upon the patient's fluid and electrolyte status while considering hyperosmolarity effects on GI function; semi-elemental or predigested formulas may be needed for malabsorption. More aggressive nutrition support such as intradialytic parenteral nutrition, tube feeding, or total parenteral nutrition may be required in some cases (Chapter 12).

Megestrol acetate, an appetite stimulant approved by the Food and Drug Administration for the treatment of HIV-associated anorexia, promotes weight gain primarily from fat, but it may be helpful in maintaining total energy intake *(89,91)*. Testosterone increases LBM in hypogonadal men with AIDS wasting, and synthetic anabolic agents such as oxandrolone, nandrolone decanoate, and oxymetholone can also increase weight and LBM; however, only oxandrolone has been approved for treating wasting associated with AIDS *(87,89,91)*. These drugs may not be appropriate for the HIV dialysis patient. Recombinant human growth hormone has been shown to improve nitrogen balance and to promote lean tissue retention in HIV patients with CKD, although long-term effects are unknown *(89,91)*. Anticytokine therapies as well as ω-3 FAs have been used to attenuate catabolic response to inflammation, but the role of these therapies are not fully understood *(87,89,91)*.

Progressive resistance exercise training can maintain or restore fitness as well as increase muscle mass and LBM when combined with adequate dietary intake. While aerobic exercise may be beneficial with fat mobilization, there is minimal increase in BCM with a possible rise in CD4 count *(88,90,91)*.

4.6. Metabolic Abnormalities and Changes in Body Composition

HAART-related metabolic abnormalities may develop, such as dyslipidemia (with hypertriglyceridemia, low HDL, increased LDL, and elevated total cholesterol), abnormal carbohydrate metabolism (ranging from IR with and without glucose intolerance to DM), and lactic acidosis/hyperlactatemia *(93,94)*. The inflammatory response to chronic HIV infection may be proatherogenic *(94)*. The Multicenter AIDS Cohort Study found increased risk of DM in HIV-infected men on HAART *(95)*. IR with HIV increases the risk for fat abnormalities and loss of SC fat, with IR often preceding this *(93,94,96)*. Lipodystrophy or fat redistribution is common in 30–50% of patients receiving antiretroviral therapy. It encompasses loss of facial, buttocks, and extremity fat, while increased fat accumulation is seen in dorsocervical fat ("buffalo hump"), breast enlargement, and greater abdominal girth due to increased visceral adipose tissue (VAT) *(94,96)*. When fat redistribution occurs, patients should be screened and monitored for glucose and lipid metabolism disorders *(88,94)*.

A simulated metabolic syndrome presents with abdominal obesity, HTN, abnormal serum lipids, and IR with concern for development of CAD, stroke, and PVD *(94,96)*. Treatment would be similar to non-HIV-infected individuals with this profile regarding diet, exercise, and medications *(86,93,95,96)*. However, there is the potential for adverse drug interactions with high doses of lipid-lowering medications *(94,97)*.

IR has been shown to decrease with the use of thiazolidinedione; however, consistent improvements in VAT or in SC lipoatrophy with thiazolidinedione have not been shown. Increases in adipogenesis in vitro suggest that these agents may be able to reverse SC fat loss *(93,94)*. Hypocaloric diets are recommended for overweight patients (i.e., BMI >27 kg/m^2), although rapid weight loss should be avoided; and both aerobic exercise and resistance training should be pursued *(94)*.

4.7. Anemia

Anemia is worsened with HIV due to direct effects on erythropoiesis, opportunistic infections and parovirus B19 infection, antiretroviral drugs, and drugs that treat infection or malignancies *(80,84,98)*. Dose requirements of epoietin alpha may be higher, and the safety of IV iron administration is not fully known *(98)*. Careful monitoring of viral load, CD4 count, and ferritin is indicated when IV iron is administered due to anemia with transferrin saturation ≤ 20% and/or ferritin ≤100 ng/ml, as studies have shown that OS in the presence of iron may activate HIV-1. Ferritin may also be elevated as an acute phase reactant, and high iron stores may adversely affect outcomes *(80,84,98)*.

5. SUMMARY

Nutrition care for patients receiving maintenance dialysis with DM, CVD, and/or HIV has the common themes of energy balance, preventing malnutrition, minimizing dyslipidemia, attaining good glycemic control, and understanding the roles of inflammation and OS. The medical nutrition therapies are similar, emphasizing adequate protein intake, appropriate energy consumption, modification of fats, and carbohydrate regulation. Each patient's nutritional care plan must be individualized in light of his/her comorbid conditions, age, psychosocial status, and rehabilitation potential. The overall impact of the disease state needs to be considered beyond the renal restrictions, if outcome and quality of life are expected to improve.

6. CASE STUDY

MT is a 73-year-old male with type 2 DM of 25 years duration who initiated HD 3.5 months ago. His past medical history is significant for diabetic neuropathy, nephropathy, and retinopathy; HTN, dementia, CAD (s/p coronary artery bypass graft 3 years ago); prior amputation of two toes and recent amputation of great toe tip. At the start of HD, MT had a urine output of approximately 500 ml/day; a 24 hours urine has not been repeated since then. He dialyzes for 4 hours on the second shift starting at 11 am. Initial nutrition education was done upon dialysis initiation by another dietitian who followed him until 1 month ago. Home glucose monitoring was being done twice per day. Diabetic medication adjustments and teaching were done by family doctor and his staff.

Laboratory tests reveal:

	BUN	*Creatinine*	*Na*	*K*	*Ca*	*PO_4*	*Alb*	*Glucose*	*HbA1C*
Initial	55	4.6	138	3.7	9.4	3.2	3.4	472	7.4
Most recent	83	6.4	136	4.7	8.2	3.3	3.7	347	9.0

His medications include Ancef® every treatment × 2 weeks, started 10 days ago, Augmentin®, Cardizem®, Cardura®, Coreg®, Epogen® 10400 units, 70/30 insulin: 15 units AM, 10 units PM; initiated 2 weeks after starting HD and changed 2 weeks ago to 30 units AM, 15 units PM.

His wife states that two nights ago he had episode of severe hypoglycemia at 9.30 pm. He was asleep and appeared sweaty, so she checked glucose which was 40. She treated the hypoglycemia but called paramedics as blood sugar was still low. MT earlier that day had been hypoglycemic at end of HD treatment for which he received D50, and he was discharged from the dialysis unit in stable condition.

6.1. Key Questions

1. What questions would you ask her to problem solve and help deter reoccurrence?
2. Based on her answers and medical history, what factors may be influencing his glycemic control?
3. What interventions would you advise?

REFERENCES

1. Zoccali C, Tripepi G, Mallamaci F. Predictors of cardiovascular death in ESRD. Semin Nephrol 2005;25:358–362.
2. National Center for Chronic Disease Prevention and Health Promotion. National diabetes fact sheet. Available at: http://www.cdc.gov/diabetes/pubs/estimates05.htm. Accessed September 17, 2006.
3. United States Renal Data System. Annual data report 2005. Available at: http://www.usrds.org/adr.htm. Accessed on September 7, 2006.
4. Tone A, Shikata K, Matsuda M, et al. Clinical features of non-diabetic renal diseases in patients with type 2 diabetes. Diabetes Res Clin Pract 2005;69: 237–242.
5. Locatelli F, Pozzoni P, DelVecchio L. Renal replacement therapy in patients with diabetes and ESRD. J Am Soc Nephrol 2004;15:S25–S29.
6. Akmal M. Hemodialysis in diabetic patients. Am J Kidney Dis 2001;38:S195–S199.
7. Lok CE, Oliver MJ, Rothwell DM, Hux JE. The growing volume of diabetes-related dialysis: a population based study. Nephrol Dial Transplant 2004;19: 3098–3103.
8. American Diabetes Association. Nutrition recommendations and interventions for diabetes. Diabetes Care 2007; 30:S48–S65.
9. Pupim LB, Heimbürger O, Qureshi AR, et al. Accelerated lean body mass loss in incident chronic dialysis patients with diabetes mellitus. Kidney Int 2005; 2368–2374.
10. Morioka T, Emoto M, Tabata T, et al. Glycemic control is a predictor of survival for diabetic patients on hemodialysis. Diabetes Care 2001;24:909–913.
11. American Association of Clinical Endocrinologists and the American College of Endocrinology. The American Association of Clinical Endocrinologists

medical guidelines for the management of diabetes mellitus: The AACE system of intensive diabetes self-management – 2002 update. Endocr Pract 2002;8: S40–S82.

12. Tzamaloukas AH, Friedman EA. Diabetes. In: Daugirdes JT, Blake PG, Ing TS, ed. Handbook of Dialysis, 3rd edition. Philadelphia: Lippincott, Williams, Wilkins, 2001:453–465.
13. Bloomgarden ZT. The European association for the study of diabetes. Diabetes Care 2005;28:1250–1257.
14. Vonesh EF, Snyder JJ, Foley RN, Collins AJ. The differential impact of risk factors on mortality in hemodialysis and peritoneal dialysis. Kidney Int 2004;66: 2389–2401.
15. Termorshuizen F, Korevaar JC, Dekker FW, et al. Hemodialysis and peritoneal dialysis: comparison of adjusted mortality rates according to the duration of dialysis: analysis of the Netherlands cooperative study on the adequacy of dialysis. J Am Soc Nephrol 2003;14:2851–2860.
16. Sitter T, Sauter M. Impact of glucose in PD: Saint or sinner? Perit Dial Int 2005;25:415–425.
17. Torun D, Ogurzkurt L, Sezer S, et al. Hepatic subcapsular steatosis as a complication associated with intraperitoneal insulin treatment in diabetic peritoneal dialysis patients. Perit Dial Int 2005;25:595–600.
18. Strid H, Simren M, Johansson A, et al. The prevalence of gastrointestinal symptoms in patients with chronic renal failure is increased and associated with impaired psychological well-being. Nephrol Dial Transplant 2002;17:1434–1439.
19. Gokal R. Peritoneal dialysis in the 21st century: an analysis of current problems and future developments. J Am Soc Nephrol 2002;13:S104–S116.
20. Yao Q, Lindholm B, and Heimburger O. Peritoneal dialysis prescription for diabetic patients. Perit Dial Int 2005;25:S76–S79.
21. Liu J, Rosner MH. Lipid abnormalities associated with end-stage renal disease. Semin Dial 2006;19:32–40.
22. Ansari A, Thomas S, Goldsmith D. Assessing glycemic control with diabetes and end-stage renal disease. Am J Kidney Dis 2004;41:523–531.
23. Wong Y-H, Szeto C-C, Chow K-M, et al. Rosiglitazone reduces insulin requirement and C-reactive protein levels in type 2 diabetes patients receiving peritoneal dialysis. Am J Kidney Dis 2005;46:713–719.
24. Quellhorst E. Insulin therapy during peritoneal dialysis: pros and cons of various forms of administration. J Am Soc Nephrol 2002;13:S92–S96.
25. Chow KM, Szeto CC, Leung CB, Kwan BC, et al. A risk analysis of continuous ambulatory peritoneal dialysis-related peritonitis. Perit Dial Int 2005;25:374–379.
26. Khalili K, Lan FP, Hanbridge AE, et al. Hepatic subcapsular steatosis in response to intraperitoneal insulin delivery: CT findings and prevalence. AJR Am J Roentgenol 2003;180:1601–1604.
27. Van Vliem BA, Schoonjans RS, Strujik DG, et al. Influence of dialysate on gastric emptying time in peritoneal dialysis patients. Perit Dial Int 2002;22:32–38.
28. Uhlig K, Levey AS, Sarnak MJ. Traditional cardiac risk factors in individuals and chronic kidney disease. Semin Dial 2003;16:116–127.
29. KDOQI Clinical Pratice Guidelines and Clinical Recommendations for Diabetes and Chronic Kidney Disease. AJKD 2007; 49:S62–S73.

30. American Diabetes Association. Standards of medical care in diabetes. Diabetes Care 2007; 30(Suppl 1):S4–S41.
31. Joy MS, Cefalu WT, Hogan SL, Nachman PH. Long-term glycemic control measurements in diabetic patients receiving hemodialysis. Am J Kidney Dis 2002;39:297–307.
32. Little RR, Tennhill AL, Rohlfing C, et al. Can glycohemoglobin be used to assess glycemic control in patients with chronic renal failure? Clin Chem 2002;48: 785–786.
33. Cornell S, Briggs A. Newer treatment strategies for the management of type 2 diabetes mellitus. J Pharm Pract 2004;17:49–54.
34. Gerich JE. The importance of controlling postprandial hyperglycemia. Practical Diabetol 2005;24:22–26.
35. Hirsch IB, Bergenstal RM, Parkin CG, et al. A real-world approach to insulin therapy in primary care practice. Clin Diabetes 2005;23:78–86.
36. Klonoff DC. Continuous glucose monitoring. Diabetes Care 2005;28:1231–1239.
37. Pupim LB, Flakoll PJ, Majchrzak KM, et al. Increased muscle protein breakdown in chronic hemodialysis patients with type 2 diabetes mellitus. Kidney Int 2005;68:1857–1865.
38. Cano NJM, Roth H, Aparicio M, et al. Malnutrition in hemodialysis diabetic patients: evaluation and prognostic influence. Kidney Int 2002;62:593–601.
39. Chavez EM, Taylor GW, Borrel LN, Ship JA. Salivary function and glycemic control in older persons with diabetes. Oral Surg Oral Med Oral Pathol Oral Radiol Endod 2000;89:305–311.
40. Moore PA, Guggenheimer J, Etzel KR, et al. Type 1 diabetes mellitus, xerostomia, and salivary flow rates. Oral Surg Oral Med Oral Pathol Oral Radiol Endod 2001;92:281–291.
41. Meurman JH, Collin H-L, Niskanen L, et al. Saliva in non-insulin-dependent diabetic patients and control subjects. Oral Surg Oral Med Oral Pathol Oral Radiol Endod 1998;86:69–76.
42. Chavez EM, Borrell LN, Taylor GW, Ship JA. A longitudinal analysis of salivary flow in control subjects and older adults with type 2 diabetes. Oral Surg Oral Med Oral Pathol Oral Radiol Endod 2001;91:166–173.
43. Sreebny LM, Yu A, Green A, Valdini A. Xerostomia in diabetes mellitus. Diabetes Care 1992;15:900–904.
44. Vinik AI, Freeman R, Erbas T. Diabetic autonomic neuropathy. Semin Neurol 2003;23:365–372.
45. Duby JJ, Campbell RK, Setter SM, et al. Diabetic neuropathy: an intensive review. Am J Health Syst Pharm 2004;61:160–176.
46. Vinik AI, Maser R, Mitchell BD, Freeman R. Diabetic autonomic neuropathy. Diabetes Care 2003;26:1553–1574.
47. Syed AA, Rattansingh A, Furtado SD. Current perspectives on the management of gastroparesis. J Postgrad Med 2005;51:54–60.
48. Pagenkemper JJ. Nutrition management of diabetes in chronic kidney disease. In: Byham-Gray L, Wiesen K, ed. A Clinical Guide to Nutrition Care in Kidney Disease. Chicago: American Dietetic Association, 2004:93–106.
49. Lin Z, Forster J, Sarosiek I, McCallum RW. Treatment of diabetic gastroparesis by high-frequency gastric electrical stimulation. Diabetes Care 2004;27: 1071–1076.

50. Rayner CK, Samson, M, Jones KL, Horowitz M. Relationships of upper gastrointestinal motor and sensory function with glycemic control. Diabetes Care 2001;24:371–381.
51. Funnell MM, Greene DA. Diabetic neuropath. In: Funnell MM, Hunt C, Kulkarni K, Rubin RR, and Yarborough PC, ed. A Core Curriculum for Diabetes Education, 3rd edition. Chicago: American Association of Diabetes Educators, 1998: 709–743.
52. Vaziri ND. Oxidative stress in uremia: nature mechanisms and potential consequences. Semin Nephrol 2004;24:469–473.
53. Dursun E, Dursun B, Suleymanlar G, Ozben T. Effect of haemodialysis on the oxidative stress and antioxidants in diabetes mellitus. Acta Diabetol 2005;42: 123–128.
54. Evans N, Forsyth E. End-stage renal disease in people with type 2 diabetes: systemic manifestations and exercise implications. Phys Ther 2004;84:454–463.
55. Sarnak M, Levey AS, Schoolwerth AC, et al. Kidney disease as a risk factor for development of cardiovascular disease: a statement from the American Heart Association Councils on Kidney in Cardiovascular Disease, High Blood Pressure Research, Clinical Cardiology and Epidemiology and Prevention. Hypertension 2003;42:1050–1065.
56. Wright RS, Reeder GS, Herzog CA, et al. Acute myocardial infarction and renal dysfunction: a high-risk combination. Ann Intern Med 2002;137:563–570.
57. Longnecker JC, Coresh J, Powe NR. Traditional cardiovascular risk factors in dialysis patients compared with the general population: the CHOICE study. J Am Soc Nephrol 2002;13:1918–1927.
58. Kundhal K, Lok CE. Clinical epidemiology of cardiovascular disease in chronic kidney disease. Nephron Clin Pract 2005;101:c47–c52.
59. McCullough PA. Cardiovascular disease in chronic kidney disease from a cardiologist's perspective. Curr Opin Nephrol Hypertens 2004;18:591–600.
60. Qureshi AR, Alvestrand A, Divino-Filho JC. Inflammation, malnutrition, and cardiac disease as predictors of mortality in hemodialysis patients. J Am Soc Nephrol 2002;13:S28–S36.
61. Nicholls AJ. Heart and circulation. In: Daugirdes JT, Blake PG, Ing TS, ed. Handbook of Dialysis, 3rd edition. Philadelphia: Lippincott, Williams, Wilkins, 2001:583–600.
62. Fox CS, Longnecker JC, Powe NR, et al. Undertreatment of hyperlipidemia in a cohort of United States kidney dialysis patients. Clin Nephrol 2004;61: 299–307.
63. National Kidney Foundation. Kidney disease outcomes quality initiative clinical practice guidelines for cardiovascular disease in dialysis patients. Available at: http://www.kidney.org/professionals/kdoqi/guidelines_cvd/index.htm. Accessed on November 10, 2007.
64. Desmeules S, Arcand-Bossé J-F, Bergeron J, et al. Nonfasting non-high-density lipoprotein cholesterol is adequate for lipid management in hemodialysis patients. Am J Kidney Dis 2005;45:1067–1072.
65. Shinohara K, Shoji T, Emoto M, et al. Insulin resistance as an independent predictor of cardiovascular mortality in patients with ESRD. J Am Soc Nephrol 2002;13:1894–1900.

66. American Diabetes Association. Diagnosis and classification of diabetes mellitus. Diabetes Care 2007;30(Suppl 1):S42–S47.
67. Nair AP, Nemirovsky D, Kim M, et al. Elevated homocysteine levels in patients with end-stage renal disease. Mt Sinai J Med 2005;72:365–373.
68. Böger CA, Götz A, Stubanus M, et al. C-Reactive protein as predictor of death in end-stage diabetic nephropathy: role of peripheral artery disease. Kidney Int 2005;68:217–227.
69. Stevinkel P. Interactions between inflammation, oxidative stress, and endothelial dysfunction in end-stage renal disease. J Ren Nutr 2003;13:144–148.
70. Higdon J. Phytosterols. The Linus Pauling Institute Micronutrient Information Center. Last updated 8/11/2005. Available at: http://lpi.oregonstate.edu/infocenter/phytochemicals/sterols/index.html. Accessed September 4, 2006.
71. National Heart Foundation of Australia Executive Summary. Position statement on the relationship between carbohydrates, dietary fibre, glycaemic index/glycaemic load and cardiovascular disease. March 2006. Available at: http://www.heartfoundation.com.au/downloads/PP-585_Exec_Summ_Nutr_PS-CHO_06_Feb_Final.pdf. Accessed September 4, 2006.
72. Berger A, Jones PJH, Abumweis SS. Plant sterols: factors affecting their efficacy and safety as functional food ingredients. Lipids Health Dis 2004;3:5–24.
73. National Heart Foundation of Australia's Nutrition and Metabolic Advisory Committee Position Statement. Plant sterols and stanols. National Heart Foundation of Australia. August 2003. Available at: http://www.heartfoundation.com.au/downloads/PlantSterol_Aug03_final.pdf. Accessed September 4, 2006.
74. American Heart Association. Fish and omega-3 fatty acids. Available at: http://www.americanheart.org/presenter.jhtml?identifier=4632. Accessed September 4, 2006.
75. Higdon J. Essential (omega-3 and omega-6) fatty acids. The Linus Pauling Institute Micronutrient Information Center. Last updated 12/07/2005. Available at: http://lpi.oregonstate.edu/infocenter/othernuts/omega3fa/index.html. Accessed September 4, 2006.
76. Martin CJ. Fats: the good, the bad, and the ugly. J Ren Nutr 2004;14:E1–E3.
77. Centers for Disease Control and Prevention. HIV/AIDS. Available at: http://www.cdc.gov/hiv/. Accessed September 10, 2006.
78. National Institute of Allergy and Infectious Diseases. HIV infection and AIDS: an overview. Available at: http://www.niaid.nih.gov/factsheets/hivinf.htm. Accessed September 17, 2006.
79. Rao TKS. Human immunodeficiency virus infection in end-stage renal disease patients. Semin Dial 2003;16:233–244.
80. Mandayam S, Ahuja TS. Dialyzing a patient with human immunodeficiency virus infection: what a nephrologists needs to know. Am J Nephrol 2004;24:511–521.
81. Gupta SK, Eustace JA, Winston JA, et al. Guidelines for the management of chronic kidney disease in HIV-infected patients: recommendations of the HIV Medicine Association of the Infectious Disease Society of America. Clin Infect Dis 2005;40:1559–1580.
82. Brook MG, Miller RF. HIV associated nephropathy: a treatable condition. Sex Transm Infect 2001;77:97–100.
83. Ahuja TS, Collinge N, Grady J, Khan S. Is dialysis modality a factor in survival of patients with ESRD and HIV-associated nephropathy. Am J Kidney Dis 2003;41:1060–1064.

84. Ahuja TS, O'Brien WA. Special issues in the management of patients with ESRD and HIV infection. Am J Kidney Dis 2003;41:279–291.
85. AidsInfo. HIV and its treatment: what you should know. September 23, 2005. Available at: http://aidsinfo.nih.gov/contentfiles/HIVandItsTreatment_cbrochure_en.pdf. Accessed September 4, 2006.
86. Izzedine H, Launay-Vacher V, Baumelou A, Deray G. An appraisal of antiretroviral drugs in hemodialysis. Kidney Int 2001;60:821–830.
87. Nerad J, Romeyn M, Silverman E, et al. General nutrition management in patients infected with human immunodeficiency virus. Clin Infect Dis 2003;36:S52–S62.
88. New York State Department of Health AIDS Institute. Clinical guidelines. Available at: http://www.hivguidelines.org/Content.aspx?Page ID=257. Accessed September 10, 2006.
89. Grinspoon S, Mulligan K. Weight loss and wasting in patients infected with human immunodeficiency virus. Clin Infect Dis 2003;36:S69–S78.
90. Shevitz AH, Knox TA. Nutrition in the era of highly active antiretroviral therapy. Clin Infect Dis 2001;32:1769–1775.
91. Mulligan K, Schambelan M. HIV-associated wasting. In: Peiperl L, Coffey S, Volberding P, ed. HIV InSite Knowledge Base. November 2003. Available at: http://hivinsite.ucsf.edu/InSite?page=kb-04-01-08. Accessed September 10, 2006.
92. Koch J, Kim LS, Friedman S. Gastrointestinal manifestations of HIV. In: Peiperl L, Coffey S, Volberding P, ed. HIV InSite Knowledge Base. June 1998. Available at: http://hivinsite.ucsf.edu/InSite?page=kb-04&doc=kb-04-01-11. Accessed September 10, 2006.
93. Gelato MC. Insulin and carbohydrate dysregulation. Clin Infect Dis 2003;36: S91–S95.
94. Chow DC, Day LJ, Souza, SA, Shikuma CM. Metabolic complications of HIV therapy. In: Peiperl L, Coffey S, Volberding P, ed. HIV InSite Knowledge Base. May 2003. Available at: http://hivinsite.ucsf.edu/InSite?page=kb-03-02-10#. Accessed September 10, 2006.
95. Wlodarczyk D. Managing medical conditions associated with cardiac risk with HIV. In: Peiperl L, Coffey S, Volberding P, ed. HIV InSite Knowledge Base. August 2004. Available at: http://hivinsite.ucsf.edu/InSite?page=kb-03&doc=kb-03-01-20. Accessed September 10, 2006.
96. Sattler F. Body habitus changes related to lipodystrophy. Clin Infect Dis 2003;36:S84–S90.
97. Dube M, Fenton M. Lipid abnormalities. Clin Infect Dis 2003;36:S79–S83.
98. Wingard R. Increased risk of anemia in dialysis patients with comorbid diseases. Nephrol Nurs J 2004;31:211–214.

IV

Nutrition in Chronic Disease in Special Needs Populations

1 Over the Lifespan

16 Pregnancy

Jean Stover

LEARNING OBJECTIVES

1. Identify modifications in dialysis therapy and management for the pregnant chronic kidney disease patient.
2. Describe nutritional concerns related to protein, energy and micronutrients.
3. Discuss medical and nutritional management of pregnant chronic kidney disease patients.

Abstract

Women with chronic kidney disease who become pregnant during early stages of the disease, while undergoing chronic dialysis or after renal transplantation, are all considered to be at high risk for complications. Fertility is decreased in this population prior to a well functioning transplant, especially for women undergoing dialysis. Although outcomes are improving, care of the pregnant dialysis patient presents a challenge to the renal healthcare team, and requires a multidisciplinary team approach including coordination with a high-risk obstetrics team. More intensive dialysis therapy, modifications in oral and intravenous medications and emphasis on increased intake of protein, calories and specific vitamins and minerals are necessary to improve survival of the fetus.

Key Words: Pregnancy; intensive dialysis; medications; vitamins; minerals; nutrition.

From: *Nutrition and Health: Nutrition in Kidney Disease*
Edited by: L. D. Byham-Gray, J. D. Burrowes, and G. M. Chertow

1. INTRODUCTION

Women with chronic kidney disease (CKD) who become pregnant during the early stages of the disease, while undergoing chronic dialysis or after renal transplantation, are all considered to be at high risk for complications. Hypertension is the most prevalent life-threatening complication during pregnancy in all stages of CKD. There is also a greater risk for more rapid decline in kidney function for women who become pregnant with a serum creatinine greater than or equal to 1.4 mg/dL. Women who become pregnant after a kidney transplant do not have an increased risk for loss of kidney function if the function is well preserved, but immunosuppressive medications (especially cyclosporine) have been known to contribute to infants born small for gestational age. The incidence of premature birth also remains high for women during all stages of CKD *(1,2)*.

Fertility generally returns for women who have a good functioning kidney transplant, but otherwise women with CKD tend to become pregnant less frequently than those with normal kidney function. There is also found to be a significant decrease in conception for women undergoing dialysis *(1,2)*. Although it has been reported that the occurrence of pregnancy in the dialysis population has increased from 0.9% in 1980 to 1–7% between 1992 and 2003, pregnancy is still considered relatively uncommon *(3)*.

One recent report in the literature has noted that 40–85% of infants born to women on dialysis now survive *(4)*. In efforts to continue these advances in pregnancy outcome, many challenges are presented to the team of nephrology professionals caring for pregnant women with CKD undergoing chronic dialysis therapy. The following discussion will focus on the care of this specific population, including changes in the dialysis regimen, medications and nutrition management, as they all impact on each other as well as on pregnancy outcome.

2. CONFIRMATION OF PREGNANCY

The confirmation of pregnancy in women undergoing dialysis generally requires a pelvic ultrasound in addition to the blood test that measures levels of the β subunit of human chorionic gonadotropin (hCG). This is because the kidney excretes the small amounts of hCG produced by somatic cells, and in renal failure, this test can appear positive by usual standards *(5)*. Once pregnancy is confirmed, and the individual wishes to proceed with it, she should be referred to a high-risk obstetrics practice.

3. DIALYSIS REGIMEN

Intensive dialysis is a key component to successful pregnancy outcome for women undergoing dialysis. The amount of dialysis is increased in efforts to mimic more normal kidney function during fetal development. It has been shown that 16–24 hours of hemodialysis per week is associated with improved infant survival *(3)*. This is an important message to convey to women in this population as soon as a pregnancy is confirmed, as it involves an alteration in lifestyle with a much greater time commitment to dialysis. It is also important to inform the patient at this time that she will be transferred to a hospital-based dialysis setting for fetal monitoring during the treatment once she reaches approximately 24-week gestation.

The prescribed duration and frequency of dialysis seems to vary, as noted in a review of the literature *(3–6)*. Over the years, the goals for determining adequate dialysis during pregnancy have generally included the parameter of maintaining the pre-dialysis blood urea nitrogen below 50 mg/dL. As adequate protein and energy intakes are of high importance during pregnancy, more frequent dialysis allows for more liberal dietary intakes while achieving this goal. Most dialysis regimens are now based on case reports of apparent improved infant survival with at least 20 hours/week of hemodialysis *(5)*. Although Kt/V measurements may be difficult to interpret due to urea generation by the fetus, they probably should be evaluated. One author recommends a target Kt/V in the range of 1.5–1.7 per treatment during pregnancy, while others report positive outcomes with retrospective measurements of a mean weekly Kt/V ranging from 5.0 to 9.6 ± 1.4 *(4–6)*.

As previously mentioned, hypertension can be a serious complication of pregnancy for women with CKD. Severe hypotension on the other hand may promote fetal distress *(5)*. In efforts to avoid potential hypertension due to volume overload, and potential hypotension with the need to remove large volumes of fluid during the treatment, more frequent dialysis is recommended. Most pregnant patients are dialyzed four to six times per week. Now, with nocturnal hemodialysis being a treatment option, this may be an ideal regimen for the pregnant dialysis patient. A reported case from Toronto, Canada using this method of treatment for 7.5 hours, 7 nights/week resulted in good blood pressure control for the mother during her pregnancy, and the delivery of a healthy infant weighing 3025 g at 38-week gestation. The case study illustrates an example of a patient who had already been undergoing nocturnal dialysis, but during her pregnancy the treatment time and frequency were increased *(7)*.

The content of the dialysate used for hemodialysis during pregnancy will vary depending on the amount of dialysis given as well as the mother's dietary intake and levels of serum electrolytes, calcium and bicarbonate. It is recommended that frequent monitoring of all of these levels be done during the pregnancy. With more dialysis, a 3.0 potassium (K+) dialysate concentration may be required to maintain normal serum potassium levels. Even though the fetus requires 25–30 g of calcium for proper skeletal development, it is usually not necessary to increase the dialysate calcium content when calcium-containing medications are taken and more frequent dialysis is given *(4)*. A standard 2.5 mEq/L calcium dialysate concentration is frequently used. There is also some production of calcitriol by the placenta, which makes it important to frequently monitor serum calcium levels to avoid hypercalcemia *(5)*. As goals for serum calcium in the general dialysis population are now lower *(8)*, it seems reasonable to maintain serum calcium levels for pregnant women on dialysis in the higher end of this range, although there is no literature to support this assumption.

There are case reports in the literature that discuss successful pregnancy outcomes for women undergoing peritoneal dialysis. More frequent exchanges with lesser volumes of instilled peritoneal fluid are necessary as the pregnancy progresses to allow more intense dialysis with less abdominal discomfort *(4)*. One report utilizes tidal dialysis with the automated cycler machine to promote both comfort and increased dialysis clearance *(9)*.

4. ENERGY AND PROTEIN NEEDS

Initial and ongoing nutrition assessment and counseling of the pregnant dialysis patient is very important due to increased energy, protein, vitamin and mineral needs for this population. It is recommended that the dietitian meet with the patient to discuss an overview of nutritional needs as soon as possible after the pregnancy is confirmed and she has agreed to follow through with it. Weekly follow-up using dietary recalls and/or food intake records to evaluate nutrition adequacy is suggested as well *(10)*. Also, collaboration with the dietitian at the hospital-based center where the patient will dialyze for the last few months of her pregnancy is recommended.

Generally, 35 kcal per kg/day of pregravida ideal body weight (IBW) or adjusted IBW are prescribed in the first trimester, and 300 kcal/day are added to this value for the second and third trimesters (Table 1).

Table 1
Nutrient Recommendations for the Pregnant Dialysis Patient *(6,10–13)*

Energy	35 kcal/kg pregravida IBW + 300 kcal/day in second and third trimesters; may need a nutritional supplement to meet needs.
Protein	1.2 g/kg pregravida IBW + 10 g/day (HD); 1.2–1.3 g/kg pregravida IBW + 10 g/day (PD); May need a nutritional supplement to meet needs.
Vitamins	Folic acid—at least 2 mg/day—doubling a standard renal vitamin should provide adequate folate and other water-soluble vitamins needed during pregnancy.
	Vitamin D—analogs have been given, but not enough information on safety during pregnancy.
	Vitamin A—not usually given, thus renal vitamins are generally given instead of prenatal vitamins.
Minerals	Iron—usually given IV during dialysis (generally iron sucrose or gluconate) to achieve iron studies in goal range for general dialysis population; oral iron has been used, but not as well absorbed.
	Calcium—given as calcium acetate or carbonate to bind phosphorus, or as calcium carbonate for a calcium supplement; keep in mind that there is increased absorption of calcium from dialysate with more frequent dialysis.
	Sodium, potassium and phosphorus—can often be liberalized in the diet with more dialysis; phosphate binders may not be needed.
	Zinc—at least 15 mg/day recommended.

IBW, ideal body weight; HD, hemodialysis; PD, peritoneal dialysis, IV, intravenously.

Daily protein needs are generally 1.2 g/kg IBW plus 10 g/day for women undergoing hemodialysis and 1.2–1.3 g/kg IBW plus 10 g/day for those receiving peritoneal dialysis *(10–12)*. It may be easier to meet these needs, as the diet can frequently be liberalized for sodium, potassium and phosphorus content due to the increased amount of solute removal with more dialysis. At times, though, the women may require protein or calorie/protein supplements to attain her estimated energy and protein requirements. Generally, a regular commercial supplement may be used with increased dialysis time.

5. VITAMINS AND MINERALS

Water-soluble vitamins are usually preferred over prenatal vitamins due to the need to avoid excess vitamin A for all individuals with CKD undergoing dialysis. With increased requirements for water-soluble vitamins during pregnancy, as well as increased losses anticipated with more intensive dialysis, a standard renal vitamin containing 1 mg folic acid is often doubled. Folate deficiency has been linked to neural tube defects in infants born to women without CKD, thus assuring at least 2 mg of folic acid per day for pregnant women undergoing dialysis has been encouraged *(2,6)*. Presently, there are renal vitamin preparations already containing greater than 1 mg of folic acid with adequate amounts of other needed water-soluble vitamins and even added zinc. These may be utilized as well.

Vitamin D analogs have been given intravenously during dialysis to pregnant women needing suppression of the parathyroid hormone (PTH) and to maintain normal serum levels of calcium. There does not seem to be definitive information available concerning whether these forms of vitamin D cross the placental barrier and, if so, whether they are safe relative to fetal development *(2,13)*.

Intravenous iron has been given safely to pregnant patients during hemodialysis, based on goal ranges for serum levels of transferrin saturation and ferritin used in the general dialysis population. Iron dextran has been reportedly used during pregnancy, but iron sucrose and iron gluconate are generally used at present *(13)*. Although not as well absorbed, oral iron preparations have also be used instead of intravenous iron, either alone or in combination with a vitamin.

Zinc supplements are prescribed in the amount of at least 15 mg/day to prevent increased risks of fetal malformation, preterm delivery, low birth weight and pregnancy-induced hypertension *(2,10)*. As mentioned above, zinc may be included in the renal vitamin prescribed.

Calcium-containing phosphate binders are generally given to the pregnant dialysis patient due to increased calcium needs of the fetus. There are no studies to evaluate the safety of using calcium acetate during pregnancy; however, this preparation as well as calcium carbonate have been utilized during this time *(13–15)*. They may be given with meals for phosphate binding, or apart from meals primarily for calcium supplementation if the serum phosphorus level is below goal range.

6. WEIGHT GAIN AND SERUM ALBUMIN

Due to fluid retention with CKD, it is difficult to determine actual solid body weight gain during pregnancy for women undergoing dialysis. It has been suggested that during the second and third trimesters, when most weight gain occurs, that the pregnant dialysis patient's estimated dry weight (EDW) be increased by 0.5 kg/week *(16)*. More frequent treatments with gentle fluid removal may help with this assessment, but a team approach involving regular collaboration with the dialysis technicians, nurses, physicians, dietitian and patient is very important when determining true weight gain.

The evaluation of adequate protein intake is also difficult, as the expected decrease of serum albumin during pregnancy is about 1 g/dL for women without CKD *(17)*. Recommendations are to continue weekly dietary recalls or records to assess daily protein intake and to continue to strive for a serum albumin ≥3.5 g/dL *(10)*.

7. MEDICATIONS

Because blood pressure control is very important for the pregnant dialysis patient, goals are to keep measurements ≤140/90 mmHg. If there is no apparent fluid overload but hypertension exists, medications are utilized. There are several antihypertensive agents considered safe during pregnancy including methyldopa, β blockers and labetelol. There is less experience using calcium channel blockers and clonidine, but these are likely to be safe as well. Angiotensin converting enzyme inhibitors (ACEi) and angiotensin receptive blockers, on the other hand, are contraindicated during pregnancy. Human studies have linked these drugs to an ossification defect in the fetal skull, dysplastic kidneys, neonatal anuria and death from hypoplastic lungs *(2,4,6)*.

Anemia is another complication during pregnancy, especially for women with CKD undergoing dialysis treatment. Epoetin alfa (Epogen, Amgen, Inc., Thousand Oaks, CA) given during dialysis has been used safely for this population, with no known congenital anomalies reported. The dose frequently needs to be increased as the pregnancy progresses to maintain hemoglobin ≥10–11 g/dL *(4)*. The need for blood transfusions during pregnancy for women undergoing dialysis has significantly declined with the use of this medication *(5)*.

8. SUMMARY

Pregnancy for women who have CKD, especially for those undergoing dialysis, is a complex medical condition. The dietitian must realize the importance of ensuring that the patient is counseled and evaluated regularly to increase energy and protein in her diet to meet the needs of the developing fetus. The patient must also be guided to change her usual renal vitamin regimen to include adequate amounts of folate and other water-soluble vitamins that have significance during pregnancy. The management of calcium, phosphorus and vitamin D required may need to be modified for the patient's safety, and zinc will need to be supplemented as well.

As mentioned previously, the management of a pregnant patient with CKD, especially if she is undergoing dialysis, requires a team approach involving nephrology and high-risk obstetrics health care professionals. Regular follow-up and communication are important to promote positive outcomes.

9. CASE STUDY

TW is a 34-year-old female with End-Stage Renal Disease (ESRD) due to glomerulonephritis and a history of hypertension as well. She initiated hemodialysis for 3.5 hours, three times per week in 7/01. At the end of 1/02, she reported that her last menses was in mid-12/01. A pregnancy test (beta-human chorionic globulin (hCG) was done and appeared to be positive. A pelvic ultrasound done 2/13/02 showed a live intrauterine gestation corresponding to 9 weeks, 5 days. TW was then counseled about her potential dialysis regimen if she decided to proceed with the pregnancy. When she agreed to proceed, she was referred to the high-risk obstetrics department at a university hospital associated with the dialysis facility.

TW brought a 24-h urine collection in 2/03, which was twice the volume of her collections in 10/01 and 1/02. As a total Kt/V of 1.97 was calculated, no immediate change was made in her dialysis prescription. She continued 3.5 hours, three times per week with a PSN 170 dialyzer and a 3.0 K+/2.5 Ca dialysate solution. Also, for safety, her antihypertensive medication was changed from amlodipine to methldopa when the pregnancy was confirmed.

In 4/02, when her Kt/V decreased to 1.29, TW was convinced to increase her dialysis time to at least 3.5 hours four times per week. She did not want to do any more time due to other appointments and care of her four other children.

TW is 71 inches tall with a IBW of 70 kg based on the Hamwi method. Her EDW at the approximate time of conception in 12/01 was 98.5 kg, which was 140% of her IBW. Her body mass index (BMI) was calculated to be 30.3. Her estimated kcal/protein needs were 2800/day (35/kg IBW + 300 kcals as she was nearing her second trimester) and

94 g/day (1.2 g/kg IBW + 10 g). Intakes were evaluated by frequent 24-h recalls, and she seemed to be eating well.

TW's EDW was increased from 98.5 kg in 12/01 to 101.5 kg in 2/02, to 103.5 kg in 3/02, to 104.5 kg in 4/02 and to 106 kg in 5/02 when she had reached 24-week gestation. Her serum albumin did decrease (from 3.8 g/dL) to 3.6 g/dL in 4/02 and 3.4 g/dL in 5/02 just before she was transferred to the inpatient hospital dialysis unit. TW's diet was only mildly restricted in potassium even before she increased her dialysis time. When the dialysis regimen was increased to 15 hours/week in 4/02, the diet was further liberalized. Her K+ level was 4.8 mEq/L in early 5/02.

TW was prescribed calcium carbonate (500 mg), 2 tid w/ meals. Calcium/phosphorus levels remained acceptable, but when her calcium was 8.4 mg/dL and then 8.5 mg/dL, the dialysate was increased to 3.5 mEq/L in 5/02. No active vitamin D was given, as her intact PTH was 67 pg/mL in 1/02 and the safety of giving it during pregnancy was uncertain. Even when the intact PTH in 4/02 was 544 pg/mL, none was given until after her delivery.

Other vitamins, minerals and medications prescribed for TW during pregnancy were renal vitamins (Nephrocaps, Fleming and Company Pharmaceuticals, St. Louis, MO)—2/day, zinc—50 mg/day (15 mg recommended, but the zinc gluconate obtained was easily tolerated and inexpensive), epoetin alfa (Epogen, Amgen, Inc.) adjusted as needed and increased at 22–23 weeks due to decreasing hemoglobin, and intravenous iron sucrose (Venofer, American Regent, Shirley, NY). TW was transferred to the hospital-based dialysis unit at 24-week gestation for fetal monitoring during dialysis. She remained on hemodialysis for 3.5 hours, 4 times per week. Because her intakes were decreased prior to transfer, a commercial nutrititional supplement (Boost Plus, Novartis, St. Louis Park, MI) was recommended.

She was admitted to the hospital at approximately 30-week gestation due to preterm labor. Per TW, her weight dropped to 101 kg (not confirmed) during her hospitalization prior to delivery. This may have been because she did not eat well in the hospital and did not take the nutrition supplements prescribed. Also, her weight may have been overestimated initially and her hospital weights were on a different scale wearing only a gown. The suggested weight gain for women with a BMI >30 is approximately 6 kg for a full-term pregnancy *(10)*.

TW was discharged from the hospital at 34-week gestation and readmitted when she went into labor at 34.5 weeks. She delivered a 4 lb. 6 oz. baby girl at that time. Presently, TW who had a kidney transplant in 8/05, and her daughter, who is now 4 years old, are doing well. It is speculated that this patient may have had a successful pregnancy outcome while dialyzing for less time than recommended because she did have residual kidney function *(2)*.

9.1. Questions

1. What lifestyle changes must be considered when counseling women with CKD undergoing dialysis once pregnancy is confirmed?
2. How are estimated energy needs calculated for the pregnant dialysis patient?
3. How are protein needs calculated for the pregnant hemodialysis patient?
4. Which vitamin is supplemented in the diet of pregnant dialysis patients to prevent neural tube defects?
5. Why is the dialysis regimen for a pregnant dialysis patient intensified?
6. How is anemia treated for pregnant women undergoing hemodialysis?

REFERENCES

1. Hou S. Pregnancy in chronic renal insufficiency and end-stage renal disease. Am J Kidney Dis 1999; 33(2):235–52.
2. Stover J. Pregnancy and dialysis. In: Byham-Gray L and Wiesen K, eds. Nutrition Care in Kidney Disease. Chicago, IL: The American Dietetic Association, 2004:121–6.
3. Holley JL and Reddy SS. Pregnancy in dialysis patients; a review of outcomes, complications and management. Semin Dial 2003; 16(5):384–88.
4. Grossman S and Hou S. Obstetrics and gynecology. In: Daugirdas JT, Blake PG, Ing TS, eds. Handbook of Dialysis, Third Edition. Philadelphia: Lippincott Williams & Wilkins, 2000:624–36.
5. Haase M, et al. A systematic approach to managing pregnant dialysis patients – the importance of an intensified haemodiafiltration protocol. Nephrol Dial Transplant 2005; 20(11):2537–42.
6. Shemin D. Dialysis in pregnant women with chronic kidney disease. Semin Dial 2003; 16(5):379–83.
7. Gangji AS, Windrim R, Gandhi S, Silverman JA, and Chan CTM. Successful pregnancy with nocturnal hemodialysis. Am J Kidney Dis 2004; 44(5):912–6.
8. The National Kidney Foundation Kidney Disease Outcomes Quality Initiative (KDOQI) clinical practice guidelines for bone metabolism and disease in chronic kidney disease. Am J Kidney Dis 2003; 42 (suppl 3):S1–S202.
9. Chang H, Miller MA, and Bruns FJ. Tidal peritoneal dialysis during pregnancy improves clearance and abdominal symptoms. Perit Dial Int 2002; 22(2):272–4.
10. Nutrition care of adult pregnant ESRD patients. In: Wiggins KL., eds. Nutrition Care of Renal Patients, Third Edition. Chicago, IL: The American Dietetic Association; 2002:105.
11. The National Kidney Foundation Kidney Disease Outcomes Quality Initiative (KDOQI) clinical practice guidelines for nutrition in chronic renal failure. Am J Kidney Dis 2000; 35(6).

12. Recommended Dietary Allowances, 10th edition. J Am Diet Assoc 1989; 89:174.
13. Vidal LV, et al. Nutritional control of pregnant women on chronic hemodialysis. J Ren Nutr 1998; 8(3):150–6.
14. Drugs.com. Available at: http://www.drugs.com. Accessed September 5, 2006.
15. Molaison EF, Baker K, Bordelon MA, Brodie P, and Powell K. Successful management of pregnancy in a patient receiving hemodialysis. J Ren Nutr 2003; 13(3):229–32.
16. Hou S. Modification of dialysis regimens for pregnancy. J Artif Organs 2002; 25:823–6.
17. Hyten FE. Nutrition and metabolism. In: Hyten F, ed. Clinical Physiology in Obstetrics. Oxford, England: Blackwell Scientific Publications; 1988:177.

17 Infancy, Childhood and Adolescence

Donna Secker

LEARNING OBJECTIVES

1. Identify three key differences in nutritional assessment, management and monitoring for children with chronic kidney disease compared to adults.
2. Describe three important nutritional issues unique to children of different ages and stages of development (i.e., infants, children and adolescents).
3. Discuss the frequent monitoring of nutritional status in children as predicated by changes in the nutrition care plan.
4. Identify the current recommendations for nutritional intake of macro- and micro-nutrients for children throughout all stages of chronic kidney disease.
5. Describe various methods of supplemental nutrition used in children with chronic kidney disease, including increasing energy density of expressed breast milk or infant formula, tube feeding, intradialytic parenteral nutrition, intraperitoneal amino acids, and vitamin and mineral supplementation.

Summary

Malnutrition in children adversely affects growth and neurocognitive and sexual development; therefore, nutritional management for children with chronic kidney disease focuses on promoting optimal growth and development through maintenance of good nutritional status and prevention of malnutrition, uremic

From: *Nutrition and Health: Nutrition in Kidney Disease*
Edited by: L. D. Byham-Gray, J. D. Burrowes, and G. M. Chertow

toxicity and metabolic abnormalities. Dietary restrictions are imposed only when clearly needed and are individualized according to age, development and food preferences. Frequent monitoring and adjustments to the nutrition care plan are required in response to changes in the child's nutritional status, age, development, anthropometrics, food preferences, residual renal function, biochemistries, renal replacement therapy, medications and psychosocial status.

Key Words: Children; chronic kidney disease; nutritional management; growth; dietary modification; enteral nutrition.

1. INTRODUCTION

Infants and young children withstand nutritional deprivation less well than adults because they have low nutritional stores and high nutritional demands for rapid physical and brain growth. Adolescents are also at greater risk because of the high demands of growth during puberty. Nutritional management for children with chronic kidney disease (CKD) focuses on promoting optimal growth and development through maintenance of good nutritional status and preventing malnutrition, uremic toxicity and metabolic abnormalities.

2. ETIOLOGY OF CKD

CKD in children can be caused by congenital, hereditary, acquired or metabolic disorders. Congenital causes are the most common and include abnormally developed kidneys (e.g., aplastic, hypoplastic and dysplastic) and obstructive uropathy [e.g., posterior urethral valves (PUV)] *(1,2)*. The second most common cause is acquired conditions, such as chronic glomerulonephritis (GN), membranoproliferative GN or focal segmental sclerosing GN. In contrast to adults, diabetic kidney disease and hypertensive nephrosclerosis are uncommon in children.

3. TREATMENT MODALITIES

Traditionally, 60–65% of pediatric patients have been treated with peritoneal dialysis (PD); however, a recent increase in the proportion of children receiving hemodialysis (HD) has occurred in North America *(1)*. Infants are particularly unstable on HD, so PD is the preferred renal replacement therapy (RRT) for this age group. Automated PD is used more often than continuous ambulatory peritoneal dialysis (CAPD), with target dwell volumes of 0.6 to 0.8 L/m^2 body surface area (BSA) for infants and 1.0 to 1.2 L/m^2 for children >2 years *(3)*. Use of daily in-center or daily nocturnal home HD in children has just begun; results have been favorable *(4–7)*. Transplantation is the preferred treatment for children with stage 5 CKD as it offers the best opportunities for rehabilitation in terms of educational and psychosocial functioning. Although not common, and depending on the treatment center, transplantation may occur as early as 1 year of age or when an infant has reached a weight of around 10 kg. Waiting times for transplantation are typically shorter for children than adults; approximately 50% of children receive a transplant within 18 months on dialysis *(1)*.

4. MALNUTRITION IN CHILDREN WITH CKD

Malnutrition is a common complication among children with CKD; its exact prevalence is unknown. Using a variety of nutritional parameters to define malnutrition, three small studies of children on dialysis who were not receiving growth hormone have documented that the prevalence of malnutrition is around 40–45% *(8–10)*. The effects of malnutrition in children are the same as in adults, but in children, it may result in growth failure and neurodevelopmental delay. Malnutrition, identified as growth delay, extremes in body mass index (BMI) for age and hypoalbuminemia at the start of dialysis are associated with an increased risk for morbidity and mortality in children with stage 5 CKD *(11–14)*. In addition, chronically malnourished children display behavioral changes such as irritability, apathy and attention deficits, and those that are stunted may experience social disadvantages that adversely affect development and quality of life *(15)*. Avoidance of malnutrition is one of the most important goals in managing children with CKD *(16,17)*.

5. GROWTH

Growth, not mortality, is the main prognostic variable in children with CKD. Fetal, infant, environmental and maternal factors can interact to impair intrauterine and postnatal growth *(18)*. Observed genetic differences in birth weight among various populations are small, and racial/ethnic differences in growth are primarily due to health and environmental influences (e.g., poor nutrition, infectious disease and low socioeconomic status), rather than ethnic differences in growth potential *(19–21)*. Adequate nutritional intake is the most important precondition for growth during the first 2 years of life where growth and neurological development are recognized to be more directly dependent on caloric intake than in older age groups. Later in childhood and adolescence, growth depends mainly on hormonal factors, with nutrition providing a supportive influence.

In addition to malnutrition and hormonal disturbances, many other factors combine to influence growth in children with CKD, including age at onset of the disease, etiology of CKD, renal osteodystrophy (ROD), water and electrolyte disturbances, metabolic acidosis, inflammation and anemia. Growth failure typically occurs once the glomerular filtration rate (GFR) drops below 40 ml/min/1.73 m^2. Bone maturation is also delayed and target adult height is often reached 2 years later than normal. Half of the children with childhood onset CKD reach an adult height that is below the 3rd percentile. Despite normal or elevated serum growth hormone levels in uremic children, administration of recombinant human growth hormone (rhGH) has successfully stimulated and improved linear growth in children with CKD. Eligibility criteria include growth potential documented by open epiphyses, height for chronological age below –2.0 standard deviation scores (SDS) or height velocity for chronological age below –2.0 SDS, and no other contraindications for rhGH use *(22)*. Nutritional intake should be optimized before starting rhGH.

6. ASSESSING NUTRITIONAL STATUS

Because there is no single parameter or gold standard measure of nutritional status for children with CKD, a variety of parameters are considered together to develop an overall impression. Each of these parameters has associated limitations in CKD and no single measure adequately reflects nutritional status *(23)*. In addition to a panel of select anthropometric, biochemical and dietary measures (Table 1), the clinician should also consider medical history, other biochemical parameters, medications, comorbid conditions, bowel habits, urine output/fluid balance and activity level.

The recommended frequency for nutritional assessment varies by age and stage of CKD *(22)*. Although appetite and growth can be impaired as early as CKD stage 1 or 2, currently there are only guidelines for children on dialysis (Table 1). Supplementation can improve weight gain, and growth in infants and toddlers; therefore, it seems prudent to assess and monitor nutritional status more often in CKD 1–4 than national guidelines for healthy children (Table 2). At all stages of CKD, more frequent assessment may be required for infants and children with malnutrition, concomitant comorbid conditions, or needing a change in RRT. Subjective global assessment, a method of nutritional assessment commonly used in the adult renal population, has not been used in children; however, a recent study in the pediatric surgical population has demonstrated its validity *(24)*, although its use in children with CKD has not yet been studied. Additional techniques

Table 1
Recommended Measures of Nutritional Status for Children Receiving Maintenance Dialysis

Parameter	*Infants and Toddlers <2 years*	*Children ≥ 2 years*
Anthropometric		
Head circumference	Monthly	Every 3–4 months until 36 months
Recumbent length	Monthly	Not applicable
Standing height	Not appropriate	Every 3–4 months
Length or Height-for-age Z-score	Monthly	Every 3–4 months
Dry weight	Monthly	Every 3–4 months
Index of weight for height i.e.,% IBW, BMI)	Monthly	Every 3–4 months
Triceps skinfold thickness	Undetermined	Every 3–4 months
Midarm muscle circumference or area	Every 3–4 months	Every 3–4 months
Biochemical		
Serum albumin	Monthly	Monthly
Serum bicarbonate	Monthly	Monthly
Dietary		
Dietary interview (i.e., food records or 24-h food recall)	Monthly	Every 3–4 months

BMI, body mass index; IBW, ideal body weight.

Table 2
Suggested Minimum Interval (months) for Assessing Nutritional Status of Children with CKD Stages 3 and 4

Age (years)	*Stage 3 CKD*	*Stage 4 CKD*
<1y	3	1
1–2y	3	2
>2y	6	3

for assessing body composition have been used sporadically in both clinical practice (dual energy X-ray absorptiometry and bioelectrical impedance analysis) and the research setting (isotope dilution, total body potassium measurement and densitometry) *(25)*.

6.1. Growth and Body Composition

Growth assessment is the single most useful tool for defining health and nutritional status in children *(26)*. The most common physical measurements for evaluating growth are recumbent length (birth to 2–3 years old) or standing height (children ≥2 years old who are able to stand upright), weight and head circumference (until 3 years of age). Weight and height reflect the size of a child (i.e., large or small) and head circumference reflects brain size. Assessment of weight alone (i.e., weight for age) is not useful because it cannot distinguish a tall, thin child from one who is short but well-proportioned. A much better anthropometric index for determining nutritional status considers a child's weight relative to his/her height, which is a measure of shape (i.e., fatness, thinness or wasting). There are several such indices [percent ideal body weight (% IBW)[1], weight-for-length/stature, and BMI] which are better associated with body composition and nutritional status than weight alone. A second preferred anthropometric index is length/height-for-age, an indicator of tallness, shortness or stunting. Suboptimal length/height-for-age is frequently associated with chronic malnutrition, organic disorders or chronic disease.

Failure to consider the child's hydration status (e.g., overhydration or a hypervolemic state) can lead to misinterpretation of measures of weight, body composition and some biochemistries. Estimation of dry weight (i.e., weight without excess fluid and at which the child is normotensive) is difficult to determine because weight gain is expected in growing children. Consideration of dietary intake, laboratory values used as markers of hydration (e.g., serum sodium and albumin), signs of edema, blood pressure and tolerance of ultrafiltration can help to differentiate true accretion of tissue mass from fluid accumulation *(27)*.

6.2. Nutrient Requirements

Nutrient recommendations for children with CKD are summarized in Tables 3–8. Recommendations should be used as a starting point

[1]Percent ideal body weight: plot length or height on growth chart to identify length-for-age or height-for-age percentile. Locate IBW as the weight at the same percentile as the height, for the same age and gender. Calculate % IBW by dividing actual weight by IBW and multiplying by 100.

(22). Modifications may be indicated during periods of catabolic stress, for children significantly above or below their IBW, or when a child's response to these recommendations is suboptimal. Chronological age should be used to determine requirements *(22)*. When chronological and height age are vastly different and a child has failed to meet goals for weight gain despite achieving requirements for chronological age, basing requirements for energy and protein on height age may provide the additional "boost" needed. Requirements for preterm infants should be based on corrected age until 2–3 years old *(28,29,30)*.

6.2.1. Energy

Adequate intake of calories is important not only for weight gain and growth but also to avoid using protein as an energy source through gluconeogenesis (Tables 6–8). Energy recommendations have traditionally been based on requirements for healthy children, because there is a lack of evidence to suggest that requirements with CKD are greater, or that children with CKD will grow better if their intake exceeds recommended amounts for healthy children. Early studies demonstrated that growth of infants and toddlers with CKD is compromised when energy intake falls below 80% of the recommended dietary allowance (RDA) *(31)*. Subsequent studies have shown that

Table 3
Equations to Estimate Energy Requirements (EER) for Children at Healthy Weights *(89)*

Age	*Estimated energy requirement (kcal/d) = total energy expenditure + energy deposition*
0–3 months	EER = [89 × weight (kg) – 100] + 175
4–6 months	EER = [89 × weight (kg) – 100] + 56
7–12 months	EER = [89 × weight (kg) – 100] + 22
13–35 months	EER = [89 × weight (kg) – 100] + 20
3–8 years	Boys: EER = 88.5 – (61.9 × age [y]) + PA × {(26.7 × weight [kg]) + (903 × height [m])} + 20
	Girls: EER = 135.3 – (30.8 × age [y]) + PA × {(10 × weight [kg]) + (934 × height [m])} + 20
9–18 years	Boys: EER = 88.5 – (61.9 × age [y]) + PA × {(26.7 × weight [kg]) + (903 × height [m])} + 25
	Girls: EER = 135.3 – (30.8 × age [y]) + PA × {(10 × weight [kg]) + (934 × height [m])} + 25

EER, estimated energy requirements; PA, Physical activity (see Table 5 for PA values).

Table 4
Equations to Estimate Energy Requirements for Children 3–18 years Who are Overweight *(89)*

Age	*Weight maintenance TEE in overweight children*
3–18 years	Boys: TEE = 114 – [50.9 × age (years)] + PA × [19.5 × weight (kg) + 1161.4 × height (m)] Girls: TEE = 389 – [41.2 × age (years)] + PA × [15.0 × weight (kg) + 701.6 × height (m)]

TEE, total energy expenditure; PA, physical activity. (See Table 5 for PA values).

the average caloric intake of children with CKD is below 80% of the RDA and that low intakes occur early and continue through all stages of the disease *(32–34)*. Early anticipation and correction of energy malnutrition, especially in infants, is therefore important in order to avoid loss of growth potential during this vulnerable period. The 2000 Kidney Disease Outcomes Quality Initiative (KDOQI) Nutrition

Table 5
Physical Activity Coefficients for Determination of Energy Requirements in Children Ages 3–18 years *(89)*

	Level of physical activity			
Gender	*Sedentary*	*Low active*	*Active*	*Very active*
	Typical ADL only	ADL + 30–60 min of daily moderate activity (e.g., walking at 5–7 km/h)	ADL + ≥60 min of daily moderate activity	ADL + ≥60 min of daily moderate activity + an additional 60 min of vigorous activity or 120 min of moderate activity
Boys	1.0	1.13	1.26	1.42
Girls	1.0	1.16	1.31	1.56

ADL, activities of daily living.

Table 6
Daily Nutrient Recommendations for Children with Chronic Kidney Disease Stages 3 and 4 ***(89,79,66,84)***

Nutrient	*Birth–6 months*	*7–12 months*	*1–3 years*	*4–8 years*	*9–13 years*	*14–18 years*
Energy	100% of the DRI (see Table 3)			100% of the DRI (see Tables 3–5)		
Protein (g/kg/day)	1.52	1.2	1.05	0.95	0.95	0.85
Sodium (mg/day) If individual has hypertension or ↑risk of fluid retention: Note: polyuric infants may need sodium supplements	Commercial or homemade baby foods made from fresh ingredients without added salt		Fresh foods without added salt	≤ 2000–2500		
Potassium If hyperkalemic:	Avoid potassium-rich foods Infants and toddlers: ≤1–3 mmol/kg/day Children and adolescents: ≤51–103mmol/day (~2–4 g/day)					
Calcium (mg/day) Total from diet + PO_4binders	≤ 2500					

(Continued)

Table 6
(Continued)

Nutrient	*Birth–6 months*	*7–12 mo*	*1–3 years*	*4–8 years*	*9–13 years*	*14–18 years*
Phosphorus (mg/d)						
PTH ↑, serum PO_4 normal	≤100	≤ 275	≤ 460	≤ 500	≤ 1250	
PTH ↑, serum PO_4 ↑	≤ 80	≤ 220	≤ 370	≤ 400	≤ 1000	
Vitamins	Routinely supplement vitamin D for exclusively breastfed infants					
If intake low or evidence of deficiency, supplement to DRI for age						
Trace minerals	Iron supplement usually required during erythropoietin therapy					
If intake low or evidence of deficiency, supplement to DRI for age						
Fluids	Note: polyuric infants may need supplemental fluids					
Restriction not usually needed, but if retaining fluid restrict to DRI for age	Total fluid intake = insensible losses + urine output + other losses + amount to be deficited					

DRI, dietary reference intake.

Table 7
Daily Nutrient Recommendations for Children Receiving Maintenance Dialysis *(22,79,66,84)*

Nutrient	*Birth–6 months*	*7–12 months*	*1–3 years*	*4–10 years*	*11–14 years*	*15–18 years*
Energy (kcal/kg/day)	108	98	102	4–6 years: 90 7–10 years: 70	Boys: 55 Girls: 47	Boys: 45 Girls: 40
Protein (g/kg/day) HD	2.6	2	1.6	4–6 years: 1.6 7–10: 1.4	Boys: 1.4 Girls: 1.4	Boys: 1.3 Girls: 1.2
PD	2.9–3	2.3–2.4	1.9-2	4–6 years: 1.9–2 7–10: 1.7–1.8	Boys: 1.7–1.8 Girls: 1.7–1.8	Boys: 1.4–1.5 Girls: 1.4–1.5
Sodium (mg/day) If individual has hypertension or ↑risk of fluid retention	Commercial or homemade baby foods made from fresh ingredients without added salt		Fresh foods without added salt	≤ 2000–2500		
Potassium If hyperkalemic	Avoid potassium-rich foods Infants and toddlers: ≤1–3 mmol/kg/day Children and adolescents: ≤51–103mmol/day (~2–4 g/day)					
Calcium (mg/day) Total from diet + PO_4 binders	≤ 2500					
Phosphorus (mg/day) PTH ↑, serum PO_4 normal	≤ 100	≤ 275	≤ 460	≤ 500	≤ 1250	

(Continued)

Table 7
(Continued)

Nutrient	*Birth–6 months*	*7–12 months*	*1–3 years*	*4–10 years*	*11–14 years*	*15–18 years*
PTH ↑, serum PO_4 ↑	≤ 80	≤ 220	≤ 370	≤ 400	≤ 1000	
Vitamins and Minerals	Iron supplement usually required during erythropoietin therapy					
	Vitamin D metabolite supplementation to maintain desirable serum levels of PTH and calcium					
If intake of water-soluble vitamins is low or evidence of deficiency, supplement to	100% of DRI for age					
Trace minerals						
If intake of zinc or copper is low or evidence of efficiency, supplement to	100% of the RDA for copper and zinc					
Fluids						
If problems with positive fluid balance, restrict to	Total fluid intake = insensible losses + urine output + other losses					

DRI, dietary reference intake; HD, hemodialysis; PD, peritoneal dialysis; PTH, parathyroid hormone; RDA, recommended dietary allowance.

Table 8
Daily Nutrient Recommendations for Children Post-Transplantation

Nutrient	*Birth–6 months*	*7–12 months*	*1–3 years*	*4–8 years*	*9–13 years*	*14–18 years*
Energy	100% of the DRI (see Table 3)			100% of the DRI (see Tables 3–5)		
Protein (g/kg/day) Note:for the first 3 months post-transplant provide 120-150% of these amounts (DRI)	1.52	1.2	1.05	0.95	0.95	0.85
Sodium (mg/d) If individual has hypertension:	Commercial or homemade baby foods made from fresh ingredients without added salt		Fresh foods without added salt	≤ 2000–2500		
Potassium						
If hyperkalemic:	Avoid potassium-rich foods Infants and toddlers: ≤ 1–3 mol/kg/day Children and adolescents: ≤ 51–103mmol/day (~2–4 g/day)					

(Continued)

Table 8
(Continued)

Nutrient	*Birth-6months*	*7–12 months*	*1–2 years*	*3–8 years*	*9–13 years*	*14–18 years*
Calcium (mg/day)	100% of DRI					
	Supplement if necessary to prevent glucocorticosteroid-induced osteoporosis					
Phosphorus (mg/day)	100% of DRI					
	Supplementation often needed in early post-transplant period to manage hypophosphatemia					
Vitamins	100% of DRI					
	Supplementation generally not needed					
Trace minerals	100% of DRI					
	Magnesium supplementation often needed in early post-transplant period to manage hypomagnesemia					
Fluids	High fluid intake required to maintain good perfusion to transplant kidney					

DRI, dietary reference intake.

Guidelines *(22)* refer to the 1989 US RDA that have since been replaced by the 2002 Dietary Reference Intakes (DRI) that include predictive equations for normal and overweight children (for weight maintenance) and include physical activity factors for children 3 years and older. The new DRI for energy are approximately 10 kcal/kg or 15–20% lower for infants and toddlers and similar or slightly higher for school-aged children and adolescents; hence, adverse effects on growth may occur if caloric intake of infants or toddlers is below 90–95% of the DRI.

Glucose absorption from peritoneal dialysate solutions provides an additional 7–10 kcal/kg for children on PD and should be included when assessing caloric intake. Following successful transplantation, immunosuppressive therapy using high dose prednisone stimulates appetite, and calorie control is needed to prevent excessive weight gain. This becomes a new issue for children who have previously struggled to consume sufficient calories. In addition, glucocorticosteroids and immunosuppressive agents such as tacrolimus (FK506) increase insulin secretion, causing impaired glucose tolerance, glycosuria and a relative resistance to insulin leading to diabetes in approximately 5–20% of children *(35)*. Avoidance of simple carbohydrates, weight control and physical exercise are prescribed for management of steroid-induced diabetes.

6.2.2. Protein

All pediatric patients need to be in positive nitrogen balance to support growth. Protein recommendations for children with stages 1–4 CKD are based on requirements for healthy children. Requirements are increased with proteinuria, catabolism, acidosis, glucocorticosteroids and peritonitis. Limited data are available on the optimal amount of protein for children on dialysis. Protein malnutrition is complicated by dialysate protein losses ranging from 100 to 300 mg protein/kg/day ($\sim$10% of daily protein intake). Losses are similar for CAPD, continuous cyclic peritoneal dialysis and tidal dialysis, but vary widely between individuals. Protein requirements are highest on a g/kg basis for infants and toddlers because protein losses are inversely related to body weight and peritoneal surface area. The KDOQI guidelines recommend an additional 0.4 g protein/kg/day to compensate for protein losses on HD and an additional 0.7–0.8 g/kg/day for losses on PD *(22)*. Similar to energy recommendations, the KDOQI guidelines refer to the 1989 US Recommended Dietary Allowances that have since been replaced with the 2002 DRI. The new DRI are lower than the RDA across all age groups. In the immediate post-transplant period, protein needs are thought to be increased by approximately 50% in association with surgical stress and the catabolic effects of steroids, decreasing back to normal recommendations for age around 3 months after transplantation.

7. NUTRITIONAL INTERVENTION AND MONITORING

Dietary restrictions are imposed only when clearly needed and should be individualized according to age, development and food preferences. Depending on the response in blood pressure or laboratory data, the dietary restrictions can be liberalized or tightened *(22)*. Dietary modifications in pediatrics are typically less restrictive than for adults in order to gain adherence, increase caloric intake and achieve optimal growth. Children who have polyuria, residual renal function (RRF) or are on PD typically require less stringent adherence to diet restrictions. Frequent monitoring and adjustments to the nutrition care plan are required in response to changes in the child's nutritional status, age, development, anthropometrics, food preferences, RRF, biochemistries, type of RRT, medications and psychosocial status.

7.1. Age-Related Feeding Problems

7.1.1. Infants

Most infants with CKD are satisfied with small volumes of oral feedings and often have slow or no progression through normal stages of acquired feeding skills. CKD is often associated with feeding problems, for example, post-traumatic feeding disorder, an oral hypersensitivity and aversion to fluids and/or foods that typically occurs in infants and toddlers with a history of previous invasive medical procedures around their oral-nasal cavities, such as intubation, oral or nasal suctioning, or the placement of nasogastric (NG) or orogastric tubes *(36,37)*. Memories of these unpleasant procedures lead the child to perceive that anything approaching their mouth is potentially painful. Other children may be able to swallow but are unwilling to eat. They may turn their head away and refuse to open their mouth when food is offered, may spit out the food or store it in their cheeks for long periods of time. Nausea and vomiting are common and contribute to feeding dysfunction, undernutrition and growth failure. Caregivers of infants and young children should be questioned about age-appropriate eating/feeding skills (e.g., sucking, chewing, swallowing and self-feeding) and related problems (e.g., storing food in cheeks, inability to advance textures and food refusal).

Delayed gastric emptying and gastric dysrhythmias occur alone or with gastroesophageal reflux disease (GERD) *(38–40)*. Delayed gastric emptying causes sustained gastric distension and a sense of satiety. GERD has been found in over 70% of infants and children with CKD who are experiencing vomiting and feeding problems *(38–40)*. Symptoms may diminish with time or persist until transplantation. Common empirical therapies for GERD include thickened feeds, maintaining the infant in an upright position during and after feedings, use of whey-predominant infant formula to improve gastric emptying, prokinetic medications (e.g., domperidone and metoclopramide), H_2-antagonists (e.g., cimetidine and ranitidine) or proton pump inhibitors (e.g., omeprazole) to prevent reflux esophagitis. When changes to feeding content and/or administration fail to improve symptoms and medical therapy for GERD is unsuccessful, continuous tube feedings, jejunal feeding or fundoplication are indicated.

With its low mineral and electrolyte content, human milk is the ideal feeding for infants with CKD. Breastfeeding should be supported; however, for a variety of reasons, infants with stage 5 CKD are often unable to meet their nutritional requirements by exclusive feeding at the breast. Motivated mothers should be supported to pump breast milk for supplemental feeding; however, many mothers neglect their own health in order to meet the demands of caring for their unwell infant and are unable to continue expressing good quantities of breast milk. Iron-fortified infant formulas containing lower amounts of phosphorus and potassium (Similac PM 60/40® Ross products Columbus, Ohio; or Good Start® Nestle, Glendale California), are indicated when relevant biochemical abnormalities begin to appear *(41)*. Soy and elemental formulas are higher in phosphorus and potassium and should be used only when necessary. Due to its high renal solute load, cow's milk is not suitable before 1 year of age and is often avoided well into the second year of life. Unless severe development delay or medical contraindication suggest otherwise, solids should be introduced and textures advanced at the same ages as for healthy infants.

7.1.2. Toddlers and Children

Poor, fussy appetites and a preference for fast food restaurants are common. Children can learn to self-induce vomiting through vigorous crying, coughing or retching to avoid eating food offered by caregivers. Poor appetites may respond favorably to small, frequent meals and snacks. Regular, structured times for meals and snacks are important. Caregivers need guidelines for setting limits around food and eating behavior, and they require constant support to consistently enforce them. Other caregivers (e.g., grandparents, baby-sitters and teachers) need to be aware of appropriate foods, fluids and established limits and should be asked to be consistent in providing care. As children become older, they need to be counselled about their diets so they can recognize and refuse restricted foods offered by others. Important topics to be discussed with caregivers and children include meals and snacks at school or after-school programs (e.g., milk coupons, special pizza or hot dog days), pica, and use of herbal supplements and alternative medications.

7.1.3. Adolescents

Teenagers who eat independent of their families need dietary information directly; however, their parents typically provide and prepare some meals eaten by the teen and therefore also need to be informed. Poor eating habits, skipped meals (in particular breakfast and/or lunch), high fluid intake (especially milk and sodas) and preference for salty fast foods and snacks are common issues in this population. Dietary instruction that addresses cafeteria foods, fast foods, snacks and acceptable drinks supports the teenager's ability to make relatively safe selections when eating out with friends. Relating nutritional status to physical appearance (e.g., healthy hair and skin), school or athletic performance (e.g., energy, alertness and muscle mass) and/or preparation for transplantation may motivate individual adolescents to improve adherence. Undernourished females may be pleased with their thin appearance and be refractory to efforts to promote weight gain. Post-transplant weight gain is most upsetting for this age group. Preparatory education on the benefits of caloric restriction and increased physical activity for weight control should begin well before transplantation occurs.

7.2. Supplementing Energy Intake

The KDOQI guidelines recommend supplemental nutritional support when growth is poor or calorie or protein intake is less than the RDA *(22)*. Manipulations to liquid feedings often require minimizing volume to maintain fluid balance, optimizing tolerance and keeping hours of feeding manageable for the family.

Infants and toddlers frequently need supplementation with oral or tube feedings of expressed breast milk or infant formula (0.67 kcal/mL; 20 kcal/oz) fortified to an energy density as high as 2 kcal/mL (60 kcal/oz) *(42,43)*. Gradual, stepwise increases of 2–4 kcal/oz theoretically improve tolerance. Increasing energy density by concentrating the base formula is often not possible because of the accompanying increase in sodium, potassium and phosphorus, therefore, fat and/or carbohydrate modules are used. The choice of which macronutrient to add is based on serum glucose and lipid profiles, presence or absence of malabsorption or respiratory distress (carbohydrate metabolism increases CO_2 production), and cost. When making a feeding which involves more than three stepwise increases (2–4 kcal/oz per step) in energy density, the distribution of energy from carbohydrate and fat should be kept similar to the base formula. Unless malabsorption is present, "heart-healthy" oils such as corn, canola or safflower are preferable because they are readily available and low in cost. To prevent oil from separating out during continuous tube feedings, an emulsified oil (i.e., Microlipid®, Novartis Medical Nutrition, Freemont, Michigan) can be used. As energy density increases, oral intake may decrease and a tube feeding may be needed.

Calories can be added to foods using heart-healthy margarines or oils, cream and other fats, sugars, syrups or carbohydrate modules. Commercial energy bars and homemade or commercial milkshakes/energy drinks made from whole milk and cream may be used; however, their phosphorus and potassium content may be too high. Milkshakes and desserts made from non-dairy products can be used to boost caloric intake for children who are hyperphosphatemic. Low calorie or calorie-free drinks should be avoided.

Non-renal enteral feedings designed for children older than 1 year have fairly high calcium and phosphorus content to support bone growth. These products are contraindicated in children with hyperphosphatemia and/or hyperkalemia. In the absence of a pediatric renal formula, adult renal products (e.g., Nepro®, Suplena®, NovaSource Renal®, Magnacal Renal® and Renalcal®), which have normal and lower protein content and are designed to be calorically dense and low in minerals and electrolytes, are recommended for children greater than 4 years of age, but have been used successfully at diluted strength in children younger than 1 year *(44)*. The magnesium content of these products is significantly higher than in breast milk or infant formulas; therefore, serum levels should be monitored closely when transitioning from an infant feeding to one of these products.

7.2.1. Tube Feeding

Tube feeding is recommended if oral supplementation is unsuccessful in correcting energy and/or protein deficits *(22)*. Studies have demonstrated the effectiveness of tube feeding in preventing and reversing weight loss and growth retardation in children of all ages, and achieving significant catch-up growth in children less than 2 years of age *(45,46)*. Various routes of tube feeding [i.e., NG, gastrostomy tube (GT), gastrojejunostomy (GJ), jejunostomy tube] have all been used successfully to provide additional breast milk, infant or enteral feedings by intermittent bolus or continuous infusion *(47)*. Choice of formula/feeding is guided by age, biochemistries, fluid allowances and cost. The method of delivery is dependent on the age of the patient, the quantity to be delivered, the composition of the formula, gastrointestinal tolerance (vomiting, delayed gastric emptying and GERD) and the safety of unsupervised feeding.

Intermittent bolus feeds are used for infants to maintain normal blood glucose levels. After infancy, continuous overnight feedings are preferred to offer the child an opportunity for spontaneous oral intake during the day and freedom from being attached to feeding equipment to attend school. If the total volume needed is too large for a child to tolerate overnight, daytime bolus feedings may be required in addition to overnight feedings. Continuous overnight feedings are usually not used before 12 months of age due to an increased risk of vomiting and aspiration, especially in infants with uremia and GERD. Feedings are initiated and advanced according to pediatric guidelines and tolerance. Volumes and rates based on body weight help to avoid intolerance in patients who are underweight or small for their age (Table 9). Reported complications include emesis, exit-site infection, leakage and displacement *(48,49)*. Wherever possible, placement of GT or GJ tubes should occur prior to initiation of PD to decrease the risk of peritonitis *(48–50)*. Oral stimulation and non-nutritive sucking should be provided to infants totally dependent on tube feeding to help smooth their transition to oral feeding following successful transplantation. Using a multidisciplinary approach to preventative and behavioral treatment programs, several centers have reported successful transitioning of virtually 100% of their patients over to complete oral feedings within 2–6 months after successful transplantation *(51–54)*, illustrating that tube feeding need not preclude development of normal oral feeding skills.

7.2.2. Parenteral Nutrition

Published guidelines for the use of parenteral nutrition (PN) in children with CKD are limited. Standard amino acid (AA) solutions (i.e., essential + non-essential AAs) are generally used and adjusted as needed. If no fluid restriction is needed, volumes can be based on daily maintenance fluid requirements. For patients who require fluid restrictions, more concentrated PN solutions of AA, dextrose and lipids

Table 9
Guidelines for Initiating and Advancing Tube Feedings in Infants and Children

Age	*Initial hourly infusion*	*Daily increases*	*Goal*[a]
Continuous Feedings			
0–1 years	1–2 mL/kg/h (10–20 mL/h)	1mL/kg/h (5–10 mL/8h)	6 mL/kg/h (21–54 mL/h)
1–6 years	2–3 mL/kg/h (20–30 mL/h)	1 mL/kg/h (10–15 mL/8h)	4–5 mL/kg/h (71–92 mL/h)
6–14 years	1 mL/kg/h (30–40 mL/h)	0.5 mL/kg/h (15–20 mL/8h)	3–4 mL/kg/h (108-130 mL/h)
>14 years	0.5–1 mL/kg/h (50 mL/h)	0.4–0.5 mL/kg/h (25 mL/8h)	125 mL/h
Bolus Feedings			
0–1 years	10–15 mL/kg/feed q 4 h (60–80 mL/feed)	20–40 mL q 4 h	20–30 mL/kg/feed q 4 h (80–240 mL/feed)
1–6 years	5–10 mL/kg/feed q 4 h (80–120 mL/feed)	40–60 mL q 4 h	15–20 mL/kg/feed q 4 h (280–375 mL/feed)
6–14 years	3–5 mL/kg/feed q 4 h (120–160 mL/feed)	60–80 mL q 4 h	10–20 mL/kg/feed q 4 h (430–520 mL/feed)
>14 years	3 mL/kg/feed q 4 h (200 mL/feed)	100 mL q 4 h	10 mL/kg/feed q 4 h (500 mL/feed)

Adapted from ref. *(90)*.

[a] Goal is expected maximum that child will tolerate; individual children may tolerate higher rates or volumes. Proceed cautiously for jejunal feedings. Goals for individual children should be based on energy requirements and energy density of feeding and therefore may be lower than expected maximum tolerance.

Table 10
Guidelines for Administration of Parenteral Amino Acids (g/kg/day) for Children

	Preterm	*Birth–1 year*	*1–12 years*	*>12 years*
Initial dose	1.5	1.5–2.0	1.5–2	1.5
Advance daily[a]	1	1	–	0.5
Goal	See protein recommendations specific to each renal replacement therapy (Tables 6–8)			
Maximum	3–3.5	2.5–3.5	1.5–2	1.5–2

[a] Rate of advancement is determined by protein tolerance.
Modified with permission from ref. *(91)*

are used. Unless otherwise indicated, goals for PN are 90% of energy recommendations (i.e., because there is no thermal effect of feeding) and 100% of protein recommendations specific to the child's stage of kidney disease and type of RRT (Tables 6–8). The AAs, dextrose and lipids should be initiated and advanced according to guidelines for non-renal patients, such as those in Tables 10–13. Mineral and electrolyte content should be adjusted to maintain acceptable serum levels and the acetate and chloride content adjusted to maintain acid/base balance. Standard parenteral dosages of multivitamins and trace elements can be used. The risk of vitamin A toxicity is minimal with a daily injectable multivitamin because the child is unlikely to have another exogenous source of vitamin A (i.e., no oral diet).

Table 11
Guidelines for Administration of Parenteral Dextrose (mg/kg/min) for Children

	Preterm	*Full term infants and children*	*Adolescents*
Initial dose	5–8	5–8	3–5
Advance daily[a]	1–2	1–3	1–3
Maximum[b]	10–16	11–12	5–8

[a] Rate of advancement is determined by glucose tolerance.
[b] Maximum will be dictated by tolerance and route of administration (peripheral and central).
Modified with permission from ref. *(91)*

Table 12
Guidelines for Administration of Parenteral Lipids (g/kg/day) for Children

	Preterm	*Full-term infants*	*Children and Adolescents*
Initial dose	1.0	1	1
Advance daily[a]	1.0	1	1
Maximum	3–4	4	2–4

[a] Rate of advancement and maximum is determined by lipid tolerance.
Modified with permission from ref. *(91)*

7.2.3. Intradialytic Parenteral Nutrition

There is early evidence of the safety and effectiveness of intradialytic parenteral nutrition (IDPN) in reversing malnutrition and improving weight gain in small numbers of malnourished children receiving maintenance HD *(55–57)*. Children have achieved gains in body weight and BMI and/or % IBW from receiving 100–150% of the recommended calorie intake from a combination of oral and parenteral AA, dextrose and lipids delivered at each dialysis session over a period of 6 weeks to 6 months. Reported adverse events have been minor and have included hypophosphatemia, transient hyperglycemia, lipid intolerance and mildly elevated liver function tests. Improvements in serum albumin concentrations were not observed as virtually all of the children had albumin levels in the normal range before IDPN was started.

7.3. Meeting Protein Needs

Voluntary protein intake usually exceeds recommendations but protein malnutrition can still occur if energy intake is low. Protein

Table 13
Summary of Daily Parenteral Nutrition Requirements for Children

Nutrient	*Preterm*	*Full-term*	*1–12 years*	*>12 years*
Energy (kcal/kg)	110–80	100–80	90–60	60–30
Dextrose (mg/kg/min)	10–16	11–12	11–12	5–8
Fat (g/kg)	3–4	3–4	2–4	2–4
Protein (g/kg)	See protein recommendations specific to each renal replacement therapy (Tables 6–8)			

Modified with permission from ref. *(91)*

intake may exceed recommendations as long as serum urea and phosphate levels are acceptable. The benefit of reducing protein intake on the progression of renal function in children with CKD stages 1–4 has not been clearly shown, and low protein diets may interfere with nutritional status and growth; therefore, the aim is to avoid excessive protein intake (i.e., >150–175% requirements) in order to minimize uremia. Protein of high biological value is encouraged because it minimizes urea production by reusing circulating nonessential AA for protein maintenance. Urea levels >100 mg/dL (>35 mmol/L) can occur for a variety of reasons including excessive protein intake, adequate protein but insufficient caloric intake, catabolism (e.g., infection), inadequate dialysis and/or access recirculation.

Protein intake may be low because of anorexia, low meat intake, chewing problems or a low phosphorus diet that limits protein-rich dairy foods. Protein intake can be increased via protein-rich foods or powdered protein modules added to infant formula, beverages, pureed foods, cereals or other moist foods. Minced or chopped meat, chicken, fish, egg, tofu or skim milk powder can be added to soups, pasta or casseroles. Milk, milk products or eggs can be substituted if meat is disliked; however, phosphorus intake will increase. Meeting protein requirements may be difficult for children receiving PD and for vegetarians, especially vegans, who may need specific dietary counselling. Persistently low urea levels (i.e., <50 mg/dL or <18 mmol/L) may indicate overall inadequate protein and caloric intake. Serum levels of proteins may not be decreased unless dietary intake has been inadequate for an extensive period.

To assess protein status, Edefonti et al. *(34)* performed nitrogen balance studies in children on PD and found that in only 50% of the studies nitrogen balance was adequate to meet estimated nitrogen requirements for growth and the metabolic needs of uremic children (i.e., ≥50 mg/kg/day). In 36% of the studies, results were considered relatively satisfactory (i.e., 0–50 mg/kg/day), and in 14% of the studies, children were in negative nitrogen balance. Children who had been on dialysis longer than 1 year had lower energy and protein intakes and poorer nitrogen balance results.

Normalized protein catabolic rate (nPCR) or protein nitrogen appearance (nPNA) indirectly assess dietary protein intake in dialysis patients with less measurement error than dietary diaries or 24-h food recalls. Both dialysis adequacy (Kt/V) and nPNA/nPCR are linked to patient survival and there is a positive association between them. A child who has a desirably low predialysis urea may be a well-nourished patient who is adequately dialyzed or an individual with a decreased protein intake. Normalized PCR or nPNA can help differentiate between the two possibilities. Whether an improvement in dialysis adequacy directly causes an improvement in appetite with an increase in protein intake has not been proven *(58)*. At the time of the 2000 KDOQI Nutrition Guidelines, there was insufficient evidence to recommend routine determination of nPCR for nutrition assessment and management of children treated with maintenance dialysis (MD) *(22)*; however, updated guidelines have recommended its use for monitoring adequacy of dialysis and dietary protein intake in children receiving maintenance HD *(3)*.

Pediatric target values for nPCR vary by age. For adolescents, nPCR values between 1.0 and 1.2 g/kg/day have been associated with positive outcomes *(59)*. Target values for children and infants have not yet been clearly delineated, but theoretically would be higher than for adolescents. Because nPCR fluctuates on a daily basis depending on what is eaten, a single value does not give a good picture of usual or average protein intake; therefore, monthly measurements are more informative.

Although PNA has been used to estimate dietary protein intake in children on PD *(58,60,61)*, associated outcome measures for interpreting PNA measurements are not well established and the recent update to the KDOQI Clinical Guidelines did not recommend routine determination of the PNA in this population *(3)*.

7.3.1. Intra-Peritoneal Amino Acid Dialysis

The use of a dialysate containing a mixture of essential and nonessential AA instead of glucose as the osmotic agent has improved protein malnutrition and nitrogen balance in children receiving PD who are unable to maintain adequate protein intake *(62,63)*. In most cases, one exchange/day is replaced with the AA dialysate, with 50–90% of the infused AA absorbed. To avoid using the AA for energy, the solution is typically given during the day when meals/snacks provide a source of calories; however, giving the AA overnight via the cycler, coupled with the standard glucose dialysate as an energy source, has also been successful *(62)*. However, routine use of intra-peritoneal amino acid dialysis (IPAA) is impractical due to cost.

7.4. Optimizing Fat Intake

High calorie diets or tube feedings rich in fats may influence lipid profiles; however, fats are an important source of calories for growing children. Hyperlipidemia is present in many children with CKD stages

3–5 *(64,65)* or post-transplantation as an adverse effect of immunosuppressant therapy. When additional fats are needed as a calorie source, "heart-healthy" ones should be used. Overall, healthy food preparation should be encouraged, such as using peanut, canola or olive oil in cooking, as these are high in monounsaturated fats. The KDOQI Cardiovascular Guidelines *(66)* recommend that management of dyslipidemias for prepubertal children with CKD and CKD stage 5 (including post-transplantation) should follow recommendations by the National Cholesterol Expert Panel in Children and Adolescents (NCEP-C) *(67)* and that management of postpubertal children or adolescents with CKD stages 4 and 5 (including post-transplantation) should follow recommendations provided in the KDOQI Clinical Practice Guidelines for Managing Dyslipidemias in Chronic Kidney Disease *(68)*. Key features of the KDOQI Dyslipidemia Guidelines that differ from the NCEP-C are (i) more frequent evaluation of dyslipidemias (i.e., after presentation with CKD, after a change in RRT, and annually); (ii) if low density lipoprotein (LDL) is 130–159 mg/dL, a therapeutic lifestyle change diet should be started (if nutritional status is adequate), followed by a statin drug in 6 months if LDL ≥130 mg/dL; and (iii) if LDL is ≥160 mg/dL, both a TLC diet and a statin should be started. Lipid-lowering drugs are also used in children with CKD and hyperlipidemia *(64)*.

7.5. Vitamin and Mineral Supplementation

When the volume or variety of dietary intake is limited by anorexia or dietary restrictions, the risk for vitamin and mineral deficiencies is increased. Children receiving dialysis have additional risks of deficiencies because of increased losses through dialysis and increased needs (e.g., iron on erythropoietin therapy and folic acid for hyperhomocysteinemia). Small studies have documented water-soluble vitamin intake below recommendations for children receiving dialysis *(69,70)*. No published studies have assessed blood vitamin levels of children undergoing dialysis in the absence of the use of a vitamin supplement. Several studies have shown that the combination of dietary and supplemental vitamin intake is routinely associated with blood concentrations that meet or exceed normal values. The KDOQI Nutrition Guidelines recommended supplementing water-soluble vitamins, zinc and copper when intake is below recommended levels or when monitoring reveals laboratory or clinical evidence of a deficiency *(22)*. More recently, the KDOQI Cardiovascular guidelines state that current opinion and evidence suggests that it is prudent to supplement, rather than risk deficiency, especially when supplementation is safe at the recommended levels *(66)*. These guidelines recommend a daily vitamin supplement that provides the recommended published vitamin profile for dialysis patients, with special attention to the inclusion of folic acid and vitamins B2, B6 and B12.

Supplementation of vitamins A and E is generally avoided because blood levels are usually normal or elevated without supplementation in individuals with CKD. Vitamin D supplementation should be provided for all exclusively breastfed infants, especially those at highest risk for vitamin D-deficiency rickets (i.e., infants born to vitamin D-deficient mothers, those having limited exposure to sunlight or those are dark skinned) *(71,72)*.

Multivitamin therapy is rarely needed after a successful renal transplant because dietary restrictions are not warranted and appetite significantly improves. In the early post-transplant period, supplements of magnesium, phosphorus and vitamin D may be required *(73)*.

Children of all ages usually require supplemental oral or intravenous iron during recombinant human erythropoietin therapy *(74)*, and folic acid supplementation should be considered for children with hyperhomocysteinemia *(75–78)*.

7.6. Maintaining Calcium and Phosphorus Balance

The incidence of renal osteodystrophy (ROD) is higher in children compared to adults due to high bone turnover in the growing skeleton. ROD contributes significantly to growth failure; therefore, maintenance of normal calcium (Ca), phosphorus (PO_4), CaXP product, and parathyroid hormone (PTH) are critical. A low phosphorus diet, in conjunction with phosphate binders and vitamin D supplementation, is an essential component of therapy for the prevention of renal bone disease and poor growth. Recent awareness of the harmful consequences of hypercalcemia and a high serum CaXP product has led clinicians to carefully evaluate calcium supplementation and calcium loads. To prevent or control ROD and decrease the risk of calcification, the KDOQI Bone Guidelines for Children with CKD *(79)* have made recommendations for the strict control of serum PTH levels and the CaXP product (Table 14) and maximum dietary phosphorus intake and intake of calcium from calcium-containing phosphorus binders and diet (Table 15).

Dietary modification must consider foods naturally high in phosphorus as well as foods and drinks in which phosphate salts have been added by the manufacturer for non-nutritive reasons. On a mixed diet, net absorption of total phosphorus in various reports ranges from

Table 14
Target Ranges for Serum Levels of Parathyroid Hormone and Calcium-Phosphorus Product for Children with Stages 2 to 5 Chronic Kidney Disease (CKD)

	CKD Stage			
Target range	2	3	4	5
Serum PTH (pg/mL)	35–70	35–70	70–110	200–300
CaXP (mg^2/dL^2)	$\leq$ 12 years old: <65 > 12 years old: <55			

PTH, Parathyroid hormone; CaXP, Calcium-phosphorus product.

65 to 90% in infants and children. There is no evidence that the absorption of phosphorus varies with dietary intake (i.e., phosphorus absorption does not improve at low intakes or decrease at high intakes) or that it is altered by uremia. Excretion of endogenous phosphorus is mainly through the kidneys with a smaller amount being excreted in the stool. The amounts of phosphorus removed by thrice weekly HD (~800 mg/treatment or 2400 mg/week) or daily PD (300–400 mg/treatment or 2100–2800 mg/week) is far less than that ingested by most children. The inefficiency of phosphate removal by standard HD and PD is highlighted by the experience of children receiving daily HD who have not required diet restrictions and have needed fewer or no phosphate binders *(5,6)*.

Table 15
Recommendations for Maximum Dietary Phosphorus Intake and Calcium Intake for Children with Hyperphosphatemia

	Dietary phosphorus intake (mg/day)		*Calcium intake (mg/day)*	
Age (years)	*High PTH, normal PO_4*	*High PTH and PO_4*	*From Ca-containing PO_4 binders*	*Total (diet + binders)*
0–0.5	≤100	≤80	≤420	≤2500
0.5–1.0	≤275	≤220	≤540	≤2500
1–3	≤460	≤370	≤1000	≤2500
4–8	≤500	≤400	≤1600	≤2500
9–18	≤1250	≤1000	≤2500	≤2500

PTH, Parathyroid hormone; PO_4, phosphorus.

Non-breastfed infants who are hyperphosphatemic require a low phosphorus formula (Similac PM 60/40® Ross Products, Columbus, Ohio; Good Start®, Nestle, Glendale California), which may be continued beyond 1 year of age to delay introducing phosphorus-rich cow's milk. Milk and milk products are limited to $\leq$240 mL/day, in combination with daily limits of other high phosphorus foods. Non-dairy, milk substitutes and frozen non-dairy desserts can be used in place of milk and ice cream. Phosphorus restriction complicates efforts to achieve adequate protein intake because protein and phosphorus are often found in the same foods. The lowest quantity of phosphorus intake in proportion to the quantity and quality of protein comes from animal flesh proteins (average: 11 mg phosphorus/g protein), whereas eggs, dairy products, legumes and lentils have higher phosphorus to protein ratios (average: 20 mg phosphorus/g protein). As a result, phosphorus control is more difficult in vegetarians and they may need more phosphorus binders in order to control serum phosphorus levels and meet dietary protein recommendations. Not all children eat three meals a day; hence, prescription of phosphorus binders needs to consider when the largest amounts of phosphorus is consumed.

Glucocorticosteroid therapy post-transplantation can induce osteoporosis; therefore, calcium supplementation may be necessary for children unable to meet the DRI for calcium *(80)*. Liberal dietary intake of phosphorus-rich foods and fluids is encouraged in the early transplantation period to manage hypophosphatemia due to transient impaired renal phosphate reabsorption *(73)*.

7.7. Modifying Sodium Intake

Serum sodium levels reflect water balance, not total body sodium; therefore it is not a good indicator of the need for sodium restriction. Restriction of salt is appropriate for children with CKD associated with salt and water retention, but not for children with salt-wasting syndromes such as obstructive uropathy, renal dysplasia, tubular disease, or polycystic kidney disease who may require sodium *supplementation* to prevent sodium depletion, a decrease in extracellular volume, and impaired growth *(81,82)*. Almost 80% of children are hypertensive 1 month after transplantation due to the effects of many of the immunosuppressive medications and require sodium restriction *(73)*. For these children, the KDOQI guidelines recommend limiting sodium intake to less than 2000–2500 mg sodium/day *(66,83)*. The most recent Dietary Guidelines for Americans *(84)* recommend that hypertensive individuals older than 2 years of age consume no more than 1500 mg sodium/day. Restaurant meals and salt added by manufacturers provide

75% of the daily sodium intake of most people living in North America *(85)*. To reduce sodium intake, children and caregivers are advised to limit salt used in cooking or added at the table, to rely on fresh rather than processed foods, to eat out less often, and to read ingredient lists and nutrient content tables on food labels to avoid salty foods, defined as having more than 140–200 mg sodium/serving *(84)*.

Homemade baby foods prepared from fresh ingredients, not packaged convenience foods, are lower in sodium; commercial baby foods do not contain added salt. Lunch ideas should be discussed, especially for children who eat at school. Salt substitutes are contraindicated in children with hyperkalemia because manufacturers use potassium chloride to replace some or all of the sodium chloride.

7.8. Limiting Potassium Intake

The potassium content of commercial baby foods differs from the equivalent table food; therefore, caregivers may require lists of the potassium content of both types of foods. Infant formulas or enteral feedings can be pre-treated with an ion exchange resin (e.g., sodium polystyrene sulfonate and calcium polystrene sulfonate) to lower their potassium content *(86,87)*. Many fruits and vegetables are high in potassium; children and caregivers should be advised to limit fruit juices and choose fruit drinks, beverages and punches instead.

A potassium restriction is rarely needed for patients receiving PD once full dialysate exchange volumes are reached. Not all hyperkalemia is diet-related, and there are a number of non-dietary causes (refer to Chapter 9).

7.9. Adjusting Fluid Intake

Children with polyuria need extra fluid intake to prevent chronic dehydration and poor growth. On the other hand, children who have edema or hypertension require fluid restrictions that are based on insensible fluid losses (Table 16), measured 24-h urine output, dialysis ultrafiltration capacity, other losses (diarrhea, gastric) and, if necessary, an amount to be deficited to bring the child closer to his/her estimated dry weight. Goals for intradialytic (fluid) weight gain on HD are based on body size and tolerance of fluid removal and ideally are $\leq$5% dry body weight. Children need frequent reminding that any food that is liquid at room temperature contains water and must be counted as part of their daily fluid allotment. Many fruits and vegetables contain significant amounts of water and can inconspicuously add to a child's fluid intake. Depending on an individual's diet, fluid from solid foods can contribute up to 800–1000 mL/day. Children struggling with fluid restrictions should be advised to drink from smaller cups or glasses,

Table 16
Normal Daily Urine Output and Insensible Losses for Children

Age	*Normal urine output*	*Insensible losses*
Preterm (<37 weeks)	≥1–2 mL/kg/hour	40 mL/kg/day
Infants (birth-1 year)	≥1–2 mL/kg/hour	40–20 mL/kg/day
Children (>1 years)	≥1 mL/kg/hour	400–500 mL/m^2/day

quench their thirst by sucking on crushed ice, chew gum, gargle or use breath sprays/sheets and, more importantly, limit their sodium intake. Additional dialysis is warranted if fluid restrictions make it impossible to meet nutritional goals. Nutrition should not be compromised in a growing child. High fluid intakes are required post-transplantation to maintain good perfusion of the transplanted kidney and to prevent toxicity from immunosuppressive agents due to dehydration *(73)*.

8. SUMMARY

Nutritional therapy is vital to the management of children with CKD. Children and adolescents are confronted with frequent dietary modifications that occur concurrently with significant changes in growth, development and independence *(16)*. As a result, optimizing nutritional status is a continuing process that requires frequent monitoring and adjustments to the nutritional plan based on changes in age, development, growth and body composition, laboratory values, RRF, RRT, medications and psychosocial status.

A registered dietitian with pediatric renal experience should be the central individual in dietary management *(16,22)*, and collaboration with the nephrologist, nurse, social worker, and therapists is indispensable to meet the nutritional needs of infants, children and adolescents with CKD *(88)*. Inputs from the child and caregivers are essential. Consistent promotion of the benefits of dietary modification and provision of practical information and emotional support to children and their families can positively influence adherence and clinical outcomes and minimize stress around nutritional issues.

9. CASE STUDY

AB is a 5-day-old infant who presents with oliguria, urea 25 mmol/L, CRT 350 mmol/L, K^+ 7.5 mmol/L, Ca^{++} 1.8 mmol/L, and PO_4 3.0 mmol/L. Investigations reveal that he has acute kidney failure due to PUV and dysplastic kidneys. Following surgical correction of his PUV, serum urea, creatinine and biochemistries normalized but he becomes polyuric and continues to have high urine output long term.

9.1. Questions

1. What infant feeding would you choose for him during the initial acute presentation, following surgical correction of his PUV, and what will be the most important nutritional management issues while he progresses through CKD stages 1–3?

9.2. Follow-Up

By 15 months of age, his GFR has dropped below 50 ml/min/1.73 m^2. His PTH is 350 ng/L, ionized Ca^{++} 1.09 mmol/L, and his PO_4 2.3 mmol/L. He takes very small amounts of solids and has been resistant to progress beyond jarred baby foods; however, now he is showing some interest in foods his parents eat. He drinks 1.5 L of water a day and gets continuous overnight G-tube feedings of a standard pediatric enteral feeding product (e.g., Resource Just for Kids).

2. What specific formula and diet modifications do you need to discuss with his caregivers?

9.3. Follow-Up

At age 4 years, AB will be starting preschool and at the same time needs to start on PD.

3. What nutritional interventions will you implement?

9.4. Follow-Up

After 18 months on PD, AB receives a living-related kidney transplant. His urine output increases to 5 L/day. He becomes hypertensive, hyperkalemic, hypophospatemic and hypomagnesemic on immunosuppressant therapy.

4. How will you adjust his nutritional regime?

REFERENCES

1. North American Pediatric Renal Transplant Cooperative Study (NAPRTCS) Annual Report. 2005. (Accessed at http://web.emmes.com/study/ped/annlrept/annlrept2005.pdf).
2. Chan JCM, Williams DM, Roth KS. Kidney failure in infants and children. Pediatr Rev 2002;23(2):47–60.

3. National Kidney Foundation. KDOQI Clinical Practice Guidelines and Clinical Practice Recommendations for Hemodialysis Adequacy, Peritoneal Dialysis Adequacy, and Vascular Access, Update 2006. Am J Kidney Dis 2006;48(1 Suppl 1):S1–322.
4. Simonsen O. Slow nocturnal dialysis as a rescue treatment for children and yong patients with end-stage renal failure. J Am Soc Nephrol 2000;11:327A.
5. Fischbach M, Terzic J, Laugel V, et al. Daily on line hemodiafiltration: A pilot experience in children. Nephrol Dial Transplant 2004;19:2360–7.
6. Geary DF, Piva E, Tyrrell J, et al. Home nocturnal hemodialysis in children. J Pediatr 2005;147(3):383–7.
7. Warady BA, Fischbach M, Geary D, Goldstein SL. Frequent hemodialysis in children. Adv Chronic Kid Dis 2007;14(3):297–303.
8. Besbas N, Ozaltin F, Coskun T, et al. Relationship of leptin and insulin-like growth factor 1 to nutritional status in hemodialyzed children. Pediatr Nephrol 2003;18:1255–9.
9. Pereira A, Hamani N, Nogueira P, Carvalhaes J. Oral vitamin intake in children receiving long-term dialysis. J Ren Nutr 2000;10(1):24–9.
10. Karupaiah T, Chooi C, Lim Y, Morad Z. Anthrompometric and growth assessment of children receiving renal replacement therapy in Malaysia. J Ren Nutr 2002;12(2):113–21.
11. Furth SL, Hwang W, Yang C, Neu AM, Fivush BA, Powe NR. Growth failure, risk of hospitalization and death for children with end-stage renal disease. Pediatr Nephrol 2002;17:450–55.
12. Furth SL, Stablein D, Fine RN, Powe NR, Fivush BA. Adverse clinical outcomes associated with short stature at dialysis initiation: A report of the North American Pediatric Renal Transplant Cooperative Study. Pediatrics 2002;109(5):909–13.
13. Wong C, Hingorani S, Gillen D, et al. Hypoalbuminemia and risk of death in pediatric patients with end-stage renal disease. Kidney Int 2002;61:630–7.
14. Wong CS, Gipson DS, Gillen DL, et al. Anthropometric measures and risk of death in children with end-stage renal disease. Am J Kidney Dis 2000;36(4): 811–9.
15. Morel P, Almond PS, Matas AJ, Gillingham KJ, Chau C, Brown A. Long-term quality of life after kidney transplantation in childhood. Transplantation 1991;52(1):47–53.
16. Warady BA, Alexander SR, Watkins S, Kohaut E, Harmon WE. Optimal care of the pediatric end-stage renal disease patient on dialysis. Am J Kidney Dis 1999;33(3):567–83.
17. Rock J, Secker D. Nutrition management of chronic kidney disease in the pediatric patient. In: Byham-Gray L, Wiesen K, eds. A Clinical Guide to Nutrition Care in Kidney Disease, 1st edition. Chicago, IL: Renal Dietitians Dietetic Practice Group of the American Dietetic Association and the Council on Renal Nutrition of the National Kidney Foundation; 2004:127–49.
18. Pinyerd BJ. Assessment of infant growth. J Pediatr Health Care 1992;6:302–8.
19. Habicht JP, Martorell R, Yarbrough C, Malina RM, Klein RE. Height and weight standards for preschool children: how relevant are ethnic differences in growth potential? Lancet 1974;1:611–5.
20. Martorell R, Medoza FS, Castillo RO. Genetic and environmental determinants of growth in Mexican-Americans. Pediatrics 1989;85:864–71.

21. Mei Z, Yip R, Trowbridge F. Improving trend of growth of Asian refugee children in the USA: Evidence to support the importance of environmental factors on growth. Asia Pac J Clin Nutr 1998;7(2):111–6.
22. National Kidney Foundation Dialysis Outcome Quality Initiative. Clinical Practice Guidelines for Nutrition in Chronic Renal Failure. Am J Kidney Dis 2000;35(6 Suppl 2):S105–36.
23. Foster B, Leonard M. Measuring nutritional status in children with chronic kidney disease. Am J Clin Nutr 2004;80:801–14.
24. Secker D, Jeejeebhoy K. Subjective Global Nutritional Assessment for children. Am J Clin Nutr 2007;85(4):1083–1089.
25. Foster B, Leonard M. Nutrition in children with kidney disease: Pitfalls of popular assessment methods. Perit Dial Int 2005;25(Suppl 3):S143–6.
26. de Onis M, Habicht JP. Anthropometric reference data for international use: recommendations from a World Health Organization Expert Committee. Am J Clin Nutr 1996;64:650–8.
27. Mitchell S. Estimated dry weight (EDW): aiming for accuracy. Nephr Nursing J 2002;29(5):421–30.
28. ESPGHAN Committee on Nutrition: Peter J. Aggett CA, Irene Axelsson, Mario DeCurtis, Olivier Goulet, Olle Hernell, Berthold Koletzko, Harry N. Lafeber, Kim F. Michaelsen, John W.L. Puntis, Jacques Rigo, Raanan Shamir, Hania Szajewska, Dominique Turck, Lawrence T. Weaver. Feeding Preterm Infants After Hospital Discharge: A Commentary by the ESPGHAN Committee on Nutrition. J Pediatr Gastroenterol Nutr 2006;42:596–603.
29. Nutrition Committee. Canadian Paediatric Society. Nutrient needs and feeding of premature infants. CMAJ 1995;152(11):1765–85.
30. Committee on Nutrition. Nutritional Needs of Low-Birth-Weight Infants. Pediatrics 1985;75(5):976–986.
31. Betts P, Magrath G. Growth pattern and dietary intake of children with chronic renal insufficiency. BMJ 1974;2:189–93.
32. Foreman JW, Abitol CL, Trachtman H, et al. Nutritional intake in children with renal insufficiency: a report of the growth failure in children with renal diseases study. J Am Coll Nutr 1996;15(6):579–85.
33. Canepa A, Perfumo F, Carrea A, et al. Protein and calorie intake, nitrogen losses, and nitrogen balance in children undergoing chronic peritoneal dialysis. Adv Perit Dial 1996;12:326–9.
34. Edefonti A, Picca M, Damiani B, et al. Dietary prescription based on estimated nitrogen balance during peritoneal dialysis. Pediatr Nephrol 1999;13:253–8.
35. Greenspan L, Gitelman S, Leung M, Glidden D, Mathias R. Increased incidence in post-transplant diabetes mellitus in children: a case-control analysis. Pediatr Nephrol 2002;17:1–5.
36. Benoit D, Green D, Arts-Rodas D. Posttraumatic feeding disorders. J Am Acad Child Adolesc Psychiatry 1997;36(5):577–8.
37. Benoit D, Coolbear J. Post-traumatic feeding disorders in infancy: behaviors predicting treatment outcome. Infant Ment Health J 1998;19(4):409–21.
38. Ravelli AM, Ledermann S, Bissett W, Trompeter R, Barratt T, Milla P. Foregut motor function in chronic renal failure. Arch Dis Child 1992;67:1343–7.
39. Ravelli AM. Gastrointestinal function in chronic renal failure. Pediatr Nephrol 1995;9:756-62.

40. Ruley EJ, Bock GH, Kerzner B, Abbott AW, Majd M, Chatoor I. Feeding disorders and gastroesophageal reflux in infants with chronic renal failure. Pediatr Nephrol 1989;3:424–9.
41. Fried M, Khoshoo V, Secker D, Gilday D, Ash J, Pencharz P. Decrease in gastric emptying time and episodes of regurgitation in children with spastic quadriplegia fed a whey-based formula. J Pediatr 1992;120(4 Pt 1):569–72.
42. Spinozzi NS, Nelson PA. Nutrition support in the newborn intensive care unit. J Ren Nutr 1996;6(4):188–97.
43. Yiu VW, Harmon WE, Spinozzi N, Jonas M, Kim MS. High-calorie nutrition for infants with chronic renal disease. J Ren Nutr 1996;6(4):203–6.
44. Gast T, Bunchman T, Barletta G. Nutritional management of infants with CKD/ESRD with use of "adult" renal-based formulas. Perit Dial Int 2007;27(Suppl 1):S34. (abstract)
45. Ledermann SE, Shaw V, Trompeter RS. Long-term enteral nutrition in infants and young children with chronic renal failure. Pediatr Nephrol 1999;13(9): 870–5.
46. Coleman JE, Watson AR, Rance CH, Moore E. Gastrostomy buttons for nutritional support on chronic dialysis. Nephrol Dial Transplant 1998;13(8):2041–6.
47. Coleman J, Warady B. Supplemental tube feeding. In: Warady BA, Alexander SR, Fine RN, Schaefer F, eds. Pediatric Dialysis. Dordrecht: Kluwer Academic Publishers; 2004:243–57.
48. Ramage IJ, Harvey E, Geary DF, Hebert D, Balfe JA, Balfe JW. Complications of gastrostomy feeding in children receiving peritoneal dialysis. Pediatr Nephrol 1999;13(3):249–52.
49. Ledermann SE, Spitz L, Moloney J. Gastrostomy feeding in infants and children on peritoneal dialysis. Pediatr Nephrol 2002;17:246–50.
50. Warady BA. Gastrostomy feedings in patients receiving peritoneal dialysis. Perit Dial Int 1999;19(3):204–6.
51. Warady BA, Kriley M, Belden B, Hellerstein S, Alan U. Nutritional and behavioural aspects of nasogastric feeding in infants receiving chronic peritoneal dialysis. Adv Perit Dial 1990;6:265–8.
52. Dello Strologo L, Principato F, Sinibaldi D, et al. Feeding dysfunction in infants with severe chronic renal failure after long-term nasogastric tube feeding. Pediatr Nephrol 1997;11(1):84–6.
53. Coleman JE, Watson AR. Growth posttransplantation in children previously treated with chronic dialysis and gastrostomy feeding. Adv Perit Dial 1998;14:269–73.
54. Kari JA, Gonzalez C, Ledermann SE, Shaw V, Rees L. Outcome and growth of infants with severe chronic renal failure. Kidney Int 2000;57(4):1681–7.
55. Krause I, Shamir R, Davidovits M, et al. Intradialytic parenteral nutrition in malnourished children treated with hemodialysis. J Ren Nutr 2002;12(1):55–9.
56. Goldstein SL, Baronette S, Vital Gambrell T, Currier H, Brewer ED. nPCR assessment and IDPN treatment of malnutrition in pediatric hemodialysis patients. Pediatr Nephrol 2002;17:531–4.
57. Orellana P, Juarez-Congelosi M, Goldstein S. Intradialytic parenteral nutrition and biochemical marker assessment for malnutrition in adolescent maintenance hemodialysis patients. J Ren Nutr 2005;15(3):312–7.

58. Cano F, Azocar M, Cavada G, Delucchi A, Marin V, Rodriguez E. Kt/V and nPNA in pediatric peritoneal dialysis: a clinical or a mathematical association? Pediatr Nephrol 2006;21:114–8.
59. Juarez-Congelosi M, Orellana P, Goldstein S. Normalized Protein Catabolic Rate Versus Serum Albumin as a Nutrition Status Marker in Pediatric Patients Receiving Hemodialysis. J Ren Nutr 2007;17(4):269–74.
60. Schaefer F, Klaus G, Mehls O. Peritoneal transport properties and dialysis dose affect growth and nutritional status in children on chronic peritoneal dialysis. J Am Soc Nephrol 1999;10(8):1786–92.
61. Cano F, Marin V, Azocar M, et al. Adequacy and nutrition in pediatric peritoneal dialysis. Adv Perit Dial 2003;19:273–8.
62. Canepa A, Verrina E, Perfumo F, et al. Value of intraperitoneal amino acids in children treated with chronic peritoneal dialysis. Perit Dial Int 1999;18(Suppl 2):S435–40.
63. Qamar I, Secker D, Levin L, Balfe J, Zlotkin S, Balfe J. Effects of amino acid dialysis compared to dextrose dialysis in children on continuous cycling peritoneal dialysis. Perit Dial Int 1999;19:237–47.
64. Querfeld U. Disturbance of lipid metabolism in children with chronic renal failure. Pediatr Nephrol 1993;7:749–57.
65. Saland J, Ginsberg H, Fisher E. Dyslipidemia in pediatric renal disease: epidemiology, pathophysiology, and management. Curr Opin Pediatr 2002;14:197–204.
66. National Kidney Foundation. KDOQI Clinical Practice Guidelines for Cardiovascular Disease in Dialysis Patients. Am J Kidney Dis 2005;45(4 Suppl 3):S1–154.
67. National Cholesterol Education Program. Report of the Expert Panel on Blood Cholesterol Levels in Children and Adolescents. Pediatrics 1992;89:495–584.
68. National Kidney Foundation. KDOQI Clinical Practice Guidelines for Managing Dyslipidemias in Chronic Kidney Disease. Am J Kidney Dis 2003;41(Suppl 3):S1–92.
69. Kriley M, Warady BA. Vitamin status of pediatric patients receiving long-term peritoneal dialysis. Am J Clin Nutr 1991;53:1476–9.
70. Warady BA, Kriley M, Alon U. Vitamin status of infants receiving long-term peritoneal dialysis. Pediatr Nephrol 1994;8:354–6.
71. American Academy of Pediatrics Section on Breastfeeding. Breastfeeding and the use of human milk. Policy Statement. Pediatrics 2005;115(2):496–506.
72. Canadian Paediatric Society, Dietitians of Canada, and Health Canada. Nutrition for Healthy Term Infants. Ottawa; 1998.
73. Kasiske BL, Vazquez MA, Harmon WE, et al. Recommendations for the outpatient surveillance of renal transplant recipients. J Am Soc Nephrol 2000;11(Suppl 15):S1–86.
74. National Kidney Foundation. KDOQI Clinical Practice Guidelines and Clinical Practice Recommendations for Anemia in Chronic Kidney Disease. Am J Kidney Dis 2006;47(5 Suppl 3):S1–S145.
75. Schroder CH, de Boer AW, Giesen AM, Monnens LA, Blom H. Treatment of hyperhomocysteinemia in children on dialysis by folic acid. Pediatr Nephrol 1999;13(7):583–5.
76. Merouani A, Lambert M, Delvin EE, Genest J, Jr., Robitaille P, Rozen R. Plasma homocysteine concentration in children with chronic renal failure. Pediatr Nephrol 2001;16(10):805–11.

77. Farid F, Faheem M, Heshmat N, Shaheen K, Saad S. Study of the homocysteine status in children with chronic renal failure. Am J Nephrol 2004;24:289–95.
78. Kang H, Lee B, Hahn H, et al. Reduction of plasma homocysteine by folic acid in children with chronic renal failure. Pediatr Nephrol 2002;17:511–4.
79. National Kidney Foundation. KDOQI Clinical Practice Guidelines for bone metabolism and disease in children with chronic kidney disease. Am J Kidney Dis 2005;46(4 Suppl 1):S1–121.
80. Institute of Medicine. Dietary Reference Intakes for Calcium, Phosphorous, Magnesium, Vitamin D, and Fluoride. Washington: National Academy of Sciences; 1997.
81. Rodriguez-Soriano J, Arant BS. Fluid and electrolyte imbalances in children with chronic renal failure. Am J Kidney Dis 1986;7:268–74.
82. Parekh RS, Flynn JT, Smoyer WE, et al. Improved growth in young children with severe chronic renal insufficiency who use specified nutritional therapy. J Am Soc Nephrol 2001;12(11):2418–26.
83. National Kidney Foundation. KDOQI Clinical Practice Guidelines on hypertension and antihypertensive agents in chronic kidney disease. Am J Kidney Dis 2004;43(5 Suppl 1):S1–290.
84. United States Department of Health and Human Services (HHS) and the Department of Agriculture (USDA). Dietary Guidelines for Americans; 2005.
85. Mattes RD, Donnelly D. Relative contributions of dietary sodium sources. J Am Coll Nutr 1991;10(4):383–93.
86. Bunchman TE, Wood EG, Schenck MH, Weaver KA, Klein BL, Lynch RE. Pretreatment of formula with sodium polystyrene sulfonate to reduce dietary potassium intake. Pediatr Nephrol 1991;5:29–32.
87. Rivard AL, Raup SM, Beilman GJ. Sodium polystyrene sulfonate used to reduce the potassium content of a high-protein enteral formula: a quantitative analysis. JPEN J Parenter Enteral Nutr 2004;28(2):76–8.
88. Harvey E, Secker D, Braj B, Picone G, Balfe JW. The team approach to the management of children on chronic peritoneal dialysis. Adv Ren Replace Ther 1996; 3(1):1–14.
89. Institute of Medicine. Dietary Reference Intakes for Energy, Carbohydrates, Fiber, Fat, Protein and Amino Acids (Macronutrients). Washington, DC: National Academy of Sciences; 2002.
90. Wilson, SE. Pediatric enteral feeding. In: Pediatric Nutrition, Theory and Practice. Grand, RJ, Sutphen, JL, et al., eds. Toronto, ON: Butterworth; 1987.
91. The Department of Clinical Dietetics. SickKids. Guidelines for the Administration of Enteral and Parenteral Nutrition in Paediatrics. Green G, Kean P, eds. 3rd edition. Toronto: 2007.

18 The Aging Adult

Julie Barboza

LEARNING OBJECTIVES

1. To describe the population shifts concerning older Americans and profile the psychosocial aspects and comorbid disease states of the aging adult.
2. To identify the normal physiological changes of aging and changes associated with chronic kidney disease.
3. To review risk factors associated with coupling of aging and chronic kidney disease.
4. To provide guidelines for nutritional assessment and management utilizing a geriatric focus.

Summary

The aging adult population will double to 71.5 million by the year 2010. Rates of chronic kidney disease (CKD) are increasing with a trend towards multiple comorbidities including congestive heart failure and diabetes mellitus (DM). Chronic diseases including CKD can leave the individual more frail than their peers and can increase morbidity and mortality risks. Based on these trends, a geriatric approach, meaning a focus on maintaining functional capacity and quality of life, may be beneficial for care of the older patient with CKD. This requires collaboration among the healthcare team members and referral to community resources.

Key Words: Geriatrics; frailty; older adults; aging; chronic kidney disease.

From: *Nutrition and Health: Nutrition in Kidney Disease*
Edited by: L. D. Byham-Gray, J. D. Burrowes, and G. M. Chertow

1. INTRODUCTION

With improvements in medical technology, decreased birth and mortality rates, the number of persons over the age of 65 is increasing *(1)*. By definition, any person over age 65 is defined as an older adult. An increased understanding of the needs of the older adult is necessary, as this population has increased by 9.5% or 3.2 million, from 1993 to 2003, to a total of 35.9 million people *(1)*. Such growth will continue until the year 2030. By this time, the population of older adults will have doubled to 71.5 million, with the minority group populations increasing from their 2000 level of 16.4% to 26.4% in 2030, and with women outnumbering men at 21 million versus 14.9 million *(1)*. The average life expectancy for individuals 65 years of age in 2003 was an average of an additional 18.2 years *(1)*.

2. PROFILE OF OLDER AMERICANS

Psychosocial factors impact the lives of older adults, as it can directly affect how they handle chronic disease. In 2004, the profile of older Americans revealed the following *(1)*:

Marital Status and Living Arrangements

- 71% of men are married versus 41% of women, and they live with their spouse.
- 19% of men live alone versus 40% of women.
- 10% of men have alternative living arrangements versus 19% of women.
- 11% of older Americans receive formal/informal personal care from others.

Economic Status

- 18.3% of men are employed versus 10.7% of women.
  The median income for men is $20,363 versus $11,845 for women.
- 12.5% of men live below the poverty level versus 7.3% of women.
- 35% of the total income of older Americans who own their home goes to housing costs versus 76% for those who rent.
- 12.7% of the total income of older Americans goes for out-of-pocket healthcare costs.

Older Americans have at least one chronic disease, with the highest percentages reported for heart disease, hypertension (HTN), arthritis, cancer or DM *(2)*. As the number of chronic diseases increases, the annual number of prescriptions increases by almost two-fold. In 2000, an older American with one to two chronic diseases had 23 prescriptions filled annually, while a person with three to four chronic diseases filled 42 prescriptions *(2)*. As a result, chronic disease can be deemed costly on many levels.

3. NORMAL PHYSIOLOGICAL CHANGES

To fully understand the unique challenges of treating the aging adult, it is instructive to review the common physiological changes that occur as a result of the normal aging process. Such information will provide a foundation for later when the aging adult diagnosed with chronic kidney disease (CKD) is discussed.

3.1. Sensory

Sensory changes can impact activities of daily living (ADLs), social interaction, ability to learn, as well as one's eating abilities and food choices. Decreased dentition and increased cavities are associated with aging *(3)*. Dental care is a high out-of-pocket healthcare expense and is often seen as not necessary or justifiable in the aging population. An increased risk for macular degeneration, a loss of central vision, is the most common cause of legal blindness in the aging adult *(3)*. This results in increased difficulty reading small print and the need for more lighting. Loss of high-frequency sensitivity, impairment of frequency discrimination, sound localization and speech discrimination impact hearing *(3)*. Older adults often do not complain of hearing loss, but of an inability to understand what is said *(3)*.

3.2. Skin

Skin changes can result in increased risk exposure to the aging adult. The epidermis, the outer layer, has decreased turnover resulting in poor wound healing, increased drying, decreased photo protection (i.e., the skin burns easier) and decreased vitamin D production *(4)*. Dermal, the under layer, has a 20% loss of thickness leading to transparent skin, increased risk of heat stroke and hypothermia due to decreased thermoregulation and decreased hypersensitivity reactions *(4)*. The subcutaneous layer atrophies impacting heat, insulation, caloric reserves and loss of the shock absorber feature. Nail thickening and decreased skin growth rates lead to an increased risk of skin trauma and heightened foot problems *(4)*.

3.3. Respiratory

Aging can result in increased susceptibility to pneumonia. There is decreased inflation, secretion expulsion, vital capacity with increased lung volume, and poor ciliary function, which leads to decreased cough reflex *(5)*. There is also decreased oxygen uptake by the cells, largely caused by an anterior–posterior chest wall increase with transverse decrease, which leads to kyphosis, calcification of the costal cartilage and reduced rib motility; equating to less deep breathing *(5)*.

3.4. Cardiovascular

Aging-related changes can lead to structural changes in the heart and the risk of systolic HTN. The heart muscle mass increases with left ventricle wall thickening and deposits of collagen, which alters pumping function; valve leaflets thicken and increase in circumference leading to an increased risk of valvular stenosis and regurgitation; arterial intimal thickening occurs and collagen in vessels leads to narrowing and decreased cardiac perfusion; the left ventricle systolic function is unchanged; and there is decreased compliance of left ventricle diastolic function with increased early diastolic left ventricle filling *(6)*. In the vessels, there is decreased compliance, leading to increased risk of systolic HTN *(6)*.

3.5. Gastrointestinal

Drug breakdown and metabolism can be affected by aging as well as vitamin and mineral metabolism. Decreased saliva flow, some impaired tongue movement during swallowing, decreased acid production and increased gastrin production, may alter the breakdown of drugs. An increased intestinal transit time and motility leads to a decreased absorption of calcium, vitamins D, B12, folate and iron. The gastrointestinal (GI) mucosa, musculature and transit time is decreased in colon, leading to an increased risk of diverticular disease. In the aging adult, the diminishing size of the liver along with decreased hepatic blood flow can alter drug metabolism *(7)*.

3.6. Musculoskeletal

A decrease in lean body mass, total body water and bone density, increased adipose tissue, and cartilage loss in weight bearing joints increases the risk of osteoarthritis and significant decreases in muscle strength and speed of contractility *(8)*.

3.7. The Kidneys

There are a variety of changes that occur in the kidneys which impact water and electrolyte exchange by direct changes in kidney structure and that potentially increase the risk of HTN, fluid overload and edema from sudden sodium load or sodium depletion with diuretics *(9)*. Older adults also have decreased thirst mechanisms, which can lead to dehydration and potential kidney injury *(9)*. A gradual decline in kidney weight starts at age 50, with the most marked decrease between the seventh and eighth decade of life. At this age, there is a progressive decline in the number of glomeruli due to sclerosis, an increase in interstitial fibrosis in the medulla resulting in decreased tubule length and a decrease in the proximal tubule volume of the individual nephrons. In addition, there is an increase in reduplication and focal thickening of both the glomerular and the tubular basement membrane *(10)*. There is also decreased renal plasma blood flow and diminished creatinine production; increased sodium conservation along with decreased sensitivity to antidiuretic hormone, diminished plasma renin and aldosterone levels, increased prevalence of hyperkalemia, impaired ammoniagenesis, impaired maximal urinary concentration and impaired water excretion *(10,11)*. Because the kidneys are an important site for drug metabolism and excretion in the body, certain medications should not be prescribed until after the individual's glomerular filtration rate (GFR) is determined so as to decrease the risk of drug toxicity.

4. AGING AND KIDNEY FUNCTION

Aging does not automatically mean progressive CKD. Based on the National Health and Nutrition Examination Survey (NHANES) III data, the mean GFR for individuals 60–69 years of age was 85 ml/min/1.73 m^2 and 75 ml/min/1.73 m^2 for persons 70 years of age, with 25% of all Americans over the age of 70 noted to have moderate to severe decreases in kidney function *(12)*. Serum creatinine levels can be inversely impacted by many chronic illnesses, such as cardiovascular disease (CVD), thereby affecting muscle mass, promoting malnutrition, inflammation, frailty and decreased kidney function, that is not obvious unless the GFR is calculated *(13,14)*. Women typically have lower muscle mass and hence lower serum creatinine values than men. All of these factors support the need to use GFR calculations as opposed to relying on the serum creatinine values to define CKD *(14,15)*.

5. AGING AND CKD RISK FACTORS

CKD typically arises in the presence of other chronic conditions. Diabetic nephropathy is the most common cause of CKD and accounts for 25% of all patients initiating renal replacement therapy (RRT) *(16)*. In the aging population, diabetes is a main contributor, but a growing number of persons, 17%, have CKD in presence of congestive heart failure (CHF) *(16)*. Twenty-two percent of persons over the age of 65 have all three comorbid conditions (CKD, DM, and CHF), which equates to a "triple threat" *(16)*. Such multiple risk factors are far less common in those under the age of 65 years; with less than 7% of the population experiencing these comorbidities *(16)*.

The primary focus of treatment for CKD is reversing or slowing the progression of CKD as well as the management of renal complications. Controlling the risk factors for CVD should be the major area of focus *(10,15)*. This should be done using established national guidelines such as The Seventh Report of the Joint National Committee on Prevention, Detection, Evaluation, and Treatment of High Blood Pressure (JNC 7) and the Kidney Disease Outcomes Quality Initiative (KDOQI) guidelines for blood pressure control *(17,18)*. Treatment for dyslipidemia should be based on the guidelines of KDOQI and the National Cholesterol Education Program Adult Treatment Plan III (NCEP ATP III) *(18,19)*. Persons with CKD should be considered in the highest risk category for CVD *(19)*. Because DM is a risk factor for CVD, establishing glycemic control with glycated hemoglobin (HgbA1C) <7% based on the American Diabetes Association guidelines is essential *(20)*.

The average age of dialysis recipients is increasing, with the median age for the incident population at 64.8 years with the majority choosing in-center hemodialysis. According to the U.S. Renal Data Systems, older persons in the 65- to 74-year age range have the highest prevalence rates among all other age groups, while the incidence and prevalence rates increase to 47–50% in those adults 75 years of age and older *(16)*. Three-fourths of all older persons begin RRT with five or more comorbid disease states. CVD, primarily CHF and ischemic heart disease, occur 2 years prior to RRT at the alarming rates of 90% in people with diabetes and nearly 70% in those without diabetes *(16)*.

5.1. Heightened Risks in Older Adults with CKD

Older adults with CKD tend to have multiple comorbidities, which coupled with the effect of aging can leave the individual more frail than their peers *(21)*. It is important to be aware of the increased risks in order to decrease morbidity and mortality in older persons *(22)*. Changes in the musculoskeletal system with aging as well as chronic disease can have a negative impact on physiological function and can lead to frailty, which is a precursor for disability *(15,22)*. Frailty is characterized by self-reported weakness, fatigue, balance or gait disorders, under-nutrition and an impaired ability to recover from insult leading to impaired homeostatic reserve, malnutrition, decreased mobility or functional loss related to diseases such as a cerebrovascular accident, Parkinson's disease or arthritis *(15)*. Shlipak *(22)* found that elderly persons with CKD were three times more likely to be frail than their counterparts with normal kidney function.

As the healthcare team looks at managing the risk factors for CVD, they need to be aware of the following areas: hypotension, increased falls risk, vascular dementia, malnutrition, anemia and bone disease.

5.1.1. Hypotension

There are multiple causes for hypotension including drug effects, aging in presence of DM or CHF/left ventricular hypertrophy, postprandial effects in presence of severe autonomic nerve system dysfunction, and/or the presence of anemia, as it can further impair cardiovascular function and hemodynamic response to hypotension. For those individuals receiving maintenance dialysis, there are added factors such as rapid fluid removal, inappropriate dry weights, and lower pre-treatment blood pressures that can lead to persons being under-dialyzed *(23,24)*.

5.1.2. Increased Falls Risk

Falls risk can be related to polypharmacy, impaired mobility, DM with impaired sensory neuropathy, autonomic neuropathy and/or visual impairment, orthostatic hypotension, renal osteodystrophy and/or the risk of osteoporosis *(15,23–25)*. In addition, depression can impact judgment and safety awareness, which can increase falls risk. Falls risk in elderly dialysis patients is high with a 4.4 relative risk of hip fracture and the resultant mortality at 1 year being 2.5 times greater; the fall rates in this group are close to the rates in nursing homes *(25,26)*. Falls can also lead to further decreases in mobility, which can directly impact ADLs and instrumental activities of daily living (IADLs) such as shopping and meal preparation.

5.1.3. Vascular Dementia

Vascular dementia increases in CVD with CKD, DM and increased age *(27,28)*. Cognitive impairment with vascular dementia is the leading cause of morbidity and mortality in CKD *(15)*. The individual with vascular dementia will have areas of intact function contrasted with others of profound impairment and task-specific disabilities based on the location of the vascular event in the brain *(15)*. The Kidney Disease Quality of Life Cognitive Function (KDQOL-CF) Subscale can be used to screen for cognitive function and help determine who needs further work-up *(29)*. Individuals on beta-blockers and those with higher educational backgrounds scored higher on the KDQOL-CF *(29)*.

5.1.4. Malnutrition

Older persons need to be screened and/or treated for malnutrition in stages 3–5 CKD, as it is strong predictor of morbidity and mortality *(10,15)*, with survival significantly influenced by age, level of serum albumin and pre-albumin level, body mass index (BMI) and presence of DM *(30)*. Based on validation studies of the KDQOL-CF tool, malnutrition is a strong predictor of cognitive impairment in addition to other factors such as stage 5 CKD, stroke history, presence of peripheral vascular disease (PVD), use of benzodiazepine, higher serum phosphorus levels and lower serum albumin *(29)*.

5.1.5. Anemia Management

Management of stages 3–5 CKD in older adults includes screening for and/or treating anemia *(10)*. Functional iron deficiency anemia has been shown to be less common in elderly persons on hemodialysis, though iron is not readily absorbed in the aging gut. It is important that the healthcare team evaluate causes of anemia, as older adults are at increased risk for folate and vitamin B12 deficiencies, diverticular disease and cancers and typically require lower doses of erythropoieten *(31)*.

5.1.6. Bone Disease Management

Because screening is occurring at earlier stages of CKD and is included as part of the National Kidney Foundation's Kidney Early Evaluation Program (KEEP), management of bone disease is addressed across the spectrum of kidney disease. Evaluation incorporates screening for and/or treating vitamin D deficiency/insufficiency, calcium and phosphorus metabolism and hyperparathyroidism. Coupled with the bone health risk factors associated with CKD are the additional risks for osteoporosis, vitamin D deficiency, decreased mobility and the lack of weight-bearing exercises in the older adult *(9,14,15)*. The most common type of renal osteodystrophy in the geriatric patients with CKD is osteitis fibrosa, with clinical manifestations including proximal muscle weakness in the lower extremities, vascular or soft tissue calcification, fractures or bone pain and intractable pruritis *(32)*. The symptoms can have significant impact on the individual's quality of life in terms of their ability to perform ADLs, IADLs and their freedom of movement.

6. GERIATRIC FOCUS

Because of the increasing age of the CKD population and the multiple comorbidities associated with this cohort, a geriatric approach to care may be beneficial. The goal of the geriatric focus is to maintain functional capacity and quality of life *(25,33,34)*. When assessing the nutritional status of older adults, the assessment should be based on their physiological and psychosocial state, keeping this geriatric focus in mind.

6.1. Barriers to Care

It is important to assess barriers to care including making observations about sensory deficits in order to maximize the teaching environment and to provide the right instructional medium. If the individual is hard of hearing, ask the person when it is easiest for them to hear. They will often note that it is hard to hear in crowds or with background noises, so choose a teaching site that is away from this environment. For those with visual deficits or of low-literacy levels, the educational materials need to be chosen carefully: the paper should not be glossy to minimize glare, the paragraphs should be double spaced and the font should be large (16 point) using plain block-type letters in black print *(35)*.

When assessing an individual's nutritional status, it is essential to be mindful of obstacles that may make the typical plan of care impractical. If the older person with CKD is not the primary person doing the shopping or cooking or if the person has cognitive impairment, have the spouse/caregiver present during the education session. There may be limited meal preparation/cooking skills due to role change, lack of caregiver, and change in living situation/economic status that may require the reliance on ready-to-eat, frozen entrees or community-sponsored meals/food banks. Listen for cues that suggest financial difficulties such as food avoidance, reporting medications are expensive, or history of the person not taking the medications as directed to make them last longer. In a caring, compassionate and non-threatening manner, any healthcare team member can ask about financial concerns or shortfalls, which impact the older adult's ability to appropriately take part in their self-management.

Be aware of an increased risk of depression, as it can also impact functional status. It can result in weight changes, changes in activity and sleep patterns, increased risk of falls, uncontrolled pain as well as decreased interest in following dietary modifications or taking medications as instructed. All healthcare team members have an obligation to communicate their observations/suspicions to other team members if they suspect the presence of depression.

6.2. Geriatric Assessment

Much of the focus when working with the older individual with CKD is on the development of a solid foundation for lifestyle modification to reduce risk factors for CVD. A firm foundation must be laid before it can be built upon, with specific individualized needs arising as a result of CKD progression. Much of this foundation is based on the JNC 7 criteria including weight management/loss to maintain BMI between 18.5 and 24.9 kg/m^2; assessment of metabolic syndrome/prediabetes based on the NCEP ATP III criteria; reduction in dietary sodium along with reduced total fat and saturated fat; aerobic physical activity for 30 or more minutes most days of the week; moderate alcohol consumption and smoking cessation if needed *(17)*.

The changes in the musculoskeletal system, as well as an increased risk of edema due to CKD, heart failure or PVD, can make the nutrition assessment challenging. With musculoskeletal changes and decreased activity levels, the basal metabolic rate and total energy intake level declines. The total energy intake level appears to be 30–35 kcal/kg of standard body weight/day for the older adult *(18)*. With the ultimate goal of achieving/maintaining an adequate BMI, laboratory values should be monitored to assess the risk for DM, weight and/or subjective global assessment should be monitored for changes, and nutritional interviews/diaries monitored for signs of deterioration in nutritional status *(18)*. Daily protein intake is generally less than that consumed by average population at levels of 0.6–0.75 g/kg of standard body weight/day for stages 3–4 CKD *(18)*. The challenge is to maintain a balance between symptoms of azotemia and protein malnutrition as aversions to protein occurs. Proteinuria also needs to be assessed as it serves many purposes. It substantiates the prognosis for kidney disease progression as well as premature CVD and serves as a guide to therapy, especially concerning the use of angiotensin converting enzyme inhibitors (ACEI) in patients experiencing profound proteinuria *(18)*.

A complete nutritional assessment involves a review of documented medical problems to determine how to satisfy the individual's nutritional needs and diet modifications. It is important to ask the older adult about any special diets or avoidance of any specific foods in order to get a sense of their understanding of the diet and disease needs. Such gentle probing is imperative because many older persons become overwhelmed as they try to compartmentalize all of the diet information they may have received over the years.

Dietary sodium modification is the cornerstone of managing the risk factors for CVD. Older adults tend to be more sensitive to changes in sodium load. It has a direct impact on blood pressure, fluid status and medication efficacy. Oftentimes if blood pressure or fluid status are not controlled, the first line of defense in the healthcare setting is medication adjustment/addition as opposed to an evaluation of dietary intake, cognition and behavior, or financial status. A dietary evaluation of an increased sodium and/or fluid load needs to be addressed; it is often a key contributor to uncontrolled fluid intake and blood pressure. A useful teaching technique that is easier for older persons to grasp is the rule of 200s, which breaks down the daily goal of <2000 mg of sodium by looking at individual foods choices *(36)*. The individual is taught to focus on two aspects of a food label, the serving size and the amount of sodium (mg) *(36)*. Often, the older adult gets confused by the percent listing next to the sodium on the label, so it is essential to stress amounts in milligrams. Encourage the person to choose food items that have 200 mg of sodium or less per serving. If the individual is willing to live with the allotted serving size or if they can increase serving sizes and maintain an intake less than 200 mg per serving, then it is a good food choice. Typically, patients are advised to choose frozen dinners with <600–700 mg sodium per day, if the use of frozen dinners is necessary *(36)*. Sometimes when dealing with older adults who are diagnosed with multiple chronic diseases later in life, balancing quality of life with quantity is important. The practitioner may need to assist the older individual in locating the lowest sodium version of their favorite foods or advising them when it is most safe to have these types of food.

Healthcare team members must stay abreast of the medications used to manage or treat the risk factors associated with CVD or CKD. It is important to have access to a good drug guide, whether it is an up-to-date book, on-line drug guide or a personal data assistant version. It is also important to be aware of food–drug interactions and adverse drug effects, many of which are GI-related causing increased risk of nausea, vomiting or constipation or which may impact function by causing somnolence or fatigue.

The assessment of nutritional status in older adults with CKD must include a review of systems such as oral health and its impact of food selections and GI symptoms or food aversions as it can be indicative of uremia. With constipation, be aware of foods or medicinal products used to correct it, as they can have a direct impact on overall health including lab data. Constipation can also decrease oral intake, increase fluid intake, result in use of higher potassium, foods such as prune juice, and result in elevated potassium from GI reabsorption.

It is important to encourage strengthening or at least maintenance of large muscle group strength as it can directly impact functional status. Resistance training has a beneficial impact on protein utilization and muscle mass in persons with CKD on low-protein diets (0.6 g/kg/day) by improving muscle mass and nutritional status as denoted by pre-albumin and function *(34)*. Regular exercise is an integral part of CKD management and quality of life. It should be encouraged and matched to the individual's current level of abilities. Some educational tools have been designed for older adults and incorporate seated techniques for maximum safety while working on upper and lower body stretching, strengthening and balance *(37)*.

7. SUMMARY

The older population is experiencing growth, which will continue through the year 2030. The aging adult has one or more chronic diseases. CKD is occurring in the presence of multiple comorbidities in this population. It is imperative that healthcare team members be aware of changes associated with aging and the increased risks that can occur in the CKD population as well as the psychosocial challenges that this population may experience. Because of the increasing age of the CKD population, a geriatric focus to the care of older patients with CKD may be beneficial. This requires the healthcare team to focus on maintaining functional capacity and quality of life. It should incorporate the goals of the older adult as they are guided in self-management and outlined in their coordinated plans of care. A geriatric focus requires a multi-disciplinary team approach with knowledge of both geriatrics and the available aging-related community resources.

8. CASE STUDY

MF is a 77-year-old Cape Verdean female with a history of DM, HTN, hyperlipidemia, coronary artery disease with two-vessel disease status post coronary-artery bypass grafting surgery, gout, osteoarthritis of hips and knees status post left-sided hip replacement, and atrial fibrillation (A Fib). She is a widow living alone in a second floor apartment on a very tight fixed income. She relies on public transportation, does her own shopping and cooking, cooks regularly for her youngest son who lives in the area. She walks with a cane, gets short of breath when carrying packages over 10 pounds and does not take part in any formal exercise program. Maria checks her blood sugar every morning, reporting them as good with a fasting range of 130–140 mg/dl.

MF reports following a no salt, no sugar diet. Based on her culture, rice and beans and tropical fruits are a staple. She uses Caribbean seasonings including adobo and sazon in her cooking. A typical day of eating includes one egg, two slices of white toast with butter or porridge, a banana and black coffee for breakfast. For lunch, she has homemade soup, fried plantains, bread and a cup of black tea. She may have an orange or mango as an afternoon snack. Her evening meal consists of fish, beef or pork cooked with seasonings, rice and beans or potatoes cooked in tomatoes with a variety of meats and a vegetable. In the evening, she may have a few cookies with a cup of tea.

Upon review of systems, MF reports the following areas of concern: some fuzziness in her vision, bouts of constipation for which she uses prune juice or a senna tea, bilateral leg swelling almost daily and bouts of knee pain which can make it difficult for her to leave her house, during those times she relies on her son to bring her food, mostly take-out or frozen entrees. She also reports some numbness and tingling sensation in her legs and difficulty keeping her balance. She denies falling but reports stumbling a few times. She brings her medicines which include glipizide 10 mg, furosemide 20 mg, lisinopril 5 mg, simavastatin 20 mg, aspirin 81 mg, colchicine 0.6 mg, warfarin 3 mg, atenolol 50 mg each day. The prescriptions were written for a 90-day supply, which should have run out by now, but most of the bottles are at least half full.

Some objective data includes height 64 inches, weight 188 pounds with no reported weight changes. Her BMI is calculated as 32.3. Her blood pressure sitting is 150/86 mmHG. Her available labs: HGBA1C 8.3%, BUN 27 mg/dl, Cr 1.4 mg/dl, Na+ 137 mEq/L, K+ 5.0 mEq/L, CO_2 20 mEq/l, RBC 3.82, Hgb 10.9 g/dl, Hct 32.4%, total cholesterol 199 mg/dl, TRG 185 mg/dl, HDL 36 mg/dl, LDL 128 mg/dl and INR 2.6.

8.1. Questions

1. What are Maria's cardiovascular (CV) risk factors?
2. What is Maria's estimated GFR (eGFR) and what is her CKD stage?
3. What are some areas of concern related to Maria's psychosocial situation?
4. What CKD complications are you concerned about based on the available information? What additional information would be useful?

5. The information you requested is provided. Based on this information, Ca++ 9.7 mg/dl, Phos 4.9 mg/dl, iPTH 252 pcg/ml, vitamin D 12.6 mg/dl, Microalb : Cr ratio 68 mg/dl, what additional areas of concern arise?

REFERENCES

1. A Profile of Older Americans: 2004. Administration on Aging. U.S. Department of Health and Human Services. (Accessed January 27, 2006, at: www.aoa.dhhs.gov/prof/Statisticstatus/profile/profiles.asp).
2. Older Americans 2004: Key Indicators of Well-Being. (Accessed January 27, 2006, at: http://agingstats.gov).
3. Burke MM, Laramie J. Sensory impairment. In: A Primary Care of the Older Adult: A Multidisciplinary Approach. St. Louis: Mosby, 2000: 439–452.
4. Burke MM, Laramie J. Aging Skin. In: A Primary Care of the Older Adult: A Multidisciplinary Approach. St. Louis: Mosby, 2000: 142–160.
5. Burke MM, Laramie J. Respiratory. In: A Primary Care of the Older Adult: A Multidisciplinary Approach. St. Louis: Mosby, 2000: 161–201.
6. Burke MM, Laramie J. The aging cardiovascular system. In: A Primary Care of the Older Adult: A Multidisciplinary Approach. St. Louis: Mosby, 2000: 202–253.
7. Burke MM, Laramie J. Gastrointestinal conditions. In: A Primary Care of the Older Adult: A Multidisciplinary Approach. St. Louis: Mosby, 2000: 254–268.
8. Burke MM, Laramie J. Musculoskeletal: common injuries. In: A Primary Care of the Older Adult: A Multidisciplinary Approach. St. Louis: Mosby, 2000: 302–353.
9. Bevan M. The older person with renal failure. Nurs Stand 2000; 14:48–54.
10. Yuanne FE, Anderson S. Kidney in aging. In: National Kidney Foundation. Primer on Kidney Diseases 4th Edition. Philadelphia: Saunders, 2005.
11. Hansberry MR, Whittier WL, Krause, MW. The elderly patient with chronic kidney disease. Adv Chronic Kidney Dis 2005; 12:71–77.
12. Coresh J, Astor BC, Greene T, et al. Prevalence of chronic kidney disease and decreased kidney function in adults US population. Third health and nutrition examination survey. Am J Kidney Dis 2003; 41:1–12.
13. Coresh J, Byrd-Holt D, Astor BC, et al. Chronic kidney disease awareness, prevalence, and trends among U.S. adults, 1999 to 2000. J Am Soc Nephrol 2005; 16: 180–188.
14. Stevens LA, Levey AS. Chronic kidney disease in the elderly – How to assess risk. N Engl J Med 2005; 20: 2122–2124.
15. Wiggins J. Core curriculum in nephrology: Geriatrics. Am J Kidney Dis 2005; 46:147–154.
16. U.S. Renal Data System, USRDS 2005 Annual Data Report: Atlas of End-Stage Renal Disease in the United States. Bethesda, Md. National Institutes of Health, National Institute of Diabetes and Digestive and Kidney Diseases, 2005. (Accessed January 27, 2006, at http://www.USRDS.org/adr.htm).
17. The Seventh Report of the Joint National Committee on Prevention, Detection, Evaluation, and Treatment of High Blood Pressure (JNC 7). Bethesda, Md.

National Institutes of Health, National Heart, Lung and Blood Institute, 2003. (Accessed January 27, 2006, at http://www.nhlbi.nih.gov/guidelines/hypertension/index.htm).
18. KDOQI Clinical practice guidelines. National Kidney Foundation, 2000–2004. (Accessed January 27, 2006 at http://www.kidney.org/professional/kdoqi/).
19. Third Report of the Expert Panel on Detection, Evaluation, and Treatment of High Blood Cholesterol in Adults (Adult Treatment Panel III). Bethesda, Md. Cholesterol Education Program, National Institutes of Health, National Heart, Lung and Blood Institute, 2002. (Accessed January 27, 2006, at http://www.nhlbi.nih.gov/guidelines/cholesterol/).
20. Clinical Practice Guidelines 2006. American Diabetes Association, 2005. (Accessed February 11, 2006, at http://care.diabetesjournals.org/content/vol29/suppl_1/).
21. Letourneau I, Ouimet D, Dumont M, Pichette V, Leblanc M. Renal replacement in endstage renal disease patients over 75 years old. Am J Nephrol 2003; 23:71–77.
22. Shlipak MG, Stehman-Breen C, Fried LF, et al. The presence of frailty in elderly persons with chronic renal insufficiency. Am J Kidney Dis 2004; 43:861–867.
23. Ismail N. Complications of hemodialysis in the elderly. (Accessed on February 25, 2005, at http://uptodateonline.com/application/topic/topic/topicText.asp?file=dialysis/20591).
24. Roberts RG, Kenny RA, Brierley EJ. Are elderly haemodialysis patients at risk of falls and postural hypotension? Int Urol Nephrol 2003; 35:415–421.
25. Sims RJ, Cassidy MJ, Masud T. The increasing number of older patients with renal disease. BMJ 2003; 327:463–464.
26. Desmet C, Beguin C, Swine C, Jadoul M. Falls in hemodialysis patients: Prospective study of incidence, risk factors and complications. Am J Kidney Dis 2005;45: 148–153.
27. Pereira AA, Weiner DE, Scott T, Sarnack MJ. Cognitive function in dialysis patients. Am J Kidney Dis 2005; 45:448–462.
28. Kurella M, Chertow GM, Luan J, Yaffe K. Cognitive impairment in chronic kidney disease. J Am Geri Soc 2004; 52:1863–1869.
29. Kurella M, Luan J, Yaffe K, Chertow GM. Validation of kidney disease quality of life (KDQOL) cognitive function subscale. Kidney Int 2004; 66:2361–2367.
30. Chauveau P, Combe C, Laville M, et al. Factors influencing survival in hemodialysis patients aged older than 75 years: 2.5 year outcome study. Am J Kidney Dis 2001; 37:997–1003.
31. Nicholas JC. A study of response of elderly patients with end-stage renal disease to epoetin alfa or beta. Drugs Aging 2004; 21:187–201.
32. Char R, Culp K. Renal osteodystrophy in older adults with end-stage renal disease. J Gerontol Nurs 2001; July:46–51.
33. Luke RG, Beck LH. Gerontologizing nephrology. J Am Soc Nephrol 1999; 10:1824–1827.
34. Castaneda C, Gordon P, Uhlin K, et al. Resistance training to counteract the catabolism of low-protein diet in patients with chronic renal insufficiency: A randomized, controlled trial. Ann Intern Med 2001; 135:965–976.
35. McHugh Sanner B. Are your written materials missing the mark? J Active Aging 2003; July/August:19–24.

36. Personal communication regarding Dietary Sodium Teaching Methods with Harvard Vanguard Medical Associates' Complex Chronic Care Dietitian, Marlene O'Donnell RD, in Boston MA on February 22, 2005.
37. Exercise for Life! A physical activity program for older adults San Francisco, Ca. Live Well, Live Long, American Society of Aging, 2004. (Accessed September 21, 2005, at http://www.asaging.org/cdc/module6/phase4/index.cfm).

2 Management of Other Disorders

19 Acute Renal Failure

Wilfred Druml

LEARNING OBJECTIVES

This chapter will enable the reader to:

1. Assess the complex metabolic environment in patients with acute renal failure.
2. Identify the impact of extracorporeal renal replacement therapies on nutrient balances.
3. Evaluate nutrient requirements in patients with acute renal failure.
4. Design and monitor nutrition support in acute renal failure patients to prevent evolution of metabolic complications.

Summary

Acute renal failure (ARF) is a clinical syndrome associated with a complex pattern of alterations of metabolism, with the induction of a prooxidative, proinflammatory and hypercatabolic state and impairment of immunocompetence. Metabolism and nutrient requirements are affected by the acutely uremic state per se, by the type and intensity of renal replacement therapy and by the underlying disease process leading to ARF and associated complications, respectively. A nutritional program for a patient with ARF must not only consider these complex metabolic alterations but also take into account the fact that because of the limited tolerance not only to fluids and electrolytes but also to various nutritional substrates and metabolic complications of nutritional support can frequently occur. Thus, nutrition support in ARF patients must be more closely monitored than in other disease states.

Key Words: Acute renal failure; metabolism; nutrient requirements; parenteral nutrition; enteral nutrition.

From: *Nutrition and Health: Nutrition in Kidney Disease*
Edited by: L. D. Byham-Gray, J. D. Burrowes, and G. M. Chertow

1. INTRODUCTION

Acute renal failure (ARF) is a proxidative, proinflammatory and hypermetabolic clinical syndrome, which continues to present one of the most challenging problems in clinical nutrition. Metabolism and nutrient requirements in patients with ARF are not only affected by the acutely uremic state per se, but also by the type and intensity of renal replacement therapy (RRT) and by the underlying disease process and associated complications, respectively. In designing a nutritional program for a patient with ARF, this complex metabolic environment has to be taken into consideration. Moreover, nutrient requirements may differ widely between individual patients and also during the course of disease, and nutrition therapy has to be coordinated with RRT. Finally, it must be noted that in ARF there is not only a limited tolerance to fluids and electrolytes but also to several nutritional substrates, a fact which increases the potential of inducing metabolic complications during nutritional support. Thus, in patients with ARF, nutrition therapy must be more closely monitored than in other disease states.

Nutritional support must be viewed as one of the cornerstones in the complex therapeutic strategies in the care of the often critically ill patients with ARF. ARF is associated with an excess "attributable mortality" which is tightly interrelated with the systemic immunologic and metabolic consequences of ARF and are aggravated by malnutrition *(1)*. The objectives of nutritional therapy, thus, exceed conventional goals, such as to maintain lean body mass, to stimulate immunocompetence and repair functions, but must also be aimed at mitigation of the inflammatory state, improvement of oxygen radical scavenging system and of endothelial functions.

Finally, it should be stressed in this context that adequate nutrition and maintaining a balanced metabolic environment such as normoglycemia and electrolyte homeostasis presents a crucial prerequisite both for prevention and therapy of ARF.

2. THE METABOLIC ENVIRONMENT AND NUTRITIONAL REQUIREMENTS IN PATIENTS WITH ARF

ARF presents a complication occurring in a broad spectrum of underlying pathologies. Clinical presentation of a patient with ARF, thus, may range from uncomplicated mono-organ failure in a noncatabolic patient to a critically ill patient with multiple-organ dysfunction syndrome (MODS). Thus, metabolic changes will be determined not only by ARF per se but also the underlying disease process

Table 1
Important Metabolic Abnormalities Induced by Acute Renal Failure

Activation of protein catabolism
Peripheral glucose intolerance/increased gluconeogenesis
Inhibition of lipolysis and altered fat clearance
Depletion of the antioxidant system
Induction of a proinflammatory state
Impairment of immunocompetence
Complex endocrine abnormalities: hyperparathyroidismus, insulin resistance, erythropoietin (EPO) resistance, resistance to growth factors, etc.

EPO, erythropoietin.

and/or additional complications and organ dysfunction *(2)*. Nevertheless, ARF in addition to the obvious effects on water, electrolyte and acid base metabolism affects all metabolic pathways of the body with specific alterations in protein and amino acid, carbohydrate and lipid metabolism and presents a proinflammatory, prooxidative and hypercatabolic state (Table 1). Moreover, the type and intensity of RRT exerts profound effect on nutrient balances.

The optimal intake of nutrients in patients with ARF is mainly influenced by the nature of the illness causing ARF, the extent of catabolism and type and frequency of RRT. Patients with ARF present a heterogeneous group of subjects with widely differing nutrient requirements, and it must be noted that these can considerably vary also during the course of disease.

2.1. Energy Metabolism and Energy Requirements

In patients with uncomplicated ARF, energy expenditure is within the range of healthy subjects. In the presence of sepsis or MODS, oxygen consumption may increase by approximately 25% *(3)*. Thus, energy expenditure in patients with ARF is rather determined by the underlying disease/associated complications and not by renal failure.

Intake of energy substrates during nutritional support should not exceed actual energy requirements. Complications, if any, from slightly underfeeding are less deleterious than from overfeeding. Increasing energy intake from 30 kcal/kg BW/day to 40 kcal/kg BW/day in patients with ARF merely increased the frequency of metabolic complications, such as hyperglycermia and hypertriglyceridemia, but had no beneficial effects *(4)*. Patients with ARF should receive

20–30 kcal/kg BW/day. Even in hypermetabolic conditions such as sepsis or MODS, energy expenditure rarely is higher than 130% of calculated basic energy expenditure and energy intake should not exceed 30 kcal/kg BW/day *(3)*.

2.2. Carbohydrate Metabolism

In many patients with ARF, hyperglycemia is present. The major cause of elevated blood glucose concentrations is peripheral insulin resistance *(5)*. Plasma insulin concentration is elevated, and insulin-stimulated glucose uptake by skeletal muscle is decreased. A second important feature of glucose metabolism in ARF is accelerated hepatic gluconeogenesis mainly from conversion of amino acids released during protein catabolism, which cannot be completely suppressed by exogenous glucose infusions *(6)*.

Hyperglycemia in the critically ill has been recognized as an important determinant in the evolution of complications such as infections and prognosis, but also for eliciting ARF *(7)*. Thus, normoglycemia must be strictly maintained during nutrition support also in patients with ARF *(8)*.

2.3. Lipid Metabolism

ARF is also associated with profound alterations of lipid metabolism. The triglyceride content of plasma lipoproteins, especially very low density lipoprotein (VLDL) and low density lipoprotein, is increased while total cholesterol and in particular high density lipoprotein cholesterol are decreased *(9)*. The major cause of lipid abnormalities in ARF is an impairment of lipolysis whereas oxidation of fatty acids is not affected *(10)*. Fat particles of artificial lipid emulsions for parenteral nutrition are degraded similarly to endogenous VLDL, and thus, impaired lipolysis in ARF retards also elimination of intravenously infused lipid particles *(9)*. Moreover, intestinal absorption of lipids is retarded in renal failure.

2.4. Protein and Amino Acid Metabolism/Protein Requirements in ARF

ARF is associated with an activation of protein catabolism with excessive release of amino acids from skeletal muscle and sustained negative nitrogen balance *(11,12)*. Muscular protein degradation and amino acid oxidation is stimulated, hepatic extraction of amino acids from the circulation, gluconeogenesis and ureagenesis are increased. In the liver, also protein synthesis and secretion of acute phase proteins is stimulated. As a consequence, imbalances in amino acid pools in plasma and in the intracellular compartment occur in ARF, and the utilization of amino acids is altered, and the clearance of many amino acids is enhanced *(13)*.

The causes of hypercatabolism in ARF are complex and manifold. They present a combination of unspecific mechanisms induced by the acute disease process and underlying illness/associated complications, with specific effects induced by the acute loss of renal function and finally also by the type and intensity of RRT (Table 1) *(12)*. The dominating mechanism is the stimulation of hepatic gluconeogenesis from amino acids, which, in contrast to healthy subjects but also to patients with CKD, can be decreased but not halted by exogenous substrate supply *(6)*. A major stimulus of muscle protein catabolism in ARF is insulin resistance. Moreover, acidosis was identified as an important factor in mediating muscular protein breakdown *(14)*.

Several additional catabolic factors are operative in ARF. The release of inflammatory mediators such as tumor necrosis factor-α, depletion of antioxidative factors, the secretion of catabolic hormones, hyperparathyroidism, suppression and/or decreased sensitivity of growth factors, the release of proteases from activated leucocytes all can stimulate protein breakdown. Moreover, RRT causes a loss of amino acids and protein and can stimulate protein catabolism.

Last but not least, inadequate nutrition contributes to the loss of lean body mass in ARF. Starvation can augment the catabolic response of ARF, and malnutrition was identified as a major determinant of morbidity and mortality in ARF *(15)*.

2.5. Amino Acid/Protein Requirements in Patients with ARF

The most controversial question in nutrition support in ARF patients concerns the optimal intake of amino acids or protein, and few studies only have attempted to define requirements *(2)*. In noncatabolic patients and during the polyuric recovery phase of AKF, a protein intake of 1–1.3 g/kg BW/day was required to achieve a positive nitrogen balance.

In critically ill patients with ARF on continuous RRT (CRRT), protein catabolic rate accounted for 1.5–1.7 g/kg BW/day and an amino acid/protein intake of 1.4–1.7 g/kg BW/day was recommended *(16)*. Recently, some authors have suggested an even higher amino acid intake of up to 2.5 g/kg BW/day *(17)*. However, there are no proven advantages of such excessive intakes which can increase uremic toxicity and provoke metabolic complications *(18)*.

Thus, unless renal insufficiency will be brief and there is no associated catabolic illness, the intake of protein or amino acids should not be lower than 1 g/kg BW/day. Catabolic patients with ARF should receive 1.2–1.5 g (maximally 1.7 g) protein/amino acids/kg BW/day. These calculations include amino acid/protein losses induced by intermittent hemodialysis, CRRT or by peritoneal dialysis.

2.6. Metabolism and Requirements of Micronutrients

Serum levels of water-soluble vitamins usually are low in ARF patients mainly because of losses induced by RRT, and thus, requirements are increased in patients with ARF. An exception is ascorbic acid: as a precursor of oxalic acid, the intake should be kept below 250 mg/day because any excessive supply may cause secondary oxalosis and may even precipitate ARF.

As in CKD, vitamin D activation and plasma levels of 25(OH) vitamin D3 and 1, 25-(OH) vitamin D_3 are severely depressed in ARF *(19)*. In contrast to CKD, serum levels of vitamin A and vitamin E are decreased in patients with ARF; vitamin K levels, however, are normal or even elevated.

Many of the findings on trace element metabolism in ARF represent unspecific alterations within the spectrum of "acute phase reaction" and do not necessarily reflect specific effects induced by ARF. However, selenium loss is increased during CRRT *(20)*, and selenium concentrations in plasma and erythrocytes are profoundly decreased in patients with ARF *(21)*. In critically ill patients, selenium replacement improved clinical outcome and reduced the incidence of ARF requiring RRT *(22)*.

Several micronutrients are important components of the organism's defense mechanisms against oxygen free radical-induced injury. A profound depression in antioxidant status has been documented in patients with ARF, and an adequate supplementation of micronutrients must be observed *(21)*.

3. ELECTROLYTES

Derangements in electrolyte balance in patients with ARF are affected by a broad spectrum of factors in addition to renal failure including the type of underlying disease and degree of hypercatabolism, type and intensity of RRT, drug therapy and also the timing, type and composition of nutritional support *(2)*.

Electrolyte requirements do not only vary considerably between patients, but it must be noted that these can fundamentally change during the course of the disease. In non-oliguric patients, in subjects on

CRRT, and during the polyuric phase of ARF, electrolyte requirements can be considerably increased. Nutrition support, especially parenteral nutrition with low electrolyte contents, can induce hypophosphatemia and hypokalemia, respectively ("refeeding syndrome"). Thus, even more than with other substrates, electrolyte requirements have to be evaluated in patients with ARF on a day-to-day basis and intakes have to be adjusted frequently.

4. METABOLIC IMPACT OF EXTRACORPORAL THERAPY

The impact of RRT on metabolism is manifold. Several water-soluble substances, such as amino acids, vitamins and carnitine, are lost during hemodialysis. Protein catabolism is caused not only by amino

Table 2
Metabolic Effects of Renal Replacement Therapy in Acute Renal Failure

Intermittent hemodialysis
Loss of water-soluble molecules:
Amino acids
Water-soluble vitamins
L-carnitine etc.
Activation of protein catabolism:
Loss of amino acids
Loss of proteins and blood
Induction of cytokine release (TNF-α, etc.)
Inhibition of protein synthesis
Increase in ROS production
Continuous renal replacement therapy
Heat loss
Excessive load of substrates:
Lactate, citrate, glucose, etc.
Loss of nutrients:
Amino acids, vitamins, selenium, etc.
Loss of electrolytes: phosphate, magnesium
Elimination of short chain proteins:
Hormones, mediators, but also of albumin
Metabolic consequences of bioincompatiblity:
Induction/ activation of mediator-cascades, of an inflammatory reaction, stimulation of protein catabolism

TNF-$^{\alpha}$, tumor necrosis factor α; Ros, reactive oxygen species.

acid losses, but also by activation of protein breakdown. Moreover, it has been suggested that generation of reactive oxygen species is augmented during hemodialysis.

Recently, CRRT such as continuous hemofiltration and/or continuous hemodialysis have gained wide application in the management of critically ill patients with ARF. The metabolic consequences of these modalities may become especially relevant because of the continuous mode of therapy and associated high fluid turnover of up to more than 50 L/day (Table 2) *(23)*.

5. NUTRIENT ADMINISTRATION

5.1. Patient Selection

In clinical practice, it has proved useful to distinguish three groups of ARF patients based on the extent of protein catabolism/severity of the underlying disease and resulting levels of dietary requirements (Table 3) *(2)*.

1. Group I includes patients without excess catabolism. ARF is usually caused by nephrotoxins. These patients rarely will present major nutritional problems, and in most cases, they can be fed orally and the prognosis is excellent.
2. Group II consists of patients with moderate hypercatabolism, who frequently suffer from complicating infections such as peritonitis. Tube feeding and/or intravenous nutrition are generally required, and dialysis/CRRT can become necessary.
3. Group III are patients in whom ARF occurs in association with severe trauma, burns, or sepsis in the context of MODS. Treatment is complex and includes enteral and/or parenteral nutrition, hemodialysis or CRRT plus blood pressure and ventilatory support. In these patients in addition to the severity of underlying illness, ARF per se is a major independent contributor to the poor prognosis *(1)*.

5.2. Oral Feeding

In all patients who can tolerate them, oral feedings should be used, but usually, this will be restricted to non-hypercatabolic patients (Group I). Initially, 40 g of high quality protein per day is given (0.6 g/kg BW/day) and subsequently is gradually increased to 0.8 g/kg BW/day as long as blood urea nitrogen (BUN) remains below 100 mg/dl. For patients treated by hemodialysis/peritoneal dialysis, protein intake should be increased to 1.0–1.4 g/kg/day. A supplement of water-soluble vitamins is recommended.

Table 3
Patient Classification and Nutrient Requirements in Patients with Acute Renal Failure

	Extent of catabolism		
	Mild	*Moderate*	*Severe*
Excess urea appearance (above N intake)	>5 g	5–10 g	>10 g
Clinical setting (examples)	Drug toxicity	Elective surgery ± infection	Sepsis, ARDS MODS
Mortality	20%	60%	>80%
Dialysis/CRRT–frequency	Rare	As needed	Frequent
Route of nutrient administration	Oral	Enteral and/or parenteral	Enteral and/or parenteral
Energy recommendations (kcal/kg BW/day)	20–25	20–30	25–35
Energy substrates	Glucose	Glucose+fat	Glucose+fat
Glucose (g/kg BW/day)	3.0–5.0	3.0–5.0	3.0–5.0
Fat (g/kg BW/day)	–	0.6–1.0	0.8–1.2
Amino acids/ protein (g/kg/day)	0.6–1.0 EAA+NEAA	1.0–1.4 EAA+NEAA	1.2–1.5 (1.7) EAA+NEAA
Nutrients used oral/enteral parenteral	Food	Enteral formulas Glucose 50–70% Lipids 10–20% AA 6.5–10%[a] Micronutrients[b]	Enteral formulas Glucose 50–70% Lipids 10–20% AA 6.5–10%[a] Micronutrients[b]

ARDS, acute respiratory distress syndrome; CRRT, continuous renal replacement therapy; EAA, essential amino acids, MODS, Multiple organ dysfunction syndrome, NEAA, non-essential amino acids.

[a] AA, amino acid solution: general or special "nephro" solutions (EAA + specific NEAA).

[b] Multi-vitamin and multi-trace element preparations.

5.3. *Enteral Nutrition (Tube Feeding)*

In the past, parenteral nutrition was the preferred route of nutritional support in patients with ARF. During recent years, enteral nutrition has become the primary type of nutritional support for all critically ill patients and also for patients with ARF. Even small amounts of luminally provided diets can help to support intestinal defense functions, support the intestinal immune system and reduce infectious complications. Moreover, enteral nutrition might exert specific advantages in ARF. In experimental ARF, enteral nutrition can augment renal plasma flow and improve renal function *(24)*. Enteral nutrition was a factor associated with an improved prognosis in critically ill patients with ARF *(17,25)*.

Nevertheless, gastrointestinal motility is impaired in many patients with ARF, and frequently, it is not possible to meet requirements by the enteral route alone, and parenteral nutrition at least supplementary and/or temporarily may become necessary *(26)*.

Unfortunately, few systematic studies on enteral nutrition have been conducted in patients with ARF. In the largest study to date, nutritional effects, feasibility and tolerance of enteral nutrition using either a conventional diet or a preparation adapted to the metabolic needs of hemodialysis patients was performed in 182 patients with ARF *(27)*. Side effects of enteral nutrient supply were higher and the amount of nutrients provided was lower in patients with ARF as compared with normal renal function, but in general, enteral nutrition was well tolerated, safe and effective.

5.3.1. Enteral Formulas

Essentially, three types of enteral formulas can be used in ARF patients, but none of these diets have been developed specifically for nutrition in ARF *(28)*.

1. Elemental powder diets: These formulas conform to the concept of a low protein diet supplemented with essential amino acids (EAA) in CKD. These diets are not complete and should be replaced by more complete ready-to-use liquid products.
2. Standard enteral formulas designed for non-uremic patients: In many intensive care patients with ARF, these standard enteral formulas are used. Disadvantages are the amount and type of protein, the high content of electrolytes. Whether diets enriched in specific substrates such as glutamine, arginine, nucleotides or omega-3-fatty acids ("Immunonutrition") might exert beneficial effects also in patients with ARF remains to be shown.

3. Specific enteral formulas adapted to the metabolic alterations of uremia: Ready-to-use liquid diets adapted to nutrient requirements of patients on regular hemodialysis therapy for the moment present the most reasonable approach in enteral nutrition of hypercatabolic intensive care patients with ARF *(27)*. However, these preparations generally are lower in potassium, magnesium and phosphate, and their use may create refeeding electrolyte deficiencies.

5.4. Parenteral Nutrition

In the critically ill patient with ARF, it is frequently impossible to cover nutrient requirements exclusively by the enteral route alone, and a supplementary or even total parenteral nutrition may become necessary.

5.4.1. Composition of Parenteral Nutrition Solution

5.4.1.1. Glucose. Glucose should be used as the main energy substrate. In contrast to earlier recommendations, glucose intake must be restricted to <3–5 g/kg BW/day as higher intakes are not used for energy but will promote lipogenesis with fatty infiltration of the liver, excessive carbon dioxide production and impair immunocompetence. Glucose tolerance is decreased in ARF, and infusion of insulin is frequently necessary to maintain normoglycemia. Moreover, insulin requirements are approximately 25% higher in parenteral than during enteral nutrition. Strictly maintaining normoglycemia by "intensive insulin therapy" must be observed to prevent the evolution of ARF and other complications *(29)*. By limiting energy intake and providing a portion of the energy by lipid emulsions, the risk of developing hyperglycemia can be reduced.

5.4.1.2. Lipid Emulsions. Changes in lipid metabolism associated with ARF should not prevent the use of lipid emulsions, but the amount infused must be adjusted to the patient's capacity to utilize lipids, and plasma triglyceride concentrations have to be monitored regularly. Usually, 1 g fat/kg BW/day will not increase plasma triglycerides substantially. Whether lipid emulsions with a lower content of polyunsaturated fatty acids (replacing soybean oil by olive oil and/or fish oil and/or medium chain triglycerides) to reduce potential proinflammatory side effects should be preferred in patients with ARF remains to be shown *(30)*.

5.4.1.3. Amino Acid Solutions. Three types of amino acid solutions for parenteral nutrition in patients with ARF have been used: exclusively EAA, standard solutions of EAA plus nonessential amino acid (NEAA) and specifically designed "nephro" solutions of adapted proportions of EAA and specific NEAA that might become "conditionally essential" in ARF.

The use of solutions of EAA alone was based on principles established for treating CKD patients with a low protein diet and an EAA supplement. These solutions should no longer be used because they are incomplete (several amino acids designated as NEAA such as histidine, arginine, tyrosine, serine and cysteine may become indispensable in ARF patients) and have an unbalanced composition *(2)*.

Thus, solutions including both EAA and NEAA in standard proportions or in special proportions ("nephro" solutions) should be used in patients with ARF. Because of the low water solubility of tyrosine, dipeptides containing tyrosine (such as glycyl-tyrosine) are contained in modern "nephro" solutions *(2,31)*.

Table 4
"Parenteral Nutrition in Acute Renal Failue: "Renal Failure Fluid" (All-in-One Solution)[a]

Component	*Quantity*	*Remarks*
Glucose (40–70%)	500 ml	In the presence of severe insulin resistance switch to D30 W
Fat emulsion (10–20%)	500 ml	Start with 10%, switch to 20% if triglycerides are <350 mg/dl
Amino acids (6.5–10%)	500 ml	General or special "nephro" amino acid solutions including EAA and NEAA
Water soluble vitamins[b]	2 × RDA	Limit vitamin C intake <250 mg/day
Fat soluble vitamins[b]	RDA	Increased requirements of vitamin E
Trace elements[b]	RDA	Plus selenium 100–300 μg/day
Electrolytes	As required	CAVH: hypophosphatemia or hypokalemia after initiation of TPN
Insulin	As required	Added directly to the solution or given separately

EAA,essential amino acid; NEAA, nonessential amino acid; RDA, recommended daily allowance; TPN, total parenteral nutrition.

[a] "All-in-one solution" with all components contained in a single bag. Infusion rate, initially 50% of requirements, to be increased over a period of 3 days to satisfy requirements.

[b] Combination products containing the RDA.

Recently, it was suggested that glutamine exerts important metabolic functions in catabolic patients. Glutamine exerts beneficial effects on renal function and can improve survival in critically ill patients *(32)*. In a post hoc analysis of this study, this effect was most pronounced in patients with ARF (4/24 survivors without, 14/23 with glutamine, $p < 0.02$). Because free glutamine is not stable in aqueous solutions, glutamine-containing dipeptides are used as glutamine source in parenteral nutrition.

5.4.1.4. "Renal Failure Fluid." Standard solutions with amino acids, glucose and lipids contained in a single bag are available ("all-in-one" solutions) (Table 4). As required, vitamins, trace elements and electrolytes can be added to these solutions. To ensure maximal nutrient utilization and prevent metabolic derangements, the infusion must be started at a low rate (providing about 50% of requirements) and gradually increased over several days.

6. COMPLICATIONS OF NUTRITIONAL SUPPORT

Technical problems and infectious complications originating from central venous catheters or enteral feeding tubes, metabolic complications of artificial nutrition and gastrointestinal side effects of enteral nutrition are similar in ARF patients and in nonuremic subjects. However, both metabolic and gastrointestinal complications are far more pronounced and occur more frequently in ARF because the utilization of various nutrients is impaired and the tolerance to electrolytes and volume load is limited, and moreover, gastrointestinal motility is impaired in ARF. By gradually increasing the infusion rate and avoiding any infusion above requirements, many side effects can be minimized. In patients with ARF, nutrition therapy must be more closely monitored than in other disease states.

7. SUMMARY

Nutrition support must be viewed as a cornerstone in the treatment of patients with ARF, a type of metabolic intervention which must be viewed together with RRT and fluid and electrolyte management, respectively.

In a patient with ARF, it is not the impairment of renal function that determines the decision to initiate nutrition support but as in any other patient groups the nutritional state, type and severity of underlying diseases and associated complications, and the extent of hypercatabolism.

However, in a patient who has acquired ARF and needs nutrition support, the nutritional regimen must take into account the multiple metabolic consequences of ARF, of RRT and of the underlying disease process and/or associated complications.

Modified by the clinical context in which it is occurring, ARF presents a catabolic, prooxidative and proinflammatory syndrome with pronounced increases in micronutrient requirements.

Modern CRRT, because of the continuous mode of therapy and the associated high fluid turnover, exerts a massive impact on electrolyte and nutrient balances.

Nutritional strategies in ARF differ fundamentally from those in CKD. Nutritional regimens that satisfy minimal requirements in metabolically stable CKD are not sufficient for the mostly hypermetabolic patients with ARF.

In the critically ill patient with ARF, nutrition support must be initiated early and both qualitatively and quantitatively sufficient *(33,34)*.

However, because of the potential of creating serious complications, also any type of "hyperalimentation" with substrate infusions exceeding actual requirements should be carefully avoided in ARF patients.

Enteral nutrition has become the preferred type of nutrition support also in patients with ARF. Nevertheless, many patients have severe limitations to enteral nutrition and will require supplementary or even total parenteral nutrition.

Especially in patients with ARF, nutrition therapy must leave a merely quantitatively oriented approach in covering nitrogen and energy requirements, but must move toward a more qualitative type of metabolic intervention aimed at modulating the inflammatory state, the oxygen radical scavenger system, immunocompetence and endothelial functions and taking advantage of specific pharmacologic effects of various nutrients.

ARF is associated with a high "attributable" mortality, patients not only die with but also from ARF. A reduction of the distressingly high mortality of patients with ARF will also depend on further improvements in metabolic care.

REFERENCES

1. Druml W: Acute renal failure is not a "cute" renal failure! Intensive Care Med 2004; 30(10): 1886–90.
2. Druml W: Nutritional support in patients with acute renal failure. In: Molitoris B, Finn W, eds. Acute Renal Failure. A Companion to Brenner & Rector's The Kidney. Philadelphia: WB Saunders Company, 2001.

3. Schneeweiss B, Graninger W, Stockenhuber F, et al.: Energy metabolism in acute and chronic renal failure. Am J Clin Nutr 1990; 52(4): 596–601.
4. Fiaccadori E, Maggiore U, Rotelli C, et al.: Effects of different energy intakes on nitrogen balance in patients with acute renal failure: a pilot study. Nephrol Dial Transplant 2005; 20(9): 1976–80.
5. May RC, Clark AS, Goheer MA, Mitch WE: Specific defects in insulin-mediated muscle metabolism in acute uremia. Kidney Int 1985; 28(3): 490–7.
6. Cianciaruso B, Bellizzi V, Napoli R, Sacca L, Kopple JD: Hepatic uptake and release of glucose, lactate, and amino acids in acutely uremic dogs. Metabolism 1991; 40(3): 261–9.
7. van den Berghe G, Wouters P, Weekers F, et al.: Intensive insulin therapy in the critically ill patients. N Engl J Med 2001; 345(19): 1359–67.
8. Van den Berghe G, Wouters PJ, Bouillon R, et al.: Outcome benefit of intensive insulin therapy in the critically ill: insulin dose versus glycemic control. Crit Care Med 2003; 31(2): 359–66.
9. Druml W, Fischer M, Sertl S, Schneeweiss B, Lenz K, Widhalm K: Fat elimination in acute renal failure: long-chain vs medium-chain triglycerides. Am J Clin Nutr 1992; 55(2): 468–72.
10. Druml W, Zechner R, Magometschnigg D, et al.: Post-heparin lipolytic activity in acute renal failure. Clin Nephrol 1985; 23(6): 289–93.
11. Price SR, Reaich D, Marinovic AC, et al.: Mechanisms contributing to muscle-wasting in acute uremia: activation of amino acid catabolism. J Am Soc Nephrol 1998; 9(3): 439–43.
12. Druml W: Protein metabolism in acute renal failure. Miner Electrolyte Metab 1998; 24(1): 47–54.
13. Druml W, Fischer M, Liebisch B, Lenz K, Roth E: Elimination of amino acids in renal failure. Am J Clin Nutr 1994; 60(3): 418–23.
14. Mitch WE: Robert H Herman Memorial Award in Clinical Nutrition Lecture, 1997. Mechanisms causing loss of lean body mass in kidney disease. Am J Clin Nutr 1998; 67(3): 359–66.
15. Fiaccadori E, Lombardi M, Leonardi S, Rotelli CF, Tortorella G, Borghetti A: Prevalence and clinical outcome associated with preexisting malnutrition in acute renal failure: a prospective cohort study. J Am Soc Nephrol 1999; 10(3): 581–93.
16. Chima CS, Meyer L, Hummell AC, et al.: Protein catabolic rate in patients with acute renal failure on continuous arteriovenous hemofiltration and total parenteral nutrition. J Am Soc Nephrol 1993; 3(8): 1516–21.
17. Scheinkestel CD, Kar L, Marshall K, et al.: Prospective randomized trial to assess caloric and protein needs of critically Ill, anuric, ventilated patients requiring continuous renal replacement therapy. Nutrition 2003; 19(11–12): 909–16.
18. Macias WL, Alaka KJ, Murphy MH, Miller ME, Clark WR, Mueller BA: Impact of the nutritional regimen on protein catabolism and nitrogen balance in patients with acute renal failure. JPEN J Parenter Enteral Nutr 1996; 20(1): 56–62.
19. Druml W, Schwarzenhofer M, Apsner R, Horl WH: Fat-soluble vitamins in patients with acute renal failure. Miner Electrolyte Metab 1998; 24(4): 220–6.
20. Berger MM, Shenkin A, Revelly JP, et al.: Copper, selenium, zinc, and thiamine balances during continuous venovenous hemodiafiltration in critically ill patients. Am J Clin Nutr 2004; 80(2): 410–6.

21. Metnitz GH, Fischer M, Bartens C, Steltzer H, Lang T, Druml W: Impact of acute renal failure on antioxidant status in multiple organ failure. Acta Anaesthesiol Scand 2000; 44(3): 236–40.
22. Angstwurm MW, Schottdorf J, Schopohl J, Gaertner R: Selenium replacement in patients with severe systemic inflammatory response syndrome improves clinical outcome. Crit Care Med 1999; 27(9): 1807–13.
23. Druml W: Metabolic aspects of continuous renal replacement therapies. Kidney Int Suppl 1999; 72: S56–61.
24. Mouser JF, Hak EB, Kuhl DA, Dickerson RN, Gaber LW, Hak LJ: Recovery from ischemic acute renal failure is improved with enteral compared with parenteral nutrition. Crit Care Med 1997; 25(10): 1748–54.
25. Metnitz PG, Krenn CG, Steltzer H, et al.: Effect of acute renal failure requiring renal replacement therapy on outcome in critically ill patients. Crit Care Med 2002; 30(9): 2051–8.
26. Lefebvre HP, Ferre JP, Watson AD, et al.: Small bowel motility and colonic transit are altered in dogs with moderate renal failure. Am J Physiol Regul Integr Comp Physiol 2001; 281(1): R230–8.
27. Fiaccadori E, Maggiore U, Giacosa R, et al.: Enteral nutrition in patients with acute renal failure. Kidney Int 2004; 65(3): 999–1008.
28. Druml WM, Mitch WE: Enteral nutrition in renal disease. In: RH R, ed. Clinical Nutrition: Enteral and Tube Feeding. Philadelphia: Elsevier, Saunders, 2004.
29. Van den Berghe G, Wilmer A, Hermans G, et al.: Intensive insulin therapy in the medical ICU. N Engl J Med 2006; 354(5): 449–61.
30. Mayer K, Schaefer MB, Seeger W: Fish oil in the critically ill: from experimental to clinical data. Curr Opin Clin Nutr Metab Care 2006; 9(2): 140–8.
31. Smolle KH, Kaufmann P, Fleck S, et al.: Influence of a novel amino acids solution (enriched with the dipeptide glycyl-tyrosine) on plasma amino acid concentration of patients with acute renal failure. Clin Nutr 1997; 16: 239–46.
32. Griffiths RD, Jones C, Palmer TE: Six-month outcome of critically ill patients given glutamine-supplemented parenteral nutrition. Nutrition 1997; 13(4): 295–302.
33. Artinian V, Krayem H, DiGiovine B: Effects of early enteral feeding on the outcome of critically ill mechanically ventilated medical patients. Chest 2006; 129(4): 960–7.
34. Dvir D, Cohen J, Singer P. Computerized energy balance and complications in critically ill patients: an observational study. Clin Nutr 2006; 25(1): 37–44.

20 Nephrotic Syndrome

Jane Y. Yeun
and George A. Kaysen

LEARNING OBJECTIVES

1. Identify the nephrotic syndrome, its causes, and its complications.
2. Describe the pharmacologic management of nephrotic syndrome.
3. Discuss the nutritional management of nephrotic syndrome.

Summary

The nephrotic syndrome is defined as proteinuria >3.5 g/dL, hypoalbuminemia, edema, hyperlipidemia, and lipiduria. While there are a myriad of causes, the complications result from the severity of the proteinuria. Complications include atherosclerosis, vascular thrombosis, anasarca, infection, nutritional depletion, and progressive kidney injury. Reducing the proteinuria is critical. When specific therapy targeting the underlying etiology fails, blocking the renin-angiotensin system will reduce the proteinuria, enhanced by concomitant moderate protein restriction. Plant sources of protein may offer additional benefit in reducing proteinuria and hyperlipidemia. Vitamin D, iron, and zinc deficiency may occur due to urinary loss of carrier proteins and are treated with appropriate dietary supplementation.

Key Words: Nephrotic syndrome; complications; treatment; protein restriction; renin-angiotensin blockade; dietary treatment.

From: *Nutrition and Health: Nutrition in Kidney Disease*
Edited by: L. D. Byham-Gray, J. D. Burrowes, and G. M. Chertow

Table 1
Manifestations of Nephrotic Syndrome

3.5 g Proteinuria 3.5g/1.73 m^2/day
Hypoalbuminemia
Edema
Hyperlipidemia
Lipiduria

1. INTRODUCTION

Nephrotic syndrome results from excessive urinary losses of albumin and other plasma proteins of similar mass and presents clinically as a syndrome complex of low serum albumin levels, edema, high blood lipid levels, and lipids in the urine (Table 1). At least 3.5 g of protein must be present in a 24-h urine collection to make the diagnosis, although most patients have, on average, 6–8 g of proteinuria a day. Over 80% of the urinary protein is albumin, reflecting plasma protein composition.

2. CAUSES OF NEPHROTIC SYNDROME

Many glomerular diseases can cause the nephrotic syndrome. Heavy proteinuria is the major manifestation of some glomerular diseases, while others present with predominantly hematuria (nephritic syndrome) or a combination of proteinuria and hematuria *(1,2)*. Primary or idiopathic glomerular diseases that cause nephrotic syndrome include minimal change disease, membranous nephropathy, and focal segmental glomerulosclerosis *(1,2)* (Table 2). The diagnosis is made on the basis of pathologic appearance of the kidney tissue. No serologic markers are available. The underlying cause of these diseases is poorly understood, although the immune system is almost certainly responsible *(2)*.

Systemic diseases also can cause the nephrotic syndrome (Table 2). Diabetic nephropathy is the most common. Certain connective tissue diseases, infections and chronic inflammatory states, malignancies, drugs, and plasma cell dyscrasias also can result in the nephrotic syndrome *(1,2)*. Diagnosis is made on the basis of a thorough history, careful physical examination, and selected tests guided by the history and examination.

Table 2
Causes of Nephrotic Syndrome

Primary or idiopathic	Secondary to systemic diseases	
Membranous nephropathy (33%) Focal segmental glomerulosclerosis (33%) Minimal change disease (15%) IgA nephropathy (10%) Membranoproliferative glomerulonephritis (2–5%) Other (5–7%)	Diabetes mellitus Systemic lupus erythematosus Amyloidosis Dysproteinemias • Multiple myeloma • Other light chain mediated disease Malignancy • Breast, colon, lung • Lymphoma	Infections • Human immunodeficiency virus • Hepatitis B • Hepatitis C • Syphilis • Malaria Drugs • Non-steroidal anti-inflammatory drug • Gold, penicillamine

3. COMPLICATIONS OF NEPHROTIC SYNDROME

Regardless of the etiology of nephrotic syndrome, the clinical sequelae are identical (Table 3). All of the adverse effects discussed below result directly or indirectly (through decreased plasma albumin levels and/or oncotic pressure) from urinary protein losses *(2,3)*. Therefore, management of nephrotic patients targets reduction of proteinuria to modify the complications of nephrotic syndrome.

3.1. Sodium Retention

Sodium and water retention results from avid renal sodium retention because of loss of response to atrial natriuretic peptide *(2,4,5)* and, when serum albumin levels fall below 1.5–2.0 g/dL, decrease in plasma oncotic pressure *(2,4,5)* (Table 2). The consequences are edema, pleural effusions, and ascites. Skin breakdown from tense edema and the presence of ascites predispose to infection.

3.2. Hypercoagulability

Because of urinary loss of anti-thrombin III, decreased plasma proteins C and S, increased plasma fibrinogen, thrombocytosis, and increased platelet adherence (Table 2), nephrotic patients may be hypercoagulable *(2,5)*. They are prone to develop deep venous thrombosis, pulmonary embolism, renal vein thrombosis, and, occasionally, arterial thrombosis. The incidence of thromboembolic disease is highest in patients with membranous nephropathy, up to 50%. The higher the protein excretion and the lower the serum albumin level, the higher is the risk of thromboembolism *(2,5)*.

3.3. Hyperlipidemia

Levels of pro-atherogenic lipoproteins, such as very low density lipoprotein (VLDL), intermediate density lipoprotein (IDL), low density lipoprotein, and lipoprotein (a), are increased in nephrotic syndrome *(2,6)*. While high density lipoprotein (HDL) levels are either normal or slightly decreased, it is the small, dense, and less protective HDL particles that accumulate *(6)*. Triglyceride levels also may be increased markedly *(2,6)*, resulting both from delayed clearance of VLDL and reduced hepatic uptake of highly atherogenic IDL remnant particles. The combination of increased synthesis and impaired catabolism of lipoproteins confers a relative risk of 5.5 for myocardial infarction and of 2.8 for death from coronary artery disease *(7)*.

Table 3
Complications and Clinical Sequelae of Nephrotic Syndrome

Complication	*Mechanism*	*Clinical sequelae*
Sodium retention	Atrial natriuretic peptide resistance; ↓Plasma oncotic pressure	Edema → Skin breakdown → Cellulitis Pleural effusion → Shortness of breath Ascites → Spontaneous bacterial peritonitis
Hypercoagulable state	Loss of anti-thrombin III; ↓Plasma proteins C and S Thrombocytosis; ↑Plasma fibrinogen	Deep venous thrombosis; Pulmonary embolism; Renal vein thrombosis
Infection	Loss of immunoglobulins Loss of complement	Spontaneous bacterial peritonitis Other infections with encapsulated organisms
Hyperlipidemia	Altered lipoprotein metabolism: ↑Proatherogenic lipoproteins; Oxidized high density lipoprotein; ↑Triglycerides	Accelerated atherosclerosis
Progressive renal injury	Iron-induced oxidative injury Lipid peroxidation Complement-mediated injury	Interstitial fibrosis; Chronic kidney disease
Nutritional depletion	Loss of tissue proteins Loss of erythropoietin Loss of plasma binding proteins	Tissue proteins → Muscle wasting Anemia Transferrin → Iron deficiency anemia Thyroglobulin-binding protein → Hypothyroidism Vitamin D-binding protein → Hypocalcemia, rickets; Zinc (bound to albumin) → Zinc deficiency

3.4. Progressive Renal Injury

Prolonged and massive proteinuria leads to progressive renal injury with interstitial fibrosis and glomerular sclerosis *(8,9)*. The proteins and lipids in the urine are taken up by the proximal renal tubular cells and lead to oxidative injury of the cells. Once taken up, the oxidized lipids, iron, and complement also may act as a chemoattractant for monocytes and stimulate production of cytokines, resulting in further injury *(8,9)*.

3.5. Infection

Patients with nephrotic syndrome may develop cellulitis because of skin breakdown from tense edema, pneumonia, and spontaneous bacterial peritonitis when ascites is present (especially in children). Immunoglobulin G (IgG) and complements are lost in the urine *(2,5)*. Unlike liver-derived proteins, the synthesis rate of IgG does not increase. Encapsulated organisms are a particular threat because such organisms require either opsonization with specific antibodies or complement fixation for killing.

3.6. Nutritional Depletion

Urinary protein losses in nephrotic syndrome lead to muscle wasting presumably due to shunting of amino acid building blocks to the liver to enhance plasma protein synthesis, in the absence of a compensatory decrease in total body protein turnover *(3,10)*. Loss of erythropoietin and binding proteins that transport iron, vitamin D, and thyroxine may result in anemia, iron deficiency, hypocalcemia and rickets, and hypothyroidism, respectively *(5,11)*. Sustained and massive proteinuria may also lead to zinc deficiency, because two-thirds of circulating zinc is bound to albumin *(5)*.

4. TREATMENT OF NEPHROTIC SYNDROME

The main goal in treating nephrotic syndrome is to reduce or eliminate proteinuria to blunt or prevent the development of associated complications, to protect kidney function, and to reduce the risk for accelerated atherosclerosis *(2,5)*. Dietary management and pharmacologic management (Table 4) each play a major and complementary role in this endeavor.

Table 4
Treatment of Nephrotic Patients

	Dietary	*Pharmacologic/other*
Calorie	35 kcal/kg/day	Remove underlying cause
Protein	0.8 g/kg/day	Start immunosuppressive drugs
	Soy protein – more beneficial	Reduce proteinuria
		Angiotensin converting enzyme inhibitor
Fat	<30% of total calories	Angiotensin receptor blocker
	Cholesterol <200 mg/d	Cyclosporine
Minerals	Sodium <2 g/day	
	Iron if clearly iron deficient	Non-steroidal anti-inflammatory drug
	Calcium if vitamin D deficient (2 g/day)	Statin therapy for hyperlipidemia
	Zinc if zinc deficient (220 mg/day)	Anticoagulation for hypercoagulability
Vitamins	Calcitriol, if vitamin D deficient	Antibiotics for infection
		Diuretics for edema
Fluid	1 L/day	

4.1. Specific Treatment

If possible, treatment of the nephrotic syndrome is directed at the underlying cause (Table 4). In the cases of drug-induced nephrotic syndrome, the offending drug is stopped. Chemotherapy, radiation therapy, and/or surgical resection of the cancer occur with malignancy-related nephrotic syndrome. Antibiotics are used when the suspect is an infection. For idiopathic nephrotic syndrome, suppressing the immune system with drugs such as steroids, cyclophosphamide, or mycophenolate may result in complete resolution of the nephrotic syndrome. A full discussion of the immunosuppressive treatment of glomerular diseases is beyond the scope of this chapter.

4.2. Non-Specific Treatment

If the nephrotic syndrome does not respond to removal of the offending agent or immunosuppressive therapy, then non-specific measures are employed to reduce the proteinuria.

4.2.1. Pharmacologic Management

Angiotensin converting enzyme inhibitors (ACEI), angiotensin receptor blockers (ARBs), cyclosporine, and non-steroidal anti-inflammatory drugs (NSAID) all cause renal vasoconstriction, reduce glomerular capillary pressure, and, therefore, proteinuria *(2,12)* (Table 4). In addition, ACEI and ARB, presumably, also reduce the defect in the filtration barrier of the glomerular basement membrane *(12,13)*. Combining ACEI with ARB has additive effects on the reduction of proteinuria *(12,13)*, and these are the mainstays of pharmacologic treatment. NSAID rarely are used in treating nephrotic syndrome because of the increased risk of acute renal failure and gastrointestinal bleeding.

Statins will improve the hyperlipidemia seen in nephrotic syndrome *(2,12)* (Table 4). They also may protect nephrotic kidneys from progressive injury because of their lipid lowering and anti-inflammatory effects *(12,14)*. However, despite the favorable data in nephrotic animals, human data for cardiac *(2,12)* and renal *(12,14)* protection are scant. If other complications of nephrotic syndrome develop (hypothyroidism, thromboembolic disease, and infection), therapy targeted at the complication is started (Table 4).

4.2.2. Nutritional Management

A high protein diet in nephrotic syndrome will increase urinary protein excretion and lead to a decline in serum albumin concentration through its adverse effects on glomerular hemodynamics *(15)*. In contrast, protein restriction, especially when combined with ACEI and/or ARB therapy, will reduce proteinuria *(12,16)* (Table 4). Although some studies suggest that severe protein restriction to 0.3 g/kg/day supplemented with amino acids is of additional benefit *(17)*, most experts recommend moderate (0.7–0.8 g/kg/day) protein restriction because of concern about precipitating malnutrition *(12)*.

The type of dietary protein also seems to make a difference. Recent studies suggest that chicken and fish sources of protein may be of more benefit than pork or beef *(18)*. Vegetarian sources of protein such as soy *(18–23)* and flaxseed *(24)* reduces proteinuria and hyperlipidemia more than animal proteins. Studies of nephrotic rats suggest that the benefit derived from soy protein is due to a direct effect on the kidneys possibly to reduce nitrotyrosine formation rather than through changes in hepatic lipid metabolism *(22,23)*. The types of amino acids present in plant proteins may be responsible for their beneficial effects, rather than the change in dietary lipid content, because branched chain and gluconeogenic amino acids (such as arginine and glutamate) do not increase proteinuria in animal models while other amino acids do *(25,26)*.

If iron deficiency is clearly present, then cautious oral iron repletion is indicated, keeping in mind that iron may exacerbate renal injury (Table 4). Patients who are hypocalcemic because of vitamin D deficiency should receive oral vitamin D and calcium (Table 4). Sodium and fluid restrictions will help reduce edema and hyponatremia, especially when used in conjunction with diuretics. Although lowering dietary lipids alone will not correct the observed hyperlipidemia, its effect on lipids is additive when used with statins.

5. SUMMARY

Although the nephrotic syndrome may begin with severe proteinuria, it quickly becomes a multisystem disease. Despite the diverse causes of nephrotic syndrome, the common thread is massive urinary loss of proteins leading to an increased risk for cardiovascular disease, vascular thrombosis, anasarca, infection, nutritional depletion, and progressive kidney injury. Treatment is targeted at reducing the proteinuria in order to prevent progressive kidney injury and to reduce the associated complications. Pharmacologic and dietary management of the nephrotic syndrome are complementary, with the mainstay of therapy being the use of ACEI, ARB, statin, and moderate protein restriction (preferably with plant or soy protein) to reduce proteinuria and hyperlipidemia, while waiting for immunosuppression to control the underlying cause.

REFERENCES

1. Madaio MP, Harrington JT. The diagnosis of glomerular diseases: Acute glomerulonephritis and the nephrotic syndrome. Arch Intern Med 2001;161:25–34.
2. Orth SR, Ritz E. The nephrotic syndrome. N Engl J Med 1998;338:1202–1211.
3. de Sain-van der Velden M, de Meer K, Kulik W, et al. Nephrotic proteinuria has no net effect on total body protein synthesis: Measurements with ^{13}C valine. Am J Kidney Dis 2000;35:1149–1154.
4. Koomans H. Pathophysiology of oedema in idiopathic nephrotic syndrome. Nephrol Dial Transplant 2003;18(Suppl 6):vi30–vi32.
5. Harris RC, Ismail N. Extrarenal complications of the nephrotic syndrome. Am J Kidney Dis 1994;23:477–497.
6. Varizi ND. Molecular mechanisms of lipid disorders in nephrotic syndrome. Kidney Int 2003;63(5):1964–1976.
7. Ordonez JD, Hiatt RA, Killebrew EJ, Fireman BH. The increased risk of coronary heart disease associated with nephrotic syndrome. Kidney Int 1993;44:638–642.

8. Keane WF. Proteinuria: Its clinical importance and role in progressive renal disease. Am J Kidney Dis 2000;35(Suppl 1):S97–S105.
9. Walls J. Relationship between proteinuria and progressive renal disease. Am J Kidney Dis 2001;37:S13–S16.
10. Castellino P, Cataliotti A. Changes in protein kinetics in nephrotic patients. Curr Opin Clin Nutr Metab Care 2002;5:51–54.
11. Vaziri ND. Erythropoietin and transferrin metabolism in nephrotic syndrome. Am J Kidney Dis 2001;38:1–8.
12. Wilmer WA, Rovin BH, Hebert CJ, Rao SV, et al. Management of glomerular proteinuria: A commentary. J Am Soc Nephrol 2003;14:3217–3232.
13. Codreanu I, Perico N, Remuzzi G. Dual blockade of the renin-angiotensin system: The ultimate treatment for renal protection? J Am Soc Nephrol 2005;16:S34–S38.
14. Campese VM, Nadim MK, Epstein M. Are 3-hydroxy-3-methylglutaryl-CoA reductase inhibitors renoprotective? J Am Soc Nephrol 2005;16:S11–S17.
15. Kaysen GA. Albumin metabolism in the nephrotic syndrome: The effect of dietary protein intake. Am J Kidney Dis 1988;12:461–480.
16. Don BR, Kaysen GA, Hutchison FN, Schambelan M. The effect of angiotensin-converting enzyme inhibition and dietary protein restriction in the treatment of proteinuria. Am J Kidney Dis 1991;17:10–17.
17. Walser M, Hill S, Tomalis EA. Treatment of nephrotic adults with a supplemented, very low-protein diet. Am J Kidney Dis 1996;28:354–364.
18. Jenkins DJA, Kendall CWC, Marchie A, Jenkins AL, et al. Type 2 diabetes and the vegetarian diet. Am J Clin Nutr 2003;78:S610–S616.
19. D'Amico G, Gentile MG, Manna G, Fellin G, et al. Effect of vegetarian soy diet on hyperlipidaemia in nephrotic syndrome. Lancet 1992;339:1131–1134.
20. Dwyer J. Vegetarian diets for treating nephrotic syndrome. Nutr Rev 1993;51: 44–46.
21. Velasquez MT, Bhathena SJ. Dietary phytoestrogens: A possible role in renal disease protection. Am J Kidney Dis 2001;37:1056–1068.
22. Pedraza-Chaverri J, Barrera D, Hernandez-Pando R, et al. Soy protein diet ameliorates renal nitrotyrosine formation and chronic nephropathy induced by puromycin aminonucleoside. Life Sci 2004;74:987–999.
23. Tovar AR, Murguia F, Cruz C, et al. A soy protein diet alters hepatic lipid metabolism gene expression and reduces serum lipids and renal fibrogenic cytokines in rats with chronic nephrotic syndrome. J Nutr 2002;132:2562–2569.
24. Velasquez MT, Bhathena SJ, Ranich T, et al. Dietary flaxseed meal reduces proteinuria and ameliorates nephropathy in an animal model of type II diabetes mellitus. Kidney Int 2003;64:2100–2107.
25. Kaysen GA, al-Bander H, Martin VI, Jones H Jr, et al. Related branched-chain amino acids augment neither albuminuria nor albumin synthesis in nephrotic rats. Am J Physiol 1991;260(2 Pt 2):R177–R184.
26. Kaysen GA, Martin VI, Jones H Jr. Arginine augments neither albuminuria nor albumin synthesis caused by high-protein diets in nephrosis. Am J Physiol 1992;263(5 Pt 2):F907–f914.

21 Kidney Stones

Orfeas Liangos
and Bertrand L. Jaber

LEARNING OBJECTIVES

1. Recognize the risk factors associated with kidney stones including dietary aberrations, stone-provoking drugs and chronic diseases.
2. Identify the most common causes for kidney stones.
3. Adopt a simplified or extensive diagnostic evaluation of kidney stones.
4. Describe general and specific measures to prevent kidney stone recurrence.

Summary

This chapter provides a concise summary of the epidemiology, clinical significance, pathophysiology, diagnosis, and secondary preventive, medical management of kidney stone disease. Specific emphasis is placed on the contribution of diet on kidney stone risk and examples are given for potential dietary interventions to prevent kidney stones. Specific pharmacological therapies with their corresponding pathophysiological mechanism of action are reviewed. A practical diagnostic and therapeutic approach to the first and recurrent stone former is also discussed. Finally, a brief outlook into potential future directions for the prevention of kidney stone disease is provided, followed by a brief summary and concluding remarks.

Key Words: Kidney; renal; stones; calculi; nutrition; diet; prevention.

From: *Nutrition and Health: Nutrition in Kidney Disease*
Edited by: L. D. Byham-Gray, J. D. Burrowes, and G. M. Chertow

1. INTRODUCTION

Kidney stone disease is a common disorder that, due to its symptoms, can inflict recurrent and significant pain and suffering onto a substantial fraction of the population. Kidney stone disease affects many young and otherwise healthy individuals during their productive years, frequently leads to hospitalizations and surgical instrumentations, and therefore may have adverse economical consequences due to healthcare resource utilization as well as absenteeism from work.

The acute management of symptomatic kidney stones is not the subject of this chapter and will only be briefly discussed. The symptomatic treatment includes the administration of intravenous or oral fluids if tolerated, adequate pain control, with the use of opioids, antiemetics, nonsteroidal anti-inflammatory drugs, and potentially antispasmodic medications, and should include straining the urine to obtain a stone specimen for analysis.

2. EPIDEMIOLOGY OF KIDNEY STONES

A recent study over a 2-year period based on the National Health and Nutrition Examination Survey (NHANES) III reported a 5% prevalence of kidney stones among adults in the United States (1988–1994 period), which represented an increase from approximately 4%, when compared to an earlier period (1976–1980) *(1)*. An estimated 1 million cases of symptomatic kidney stones occur per year in the United States and comprise approximately 1% of all hospital admissions. The lifetime risk of symptomatic kidney stone disease is high. Indeed, approximately 10% of men and 5% of women will experience an episode of symptomatic kidney stone disease by age 70 *(2)*. The condition is not equally distributed among genders and ethnic groups, as 80% of all kidney stone cases occur in men and kidney stone disease is more common among Caucasians than African Americans *(1)*.

Regional variations in the frequency and nature of kidney stone disease exist within the United States *(3)*, with an increased prevalence in the southeastern region of the country *(1)*. This variation might be related to differences in temperature and sunlight exposure as well as dietary habits and beverage consumption *(4)*. Stones in the upper urinary tract are frequently seen in developed countries and are associated with a more affluent life-style that includes high animal protein consumption, hypertension and a high body mass index. Bladder stones, however, are more commonly seen in developing countries and more frequently affect individuals with a poor socioeconomic status.

3. CONSEQUENCES OF KIDNEY STONE FORMATION

Patients with kidney stones typically present with an episode of severe pain, which is frequently accompanied by autonomic symptoms such as lightheadedness, diaphoresis, nausea and vomiting. The severity of symptoms often causes the affected individual to visit an emergency department, frequently requiring hospitalization. Accordingly, absenteeism from work is a common result of this disorder.

Ureteral obstruction, upper urinary tract infection, need for stone removal by instrumentation, extracorporeal shock wave lithotripsy or surgery are typical sequelae of symptomatic stone disease. In addition, patients with recurrent stone disease carry an increased risk of chronic kidney disease *(5)*. The cumulative 10-year recurrence rate of kidney stone disease is estimated at 60% for men and 30% for women *(2)*.

4. PATHOGENESIS

4.1. Stone Chemistry

Kidney stones can form from a variety of substances excreted in the urine and frequently consist of two or more different substances (Table 1). Calcareous (calcium oxalate, or calcium phosphate) stones are by far the most common and make up over 80% of kidney stones. Metabolic defects leading to their formation include hypercalciuria in over 65% of cases, and less frequently, hyperuricosuria, hyperoxaluria, hypocitraturia or a combination thereof.

Table 1
Composition of Kidney Stones

Type	*Frequency (%)*
Calcium	70–88
Calcium oxalate	36–70
Calcium phosphate	6–20
Mixed	11–31
Magnesium ammonium phosphate (struvite)	6–20
Uric acid	6–17
Cystine	0.5–3
Miscellaneous	1–4

4.2. Crystallization of Stone-Forming Salts

For a kidney stone to form, the concentration of a dissolved salt has to exceed its solubility in the urine, a condition referred to as supersaturation *(5)*. The development of supersaturation depends on the solubility of the dissolved substance, its amount excreted in the urine and the urine volume and pH. Changes in any one of these conditions individually affect stone formation. In addition, substances may remain in solution even after their concentration has exceeded their solubility. This phenomenon is referred to as metastability *(5)* and is mediated by various substances that are present in the urine such as citrate, osteopontin and Tamm-Horsfall Protein.

4.3. Phases of Stone Formation and Growth

Several steps are still required for a kidney stone to form, after crystallization has occurred. In the nucleation phase, crystals form around a nucleus that may be chemically distinct from the crystals themselves, such as a cell membrane component, protein or a different mineral (heterogeneous nucleation), or the crystals form a nucleus around preexisting crystals of the same substance (homogeneous nucleation). In calcium-oxalate stone formers, for example, calcium-oxalate crystals form on a plaque consisting mainly of calcium phosphate, based at the tip of the renal papillae, called the Randall plaque *(6)*. More recently, the basement membrane of the thin loop of Henle was identified as the site of origin for this plaque *(7)*. Nucleation is followed by an aggregation phase and appositional stone growth.

5. RISK FACTORS FOR STONE FORMATION

Conditions favoring stone formation are supersaturation of urine, increased excretion of urinary stone-forming salts (calcium, oxalate, uric acid and phosphate), urinary pH, decreased urinary volume, the presence of crystallization facilitators such as uric acid, and the absence of urinary crystallization inhibitors such as citrate. The effect of urine pH is not uniform, and an acid or alkaline urine pH increases the solubility of some and decreases the solubility of other salts. Magnesium ammonium phosphate or struvite stones, due to their infectious etiology, are more common in patients with chronically infected urine due to neurologic disorders, such as paraplegia and neurogenic bladder, and also in patients with recurrent urinary tract infections without anatomical or functional urologic disturbances. Additional risk factors are stone provoking drugs, dietary aberrations, positive family history and underlying metabolic disorders. Stone provoking drugs include vitamin C, which can increase oxaluria, vitamin D, which causes hypercalciuria, uricosuric agents, acetazolamide, which can alkalinize the urine resulting in calcium phosphate precipitation, and triamterene, sulfadiazine, and more recently indinavir *(8)*, which can precipitate and form into stones when excreted in the urine.

5.1. Hypercalciuria

Potential mechanisms of hypercalciuria are increased intestinal calcium absorption that could be intrinsic or 1, 25-dihydroxy-vitamin D_3 mediated (absorptive hypercalciuria); decreased renal tubular reabsorption of calcium (renal hypercalciuria); or enhanced excretion due to bone demineralization (resorptive hypercalciuria). Absorptive hypercalciuria is the most common form and represents 90% of all hypercalciuric states. In this disorder, increased intestinal calcium absorption is the primary defect and the serum calcium level and fasting urine calcium excretion are typically normal. Secondarily, the parathyroid function is suppressed. Potential secondary causes of absorptive hyercalciuria include the milk alkali syndrome and sarcoidosis. In renal hypercalciuria, which is second in frequency (8%), fasting urinary calcium excretion is primarily elevated due to a renal defect, leading to secondary hyperparathyroidism and increased intestinal calcium absorption. In this disorder, the serum calcium level is usually normal. Resorptive hypercalciuria is third in frequency (2%) and is often a result of primary hyperparathyroidism or prolonged immobilization, leading to an elevated serum calcium level, elevated intestinal absorption and hypercalciuria.

5.2. Hyperoxaluria

Mechanisms of hyperoxaluria can be increased dietary intake of oxalate, intake of oxalate precursors (vitamin C) or increased intestinal oxalate absorption due to decreased intestinal calcium availability. The latter can be due to a low calcium diet (e.g., in lactose intolerance or vegetarian diet), calcium hyper-absorptive states such as in 1, 25-dihydroxy-vitamin D_3 hyper-vitaminosis, renal phosphate leak (resulting in secondary hyper-vitaminosis D), and use of calcium binders such as cellulose phosphate or intestinal free fatty acids (e.g., in malabsorption syndromes). A variety of gastrointestinal diseases including inflammatory bowel disease and small bowel surgery (e.g., resection or bypass) with or without ileostomy may lead to hyperoxaluria combined with hypocitraturia *(9)*. An increasingly recognized risk factor for calcium oxalate stone formation is the increased bioavailability of oxalate due to a decrease in intestinal oxalate degradation.

This might be due to the absence of intestinal colonization with *Oxalobacter formigenes*, an oxalate-degrading microorganism that acts through the enzyme oxalyl-CoA decarboxylase *(10,11)*. Of note, this intestinal bacterium is antibiotic sensitive, and repeated antibiotic therapies can eradicate it *(12)*.

5.3. Hypocitraturia

Because urinary citrate binds calcium by chelation and reduces the risk of calcium salt precipitation, decreased urinary citrate excretion leads to an increased risk for calcium stone formation. Hypocitraturia can be caused by chronic metabolic acidosis due to bicarbonate loss from chronic diarrhea or in laxative abuse, small bowel pathology, metabolic acid loads from high animal protein intake *(13)*, renal tubular acidosis, polycystic kidney disease *(14)*, medullary sponge kidney, ureteral diversion and starvation. Drugs such as ethacrynic acid, angiotensin converting enzyme inhibitors (ACEI) *(15)* and acetazolamide can also reduce urinary citrate excretion.

5.4. Hyperuricosuria

Gouty diathesis is not only a risk factor for the development of uric acid stones, but for kidney stone disease in general *(16)*. Uric acid stones are also more common in patients with myeloproliferative disorders or ulcerative colitis. In addition, bowel surgery that involves the colon is a risk factor for hyperuricosuria and uric acid stone formation *(17)*.

5.5. Dietary Factors and Risk of Stone Formation

Dietary risk factors for stone formation are numerous and can potentially be influenced positively. A low fluid intake results in low urine volume and an increased concentration of urinary solutes. A high sodium intake increases urinary calcium excretion and facilitates calcium stone formation. A diet high in animal proteins increases the amount of inorganic acids produced by amino acid metabolism, whereby decreasing urinary pH. In addition, urinary calcium excretion increases as a consequence of the acidifying effect of oxidation of sulfur in the amino acids methionine, cysteine and cystine to sulfate, which is excreted with calcium to maintain electroneutrality *(18)*. A high oxalate diet, for example from peanuts, tea, instant coffee, rhubarb, beets, beans, berries, chocolate, dark leafy greens, oranges, tofu, sweet potatoes and draft beer, increases oxalate absorption and subsequent urinary excretion with an increased risk of calcium oxalate stone formation. A diet high in purines, for example from the consumption of red meat or game, sweetbreads, beer or cruciferous vegetables, increases urinary uric acid excretion.

Dietary calcium intake can have variable effects. Excessive calcium intake, especially if combined with vitamin D_3 supplementation, will invariably lead to increased calcium absorption and hypercalciuria. Dietary calcium restriction on the other hand reduces enteral chelation of dietary oxalate, thereby facilitating oxalate absorption and subsequently urinary excretion. The effect of dietary calcium intake on kidney stone formation has been examined in several clinical studies. In a prospective study of the relation between dietary calcium intake and the risk of symptomatic kidney stones in a large cohort of men (40–75 years of age), with no prior history of kidney stones, after adjusting for relevant risk factors, a higher dietary calcium intake was associated with a reduced risk of kidney stones *(19)*. In this study, higher intake of animal protein was also found to promote kidney stone formation, and high intake of dietary potassium and fluid was found to be protective. These findings were also confirmed for women, in a similar, large, observational study *(20)*; however, in contrast to a protective effect of an increased intake of calcium-containing foods, the intake of calcium supplements actually increased kidney stone risk. In addition, sucrose and sodium intake increased, and potassium and fluid intake decreased the risk of stone development. Most of these findings were confirmed in a subsequent prospective cohort study of younger women (27–44 years of age) *(21)*. In this study, however, no stone promoting effect of dietary calcium supplements was demonstrable, but a protective effect of phytate intake, a plant-based chelating agent present in rice and other whole grains, was observed. An additional analysis of the same cohort demonstrated a protective effect of vitamin B_6 supplements, which can decrease oxalate production, and no deleterious effect of vitamin C supplements, up to a dose of 1500 mg/day *(22)*. The latter finding is rather unexpected because vitamin C is metabolized to oxalate and should theoretically increase oxaluria.

A more recent, 5-year prospective randomized trial compared the effect of two diets in 120 men with recurrent calcium-oxalate stones and hypercalciuria. The group that followed a normal calcium diet, along with animal protein and sodium restriction, had less frequent recurrence of stones compared to those who restricted dietary calcium intake *(23)*. Due to the controlled design of this trial, these results confirm and corroborate the data obtained in the prior observational cohort studies.

5.6. Effect of Beverage Consumption on Kidney Stone Risk

In addition to the total amount of fluid consumed, the type of beverage might also play a role. In a study of middle-aged women, coffee, tea and wine were independently protective, whereas grapefruit juice promoted incident kidney stone disease *(24)*. These findings were also confirmed in men; however, in this cohort, beer consumption reduced, and apple juice increased the risk of kidney stones *(25)*. Some beverages, for example stronger hopped beers such as European lagers, might contain significant amounts of oxalate but their consumption also increases urinary flow and dilution, which likely explains their neutral or even beneficial effects in observational studies *(25,26)*. A general recommendation for the consumption of alcoholic beverages to reduce the risk of stone disease cannot be given at the present time. In addition, coffee, as opposed to tea consumption, has repeatedly been shown to be protective in a dose-dependent fashion, which likely relates to its effects on urine flow *(25,26)*.

5.7. Genetic Risk Factors for Kidney Stones

The importance of genetic risk factors has been documented in a series of observational studies. In one study, a family history of kidney stones substantially increased the risk of incident stone formation in men *(27)*. In a sibling study in Vietnam era veterans, a higher concordance rate for urinary calculi in monozygotic twins compared with dizygotic twins was found and a heritability of 56% was estimated compared with an environmental effect of 44% *(26)*. Specific monogenic diseases associated with kidney stone disease and nephrocalcinosis have also been described *(5)*. However, the exact genetic mechanisms for the more common idiopathic calcium oxalate stone disease remain poorly understood.

6. DIAGNOSTIC EVALUATION

Not every patient with one single episode of nephrolithiasis will require undergoing an extensive metabolic evaluation. At an initial clinic visit, after the first symptomatic kidney stone episode has occurred in a patient who is not considered to be at risk for recurrent stone formation, a simplified evaluation is usually sufficient. This consists of a dietary and medical history, a urinalysis and urine culture, a concise blood test panel, and a stone analysis if available. However, patients with recurrent nephrolithiasis or those considered at risk for recurrence might benefit from a more comprehensive metabolic and

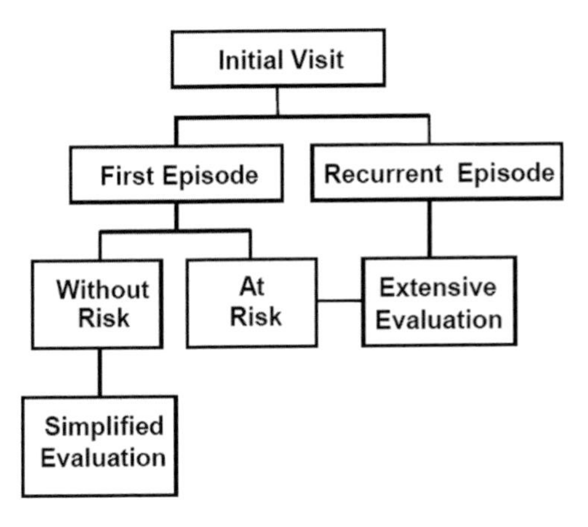

Fig. 1. Scheme for the diagnostic evaluation of patients with kidney stones.

dietary evaluation (Fig. 1). The extensive evaluation should consist of a dietary and medical history, a urinalysis and urine culture, blood work, stone analysis (if available), a spiral CT scan, a 24-h urine collection, and the completion of a dietary journal and food frequency questionnaire. The diagnostic value of each of these studies is outlined in the section below.

The urinalysis can identify several stone risk factors. A urine pH below 5.5, for example, promotes the formation of uric acid and cystine stones. A urinary pH of 7.5 can be associated with a concomitant infection with urea-splitting bacteria and struvite stone formation or can be associated with renal tubular acidosis.

A careful microscopic examination of the sediment of a fresh urine sample can lead to specific crystal identification and help with the differential diagnosis *(28)*. Additional urine acidification or alkalinization during microscopy might improve the morphologic identification of crystals according to their formation or dissolution in acid or alkaline environments. Urate crystals can be rhomboid, needle shaped or amorphous in shape. Oxalate crystals can be dumbbell or needle shaped (calcium oxalate monohydrate) or envelope shaped (calcium oxalate dihydrate). Cystine crystals are hexaedric and flat and, if discovered, pathognomonic for the disease. A urine cyanide-nitroprusside test can be performed to confirm the presence of cystine crystals, during which the urine would undergo purple red discoloration. Phosphate crystals can be in the form of triple phosphate (typical "coffin lid" structure) or amorphous phosphate. The urine culture rules out concomitant urinary tract infection particularly with urea-splitting bacteria.

A limited blood work panel for the evaluation of kidney stone disease includes serum electrolytes (to evaluate for renal tubular acidosis), blood urea nitrogen, creatinine, calcium and phosphorus (to evaluate for primary hyperparathyroidism), albumin, and uric acid (to evaluate for gouty diathesis).

Imaging with non-contrast helical CT scanning provides information about the extent and complications of the stones. A urological consultation is warranted if the stone does not pass spontaneously, is larger than 5 mm in diameter *(29)*, causes urinary obstruction, or is associated with urosepsis.

7. SECONDARY PREVENTION

General dietary advice should be given to every stone former, consisting of high fluid intake, usually a minimum of 3 L/day *(30)*, which would ideally allow for excretion of a dilute urine throughout the entire daytime and nighttime period. Of note, to ensure urine dilution throughout the night requires the intake of one to two 8 ounce glasses of water before bedtime, followed by an additional 8 ounce glass during the night, for example, after a void. This would inadvertently lead to frequent nocturia and thus can limit adherence. In addition, access to water during daytime work hours might be difficult for some patients. In addition, a dietary sodium restriction of less than 150 mEq/day should also be recommended to all patients. Patients with calcium oxalate stones should also limit daily intake of animal proteins to 0.8 g per kg of body weight, which is the recommended daily allowance (RDA) as determined by the Institute of Medicine, Food and Nutrition Board. It is now well documented that, especially among patients with calcium oxalate stones, dietary calcium restriction increases the risk of recurrent stone formation *(19,20,23)*. Dietary calcium restriction might also lead to an increased loss of bone mass and increases the likelihood for osteoporosis in individuals at risk for this condition. On the other hand, excessive intake of calcium, especially in form of supplements, might still increase calciuria and therefore calcium stone risk. Therefore, the recommendation should be to avoid extremes of calcium intake and remain within the RDA of 800–1000 mg/day. Dietary oxalate restriction is recommended in oxalate stone formers.

7.1. Pharmacological Prevention

In recurrent stone formers, specific pharmacological therapies should be added for secondary prevention. If hypercalciuria is present, thiazide diuretics are effective *(31)*, however, always in conjunction with a low sodium diet. Other pharmacological options include hydrochlorothiazide, chlortalidone, amiloride combined with hydrochlorothiazide, and amiloride alone.

In enteric hyperoxaluria, in which dietary restriction is not sufficient, a low fat diet, calcium carbonate supplements, medium chain fatty acids supplementation and enteral binding therapy with cholestyramine can all be attempted alone or in combination. Hypocitraturia warrants potassium citrate (Uro-Cit-K®) supplementation at a minimum dose of 40–60 mEq/day in two to three divided doses *(32)*. Other citrate formulations such as Bi-Citra® (sodium citrate) or Poly-Citra® (sodium/potassium citrate) are to be avoided due to their high sodium content.

In uric acid stone disease, urinary pH should be maintained above 7, which might warrant higher doses of potassium citrate. If citrate alone cannot raise the urine pH consistently, especially during the night, an evening dose of acetazolamide may be added. In calcium phosphate stones, however, citrate supplementation should be used with caution as excessive urinary alkalinization might promote calcium phosphate precipitation *(5)*. In hyperuricosuria, in addition to reducing purine intake, oral allopurinol has been shown to be effective.

Cystinuria, which is characterized by urinary cystine excretion of greater than 250 mg/day, is an uncommon inherited tubular disorder of amino acid transport that leads to an increased excretion of cystine, ornithine, lysine and arginine. This autosomal recessive disorder has a prevalence of 1 in 7000. Symptoms prevail in the second and third decade of life, and mixed stones account for half of the stones formed in patients with cystinuria. Common associated risk factors include hypercalciuria (19%), hyperuricosuria (22%) and hypocitraturia (44%). Urinary cystine solubility is pH dependent, and urinary alkalinization is the mainstay of treatment in this disorder. In mild to moderate cystinuria of 250–500 mg/day, treatment consists of mild sodium restriction, potassium citrate supplementation to alkalinize the urine pH, and increase in fluid intake with the goal of generating 1 L of urine per 250 mg of cystine excreted per day to maximize dilution. Of note, the patient should be instructed to maintain a log of urinary pH results to self-monitor therapy. In more severe forms of cystinuria with excretion rates of over 500 mg/day, chelating agents such as captopril or tiopronin (Thiola®) should be added.

In recurrent struvite stones, in addition to a general dietary prescription, sodium restriction and maintenance of a minimum fluid intake of 3.5 L/day, specific therapies such as the chronic use of suppressive antibiotics or urease inhibitors such as acetohydroxamic acid (Lithostat®) can be used. All dietary and pharmacological interventions should be followed up with a repeat 24-h urine collection 3 months after the intervention, in order to monitor treatment success and adherence.

8. FUTURE DIRECTIONS

Because magnesium might also increase the solubility of urinary oxalate, the additional use of magnesium supplements for the prevention of calcium oxalate stones has been an area of debate. Co-administration of magnesium oxide with citrate preparations has been shown to increase solubility indices of calcium oxalate *(33)* and decrease the relative supersaturation index of brushite *(34)*. However, the administration of magnesium oxide and hydroxide has been shown to be ineffective in preventing recurrent stone formation, and only potassium magnesium citrate was found to be beneficial *(35)*. The utility of magnesium in combination with potassium citrate therefore merits further investigation.

Pyridoxine (Vitamin B_6) supplementation can reduce oxalate production, and observational studies have shown that it might reduce the risk of stone formation in women, if given in large doses *(22)*. Before high dose pyridoxine can be introduced into the therapeutic armamentarium, these observations need further corroboration in prospective randomized controlled trials as well as expansion to other segments of the population.

In calcium oxalate stone formers with mild to moderate hyperoxaluria, the enteral administration of probiotic agents in the form of oxalate-metabolizing microorganisms such as bacteria of the *Lactobacillus* species or *O. formigenes* showed promising results. In one randomized controlled study, administration of freeze-dried *Lactobacillus* species resulted in significant reduction of urinary and fecal oxalate excretion. In addition, the administered bacteria had demonstrable in vitro oxalate-metabolizing activity *(36)*. The use of a probiotic consisting of a lactic acid bacteria mixture has also been shown to decrease enteric oxaluria in patients with malabsorption syndromes *(37)*.

Positive health economical effects of a more proactive approach to secondary prevention of kidney stones have increased the emphasis on prevention and strengthened the role of a dietary and metabolic approach compared to the more traditional surgical therapeutic approaches to kidney stone management. The importance of preventing hospitalizations and procedures does not only reduce healthcare costs *(38)*, but might also reduce the risk of chronic kidney parenchymal injury that is sustained with each symptomatic episode, and procedure such as shock wave lithotripsy *(39–41)*.

9. SUMMARY

Recurrent kidney stone disease is a common and debilitating disorder with an individual lifetime risk of approximately 10% and a recurrence rate of 50%. If a patient presents after a single episode and does not display risk factors for recurrence, only a limited evaluation is necessary, and a general internist can perform continuous management. If two or more episodes occur or in the presence of risk factors for recurrence, an extensive evaluation is warranted, which should be performed by a nephrologist. While some dietary interventions, such as an increase in fluid intake and sodium restriction can be uniformly recommended, the remainder of the therapy should be tailored to the specific metabolic and urinary abnormalities that are identified in the diagnostic evaluation. In the most common calcium oxalate stones, restriction of oxalate, purine and animal protein intake is in most cases appropriate. Finally, the treatment of kidney stone disease remains a challenging task, as it requires the patient's long-term commitment and adherence to significant dietary and lifestyle changes and compliance with, frequently complex, medical therapies.

REFERENCES

1. Stamatelou KK, Francis ME, Jones CA, Nyberg LM, Curhan GC. Time trends in reported prevalence of kidney stones in the United States: 1976–1994. Kidney Int 2003;63:1817–23.
2. Uribarri J, Oh MS, Carroll HJ. The first kidney stone. Ann Intern Med 1989;111:1006–9.
3. Curhan GC, Rimm EB, Willett WC, Stampfer MJ. Regional variation in nephrolithiasis incidence and prevalence among United States men. J Urol 1994;151:838–41.
4. Soucie JM, Coates RJ, McClellan W, Austin H, Thun M. Relation between geographic variability in kidney stones prevalence and risk factors for stones. Am J Epidemiol 1996;143:487–95.
5. Coe FL, Evan A, Worcester E. Kidney stone disease. J Clin Invest 2005;115:2598–608.
6. Randall A. The origin and growth of renal calculi. Ann Surg 1937;105:1009–27.
7. Evan AP, Lingeman JE, Coe FL, et al. Randall's plaque of patients with nephrolithiasis begins in basement membranes of thin loops of Henle. J Clin Invest 2003;111:607–16.

8. Tashima KT, Horowitz JD, Rosen S. Indinavir nephropathy. N Engl J Med 1997;336:138–40.
9. Bambach CP, Robertson WG, Peacock M, Hill GL. Effect of intestinal surgery on the risk of urinary stone formation. Gut 1981;22:257–63.
10. Troxel SA, Sidhu H, Kaul P, Low RK. Intestinal *Oxalobacter formigenes* colonization in calcium oxalate stone formers and its relation to urinary oxalate. J Endourol 2003;17:173–6.
11. Hoppe B, von Unruh G, Laube N, Hesse A, Sidhu H. Oxalate degrading bacteria: new treatment option for patients with primary and secondary hyperoxaluria? Urol Res 2005;33:372–5.
12. Mittal RD, Kumar R, Bid HK, Mittal B. Effect of antibiotics on Oxalobacter formigenes colonization of human gastrointestinal tract. J Endourol 2005;19: 102–6.
13. Kok DJ, Iestra JA, Doorenbos CJ, Papapoulos SE. The effects of dietary excesses in animal protein and in sodium on the composition and the crystallization kinetics of calcium oxalate monohydrate in urines of healthy men. J Clin Endocrinol Metab 1990;71:861–7.
14. Grampsas SA, Chandhoke PS, Fan J, et al. Anatomic and metabolic risk factors for nephrolithiasis in patients with autosomal dominant polycystic kidney disease. Am J Kidney Dis 2000;36:53–7.
15. Melnick JZ, Preisig PA, Haynes S, Pak CY, Sakhaee K, Alpern RJ. Converting enzyme inhibition causes hypocitraturia independent of acidosis or hypokalemia. Kidney Int 1998;54:1670–4.
16. Kramer HJ, Choi HK, Atkinson K, Stampfer M, Curhan GC. The association between gout and nephrolithiasis in men: The Health Professionals' Follow-Up Study. Kidney Int 2003;64:1022–6.
17. Parks JH, Worcester EM, O'Connor RC, Coe FL. Urine stone risk factors in nephrolithiasis patients with and without bowel disease. Kidney Int 2003;63: 255–65.
18. Lemann J, Jr. Calcium and phosphate metabolism: an overview in health and in calcium stone formers. In: Coe F, Favus M, Pak C, Parks J, Preminger G, eds. Kidney Stones: Medical and Surgical Management. Philadelphia: Lippincott-Raven Publishers; 1996:159–288.
19. Curhan GC, Willett WC, Rimm EB, Stampfer MJ. A prospective study of dietary calcium and other nutrients and the risk of symptomatic kidney stones. N Engl J Med 1993;328:833–8.
20. Curhan GC, Willett WC, Speizer FE, Spiegelman D, Stampfer MJ. Comparison of dietary calcium with supplemental calcium and other nutrients as factors affecting the risk for kidney stones in women. Ann Intern Med 1997;126:497–504.
21. Curhan GC, Willett WC, Knight EL, Stampfer MJ. Dietary factors and the risk of incident kidney stones in younger women: Nurses' Health Study II. Arch Intern Med 2004;164:885–91.
22. Curhan GC, Willett WC, Speizer FE, Stampfer MJ. Intake of vitamins B6 and C and the risk of kidney stones in women. J Am Soc Nephrol 1999;10:840–5.
23. Borghi L, Schianchi T, Meschi T, et al. Comparison of two diets for the prevention of recurrent stones in idiopathic hypercalciuria. N Engl J Med 2002;346:77–84.
24. Curhan GC, Willett WC, Speizer FE, Stampfer MJ. Beverage use and risk for kidney stones in women. Ann Intern Med 1998;128:534–40.

25. Curhan GC, Willett WC, Rimm EB, Spiegelman D, Stampfer MJ. Prospective study of beverage use and the risk of kidney stones. Am J Epidemiol 1996;143:240–7.
26. Goldfarb DS, Fischer ME, Keich Y, Goldberg J. A twin study of genetic and dietary influences on nephrolithiasis: a report from the Vietnam Era Twin (VET) Registry. Kidney Int 2005;67:1053–61.
27. Curhan GC, Willett WC, Rimm EB, Stampfer MJ. Family history and risk of kidney stones. J Am Soc Nephrol 1997;8:1568–73.
28. Daudon M, Jungers P. Clinical value of crystalluria and quantitative morphoconstitutional analysis of urinary calculi. Nephron Physiol 2004;98:31–6.
29. Teichman JM. Clinical practice. Acute renal colic from ureteral calculus. N Engl J Med 2004;350(7):684–93.
30. Qiang K. Water for Preventing Urinary Calculi. Cochrane Database Syst Rev 2004;(3):CD004292.
31. Laerum E, Larsen S. Thiazide prophylaxis of urolithiasis. A double-blind study in general practice. Acta Med Scand 1984;215:383–9.
32. Barcelo P, Wuhl O, Servitge E, Rousaud A, Pak CY. Randomized double-blind study of potassium citrate in idiopathic hypocitraturic calcium nephrolithiasis. J Urol 1993;150:1761–4.
33. Kato Y, Yamaguchi S, Yachiku S, et al. Changes in urinary parameters after oral administration of potassium-sodium citrate and magnesium oxide to prevent urolithiasis. Urology 2004;63:7–11.
34. Allie S, Rodgers A. Effects of calcium carbonate, magnesium oxide and sodium citrate bicarbonate health supplements on the urinary risk factors for kidney stone formation. Clin Chem Lab Med 2003;41:39–45.
35. Massey L. Magnesium therapy for nephrolithiasis. Magnes Res 2005;18:123–6.
36. Campieri C, Campieri M, Bertuzzi V, et al. Reduction of oxaluria after an oral course of lactic acid bacteria at high concentration. Kidney Int 2001;60:1097–105.
37. Lieske JC, Goldfarb DS, De Simone C, Regnier C. Use of a probiotic to decrease enteric hyperoxaluria. Kidney Int 2005;68:1244–9.
38. Parks JH, Coe FL. The financial effects of kidney stone prevention. Kidney Int 1996;50:1706–12.
39. Karalezli G, Gogus O, Beduk Y, Kokuuslu C, Sarica K, Kutsal O. Histopathologic effects of extracorporeal shock wave lithotripsy on rabbit kidney. Urol Res 1993;21:67–70.
40. Gunasekaran S, Donovan JM, Chvapil M, Drach GW. Effects of extracorporeal shock wave lithotripsy on the structure and function of rabbit kidney. J Urol 1989;141:1250–4.
41. Jungers P, Joly D, Barbey F, Choukroun G, Daudon M. ESRD caused by nephrolithiasis: prevalence, mechanisms, and prevention. Am J Kidney Dis 2004;44:799–805.

V Additional Nutritional Considerations in Kidney Disease

22 Dietary Supplements

Diane Rigassio Radler

LEARNING OBJECTIVES

1. Define dietary supplement and delineate safety and efficacy issues.
2. Identify dietary supplements that may be used by people with kidney disease.
3. Describe dietary supplements that may be associated with kidney dysfunction.
4. Discuss considerations for health professionals regarding dietary supplements.

Summary

Dietary supplements fall under the rubric of complementary and alternative medicine (CAM). Patterns of use of CAM and dietary supplements by Americans have increased, while the dietary supplement market has concurrently grown due to the demand and the regulatory changes. People with kidney disease may seek to use dietary supplements to prevent further deterioration of kidney function or to ameliorate side effects. Practitioners must be aware of dietary supplement safety and efficacy in order to advise patients on appropriate use.

Key Words: Kidney disease; complementary medicine; dietary supplements; herbs; botanicals.

1. INTRODUCTION

Complementary and Alternative Medicine (CAM) may be thought of as "a group of diverse medical and health care systems, practices, and products that are not presently considered to be part of conventional medicine" *(1)*. Practices considered CAM may evolve over

From: *Nutrition and Health: Nutrition in Kidney Disease*
Edited by: L. D. Byham-Gray, J. D. Burrowes, and G. M. Chertow
© Humana Press, Totowa, NJ

time and be adopted into conventional health care when therapies are proven safe and effective *(1)*. "CAM" may literally be divided into therapies that are "Complementary," referring to practices that are adjunctive to conventional practice; they may be "Alternative," referring to practices that are used instead of conventional practices; or, increasing in vernacular, the term "Integrative" is used to refer to a merging of allopathic approaches and CAM therapies for which evidence on safety and efficacy exists. The integrative approach seeks to deliver health care that is superior to any one modality alone *(1)*. After the formation of the National Center for Complementary and Alternative Medicine (NCCAM) within the National Institutes of Health, NCCAM categorized CAM practices into five domains: (i) alternative medical systems, (ii) mind-body interventions, (iii) biologically based treatments, (iv) manipulative and body-based methods, and (v) energy therapies *(1)*. Dietary supplements are within the domain of biologically based treatments. While people with kidney disease may wish to explore one or more of the CAM domains, the focus of this chapter is with regard to dietary supplements and kidney disease.

2. DIETARY SUPPLEMENTS

Recent estimates suggest that there are 40,000–50,000 dietary supplements available to consumers *(2)*. The extensive availability and relative ease in obtaining dietary supplements may help explain why sales continue to escalate: dietary supplements accounted for $14.1 billion in 1998, $15.5 billion in 1999, $17.1 billion in 2000, and $19.8 billion in 2003. To put that in perspective, the nutrition industry as a whole is estimated at $62 billion, which means that dietary supplements are nearly one-third of the nutrition industry *(3)*.

The surge of dietary supplements over the last 10 years may be attributed to the Dietary Supplement Health and Education Act (DSHEA) of 1994 *(4)*. At that time, Congress acknowledged that there may be a positive relationship between dietary practices, such as dietary supplements, and health, which may translate into reduced health care burden. DSHEA amended the Food, Drug and Cosmetic Act of 1958 to exclude dietary supplements from the pre-market safety evaluations that food and food products undergo. DSHEA mandated a definition of dietary supplements (Box 1) and guidelines for product claims and labels, including a disclaimer that the FDA has not evaluated the product for safety or efficacy. DSHEA authorized FDA to establish good manufacturing practices (GMP) for the supplement industry and provided for the creation of the Office of Dietary Supplements to promote research on dietary supplements *(4)*.

Box 1. Definition of Dietary Supplement *(4)*

"A product intended to supplement the diet to enhance health that contains one or more of the following:

- vitamin, mineral, amino acid, herb or botanical
- dietary substance to supplement the diet by increasing total dietary intake
- concentrate, metabolite, constituent, extract or combination of any ingredient above
- intended for ingestion as capsule, powder, gelcap and is not represented as a conventional food or as a sole item of a meal or the diet."

3. EFFICACY AND SAFETY OF CAM THERAPIES

Among the key issues concerning health care professionals regarding CAM use is the uncertainty over efficacy and safety of many CAM practices (Box 2). Many CAM modalities are inherent in other cultures; however in the United States, a CAM practice may be used in a manner not intended (more is better) and, in the case of botanicals, may differ from those of other regions due to species, soil, water, and growing conditions.

Additionally, in certain populations, such as persons diagnosed with kidney disease, dietary supplements may be contraindicated due to impaired renal function or possible drug interactions.

Through NCCAM, researchers have started the foundation for investigating efficacy and safety of various CAM modalities. NCCAM's mission is to subject CAM practices to rigorous scientific scrutiny, train CAM researchers, and provide credible information to both consumers and health care professionals *(5)*. For those CAM modalities that are found to be health promoting, efficacious, and safe, the expectation is that they will be integrated into mainstream medicine. Until then, the rise in demand and availability, ease of obtaining products, and limited proof of efficacy and safety are cause for concern.

The FDA guidelines on GMP are not yet established; however, there are several independent monitoring agencies (Box 3). Additionally, there are independent certification programs that a manufacturer can seek to endorse the product. These independent organizations will evaluate, on a voluntary basis from the manufacturer, for purity, accuracy of ingredient labeling, and manufacturing practices. One is from the United States Pharmacopoeia (USP) who will evaluate products upon specified criteria and allow the

Box 2. Deciding for or Against the Use of Dietary Supplements

For	Against
"Natural" medicines are natural	"Natural" medicines have biologic activity
Many therapies used for centuries	Limited scientific testing
May reduce need for drugs and associated side effects when used properly	Sold without knowledge of action; Active ingredient concentrations vary May displace/enhance/interfere with current therapy – St John's Wort and indinavir in HIV – Ginkgo and anticoagulation therapies
Other countries have been prescribing herbs safely for years	Quality control – No federal regulation prior to sale – May be subject to misidentification, adulteration, contamination

manufacturers to use the designation, DSVP for Dietary Supplement Verification Program, if the product passes the rigorous testing (see http://www.usp-dsvp.org). Another similar program is set up by NSF International to obtain the right to use the NSF mark (see</http://www.nsf.org/consumerdietary_supplements/dietary-faq.asp?program=DietarySup>/).

Box 3.

Quackwatch http://www.quackwatch.com/ for health fraud and quackery
Medwatch at http://www.fda.gov/medwatch
National Council Against Health Fraud http://www.ncahf.org/
Consumerlab — independent evaluation with periodic reports — part of the report is available free, some by nominal subscription. http://www.consumerlab.com

4. DIETARY SUPPLEMENTS AND KIDNEY DISEASE

The increasing use of CAM practices within the United States has been documented since the early 1990s *(6–8)*. CAM use remains a significant issue in current health care practice with continued interest and popularity among Americans in using dietary supplements *(9,10)*. Although there a dearth of published literature on the actual patterns of use of dietary supplements by people with kidney disease *(11,12)*, people may choose to use dietary supplements in an attempt to prevent further renal deterioration, or may use CAM as an adjunct to mitigate side effects of the disease or treatments *(12,13)*. Supplement use may be classified as those with potential protective effects and those that should be avoided in kidney disease. Additional considerations include dietary supplements that may be toxic and lead to kidney dysfunction and those that may have interactions with prescribed drugs.

4.1. Dietary Supplements with Potential Protective Effects

Preventing renal deterioration by using herbs and supplements would be a fortunate asset. Chinese cultures may not have dialysis abundantly available, and traditional medicine using herbs may be one of the first choices in treatment. Thus, Chinese formulations or single herbs may be promising treatments. However, most of the published research is with animal studies; human applicability, safety, efficacy, and potential interaction with other medications must be explored. Astragalus (*Astragalus membranaceus*) an adaptogenic herb and antioxidant may reduce proteinuria in glomerulonephritis *(13–15)*. Traditional Chinese Medicine uses astragalus often in combination with other herbs for its immune-enhancing potential; hence, individuals on immunosuppressants or those with autoimmune diseases should avoid its use *(16)*. The antioxidant properties of ginger (*Zingiber officinale*) may be linked to reduced inflammation in rats *(17)*. Other herbs such as milk thistle (*Silybum marianum*) and cordyceps (*Cordyceps sinensis*) may offer protection against nephrotoxic drugs *(16,18)*.

Given that people with kidney disease often also have hypertension, diabetes, or hyperlipidemia, herbs and supplements with anti-inflammatory activity may be of interest in an attempt to mitigate cardiovascular risk factors *(13)*. Omega-3 fatty acids found in foods, mainly fish, are also available in the form of dietary supplements as anti-inflammatory agents. A review article on the benefits of fish oil supplementation in renal patients suggests that omega-3 fatty acids may also be beneficial to reduce the severity of pruritis and reduce the risk of thrombosis after placement of vascular graft *(19)*. Evening primrose (*Oenothera biennis*) and borage (*Borago officinalis*) oil are sources of gamma-linolenic acid (GLA), an omega-6 fatty acid *(16)*. GLA may be converted to compounds that have anti-inflammatory properties. Judicious use of anti-inflammatory agents by people with kidney disease looks promising. Certainly side-effects and drug interactions should be monitored and noted; fish oils can decrease platelet aggregation so large doses predispose a bleeding risk and may have additive effects with anticoagulant or antiplatelet medications *(16)*.

4.2. Dietary Supplements to Avoid in Kidney Disease

While there is still much research to be done with dietary supplements and specifically in populations with kidney disease, most supplements should be approached with caution. Published literature on case reports, with either positive or negative outcomes, should not be generalized to a larger population. Theoretical mechanisms of action need testing in vivo before affirmation of use; both demonstrated and theoretical drug–herb interactions must be heeded.

Dandelion root (*Taraxacum officinale*) and Scotch broom (*Cytisus scoparius*) may have diuretic effects *(16)*. Parsley (*Carum petroselinum*), juniper (*Juniperus communis*), lovage (*Levisticum officinale*), and goldenrod (*Solidago virgaurea*) have constituents that irritate the kidney, increase renal blood flow and glomerular filtration, thus acting as aquaretics that increase water loss but not electrolyte excretion *(16,20)*. These herbs should be considered contraindicated in kidney disease.

Kidney disease may alter the pharmacokinetics of drugs. Given that dietary supplements are natural forms of active biochemical agents, and that often the active ingredient or the mechanism of action is not fully understood in healthy individuals, people with kidney disease must use caution when considering the metabolism, distribution, and excretion of natural products. For example, it is known that St. John's Wort (*Hypericum perforatum*), used for mild depression, interferes with a metabolic pathway shared by many drugs *(16,21)*. When prescribed drugs and St. John's Wort are concomitantly taken, the blood concentration of the drugs is low and often not therapeutic.

5. DIETARY SUPPLEMENTS AND KIDNEY DYSFUNCTION

Perhaps the most infamous case of herbs causing kidney dysfunction is the case or "Chinese herb nephropathy" or "aristolochic acid nephropathy." In an attempt at weight loss, a Belgian population took an herbal supplement with a mis-identified herb containing aristolochic acid, a nephrotoxic and carcinogenic herb. It was later reported that numerous people who took the supplement needed dialysis or renal transplant and several developed urothelial carcinoma *(22)*. Other less common herbs implicated in kidney dysfunction, mainly used in Africa and China, have been reported and summarized in a literature review by Colson and De Broe *(14)*. It would be prudent for practitioners whose patient populations have strong cultural ties to Africa or China, familiarize themselves with particular regimens inherent in certain populations. Refer to Table 1 for a list of herbs with adverse effects on the kidney.

6. CONSIDERATIONS FOR HEALTHCARE PROVIDER

Conscientious health care providers understand the current science in health promotion and disease management and strive for the best possible outcomes in patient care *(23–26)*. Recognizing that patients may seek to include dietary supplements as a part of their health care regimen is vital for open communication and treatment. Healthcare providers must be willing to ask questions relative to dietary supplement and be prepared to answer questions or dialogue the pros and cons of a given regimen. Often, however, the area of dietary supplements may be intimidating when faced with numerous supplements with uncommon names. Suggested approaches would be to identify those dietary supplements common to one's practice area, for example in kidney disease, research the evidence for safety and efficacy through either scientific publications or databases that synthesize the information into a quick, sound reference such as http://www.consumerlabs.com or http://www.naturaldatabase.com. Once the provider has background knowledge on the supplement, he or she is usually more comfortable dialoging with the patient regarding dietary supplements. Likewise, a patient who perceives that his provider is willing to discuss the subject is more likely to disclose the truth about thought of use or actual use. Pertinent information to discuss is what the patient wants to use and why. The "what" is relatively straight-forward but the "why" aspect may be more nebulous. Find out the

Table 1
Herbs with Adverse Renal Effects

Herb (Scientific name)	*Effect*	*Remark*
Aristolochia (*Aristolochia auricularia*)	Renal fibrosis, carcinoma	Contains aristocholic acid which is nephrotoxic and carcinogenic FDA prohibits products containing aristocholic acid Other herbs such as asarabacca and costus root may be adulterated with aristocholic acid
Neem (*Azadirachta indica*)	Nephrotoxic	Leaf or seed oil may be nephrotoxic; flower, fruit and twigs may be safe
Licorice (*Glycyrrhiz glabra*)	Hypernatremia, hypokalemia, edema	Numerous drug interactions
Senna (*Senna alexandrina*), Cascara (*Rhamnus purshiana*)	Hypokalemia	Used as laxatives. Senna leaf is not for long term use
Noni fruit (*Morinda officialis*)	Hyperkalemia	Fruit contains high concentration of potassium
Juniper berry (*Juniperus communis*), dandelion (*Taraxacum officiale*), asparagus tea (*Asparagus officinalis*), rupturewort (*Herniaria glabra*), Scotch broom (*Cytisus scoparius*), stinging nettle (*Urtica dioica*), uva ursi (*Arctostaphylos uva-ursi*)	Diuresis, electrolyte imbalance	May increase water loss without sodium excretion

FDA, food and drug administration. Adapted from refs *(18,16)*.

source of the patient's information and identify the patient's expectations. Is the patient taking the dietary supplement to treat a side effect of the disease or hoping that the supplement may cure his disease? Understanding the patient's issues and the safety and efficacy of the supplements leads to an open discussion to support or discourage their use. If the patient understands the issues around dietary supplements and still intends to take one or some, healthcare providers may allow a trial period if the supplement is not harmful and the patient can afford the out-of-pocket expense. In that case, providers may advise the patient to start with one supplement at a time, monitor and report any side effects, allow 4–6 weeks to notice the desired effect, and report back to the provider all positive, negative, or neutral experiences. As expected with all patient contact, providers must document the dialogue and communicate the patient's actions with the healthcare team.

7. SUMMARY

With the increase in use of dietary supplements, and the relative ease with which they may be marketed by manufacturers and purchased by consumers, healthcare providers need to be aware of common supplements and cognizant of the safety concerns or interactions they may have with disease status and medication regimens. Healthcare providers can identify the notable dietary supplements that may be encountered in practice and research the safety and efficacy issues with regard to their use or misuse. Honest, open communication with patients who may wish to explore the use of dietary supplements is essential for patient – provider relations and optimal outcomes for health care.

Resources – Websites worth noting are the following:

- American Botanical Council, http://www.herbalgram.org.
- CAM on PubMed, http://www.nlm.nih.gov/nccam/camonpubmed.html.
- National Center on Complementary and Alternative Medicine. http:// nccam. nih.gov.
- Natural Medicines Comprehensive Database. http://www.natural database.com.
- Office on Dietary Supplements. http://dietary-supplements.info.nih.gov/ index.aspx.
- American Herbal Foundation (many good links), http://www.herbs.org.
- NCCAM tips for consumers. http://nccam.nih.gov/health/decisions/ index.htm.
- Natural Standards, http://www.naturalstandards.com.

Please note that when you find a reputable website, you can often find more information through links. See ref. *(16)*, (also available as an online subscription at http://www.naturalmedicines.com) and ref. *(27)*.

REFERENCES

1. What Is Complementary and Alternative Medicine (CAM)? (Accessed March 15, 2006, at http://nccam.nih.gov/health/whatiscam/).
2. Office of Dietary Supplements – Update. 2006. (Accessed March 15, 2006, at http://dietary-supplements.info.nih.gov/News/ODS_Update_-_January_2006.aspx).
3. American Dietetic Association. Practice Paper of the American Dietetic Association: Dietary supplements. J Am Diet Assoc 2005;105(3):460–70.
4. Dietary Supplement Health and Education Act of 1994. 1995. (Accessed March 15, 2006, at http://vm.cfsan.fda.gov/~dms/dietsupp.html.)
5. About the National Center for Complementary and Alternative Medicine. (Accessed March 15, 2006, at http://nccam.nih.gov/about/aboutnccam/index.htm.)
6. Eisenberg DM, Davis RB, et al. Trends in alternative medicine use in the United States, 1990–1997: results of a follow-up national survey. JAMA 1998;280(18):1569–75.
7. Eisenberg DM, Kessler RC, Foster C, Norlock FE, Calkins DR, Delbanco TL. Unconventional medicine in the United States. Prevalence, costs, and patterns of use. N Engl J Med 1993;328(4):246–52.
8. Kessler RC, Davis RB, et al. Long-term trends in the use of complementary and alternative medical therapies in the United States. Ann Intern Med 2001;135(4):262–8.
9. Barnes PM, Powell-Griner E, McFann K, Nahin RL. Complementary and alternative medicine use among adults: United States, 2002. Adv Data 2004(343): 1–19.
10. Kelly JP, Kaufman DW, Kelley K, Rosenberg L, Anderson TE, Mitchell AA. Recent trends in use of herbal and other natural products. Arch Intern Med 2005;165(3):281–6.
11. Burrowes JD, Van Houten G. Use of alternative medicine by patients with stage 5 chronic kidney disease. Adv Chronic Kidney Dis 2005;12(3):312–25.
12. Dahl NV. Herbs and supplements in dialysis patients: panacea or poison? Semin Dial 2001;14(3):186–92.
13. Markell MS. Potential benefits of complementary medicine modalities in patients with chronic kidney disease. Adv Chronic Kidney Dis 2005;12(3):292–9.
14. Colson CR, De Broe ME. Kidney injury from alternative medicines. Adv Chronic Kidney Dis 2005;12(3):261–75.
15. Li X, Wang H. Chinese herbal medicine in the treatment of chronic kidney disease. Adv Chronic Kidney Dis 2005;12(3):276–81.
16. Jellin J, Gregory P, et al. Pharmacist's Letter/Prescriber's Letter Natural Medicines Comprehensive Database. Stockton, CA: Therapeutic Research Faculty; 2002.
17. Ojewole JA. Analgesic, antiinflammatory and hypoglycaemic effects of ethanol extract of Zingiber officinale (roscoe) rhizomes (zingiberaceae) in mice and rats. Phytother Res 2006;20(9):764–72.

18. Combest W, Newton M, Combest A, Kosier JH. Effects of herbal supplements on the kidney. Urol Nurs 2005;25(5):381–6.
19. Vergili-Nelsen JM. Benefits of fish oil supplementation for hemodialysis patients. J Am Diet Assoc 2003;103(9):1174–7.
20. Robbers JE, Tyler VE. Tyler's Herbs of Choice: The Therapeutic Use of Phytomedicinals. New York: The Haworth Herbal Press; 1999.
21. Henney JE. Risk of drug interactions with St John's Wort. JAMA 2000;283(13):1679.
22. Vanherweghem LJ. Misuse of herbal remedies: the case of an outbreak of terminal renal failure in Belgium (Chinese herbs nephropathy). J Altern Complement Med 1998;4(1):9–13.
23. Bauer BA. Herbal therapy: what a clinician needs to know to counsel patients effectively. Mayo Clin Proc 2000;75(8):835–41.
24. Ernst E, Cohen MH. Informed consent in complementary and alternative medicine. Arch Intern Med 2001;161(19):2288–92.
25. Cohen MH. Legal issues in caring for patients with kidney disease by selectively integrating complementary therapies. Adv Chronic Kidney Dis 2005;12(3): 300–11.
26. Cohen MH, Eisenberg DM. Potential physician malpractice liability associated with complementary and integrative medical therapies. Ann Intern Med 2002;136(8):596–603.
27. Saubin A. The Health Professionals Guide to Dietary Supplements, 2nd edition. Chicago: The American Dietetic Association; 2002.

23 Issues Affecting Dietary Adherence

Jerrilynn D. Burrowes

LEARNING OBJECTIVES

1. Understand the factors that affect adherence.
2. Describe behavior change models and its implications for nutritional counseling.
3. Develop strategies to improve adherence.

Summary

Medical nutrition therapy is an integral part of the treatment plan for chronic kidney disease (CKD). Dietary adherence is a key component for successful treatment of CKD. However, a number of factors may inhibit or improve dietary adherence including lifestyle, attitudes toward the disease, economics, cultural barriers, and social support. This chapter reviews these factors and provides strategies for achieving dietary adherence.

Key Words: Dietary adherence; dietary compliance; chronic kidney disease; behavior change models; nutrition counseling.

1. INTRODUCTION

Medical nutrition therapy is an integral component for successful treatment outcomes in patients with chronic kidney disease (CKD). Treatment requires adherence to varied and complex dietary prescriptions which may change during the course of therapy. A number

From: *Nutrition and Health: Nutrition in Kidney Disease*
Edited by: L. D. Byham-Gray, J. D. Burrowes, and G. M. Chertow

of factors may influence a patient's ability to follow the recommended dietary treatment in both positive and negative ways. Factors that can inhibit dietary adherence include lifestyle, attitudes towards the disease, economics, and cultural barriers. Other factors can improve adherence such as social support and the health practitioner's knowledge of the patient's culture, food habits, beliefs, and practices. The health care provider needs to understand the factors that influence dietary adherence. This chapter will review these topics and will present strategies for achieving dietary adherence in patients with CKD.

2. DEFINITIONS

Adherence is the extent to which a person's behavior (e.g., taking medications, following a diet, and/or executing lifestyle changes) corresponds with agreed recommendations from a health care provider. The terms adherence and compliance are often used interchangeably in the literature. The main difference in the terms is that adherence requires the patient's agreement to the recommendations, as the patient and/or family member is an active collaborator with the health professional in the treatment process. On the other hand, compliance entails obedience to a directive (e.g., "take this medication 3 times a day") *(1)*.

Adherence is the single most important modifiable factor that compromises treatment outcome. It is also a primary determinant of the effectiveness of treatment because poor adherence attenuates optimum clinical benefit *(2,3)*. The best treatment can be rendered ineffective by poor adherence *(1)*.

3. DIETARY ADHERENCE

Dietary adherence is critical for the successful treatment of CKD. Poor adherence places individuals at risk for complications such as malnutrition, fluid overload, hyperkalemia, and hyperphosphatemia. It can also complicate the patient-health professional relationship and prevent an accurate assessment of the quality of care provided. In addition, patients with CKD usually have multiple comorbidities such as diabetes and hypertension, which make their treatment regimens more complex. These conditions are known to be characterized by poor rates of adherence and serve to further increase the likelihood of poor treatment outcomes *(1)*.

Poor adherence wastes health care resources, it causes preventable morbidity and mortality and the loss of health care dollars and productivity, and it jeopardizes patient care *(4)*. Poor adherence to dietary treatment usually goes undetected by health care providers *(5)*; it can confuse the clinical picture and diagnostic process and contribute to unnecessary testing and procedures resulting in inappropriate regimen changes, such as an increase in phosphate binders or changes in the dialysis prescription.

4. FACTORS AFFECTING DIETARY ADHERENCE

Adherence is a multidimensional phenomenon determined by the interaction of specific factors, such as social and economic factors, health care team and system-related factors, condition-related factors, therapy-related factors, and patient-related factors (Table 1). Health care providers must understand how these factors influence adherence and develop strategies which incorporate them in the education process.

4.1. Social and Economic Factors

Some of the social and economic factors which significantly affect adherence include poverty, illiteracy, a low level of education, unemployment, a lack of effective social support networks, unstable living conditions, long distance from treatment center, high cost of transportation to the clinic, high cost of medication, changing environmental situations, culture and lay beliefs about illness and treatment, and family dysfunction. Patient and family characteristics constitute additional sets of factors influencing adherence. In fact, the attitude and support of family members and significant others are possibly the most important factors or motivators for positive adherence *(1)*. The ability of the patient and/or family to understand the importance of following the prescribed treatment is an important element. Family members and/or significant others should attend nutrition counseling sessions with the patient to learn more about their condition and its treatment.

4.2. Health Care Team and System-Related Factors

Little research has been conducted on the effects of the health care team and system-related factors on adherence. A good patient-provider relationship may improve adherence; however, there are many factors that have a negative effect on adherence, such as lack of knowledge and training for health care providers on managing chronic diseases, overworked health care providers, lack of feedback on performance, short consultations, and lack of knowledge about adherence and effective interventions for improving it *(1)*.

Table 1
Factors Affecting Dietary Adherence to Medical Nutrition Therapy for the Control of Chronic Kidney Disease and Interventions for Improving Adherence

Factors affecting dietary adherence	*Interventions to improve adherence*
Social and economic factors	• Assessment of social needs • Family preparedness
Health care team and system-related factors	• Multidisciplinary care • Good patient-health care provider relationship • Training of health professionals on adherence • Identification of the treatment goals and development of strategies to meet them • Continuing education for health professionals • Continuous monitoring and reassessment of treatment • Non-judgmental attitude and assistance • Training in communication skills for health care providers
Condition-related factors	• Education on proper use of medications (e.g., phosphate binders) • Nutrition education
Therapy-related factors	• Simplification of regimens • Education on proper use of medications (e.g., phosphate binders)
Patient-related factors	• Behavior and motivational interventions • Good patient-health care provider relationship • Self-management of disease and treatment • Assessment of psychological needs

4.3. Condition-Related Factors

Illness-related demands faced by the patient represent condition-related factors. Strong determinants of adherence are those related to the severity of symptoms, level of disability (e.g., physical, psychological, social, and vocational), rate of progression and severity of the disease, and the availability of effective treatment. The impact of these factors depends on how they influence patients' risk perception, the importance of following the treatment, and the priority placed on adherence *(1)*.

4.4. Therapy-Related Factors

Therapy-related factors which affect adherence include the complexity of the medical regimen, the duration of treatment, previous treatment failures, frequent changes in treatment, the immediacy of beneficial effects, side effects, and the availability of medical support to deal with them *(1)*. A patient's ability to carry out the regimen as prescribed depends on how simple or difficult the regimen proves to be and what supports are available to assist the patient *(5)*.

Adherence is best achieved when the initial dietary regimen is simple, with complexities introduced gradually. For example, a simple dietary regimen may be restricting high potassium foods, whereas a more complex regimen may be following the renal diet exchange system. Furthermore, practical social difficulties such as not being able to coordinate the timing of phosphate binders with meals may also jeopardize adherence. Continuous reassurance may promote increased adherence when (and if) challenges arise.

4.5. Patient-Related Factors

The resources, knowledge, attitudes, beliefs, perceptions, and expectations of the patient represent patient-related factors that affect adherence. The patients' knowledge and beliefs about their illness, motivation to manage it, confidence (self-efficacy) in their ability to engage in illness management behaviors, and expectations regarding the outcome of treatment and the consequences of poor adherence influence adherence behavior *(1)*. A patient's motivation to adhere to a prescribed treatment is influenced by the value that he or she places on following the regimen (cost–benefit ratio) and the degree of confidence in being able to follow it *(6)*. Long-term and particularly lifetime regimens require commitment.

5. BEHAVIOR CHANGE MODELS

Learning new, complex patterns of behavior normally requires modifying many of the small behaviors that compose an overall complex behavior. Some behavior change models have many similarities (i.e., they address how individual factors such as knowledge, attitudes, beliefs, prior experience, and personality influence behavioral choices). These include the health belief model, the theory of

Table 2
Behavior Change Models and Theories Which Address How Individual Factors Such as Knowledge, Attitudes, Beliefs, Prior Experience, and Personality Influence Behavior Choices

Theory/model	*Focus*	*Key concepts*
Health belief model	Peoples' perceptions of the threat of a health problem and appraisal of behavior recommended to prevent or manage a problem	• Perceived susceptibility • Perceived severity • Perceived benefits of action • Perceived barriers to action • Cues to action • Self-efficacy
Theory of reasoned action/theory of planned behavior	People are rational beings whose intention to perform a behavior strongly relates to its actual performance through beliefs, attitudes, subjective norms, and perceived behavioral control	• Behavioral intention • Subjective norms • Attitudes • Perceived behavioral control
Stages of change/ transtheoretical model	Readiness to change or attempt to change a health behavior varies among individuals and within an individual over time	• Precontemplation • Contemplation • Preparation • Action • Maintenance • Relapse

reasoned action/theory of planned behavior, and the transtheoretical model/stages of change (Table 2). Each model implies that patients make rational decisions regarding future events and, given the appropriate skills, they can establish goals and modify or regulate behavior to achieve these goals *(7)*. However, counseling patients based on the stages of change model appears to be effective because the focus of the counseling session will depend on where the patient is in the stage of change.

5.1. Stages of Change

Identifying the patient's readiness to make dietary changes is useful when planning interventions. The transtheoretical model, also referred to as the stages of change model, suggests that behavioral change is a process that occurs gradually, with the patient moving back and forth through a series of fairly predictable stages [i.e., from precontemplation (not considering change), to motivation to change (contemplation), before making a commitment to change (preparation), actually making the change (action), and working to prevent relapse (maintenance)]. Relapse is a common occurrence and part of the normal process of working toward lifelong change (Table 3) *(8)*. Each patient's readiness to change must be assessed continuously, and nutrition counseling should be tailored according to the patient's stage of change.

6. STRATEGIES FOR ACHIEVING DIETARY ADHERENCE

Strategies to improve dietary adherence fit into one of three categories: educational, behavioral, and organizational *(9–11)*. Effective nutrition education is the first step in achieving dietary change. Education regarding the nutrition management of the disease should raise the patient's level of adherence to an acceptable level, but probably not to a desired level. Education is better than no intervention, but it is not sufficient to raise adherence to the desired level *(12)*.

Educational interventions rely on the transmission and dissemination of information and instructions with or without motivational appeal, with the intermediate objective of affecting a patients' knowledge and attitudes *(13)*. However, education alone is usually not sufficient to achieve long-term dietary adherence and sustain behavioral change. The use of innovative educational techniques such as games or videotapes, which increase patient knowledge and attitudes about their condition and which promote dietary adherence, should be employed.

Table 3
Stages of Change Model and Implications for Nutritional Counseling

Stage of change	*Patient characteristics*	*Implications for nutritional counseling*
Precontemplation	• Not intending to take action in the foreseeable future (up to 6 months) • Often characterized as lacking control • Denial or resignation is common • Not yet convinced of seriousness of condition	• Relationship building is key • Educate about the underlying disease process • Put minimal emphasis on specific dietary changes • Focus on overcoming individual barriers
Contemplation	• Considering advantages and disadvantages of changing behavior • Apprehensive about behavior change because it may have failed in the past • Not ready to commit to making a change	• Educate about the long-term complications of disease, the role of diet in medical management, and expected benefits • Begin to establish a timeline for change
Preparation	• Ready to change behavior (within the next 2 weeks) • Need assistance with problem-solving and social support • Taking tentative steps to change behavior	• Provide initial simplified diet instruction • Begin to set goals and encourage self-management/self-monitoring behaviors • Reinforce initial successful behavioral changes • Build self-efficacy

Action	• Taking significant steps to change behavior	• Use progressive nutrition education to develop self-management skills • Offer continued reinforcement for successful incorporation of new behaviors into the individual's lifestyle • Develop goal-setting and self-monitoring skills
Maintenance	• Incorporating new behaviors into lifestyle and sustaining them • Maintenance change in behavior	• Develop self-management and coping skills • Develop skills for managing diet in new/difficult social situations • Build self-management, goal-setting, and monitoring skills
Relapse	• Experiencing a normal part of the process of change • Usually feels demoralized	• Reinforce self-management skills and develop social supports • Address factors associated with relapse • Build self-efficacy and self-monitoring skills • Emphasize what can be learned from relapse versus focusing on failure

Nutrition information should be sensitive to the personal characteristics of the patient, including attitudes, cultural norms, beliefs, and reading skills *(14)*.

Behavioral strategies are procedures that attempt to influence specific non-adherent behaviors directly through the use of techniques such as reminders, tailoring, contracting, self-monitoring, reinforcement, and family/peer support, but with information and instruction playing a secondary role *(9)*. However, behavioral strategies fail to maintain consistent change in the long term *(10)*.

Organizational strategies focus primarily on clinic and regimen convenience and on the utilization of personnel for fostering dietary adherence *(11)*. Organizational change can prevent or reduce adherence problems, often without the need for altering provider workloads or increasing budgets. Examples include making special appointments at odd hours or telephoning patients at home with abnormal lab results.

7. SUMMARY

Dietary adherence is a critical component for successful management of CKD. Several factors can inhibit or improve dietary adherence. Educational, behavioral and organizational strategies may improve adherence. Dietitians who educate patients with CKD and/or their family members need to be aware of these factors and the strategies, which may improve adherence.

REFERENCES

1. World Health Organization. Adherence to long term therapies: Evidence for action. World Health Organization, 2003. Retrieved from: <http://www.emro.who.int//ncd/Publications/adherence_report.pdf>. Accessed March 18, 2006.
2. The World Health Report 2002. Reducing Risks, Promoting Healthy Life. Geneva, World Health Organization, 2002.
3. Dunbar-Jacob J, Erlen JA, Schlenk EA, Ryan CM, Sereika SM, Doswell WM. Adherence in chronic disease. Annu Rev Nurs Res 2000; 18:48–90.
4. DiMatteo MR. Variations in patients' adherence to medical recommendations. Med Care 2004; 42(3):200–209.
5. DiMatteo MR, Reiter RC, Gambone JC. Enhancing medication adherence through communication and informed collaborative choice. Health Commun 1994; 6(4):253–265.
6. Miller W, Rollnick S. Motivational Interviewing. New York: Guilford Press, 1999.
7. Brawley LR, Culow-Reed SN. Studying adherence to therapeutic regimens: overview, theories, recommendations. Control Clin Trials 2000;21:156S–163S.

8. Prochaska JO, Johnson S, Lee P. The transtheoretical model of behavior change. In: Schumaker SA, Schron E, Ockene JK, McBee WL, eds. The Handbook of Health Behavior Change, 2nd ed. New York: Springer Publishing, 1998:59–84.
9. Dunbar JM, Marshall GD, Hovell MF. Behavioral strategies for improving compliance. In: Haynes RB, Taylor DW, Sacket DL, eds. Compliance in Health Care. Baltimore: Johns Hopkins University Press, 1979:174–190.
10. Leventhal H, Cameron L. Behavioral theories and the problem of compliance. Patient Educ Couns 1987;10:117–138.
11. Gibson ES. Compliance and the organization of health services. In: Haynes RB, Taylor DW, Sacket DL, eds. Compliance in Health Care. Baltimore: Johns Hopkins University Press, 1979:278–285.
12. Elliott WJ. Compliance strategies. Curr Opin Nephrol Hypertens 1994;3:271–278.
13. Green LW. Educational strategies to improve compliance with therapeutic and preventive regimens: the recent evidence. In: Haynes RB, Taylor DW, Sacket DL, eds. Compliance in Health Care. Baltimore: Johns Hopkins University Press, 1979:157–173.
14. Sherman AM, Bowen DJ, Vitolins M, Perri MG, Rosal MC, Sevick MA, Ockene JK. Dietary adherence: characteristics and interventions. Control Clin Trials 2000;21:206S–211S.

24 Outcomes Research

Laura D. Byham-Gray

LEARNING OBJECTIVES

1. To define the principles and processes of outcomes research.
2. To discuss potential outcomes research projects in nutrition and kidney disease.
3. To identify the role of nutrition in patient outcomes among individuals diagnosed with kidney disease.

Summary

The morbidity and mortality rates among patients diagnosed with chronic kidney disease remain unacceptably high in spite of recent advances in medicine and technology. Differences in patient outcomes exist among the industrialized countries and are partly explained by variances in clinical practice. Outcomes research has been the primary methodology used to more fully investigate the root causes for practice variation. Research has established the relationships between nutritional status with morbidity and mortality. Although several nutrition parameters are prognostic indicators, they are complex to understand, and one specific measure that definitively diagnoses nutritional risk is lacking. Further scientific inquiry should explore the effectiveness of medical nutrition therapy on key nutrition-related outcomes, thereby answering some of the questions plaguing current clinical practice.

Key Words: Outcomes research; kidney disease; nutrition; morbidity; mortality; evidence-based practice; Dialysis Outcomes and Practice Patterns Study; Kidney Disease Outcomes Quality Initiative; Kidney Disease: Improving Global Outcomes.

From: *Nutrition and Health: Nutrition in Kidney Disease*
Edited by: L. D. Byham-Gray, J. D. Burrowes, and G. M. Chertow

1. INTRODUCTION

Despite advances in medicine and technology, clinical outcomes among patients diagnosed with chronic kidney disease (CKD) have remained suboptimal. For example, the 5-year survival rates among individuals diagnosed with end-stage kidney disease (ESKD) who were greater than 45 years of age is less than 50% *(1)*. Patients diagnosed with CKD on maintenance dialysis often experience a lower quality of life, greater risk of morbidity, higher rates for hospitalization, and increased mortality when compared with the general population *(2)*. The Dialysis Outcomes and Practice Patterns Study (DOPPS), which will be discussed in Subheading 7, has identified that the annual mortality rate among dialysis patients in the United States is approximately 22% *(3)*. Japan and Europe report substantially lower first-year crude mortality rates, 7% and 16%, respectively *(3,4)*. Although the United States dialyzes patients with higher mean ages (60.5 ± 15.5 years) and a larger number of comorbid diseases, when regression models were adjusted for such case-mix factors, the United States still had higher overall morbidity and mortality rates.

Historically, much of the focus in kidney disease had been on the burgeoning ESKD population and the respective renal replacement therapies necessary for life maintenance, that is, hemodialysis (HD), peritoneal dialysis (PD), and transplantation. For the first time in decades, the incidence of individuals advancing to ESKD has slowed and may be related to the greater emphasis on early screening for CKD in primary care settings leading to more timely and appropriate intervention *(1)*. Even though only about 2% of the 20 million individuals diagnosed with CKD progress to ESKD, they are also more likely to die than to initiate dialysis *(5)*. Obviously, newer approaches to CKD management and treatment are warranted. Capturing the patient once at stage 5 CKD may be too late to make meaningful changes in the factors associated with poorer outcome *(6)*.

This chapter will give a brief overview of the importance for studying the role that nutrition status has on key clinical outcomes in CKD, a clear description of outcomes research and its related methodology as well as a thorough discussion of clinical guidelines and their role in reducing practice variation.

2. CHALLENGES FOR NUTRITION

The importance of nutrition in the treatment and management of CKD is unquestionable. The relationship between nutritional status and morbidity and mortality has been researched extensively. There are a multitude of outcomes across the spectrum of kidney disease that could and should be measured. For example, bone disease, diabetes, dyslipidemias, hypertension, dialysis adequacy, and anemia are all either directly or indirectly related to nutrition intervention. Understanding how nutritional status impacts morbidity and mortality is seemingly more difficult than studying either dialysis adequacy or anemia management *(7)*. Malnutrition is an independent contributor for mortality risk. However, it remains unclear whether malnutrition occurs over a period of time or as the result of a suboptimal status at the time of dialysis initiation, that is, is the association between malnutrition and death secondary to changes in nutrition status experienced over time or rather the presence of abnormalities at baseline *(8)*? Multiple factors may explain the complexity of defining, treating, and reversing the malnutrition experienced and comprise both nutritional and non-nutritional components *(8)*. Nutritional status can be assessed and "diagnosed" by either using anthropometric measures, biochemical indices, clinical symptoms, or dietary intake records separately or together, therein lies the difficulty, the lack of one single measure that provides a good estimate of nutritional status. Compounding these challenges is the impact of metabolic aberrations and hemodynamic imbalances secondary to CKD that may falsely affect nutrition parameters, for example, non-nutritional factors such as inflammation or hydration status may interfere with the reliability of measuring nutritional status through conventional means.

Nutrition intervention or medical nutrition therapy (MNT) does make a positive impact on patient health outcomes. Studies have reported the effectiveness of nutrition intervention for a number of clinical outcomes related to disease states/conditions such as diabetes, hyperlipidemia, cancer, unintentional weight loss, as well as outcomes related to cost and health care utilization *(9–16)*. There have been studies that have measured the impact of nutrition intervention (e.g., counseling, oral supplementation, educational programs, and intradialytic parenteral nutrition on CKD patients), but they are limited by sample size, study duration, and no randomized controlled trials have specifically measured the impact of MNT on malnutrition and improved outcomes *(8,17–20)*. More research is needed to determine how nutrition therapy or dietetic practice patterns may affect outcomes. DOPPS may provide such key information that links "best evidence" to "best practices," thereby enabling the exploration of possible predictors and uncovering whether an improvement of poor outcomes can be assured with nutritional and/or inflammatory intervention.

3. OUTCOMES RESEARCH DEFINED

"The American health care delivery system is in need of fundamental change" is the opening statement to the Institute of Medicine's (IOM) report entitled *Crossing the Quality Chasm: A New Health Care System for the 21st Century, 2001 (21)*. It is a provocative beginning for "what works and what doesn't work in health care" *(22)*. Thus, the goals for outcomes research are really to determine *(23)* (i) which treatments are the most effective, (ii) which providers give the best care, (iii) which health plans are the most efficient, (iv) which delivery systems provide the most patient-centered care, and (v) who produces the best outcomes? Obviously, patients, providers, payers, and policymakers are all interested in the answers to the above questions.

Outcomes research is often referred to as the "third revolution in health care" and is defined as "the process of obtaining data to measure the effect of a particular intervention on patient care" *(22)*. It may sometimes be referred to as "medical effectiveness research (MER)" or "outcomes effectiveness research (OER)." Regardless of the acronym, outcomes research collects and analyzes data with the intent of aiding patients, health care professionals, insurers, and administrators in the selection of suitable medical treatment options and setting health care policy *(24)*.

The benefits for conducting outcomes research are multiple *(25,26)*: it identifies best practices and improves the knowledge base of medical and health sciences, quantifies cost-effectiveness of therapeutic interventions, supplies the basis for practice guidelines, formulates methods for continuous quality improvement (CQI), and sets benchmarking thresholds, aids in market decisions, and provides accountability.

There is a distinction to be made between outcomes management and outcomes research; both are equally important for patient care. Outcomes management is the outcomes data routinely collected by practitioners and lays the groundwork for outcomes research *(27)*. Outcomes management allows the practitioner to participate in research at a more basic level and fosters further professional development in research skills. Outcomes research employs controlled research procedures.

Seemingly, CQI or process improvement interface with outcomes management and outcomes research. CQI includes the analysis of the process(es), identification of key quality characteristics of the process, as well as the outcomes of interest. The measurements will often include data collection on key process variables rather than direct clinical care that are hypothesized to influence patient outcomes, such as, frequency of contact, length of time between encounters, or when appointment scheduling occurs, and so on. Outcomes research, on the other hand, focuses closely on the impact of the treatment on patient outcome. Generally, outcomes projects lead to changes in practice standards. Quality of care is then measured according to such standards; this aspect represents CQI.

Table 1
Topics Addressed in the ESRD Clinical Performance Measures Based on National Kidney Foundation Dialysis Outcome Quality Initiative Guidelines

Hemodialysis adequacy	Monthly measurement of delivered hemodialysis dose
	Method of measurement of delivered hemodialysis dose
	Minimum delivered hemodialysis dose
	Method of postdialysis blood urea nitrogen sampling
	Baseline total cell volume measurement of dialyzers intended for reuse
Peritoneal dialysis adequacy	Measurement of total solute clearance at regular intervals
	Calculate weekly Kt/V_{urea} and creatinine clearance in a standard way
	Delivered dose of peritoneal dialysis
Vascular access management	Maximizing placement of arteriovenous fistulae
	Minimizing use of catheters as chronic dialysis access
	Preferred/nonpreferred location of hemodialysis catheters located above the waist
	Monitoring arteriovenous grafts for stenosis
Anemia management	Target hematocrit for epoetin therapy
	Assessment of iron stores in anemic patients or those prescribed epoetin
	Maintenance of iron stores: target
	Administration of supplemental iron

Adapted from *(28)*.

One example of quality measures that serve as benchmarks is the End Stage Renal Disease (ESRD) Clinical Performance Measures (CPMs) Project funded by the Centers for Medicare and Medicaid Services (CMS) *(28)*. This project was completed secondary to the Balanced Budget Act (BBA) which mandated that "CMS develop and implement, by January 1, 2000, a method to measure and report the quality of renal dialysis services provided under the Medicare program" *(28)*. CMS formulated performance measures based on the National Kidney Foundation's (NKF) Dialysis Outcome Quality Initiative (DOQI) Clinical Practice Guidelines and changed the original benchmarks of the ESRD Core Indicators Project to include algorithms that applied the CPMs as well as generated a number of data collection instruments and their respective methodologies. Sixteen ESRD CPMs are currently rated: five for HD adequacy, three for PD adequacy, four for anemia management, and four for vascular access. The CPMs are summarized in Table 1. Such indicators although measured in the context of outcomes management can be rich sources of data for potential outcomes research projects.

4. TYPES OF OUTCOMES

Although categories in the literature may slightly vary, there are generally three "types of outcomes": clinical, patient-oriented, and economic. Clinical outcomes focus on health status outcomes and can include mortality, risk factors, changes in development or progression of symptoms, disease and its sequelae, and complications from treatment *(22)*. Some sample types related to nutrition and CKD are provided in Table 2 *(29)*. Patient-oriented outcomes give attention to the consequences of intervention that are of concern to patients/families, such as survival, symptom relief, adverse effects of condition or its treatment, functional status, quality of life, and satisfaction. Economic outcomes are related to indicators that reduce length of stay, minimize care costs, or maximize revenue generation. A number of resources are available on outcomes research that the practitioner may find useful which fully define validated measurements to use for studying these types of outcomes *(23,26,30,31)*. For example, if the practitioner wanted to study the effects of nutrition status on quality of life (patient-oriented outcome), she/he would need to consult a number of validated tools to determine which one measured the constructs of interest.

Attributes of good outcome variables are "objective, precise, quantitative and translatable" *(22)*. Dhingra and Laski *(32)* add that they should be "valid, reproducible, actionable, and comparable over geographic, demographic and temporal boundaries"; allowing for

Table 2
Main Types of Outcome Measures with Nutrition-Related Examples Cited

Clinical outcomes	*Economic outcomes*	*Patient-oriented outcomes*
Mortality rate	MNT reimbursement	Ability to live independently
Weight status	Length of stay	Perception of patient care received
Body mass index	Hospitalizations	Health-related quality of life
Subjective global assessment	Delays in CKD progression	Symptom relief from early satiety
Albumin or pre-albumin levels	Cost of enteral versus parenteral nutritional supplementation	Functional status
Lean body mass		
C-reactive protein	Cost benefits of MNT versus other adjunctive therapies	
Normalized protein catabolic rate		
Interdialytic weight gains		

CKD, chronic kidney disease; MNT, medical nutrition therapy. Adapted from *(29)*.

benchmarking to occur. Thus, one way to initiate more outcomes research in clinical practice specifically in kidney disease is to use the outcomes measures published in evidence-based practice guidelines.

5. EVIDENCE-BASED PRACTICE GUIDELINES

Variances in patient outcome and survival among industrialized countries (e.g., United States, Europe, Canada, Australia and Japan) were first recognized in 1989 at the Dallas symposium on morbidity and mortality of dialysis patients *(33)*. The United States reported the largest number of new patients with ESKD maintained on dialysis but also registered the highest crude mortality rate, ranging from 22 to 24%. Such results motivated several organizations and regulatory agencies (e.g., National Institutes of Health, Health Care Financing Administration, and Kidney Physicians Association) to focus on efforts for improving the quality of care delivered to dialysis patients in the United States. This emphasis on "medical effectiveness" led to the creation of the DOQI Project of the NKF in 1995. To reflect a broader mission of improving the health status of patients across the spectrum of kidney disease, DOQI was later named in 1999 as the Kidney Disease Outcomes Quality Initiative (KDOQI). The first DOQI guidelines were released in 1997 with subsequent updates and topics expanded from dialysis adequacy, vascular access, and anemia to PD, nutrition, CKD, dyslipidemia, bone disease, hypertension, and diabetes *(34)*.

Concomitantly, other countries initiated their own system for creating and developing practice guidelines *(35)*. Although similar recommendations were established internationally, the target ranges or values for specific outcome measures and how the evidence was rated were highly variable. A more uniform approach for evidence analysis was needed, therefore the concept of KDIGO (which stands for Kidney Disease: Improving Global Outcomes) was conceived. Its mission is "to improve care and outcomes of kidney disease patients worldwide through promoting coordination, collaboration, and integration of initiatives to develop and implement clinical practice guidelines" *(36)*.

6. PRACTICE GUIDELINES AND PATIENT CARE

There is limited research whether practice guidelines actually affect clinical practice. Nonetheless, successful implementation of evidence-based guidelines can improve outcomes *(37,38)*. As mentioned earlier, Sugarman and associates *(28)* have outlined the methods for making the CPMs measurable indicators of quality; based on the KDOQI evidence-based guidelines. The American Dietetic Association has published a number of *Evidence-Based Guides for Practice*; one is specific to non-dialysis kidney disease, and they provide a number of reliable indicators to measure nutritional status and to monitor outcomes *(39)*. A great initial outcomes research project may be to examine whether integrating such guidelines does impact care. It represents a simple study design, and no studies currently exist in nutrition and CKD.

Nonetheless, there are challenges to implementing guidelines into practice; often out of the practitioner's purview. Burrowes and colleagues *(40)* surveyed renal dietitians about whether they implemented the KDOQI Nutrition Guidelines. The vast majority (92%) had integrated at least one guideline into practice, whereas only 5% had implemented all of them. Dietitians were unable to change their clinical practice according to the best evidence, as a number of barriers existed, such as unavailable equipment or tools (e.g., computers, food models, and calipers), high patient-to-dietitian staffing ratios, or the lack of administrative support for change. Thus, it is increasingly difficult to determine the impact of practice guidelines on patient outcomes if resources are lacking for their successful implementation. Given such concerns in general clinical practice, the DOPPS has attempted to implement accepted standards indicated in KDOQI *(35)*.

7. DIALYSIS OUTCOMES AND PRACTICE PATTERNS STUDY

Discrepancies in patient outcomes are believed to be largely related to varying practice patterns. A detailed investigation entitled the Dialysis Outcomes and Practice Patterns Study was initiated in 1996. The goal of DOPPS was "to increase the longevity of patients on hemodialysis" and was very similar to KDOQI, but just used different steps at reaching the goal. DOPPs measures a large number of practice patterns to detect new evidence for modifiable treatment factors that are associated with improved outcomes *(35)*. DOPPS is a large, international, observational, prospective HD study which originally involved seven countries (Japan, United States, and five European countries, France, Germany, Italy, Spain and the United Kingdom) and was expanded to 12 with the addition of Australia/New Zealand, Belgium, Canada, and Sweden *(41)*. A more detailed discussion of the study is available elsewhere *(41,42)*. DOPPS gathers substantial data regarding the patient's demographic characteristics, medical history, laboratory values, drug therapies and prescriptions, and dialysis unit practices and outcomes *(43)*. It seeks to identify what dialysis practices reduce mortality rates, lower hospitalizations, enhance health-related quality of life, and improve vascular access outcomes after controlling for the effects of comorbid diseases and demographic variables. Approximately 70 papers have been published using this dataset, with results reported concerning key nutrition indicators *(44–46)*. Nutrition-related outcomes associated with increased mortality include low body mass index, substandard Subjective global assessment (SGA) score, and hypoalbuminemia. Poorer nutrition-related outcomes are cited among U.S. patients in comparison to other countries. In future analyses, DOPPS is expected to supply more insight about therapeutic interventions used at the dialysis facilities that may result in better nutrition-related outcomes and provide direction toward what variables to study more closely to affect clinical practice changes.

8. SUMMARY

The incidence of CKD patients is expanding in the United States, but sadly, poor outcomes prevail. The complexity of malnutrition impedes its understanding, treatment, and subsequent elimination as a contributor toward the morbidity and mortality experienced in this specific patient population. The generation of evidence-based practice guidelines assists in positively affecting change in practice to optimize outcomes and serves as sources for measurable indicators in outcomes research. Existing databases have revealed a strong link between nutrition and outcomes in CKD. More research should concentrate on these areas of nutritional intervention in order to improve outcomes, provide direction for future study, and potentially create changes in clinical practice.

REFERENCES

1. National Institutes of Health National Institute of Diabetes, Digestive and Kidney Diseases (NIoDaDaKD). USRDS 2005 Annual Data Report: Atlas of End-Stage Renal Disease in the United States. Bethesda, MD, 2005.
2. Nicolucci A, Procaccini, DA. Why do we need outcomes research in end-stage kidney disease? J Nephrol 2000;13:401–404.
3. Goodkin D, Young EW, Kurokawa K, Prutz K-G, Levin N. Mortality among hemodialysis patients in Europe, Japan, and the United States: Case-Mix effects. Am J Kidney Dis 2004;44:S16–S21.
4. Goodkin D, Bragg-Gresham JL, Koenig KG, et al. Association of comorbid conditions and mortality in hemodialysis patients in Europe, Japan, and the United States: The Dialysis Outcomes Practice Patterns Study (DOPPS). J Am Soc Nephrol 2003;14:3270–3277.
5. Keith D, Nichols GA, Gullion C, et al. Longitudinal follow-up and outcomes among a population with chronic kidney disease in a large managed care organization. Arch Intern Med 2004;164:659–663.
6. Komenda P, Levin A. Analysis of cardiovascular disease and kidney outcomes in multidisciplinary chronic kidney disease clinics: Complex disease requires complex care models. Curr Opin Nephrol Hypertens 2006;15:61–66.
7. Szczech L. What has outcomes research taught us about evidence-based treatment of end-stage kidney disease? J Clin Outcomes Manag 2002;9:509–514.
8. McClellan W, Rocco MV, Flanders WD. Epidemiologic cohort studies of critical nutritional issues in the care of the dialysis patient. Report of the Epidemiology Work Group. J Ren Nutr 1999;9:133–137.
9. Lemon C, Lacey K, Lohse B, et al. Outcomes monitoring of health, behavior, and quality of life after nutrition intervention in adults with type 2 diabetes. J Am Diet Assoc 2004;104:1805–1815.
10. Franz M, Splett PL, Monk A. Cost-effectiveness of medical nutrition therapy provided by persons with non-insulin-dependent diabetes mellitus. J Am Diet Assoc 1995;95:1018–1024.

11. McGehee M, Johnson EQ, Rasmussen HM, et al. Benefits and costs of medical nutrition therapy by registered dietitians for patients with hypercholesterolemia. J Am Diet Assoc 1995;95:1041–1043.
12. Delahanty L, Sonnenberg LA, Hayden D, et al. Clinical and cost outcomes of medical nutrition therapy in hypercholesterolemia: A controlled trial. J Am Diet Assoc 2001;101:1012–1023.
13. Tosteson A, Weinstein MC, Hunink MGM, et al. Cost-effectiveness of population wide educational approaches to reduce serum cholesterol levels. Circulation 1997;95:24–30.
14. Pavlovich W, Waters H, Weller W, et al. Systematic review of literature on cost-effectiveness of nutrition services. J Am Diet Assoc 2004;104:226–232.
15. Isenring E, Capra S, Bauer JD. Nutrition intervention is beneficial in oncology outpatients receiving radiotherapy to the gastrointestinal or head and neck area. Br J Cancer 2004;91:447–452.
16. Splett P, Roth-Yousey LL, Vogelzang JL. Medical nutrition therapy for the prevention and treatment of unintentional weight loss in residential healthcare facilities. J Am Diet Assoc 2003;103:352–362.
17. Krause I, Shamir R, Davidovits M, et al. Intradialytic parenteral nutrition in malnourished children treated with hemodialysis. J Ren Nutr 2002;12:55–59.
18. Shaw-Stuart N, Stuart A. The effect of an educational patient compliance program on serum phosphate levels in patients receiving hemodialysis. J Ren Nutr 2000;10:80–84.
19. Wilson B, Fernandez-Madrid A, Hayes A, et al. Comparison of the effects of two early intervention strategies on the health outcomes of malnourished hemodialysis patients. J Ren Nutr 2001;11:166–171.
20. Beutler K, Park GK, Wilkowski MJ. Effect of oral supplementation on nutrition indicators in hemodialysis patients. J Ren Nutr 1997;7:77–82.
21. Institute of Medicine. Crossing the Quality Chasm: A New Health System for the 21st Century. Washington, DC: National Academy Press.
22. Splett PL. Outcomes research and cost-effectiveness. In: Monsen E, ed. Research: Successful Approaches, 2nd Edition. Chicago, IL: The American Dietetic Association, 2003.
23. Iezzoni L. Risk Adjustment for Measuring Health Care Outcomes. Chicago, IL: Health Administration Press, 2003.
24. August DA. Creation of a specialized nutrition support outcomes research continuum: If not now, when? JPEN J Parenter Enteral Nutr 1996;20:394–400.
25. Inman-Felton A. The fundamentals of outcomes research: Demonstrating the effectiveness of nutrition interventions. Support Line 1997:13–17.
26. Kane R. Understanding Health Care Outcomes Reseach. Boston, MA: Jones & Bartlett Publishers, 2004.
27. Biesemeier C. Demonstrating the effectiveness of medical nutrition therapy. Top Clin Nutr 1999;14:13–24.
28. Sugarman J, Frederick PR, Frankenfield DL, et al. Developing clinical performance measures based on the Dialysis Outcomes Quality Initiative Clinical Practice Guidelines: Process, outcomes, and implications. Am J Kidney Dis 2003;42:806–812.
29. Schiller MR, Moore C. Practical approaches to outcomes evaluation. Top Clin Nutr 1999;14:1–12.

30. Block DJ. Healthcare Outcomes Management: Strategies for Planning and Evaluation. Boston, MA: Jones and Bartlett Publishers, Inc., 2006.
31. Nolan M, Mock, V. Measuring Patient Outcomes. Thousand Oaks, CA: Sage Publications Inc., 2000.
32. Dhingra H, Laski, ME. Outcomes research in dialysis. Semin Nephrol 2003;23:295–305.
33. Hull A, Parker, T. Proceedings from the Morbidity, Mortality, and Prescription of Dialysis Symposium. Dallas, Texas. September 15–17, 1989. Am J Kidney Dis 1990;15:365–383.
34. National Kidney Foundation. Clinical Practice Guidelines. Available at: http://www.kidney.org. Accessed January 7, 2007.
35. Port F, Eknoyan G. The Dialysis Outcomes and Practice Patterns Study (DOPPS) and the Kidney Disease Outcomes Quality Initiative (K/DOQI): A cooperative initiative to improve outcomes for hemodialysis patients worldwide. Am J Kidney Dis 2004;44:S1–S6.
36. Kidney Disease: Improving Global Outcomes (KDIGO). Clinical Practice Guidelines. Available at: http://www.kdigo.org. Accessed January 7, 2007.
37. Collins A, Roberts TL, St Peter WL, et al. United Renal Data System assessment of the impact of the National Kidney Foundation-Dialysis Outcomes Quality Initiative guidelines. Am J Kidney Dis 2002;39:784–795.
38. Grol R, Grimshaw, J. From best evidence to best practice: Effective implementation of change in patients' care. Lancet 2003;362:1225–1230.
39. The American Dietetic Association. Medical Nutrition Therapy Evidence-Based Guides for Practice: Chronic Kidney Disease (non-dialysis) Medical Nutrition Therapy Protocol. Chicago, Illinois: The American Dietetic Association, 2002.
40. Burrowes J, Russell GB, Rocco, MV. Multiple factors affect renal dietitians' use of the NKF K/DOQI Adult Nutrition Guidelines. J Ren Nutr 2005;15:407–426.
41. Pisoni R, Gillespie BW, Dickinson DE, et al. The Dialysis Outcomes and Practice Patterns Study: Design, data elements, and methodology. Am J Kidney Dis 2004;44:S7–S15.
42. Pisoni R, Greenwood, RN. Selected lessons learned from the Dialysis Outcomes and Practice Patterns Study (DOPPS). Contrib Nephrol 2005;149:58–68.
43. Goodkin D, Mapes DL, Held PJ. The Dialysis Outcomes Practice Patterns Study (DOPPS): How can we improve the care of hemodialysis patients? Semin Dial 2001;14:157–159.
44. Pifer T, McCullough KP, Fort FK, et al. Mortality risk in hemodialysis patients and changes in nutritional indicators: Dialysis Outcomes Practice Patterns Study (DOPPS). Kidney Int 2002;62:2238–2245.
45. Hecking E, Bragg-Gresham JL, Rayner HC, et al. Haemodialysis prescription and nutritional indicators in five European countries: Results from the Dialysis Outcomes Practice Patterns Study (DOPPS). Nephrol Dial Transplant 2004;19:100–107.
46. Combe C, McCullough KP, Asano Y, et al. Kidney Disease Outcomes Quality Initiative (K/DOQI) and the Dialysis Outcomes and Practice Patterns Study (DOPPS): Nutrition guidelines, indicators, and practices. Am J Kidney Dis 2004;44 Suppl 2:S39–S46.

25 Suggested Resources for the Practitioner

Patricia DiBenedetto-Barbá, Jerrilynn D. Burrowes, and Laura D. Byham-Gray

LEARNING OBJECTIVES

1. To find information on topics related to kidney disease and related fields.
2. To access information in a systematic and well-defined manner.
3. To share information and to distribute it to those persons who are in need and will benefit from this knowledge.
4. To identify the vast number of resources available.

Summary

Information is one of the most important resources for the practitioner. This chapter provides many of the most up-to-date resources in print and on the internet for the practitioner involved in the care of patients with kidney disease. This subject matter is constantly changing and being updated. The resources listed are of interest to professionals, clients and other interested individuals. The topics covered include information on kidney disease, diabetes, hypertension, and other related areas of these general topics. However, the sites and information listed do not constitute an endorsement by the authors or editors.

Key Words: Potassium sources; phosphorus sources; phosphorus additives; vitamins; oxalate; renal cookbooks.

From: *Nutrition and Health: Nutrition in Kidney Disease*
Edited by: L. D. Byham-Gray, J. D. Burrowes, and G. M. Chertow

1. INTRODUCTION

This chapter was developed with the practitioner in mind. Its primary goal is to be a handy reference for the enumerable resources needed and used by the practitioner in chronic kidney disease (CKD). The chapter's organization includes four basic sections: (i) evidence-based practice guidelines, (ii) diet-related resources and food lists, (iii) critical tools for conducting nutrition assessments and delivering quality care, and (iv) internet websites. Much of what is contained in this chapter may augment topics already presented and discussed in earlier chapters.

2. EVIDENCE-BASED PRACTICE GUIDELINES

Practitioners in CKD have several evidence-based practice guidelines at their disposal that will assist them in evaluating each patient's status and making appropriate clinical decisions for care delivery. The National Kidney Foundation (NKF) and their appointed Work Groups as part of the Kidney Disease Outcomes Quality Initiative (KDOQI) have published a number of practice guidelines related to topics of dialysis adequacy, anemia, bone disease, and of course, nutrition, as well as many others. The reader is encouraged to refer the NKF website (http://www.kidney.org) for routine updates and a full explanation of each practice guideline. Due to space limitations, what is addressed in the next section are practice guideline statements that discuss nutrition management in CKD for adults and children, bone disease, and diabetes only.

2.1. KDOQI Practice Guideline Statements

2.1.1. Clinical Practice Guidelines for Nutrition in Chronic Renal Failure (http://www.kidney.org/professionals/KDOQI/guidelines_updates/doqi_nut.html)

I. Adult Guidelines

A. Maintenance Dialysis

1. Evaluation of Protein-Energy Nutritional Status

Guideline 1. Use of a Panel of Nutritional Measures
Nutritional status in maintenance dialysis (MD) patients should be assessed with a combination of valid, complementary measures rather than any single measure alone. (*Opinion*)

Guideline 2. Panels of Nutritional Measures for Maintenance Dialysis Patients
For MD patients, nutritional status should be routinely assessed by predialysis or stabilized serum albumin, percent of usual body weight, percent of standard (NHANES II) body weight, subjective global assessment, dietary interviews and diaries, and nPNA. (*Opinion*)

Guideline 3. Serum Albumin
Serum albumin is a valid and clinically useful measure of protein-energy nutritional status in MD patients. (*Evidence*)

Guideline 4. Serum Prealbumin
Serum prealbumin is a valid and clinically useful measure of protein-energy nutritional status in MD patients. (*Evidence and Opinion*)

Guideline 5. Serum Creatinine and Creatinine Index
Serum creatinine and the creatinine index are valid and clinically useful measures of protein-energy nutritional status in MD patients. (*Evidence and Opinion*)

Guideline 6. Serum Cholesterol
Serum cholesterol is a valid and clinically useful measure of protein-energy nutritional status in MD patients. (*Evidence and Opinion*)

Guideline 7. Dietary interviews and Diaries
Dietary interviews and/or diaries are valid and clinically useful measures of dietary protein and dietary energy intake in MD patients. (*Evidence and Opinion*)

Guideline 8. Protein Equivalent of Total Nitrogen Appearance (PNA)
PNA or Protein Catabolic Rate (PCR) is a valid and clinically useful measure of net protein degradation and protein intake in MD patients. (*Evidence*)

Guideline 9. Subjective Global Assessment (SGA)
SGA is a valid and clinically useful measure of protein-energy nutritional status in MD patients. (*Evidence*)

Guideline 10. Anthropometry
Anthropometric measurements are valid and clinically useful indicators of protein-energy nutritional status in MD patients. (*Evidence and Opinion*)

Guideline 11. Dual Energy X-Ray Absorptiometry (DXA)
DXA is a valid and clinically useful technique for assessing protein-energy nutritional status. (*Evidence and Opinion*)

Guideline 12. Adjusted Edema-Free Body Weight (aBW_{ef})
The body weight to be used for assessing or prescribing protein or energy intake is the aBW_{ef}. For HD patients, this should be obtained postdialysis. For PD patients, this should be obtained after drainage of dialysate. (*Opinion*)

2. Management of Acid-Base Status

Guideline 13. Measurement of Serum Bicarbonate
Serum bicarbonate should be measured in MD patients once monthly. (*Opinion*)

Guideline 14. Treatment of Low Serum Bicarbonate
Predialysis or stabilized serum bicarbonate levels should be maintained at or above 22 mmol/L. (*Evidence and Opinion*)

3. Management of Protein and Energy Intake

Guideline 15. Dietary Protein Intake (DPI) in Maintenance Hemodialysis (MHD)
The recommended DPI for clinically stable MHD patients is 1.2 /kg body weight/d. (*Evidence and Opinion*)

Guideline 16. DPI in Chronic Peritoneal Dialysis (CPD)
The recommended DPI for clinically stable CPD patients is 1.2 to 1.3 /kg body weight/d. (*Evidence*)

Guideline 17. Daily Energy Intake for MD Patients
The recommended daily energy intake for MHD or CPD patients is 35 cal/kg body weight/d for those who are less than 60 years of age and 30 to 35 kcal/kg body weight/d for individuals 60 years or older. (*Evidence and Opinion*)

4. Nutritional Counseling and Follow-Up

Guideline 18. Intensive Nutritional Counseling with MD
Every MD patient should receive intensive nutritional counseling based on an individualized plan of care developed before or at the time of commencement of MD therapy. (*Opinion*)

Guideline 19. Indications for Nutrition Support
Individuals undergoing MD who are unable to meet their protein and energy requirements with food intake for an extended period of time should receive nutritional support. (*Evidence and Opinion*)

Guideline 20. Protein Intake During Acute Illness
The optimum protein intake for a MD patient who is acutely ill is at least 1.2 to 1.3 g/kg/d. (*Opinion*)

Guideline 21. Energy Intake During Acute Illness
The recommended energy intake for a MD patient who is acutely ill is at least 35 kcal/kg/d for those who are less than 60 years of age and at least 30 to 35 kcal/kg/d for those who are 60 years of age or older. (*Evidence and Opinion*)

Guideline 22. L-Carnitine for MD Patients
There are insufficient data to support the routine use of L-carnitine for MD patients. (*Evidence and Opinion*)

B. Advanced Chronic Renal Failure without Dialysis

Guideline 23. Panels of Nutritional Measures for Nondialyzed Patients
For individuals with chronic renal failure (CRF) (GFR $<$ 20 ml/min/1.73 m^2), protein-energy nutritional status should be evaluated by serial measurements of a panel of markers including at least one value from each of the following clusters: (1) serum albumin; (2) edema-free actual body weight, percent standard (NHANES II) body weight, or SGA; and (3) nPNA or dietary interviews and diaries. (*Evidence and Opinion*)

Guideline 24. Dietary Protein Intake for Nondialyzed Patients
For individuals with CRF (GFR $<$ 25 ml/min/1.73 m^2) who are not undergoing MD, the institution of a planned low protein diet providing 0.60 protein/kg/d should be considered. For individuals who will not accept such a diet or who are unable to maintain adequate dietary energy intake (DEI) with such a diet, an intake of up to 0.75 /kg/d may be prescribed. (*Evidence and Opinion*)

Guideline 25. Dietary Energy Intake for Nondialyzed Patients
The recommended DEI for individuals with CRF (GFR $<$ 25 ml/min/1.73 m^2) who are not undergoing MD is 35 kcal/d for those who are younger than 60 years old and 30 to 35 kcal/d for individuals who are 60 years of age or older. (*Evidence and Opinion*)

Guideline 26. Intensive Nutritional Counseling for CRF
The nutritional status of individuals with CRF should be monitored at regular intervals. (*Evidence*)

Guideline 27. Indications for Renal Replacement Therapy
In patients with CRF (GFR $<$ 15 to 20 ml/min/1.73 m^2) who are not undergoing MD, if protein-energy malnutrition develops or persists despite vigorous attempts to optimize protein and energy intake and there is no apparent cause for malnutrition other than low nutrient intake, initiation of MD or a renal transplant is recommended. (*Opinion*)

II. Pediatrics Guidelines

Guideline 1. Patient Evaluation of Protein-Energy Nutritional Status
The most valid measures of protein and energy nutrition status in children treated with MD include: (*Evidence and Opinion*)

- Dietary interview/diary (*Opinion*)
- Serum albumin (*Opinion*)
- Height or length (*Evidence and Opinion*)
- Estimated dry weight (*Evidence and Opinion*)
- Weight/Height Index (*Opinion)*
- Mid-arm circumference and muscle circumference or area (*Opinion*)
- Skinfold thickness (*Opinion*)
- Head circumference (3 years or less) (*Evidence and Opinion*)
- Standard deviation score (SDS or Z score) for height (*Evidence and Opinion*)

Guideline 2. Management of Acid-Base Status
Because academia exerts a detrimental effect on growth and nutritional status, serum bicarbonate levels below 22 mmol/L should be corrected with oral administration of alkali therapy and/or the use of higher sodium bicarbonate dialysate solution in patients treated with MHD. *(Evidence and Opinion*)

Guideline 3. Urea Kinetic Modeling
Urea kinetic modeling may have a role in the nutritional assessment and management of children treated with MD. Although PNA is useful to assess and follow nutritional status in adults, there is currently insufficient evidence to recommend its routine use in pediatric patients. *(Evidence and Opinion*)

Guideline 4. Interval Measurements
Scheduled, interval measurements of growth and nutrition parameters should be obtained to provide optimal care of the nutritional needs of children on maintenance PD or HD. *(Evidence and Opinion*)

Guideline 5. Energy Intake for Children Treated with MD
The initial prescribed energy intake for children treated with MHD or PD should be at the Recommended Dietary Allowance (RDA) level for chronological age. Modifications should then be made depending upon the child's response. *(Evidence and Opinion*)

Guideline 6. Protein Intake for Children Treated with MD
Children treated with MHD should have their initial DPI based on the RDA for chronological age and an additional increment of 0.4 /kg/d. *(Evidence and Opinion*)

Children treated with maintenance PD should have their initial DPI based on the RDA for their chronological age plus an additional increment based on anticipated peritoneal losses. *(Evidence and Opinion)*

Guideline 7. Vitamin and Mineral Requirements
The recommended dietary intake should achieve 100% of the Dietary Reference Intakes (DRI) for thiamin (B_1), riboflavin (B_2), pyridoxine (B_6), vitamin B_{12}, and folic acid. An intake of 100% of the RDA should be the goal for vitamins A, C, E, and K, copper, and zinc. *(Evidence and Opinion)*

Guideline 8. Nutrition Management
Every dialysis patient and appropriate family member (or caretaker) should receive intensive nutrition counseling based on an individualized plan of care, which includes relevant, standardized measurements of growth and physical development, developed prior to or at the time of initiation of MD. *(Opinion)*

The nutrition plan of care developed during the early phase of MD therapy should be reevaluated frequently and modified according to progress. The maximum time between such updates is 3 to 4 months. *(Opinion)*

Guideline 9. Nutritional Supplementation for Children Treated with MD
Supplemental nutritional support should be considered when a patient is not growing normally (e.g., does not have normal height velocity) or fails to consume the RDA for protein and/or energy. Supplementation by the oral route is preferred followed by enteral tube feeding. *(Evidence and Opinion)*

Guideline 10. Recommendations for the Use of Recombinant Human Growth Hormone (hGH) for Children Treated with MD
Treatment with recombinant hGH in dialysis patients with growth potential should be considered under the following conditions: *(Evidence and Opinion)*

- Children who have (1) a height for chronological age more negative than 2.0 SDS or (2) a height velocity for chronological age SDS more negative than 2.0 SDS, (3) growth potential documented by open epiphyses, and (4) no other contraindication for recombinant hGH use.
- Prior to consideration of the use of recombinant hGH, there should be correction of (1) insufficient intake of energy, protein, and other nutrients, (2) acidosis, (3) hyperphosphatemia (the level of serum phosphorus should be less than 1.5 times the upper limit for age), and (4) secondary hyperparathyroidism.

2.1.2. KDOQI Clinical Practice Guidelines for Bone Metabolism and Disease in Chronic Kidney Disease (http://www.kidney.org/professionals/KDOQI/guidelines_bone/index.htm)

Guideline 3: Evaluation of Serum Phosphorus Levels

3.1 In CKD patients (Stages 3 and 4), the serum level of phosphorus should be maintained at or above 2.7 mg/dL (0.87 mmol/L) (*Evidence*) and no higher than 4.6 mg/dL (1.49 mmol/L). (*Opinion*)

3.2 In CKD patients with kidney failure (Stage 5) and those treated with hemodialysis (HD) or peritoneal dialysis (PD), the serum levels of phosphorus should be maintained between 3.5 and 5.5 mg/dL (1.13 and 1.78 mmol/L). (*Evidence*)

Guideline 4: Restriction of Dietary Phosphorus in Patients with CKD

4.1 Dietary phosphorus should be restricted to 800 to 1,000 mg/day (adjusted for dietary protein needs) when the serum phosphorus levels are elevated (>4.6 mg/dL [1.49 mmol/L]) at Stages 3 and 4 of CKD, (Opinion) and >5.5 mg/dL (1.78 mmol/L) in those with kidney failure (Stage 5). (*Evidence*)

4.2 Dietary phosphorus should be restricted to 800 to 1,000 mg/day (adjusted to dietary protein needs) when the plasma levels of intact PTH are elevated above target range of the CKD stage. (*Evidence*)

4.3 The serum phosphorus levels should be monitored every month following the initiation of dietary phosphorus restriction. (*Opinion*)

Guideline 6: Serum Calcium and Calcium-Phosphorus Product

6.1 The serum levels of corrected total calcium should be maintained within the "normal" range for the laboratory used in Stages 3 and 4 CKD. (*Evidence*)

6.2 Serum levels of corrected total calcium should be maintained within the normal range for the laboratory used, preferably toward the lower end (8.4 to 9.5 mg/dL [2.10 to 2.37 mmol/L]) in Stage 5 CKD with kidney failure. (*Opinion*)

6.3 In the event corrected total serum calcium level exceeds 10.2 mg/dL (2.54 mmol/L), therapies that cause serum calcium to rise should be adjusted as follows:

6.3a In patients taking calcium-based phosphate binders, the dose should be reduced or therapy switched to a noncalcium-, nonaluminum-, nonmagnesium-containing phosphate binder. (*Opinion*)

6.3b In patients taking active vitamin D sterols, the dose should be reduced or therapy discontinued until the serum levels of corrected total calcium return to the target range (8.4 to 9.5 mg/dL [2.10 to 2.37 mmol/L]). (*Opinion*)

6.3c If hypercalcemia (serum levels of corrected total calcium >10.2 mg/dL [2.54 mmol/L]) persists despite modification of therapy with vitamin D and/or discontinuation of calcium-based phosphate binders, dialysis using low dialysate calcium (1.5 to 2.0 mEq/L) may be used for 3 to 4 weeks (*Opinion*)

In CKD Patients (Stages 3 to 5):

6.4 Total elemental calcium intake (including both dietary calcium intake and calcium-based phosphate binders) should not exceed 2,000 mg/day. (*Opinion*)

6.5 The serum calcium-phosphorus product should be maintained at <55 mg^2/dL^2. (*Evidence*) This is best achieved by controlling serum levels of phosphorus within the target range. (*Opinion*)

6.6 Patients whose serum levels of corrected total calcium are below the lower limit for the laboratory used (<8.4 mg/dL [2.10 mmol/L]) should receive therapy to increase serum calcium levels if:

6.6a There are clinical symptoms of hypocalcemia such as paresthesia, Chvostek's and Trousseau's signs, bronchospasm, laryngospasm, tetany, and/or seizures (*Opinion*); or

6.6b The plasma intact PTH level is above the target range for the CKD Stage. (*Opinion*)

6.7 Therapy for hypocalcemia should include calcium salts such as calcium carbonate (*Evidence*) and/or oral vitamin D sterols. (*Evidence*)

Guideline 8: Vitamin D Therapy in CKD Patients

Guideline 8A: Active Vitamin D Therapy in Patients with Stages 3 and 4 CKD

8A.1 In patients with CKD Stages 3 and 4, therapy with an active oral vitamin D sterol (calcitriol, alfacalcidol, or doxercalciferol) is indicated when serum levels of 25(OH)-vitamin D are >30 ng/mL (75 nmol/L), and plasma levels of intact PTH are above the target range for the CKD stage. (*Evidence*)

8A.1a Treatment with an active vitamin D sterol should be undertaken only in patients with serum levels of corrected total calcium <9.5 mg/dL (2.37 mmol/L) and serum phosphorus <4.6 mg/dL (1.49 mmol/L). (*Opinion*)

8A.1b Vitamin D sterols should not be prescribed for patients with rapidly worsening kidney function or those who are noncompliant with medications or follow-up. (*Opinion*)

8A.2 During therapy with vitamin D sterols, serum levels of calcium and phosphorus should be monitored at least every month after initiation of therapy for the first 3 months, then every 3 months thereafter. Plasma PTH levels should be measured at least every 3 months for 6 months, and every 3 months thereafter. (*Opinion*)

8A.3 Dosage adjustments for patients receiving active vitamin D sterol therapy should be made as follows:

8A.3a If plasma levels of intact PTH fall below the target range for the CKD stage, hold active vitamin D sterol therapy until plasma levels of intact PTH rise to above the target range, then resume treatment with the dose of active vitamin D sterol reduced by half. If the lowest daily dose of the active vitamin D sterol is being used, reduce to alternate-day dosing. (*Opinion*)

8A.3b If serum levels of corrected total calcium exceed 9.5 mg/dL (2.37 mmol/L), hold active vitamin D sterol therapy until serum calcium returns to <9.5 mg/dL (2.37 mmol/L), then resume treatment at half the previous dose. If the lowest daily dose of the active vitamin D sterol is being used, reduce to alternate-day dosing. (*Opinion*)

8A.3c If serum levels of phosphorus rise to >4.6 mg/dL (1.49 mmol/L), hold active vitamin D therapy, initiate or increase dose of phosphate binder until the levels of serum phosphorus fall to ≤ 4.6 mg/dL (1.49 mmol/L); then resume the prior dose of active vitamin D sterol. (*Opinion*)

Guideline 8B. Vitamin D Therapy in Patients on Dialysis (CKD Stage 5)

8B.1 Patients treated with HD or PD with serum levels of intact PTH levels >300 pg/mL (33.0 pmol/L) should receive an active vitamin D sterol (such as calcitriol, alfacalcidol, paricalcitol, or doxercalciferol) to reduce the serum levels of PTH to a target range of 150 to 300 pg/mL (16.5 to 33.0 pmol/L). (*Evidence*)

8B.1a The intermittent, intravenous administration of calcitriol is more effective than daily oral calcitriol in lowering serum PTH levels. (*Evidence*)

8B.1b In patients with corrected serum calcium and/or phosphorus levels above the target range, a trial of alternative vitamin D analogs, such as paricalcitol or doxercalciferol may be warranted. (*Opinion*)

8B.2 When therapy with vitamin D sterols is initiated or the dose is increased, serum levels of calcium and phosphorus should be monitored at least every 2 weeks for 1 month and then monthly thereafter. The plasma PTH should be measured monthly for at least 3 months and then every 3 months once target levels of PTH are achieved. (*Opinion*)

8B.3 For patients treated with peritoneal dialysis, oral doses of calcitriol (0.5 to 1.0 μg) or doxercalciferol (2.5 to 5.0 μg) can be given 2 or 3 times weekly. Alternatively, a lower dose of calcitriol (0.25 μg) can be administered daily. (*Opinion*)

8B.4 When either hemodialysis or peritoneal dialysis patients are treated with active vitamin D sterols, management should integrate the changes in serum calcium, serum phosphorus, and plasma PTH. (*Opinion*)

Guideline 13. Treatment of Bone Disease in CKD

Guideline 13c. Adynamic Bone Disease

13C.1 Adynamic bone disease in stage 5 CKD (as determined either by bone biopsy or intact PTH <100 pg/ml [11.0 pmol/L]) should be treated by allowing plasma levels of intact PTH to rise in order to increase bone turnover. (*Opinion*)

13C.1a This can be accomplished by decreasing doses of calcium-based phosphate binders and vitamin D or eliminating such therapy. (*Opinion*)

2.1.3. KDOQI Clinical Practice Guidelines and Clinical Practice Recommendations for Diabetes and Chronic Kidney Disease [*American Journal of Kidney Diseases*, 49(Suppl 2), 2007]

Guideline 5. Nutritional management in diabetes and CKD

5.1 Target dietary protein intake (DPI) for people with CKD stages 1–4 should be the recommended dietary allowance (RDA) of 0.8 g/kg body weight/day.

2.2. American Dietetic Association Medical Nutrition Therapy Protocols for Non-Dialysis Kidney Disease

As of January 2002, Medicare covers medical nutrition therapy (MNT) for the treatment of CKD [glomerular filtration rate (GFR) <60 ml/min/1.73 m^2]. This was a major breakthrough for the dietetics profession and a boon to the patients who need these services. The recognition of the importance of MNT in the treatment of CKD can make significant progress in the care of many eligible persons. The particulars of coverage can be found at the website for the Centers for Medicare and Medicaid Services (CMS), http://www.cms.hhs.gov. The American Dietetic Association has published evidence-based practice protocols that guide care for non-dialysis kidney disease and can be ordered at the website http://www.eatright.org. The following two pages provide a summary sheet of these MNT protocols.

Summary Page
Chronic Kidney Disease (non-dialysis)
Medical Nutrition Therapy Protocol

Setting: Ambulatory Care or adapted for other healthcare settings (Adult 18+ years old) — **Number of encounters: 3 to 6**

No of Encounters	Length of encounters	Time between encounters
1	60-90 minutes	3-4 weeks
2	45-60 minutes	3-4 weeks
3	30-45 minutes	3-4 weeks
4,5,6	30-45 minutes	6- 8 weeks or as identified by reassessment

Expected Outcomes of Medical Nutrition Therapy

Base Line Evaluation of Outcomes

Outcomes Assessment Factors	1	2	3	4*	Expected Outcomes	Ideal/goal Value
Clinical Assessment						
Laboratory Values:						
• Serum albumin	√		√	√	• Maintain normal range	• Serum Albumin: > 4.0 g/dL (Grade II)
• Serum CO_2	√			√	• Maintain normal range	• Serum CO_2: 24-32 mEq/L (Grade II)
• Serum potassium	√	√	√	√	• Maintain normal range	• Serum Potassium: 3.5-5.5 mmol/L (Grade II)
• Serum calcium	√	√	√	√	• Progress toward goal	• Serum Calcium: 8.5- 10.2 mg/dL (corrected)
• Serum phosphorus	√	√	√	√	• Progress toward goal	• Serum Phosphorus:3.4-5.5 mg/dL(Grade II)
• PTH	√			√	• Progress toward goal	• Intact PTH: 100-300 pg/ml (Grade II)
• Serum glucose (if diabetic)	√			√	• ↓ 10% or at goal	• Random glucose:<140-160 mg/dL(blood); <160-180 mg/dL (plasma); A1C: <7% (Grade I)
• A1C (if diabetic)	√			√		
• Serum lipids	√		√	√	• Within normal range or △ by 10% to 20% if abnormal	• Cholesterol:>160 to <200 mg/dL • LDL-chol:<100 mg/dL;TG:<150 mg/dL • HDL-chol: >45 mg/dL (M) >55 (F)
• Serum Creatinine/GFR	√	√	√	√	• Stabilize creatinine, GFR and urinary albumin excretion	• Serum creatinine/GFR: stabilizes
• Urine albumin	√	√	√	√		• Urine albumin:<30 mg/d or <3 mg/dL
• Hemoglobin	√			√	• Adequate iron, folate for erythropoiesis when rHuEPO is administered	• Hgb: 12 g/L (M); 11 g/L (F) (Grade II)
• Ferritin	√			√		• Ferritin: >100 ng/mL,
• Transferrin saturation	√			√		• Transferrin saturation: >20%
• RBC folate	√			√		• RBC folate: >200 ng/ml
Nutrition/Physical:					• Maintains height, skeletal muscle, weight, fat stores (Weight should be edema free)	• Height: Yearly heights to monitor spinal osteoporosis/bone loss
• Height	√					
• Weight/BMI	√	√	√	√		• BMI: ≥24
• Body composition/SGA	√		√	√		
• Blood pressure	√	√	√	√	• Achieves blood pressure goal	• Blood pressure: <125/75: >1 g proteinuria or diabetic nephropathy; <130/85 without proteinuria (Grade II)
Functional Status:						
• ADLs, IADLs	√	√	√	√	• Improves/maintains functional status	• Optimum functional status
Therapeutic Lifestyle Changes						**MNT Goal: Maintain kidney function, ↓ progression; maintain nutritional status**
• Food/Meal Plan	√	√	√	√	• Chooses appropriate kinds and amounts of food	• Kcal: BEE (consider stress, dietary protein, weight goals) (Grade I)
	√	√	√	√	• Chooses 50% high biological animal or plant sources of protein	• Protein: 0.6 to 1.0 g/kg/IBW based on GFR, urinary protein excretion, degree of malnutrition, stress, motivation (Grade I)
	√	√	√	√	• Limits total fat, SF, cholesterol to meet serum lipid goals	• Fat: 25-30%, <7% SFA, <200 mg cholesterol
	√	√	√	√	• Eats at consistent times if diabetic	• Carbohydrate: 50 to 60% kcal
	√	√	√	√	• Limits high sodium foods	• Sodium: individualized, 1-3 g/d
	√	√	√	√	• Consumes potassium per labs	• Potassium: Individualized based on labs
	√	√	√	√	• Limits phosphorus per labs	• Phosphorus: 8-12 mg/kg IBW; phosphate binders/vitamin D analogues may be needed
	√	√	√	√	• Consumes calcium supplements if prescribed, based on labs	• Calcium: Individualized: ~800 to 1200 mg/d
• Food/supplement intake	√	√	√	√	• Maintains adequate appetite	• Consumes >80% meals/supplements • Adequate to maintain weight, body composition
• Food label reading		√	√	√	• Accurately reads food labels	
• Recipe modification	√	√	√	√	• Modifies recipes as needed	
• Food preparation	√	√	√	√	• Uses methods to ↓ sodium	
• Self-monitoring	√	√	√	√	• Records daily food intake	• Dietary intake = prescription >80% of time
• Eating away from home			√	√	• Selects food appropriately	
• Potential food/nutrient/drug interaction	√	√	√	√	• Follows protocols for medications	• Consumes >80% phosphate binders with meals/snacks if prescribed
• Smoking/use of alcohol	√	√	√	√	• Participates in smoking cessation program; limits use of alcohol	
• Physical activity	√	√	√	√	• Participates in physical activity	• Maintains muscle stores/strength

Fig. 1. ADA MNT Protocols.

Flow Chart
Medical Nutrition Therapy Process for Chronic Kidney Disease (non-dialysis)
Medical Nutrition Therapy Protocol

Referral/Consult Information (< 30 days prior to encounter 1)	
RD to obtain pertinent clinical data from referral source or client medical record/information system:	
✓ Laboratory values (creatinine, GFR, serum albumin, Hgb, bicarbonate, potassium, phosphorus, calcium, urine albumin, PTH, cholesterol, fasting TG, A1C if diabetic)	✓ Presenting signs and symptoms (eg, poor appetite, altered taste, recent weight loss, GI symptoms)
✓ Physician goals or medical plans	✓ Medications (dose, frequency)
✓ Primary cause of chronic kidney disease	✓ Prescribed vitamins/minerals
✓ Medical history (co-morbidities, e.g., diabetes)	✓ Physical activity clearance or limitations

↓

Encounter 1: 60-90 minutes
RD to obtain clinical data from client medical record/information system and client interview:
□ **Nutrition-focused assessment**: Evaluation of height, weight, usual weight, BMI, GFR, laboratory values, other clinical data (e.g., blood pressure, ADLs, IADLs, SGA), client's knowledge of kidney disease, readiness to learn; comprehensive diet history including meal pattern and current dietary intake of calories, protein, type of fat, sodium (potassium, phosphorus, calcium as appropriate for stage of CKD and GFR), vitamin/mineral supplements, herbal supplements and over the counter medications (OTC), physical activity pattern, and psychosocial and economic issues impacting nutrition therapy. Consider co-morbid conditions and need for additional modifications in the nutrition care plan, e.g., diabetes.
□ **Intervention and self-management training**: Provide rationale for nutrition therapy to slow the progression of kidney disease and to maintain normal blood pressure and blood glucose. Nutrition therapy should include an individualized nutrition prescription appropriate to the stage of CKD and GFR to maintain optimal laboratory indices, a meal plan and self-monitoring strategies (e.g., food and medication records to be kept). The nutrition prescription may include g protein/day, type of fat, sodium, % carbohydrates (if diabetic), nutritional supplements, specific amounts of potassium, phosphorus, calcium, phosphate binders and vitamin D supplementation. Mutually established goals and outcomes for eating, following medication protocol, recording food and medication intake, participating in physical activity, and self-blood glucose monitoring (if diabetic).
□ **Plan for reassessment and follow-up:** Establish timeline for follow-up and provide client with contact information. Identify *expected outcomes* to determine response to care (e.g., ↓ blood pressure, maintain weight, glycemic control if diabetic)
□ **Communication:** Provide documentation to physician and other relevant health care team members according to organization's policy.

↓ **3 to 4 weeks between encounters**

Encounter 2: 45-60 minutes
RD to obtain clinical data from client medical record/information system and client/caregiver interview:
□ **Nutrition-focused assessment**: Reassess weight, BMI, laboratory values, other clinical data (e.g., blood pressure, blood glucose records, and medication changes). Obtain brief diet history and assess client's adherence/comprehension to nutrition prescription (e.g., intake of calories, fat, type of fat, sodium, key nutrients appropriate for stage of CKD and GFR, e.g., protein, potassium, phosphorus, calcium, and physical activity goals--refer to food/medication records). Reinforce or modify mutually established goals/outcomes for eating and keeping food and medication records.
□ **Intervention and self-management training**: Provide new nutrition information (e.g., food label reading, food preparation methods). Reinforce or modify individualized nutrition prescription, meal plan, self-monitoring and behavior strategies. Reinforce or modify mutually established goals/outcomes for eating, behavior changes, self-monitoring, and physical activity.
□ **Plan for reassessment and follow-up:** Establish timeline for follow-up and provide client with contact information. Identify *expected outcomes* to determine response to care (e.g., ↓ blood pressure, ↓ A1C, ↑ intake of protein and energy).
□ **Communication:** Provide documentation to physician and other relevant health care team members according to organization's policy.

Fig. 2. ADA MNT Protocols.

 **3 to 4 weeks between encounters**

Encounters 3: 30-45 minutes
RD to obtain clinical data from client medical record/information system and client/caregiver interview:
□ **Nutrition-focused assessment**: Reassess weight, BMI, SGA, laboratory values and medication changes. Obtain brief diet history and assess client's adherence to any diet modifications (kcal, protein, sodium, potassium, phosphorus), review labs and discuss other clinical data (e.g., blood pressure, A1C if diabetic) and comprehension to nutrition prescription (e.g., intake of calories, protein, sodium, potassium, phosphorus, calcium), and physical activity goals (review food/medication records).
□ **Intervention and self-management training**: Reinforce or modify mutually established goals/outcomes for eating, self-monitoring (food/medication records), physical activity. Review lab results (serum and urine albumin, serum potassium and phosphorus) and modify diet prescription as appropriate. Review food and medication records and reinforce/modify mutually established goals/outcomes for eating, behavior changes, self-monitoring with food/medication records. Provide new nutrition information (e.g, recipe modification for specific nutrients that need modifying e.g., sodium and dining out strategies). Reinforce or modify individualized nutrition prescription, meal plan, self-monitoring and behavior strategies.
□ **Plan for reassessment and follow-up:** Establish timeline for follow-up and provide client with contact information. Identify *expected outcomes* to determine response to care (e.g., ↑ serum albumin, ↓ blood pressure, ↓ urine albumin, ↓ A1C, ↑ dietary intake of protein and energy, ↓ serum phosphorus)
□ **Communication:** Provide documentation to physician and other relevant health care team members according to organization's policy.

 **6 to 8 weeks between encounters**

Encounters 4, 5, 6 : 30-45 minutes
RD to obtain clinical data from client medical record/information system and client/caregiver interview:
□ **Nutrition-focused assessment**: Reassess weight, BMI, SGA, laboratory values and medication changes. Obtain brief diet history and assess client's adherence to any diet modifications (kcal, protein, kind of fat, potassium, phosphorus), review labs and discuss other clinical data (e.g., blood pressure, A1C if diabetic) and comprehension to nutrition prescription (e.g., intake of calories, protein, sodium, potassium, phosphorus, calcium), and physical activity goals (review food/medication records).
□ **Intervention and self-management training**: Reinforce or modify mutually established goals/outcomes for eating, self-monitoring (food/medication records), physical activity. Review lab results (serum and urine albumin, serum potassium and phosphorus) and modify diet prescription as appropriate. Review food and medication records and reinforce/modify mutually established goals/outcomes for eating, behavior changes, self-monitoring with food/medication records. Provide new nutrition information based on changes in kidney function, medications, diet prescription. Reinforce or modify individualized nutrition prescription, meal plan, self-monitoring and behavior strategies.
□ **Plan for reassessment and follow-up:** Establish timeline for follow-up and provide client with contact information. Identify *expected outcomes* to determine response to care (e.g., ↑ serum albumin, ↓ blood pressure, ↓ urine albumin, ↓ A1C, ↑ dietary intake of protein and energy, ↓ serum phosphorus, normalize serum potassium)
□ **Communication:** Provide documentation to physician and other relevant health care team members according to organization's policy.

Fig. 2. (*Continued*).

3. DIET-RELATED RESOURCES AND FOOD LISTS

Dietitians assist in translating research in nutrition and CKD to patients and their caregivers. A number of resources are available to help with this process, and a few are provided within this chapter. The following sections comprise several tables that cover the nutrition composition of foods with various stages of CKD, food sources of potassium, phosphorus and oxylate, examples of high biologic value protein sources, and renal micronutrient supplements.

Table 1
Nutrition Composition of Foods (Per Serving) for People with Stage 3 or 4 Chronic Kidney Disease

Foods	*Protein (g)*	*Calories (kcal)*	*Sodium (mg)*	*Potassium (mg)*	*Phosphorus (mg)*	*Example*
Protein						
High	6–8	0–100	20–150	50–150	50–100	1 oz. beef, chicken, fish
High Phosphorus	6–8	50–100	20–150	50–350	100–300	1 oz. organ meats
High Sodium	6–8	50–100	200–450	50–150	50–100	¼ c cottage cheese
Vegetable						
Low K^+	2–3	10–100	0–50	20–150	10–70	1 c cabbage
Medium K^+	2–3	10–100	0–50	150–250	10–70	½ c beets
High K^+	2–3	10–100	0–50	250–550	10–70	¼ whole avocado
Fruit						
Low K^+	0–1	20–100	0–10	20–150	1–20	1 medium apple
Medium K^+	0–1	20–100	0–10	150–250	1–20	½ c medium peach
High K^+	0–1	20–100	0–10	250–550	1–20	2 apricot halves
Dairy/High PO_{4-} foods	2–3	50–200	150–400	10–100	100–200	1 c. sherbet
Breads/cereals	2–3	50–200	0–150	10–100	10–70	½ small sweet roll, no nuts
Free calories	0–1	100–150	0–100	0–100	0–100	1 (3 oz.) popsicle

c, cup; K^+, potassium, Na^{++}, sodium; PO_{4-} phosphorus; oz, ounce.
Adapted from ref. *(1)*.

Table 2
Nutrition Composition of Foods (Per Serving) for People with Stage 5 Chronic Kidney Disease

Foods	*Protein (g)*	*Calories (kcal)*	*Sodium (mg)*	*Potassium (mg)*	*Phosphorus (mg)*	*Example*
Protein						
Animal	6–8	50–100	20–150	50–150	50–100	1 oz. beef, chicken, fish
Animal (with higher Na^{++} or PO_4^- contents)	6–8	50–100	200–500	50–150	100–300	1 oz. sardines
Fruit/Vegetable						
Low K^+	0–3	10–100	1–50	20–150	0–70	1 c lettuce or ½ c grape juice
Medium K^+	0–3	10–100	1–50	150–250	0–70	½ c cooked broccoli or 2 Tbsp raisins
High K^+	0–3	1–100	0–50	250–550	0–70	1 medium tomato Or 1 small nectarine
Dairy/High PO_4^-	2–8	100–400	30–300	50–400	100–120	½ c milk
Breads/cereals	2–3	50–200	0–150	10–100	10–70	½ small bagel
Free calories	0–1	100–150	0–100	0–100	0–100	4 pieces hard candy

c, cup; K^+, Potassium; Na^{++}, Sodium; oz, ounce; Tbsp, tablespoon.
Adapted from ref. *(1)*.

3.1. Food Sources of Potassium

The following table lists foods that are high in potassium. The portion size is ½ cup unless otherwise stated. While all the foods on this list contain more than 200 mg potassium per portion, some are higher than others.

Table 4 list foods which are low in potassium. A portion is ½ cup unless otherwise noted.

Table 3

High potassium foods (>200 mg/portion)		
Fruits	*Vegetables*	*Other foods*
Apricot, raw (2 medium) dried (5 halves)	Acorn squash	Bran/Bran products
	Artichoke	
Avocado ($^1/_4$ whole)	Bamboo shoots	Chocolate (1.5–2 ounces)
Banana ($^1/_2$ whole)	Baked beans	Granola
Cantaloupe	Butternut squash	Milk, all types (1 cup)
Dates (5 whole)	Refried beans	Molasses (1 Tablespoon)
Dried fruits	Beets, fresh then boiled	Nutritional Supplements: Use only under the direction of your doctor or dietitian.
Figs, dried	Black beans	
Grapefruit juice	Broccoli, cooked	Nuts/seeds
Honeydew	Brussels sprouts	Nuts and Seeds (1 ounce)
Kiwi (1 medium)	Chinese cabbage	Peanut Butter (2 tablespoons)
Mango (1 medium)	Carrots, raw	Salt Substitutes/Lite Salt
Nectarine (1 medium)	Dried beans and peas	Salt Free Broth
Orange (1 medium)	Greens, except Kale	Snuff/Chewing Tobacco
Orange juice	Hubbard squash	Yogurt
Papaya ($^1/_2$ whole)	Kohlrabi	
Pomegranate (1 whole)	Lentils	
Pomegranate juice	Legumes	
Prunes	Mushrooms, canned	
Prune juice	Parsnips	
Raisins	Potatoes, white and sweet	
	Pumpkin	
	Rutabagas	
	Spinach, cooked	
	Tomatoes/Tomato products	
	Vegetable Juices	

Table 4

Low potassium foods		
Fruits	*Vegetables*	*Other foods*
Apple (1 medium)	Alfalfa sprouts	Rice
Apple Juice	Asparagus (6 spears)	Noodles
Applesauce	Beans, green or wax	Pasta
Apricots, canned in juice	Cabbage, green and red Carrots, cooked	Bread and bread products: (not whole grains)
Blackberries	Cauliflower	Cake: angel, yellow
Blueberries	Celery (1 stalk)	Coffee: limit to 8 ounces
Cherries	Corn, fresh (½ ear) frozen (½ cup)	Pies without chocolate or high potassium fruit
Cranberries	Cucumber	Cookies without nuts or chocolate
Fruit Cocktail	Eggplant	Tea: limit to 16 ounces
Grapes	Cucumber	
Grape Juice	Eggplant	
Grapefruit (½ whole)	Kale	
Mandarin oranges	Lettuce	
Peaches, fresh (1 small) canned (½ cup)	Mixed vegetables	
Pears, fresh (1 small) canned (½ cup)	Mushrooms, fresh	
Pineapple	Okra	
Pineapple juice	Onions	
Plums (1 whole)	Parsley	
Raspberries	Peas Green peppers	
Strawberries	Radish	
Tangerine (1 whole)	Rhubarb	
Watermelon (limit to 1 cup)	Water chestnuts, canned	
	Watercress	
	Yellow squash	
	Zucchini squash	

Adapted from ref. *(2)*.

Table 5
Food Sources of Phosphorus

Food	*Amount*	*Phosphorus (mg)*	*Protein (g)*	*mg Phos/g Protein*
Beans, legumes, Tofu:				
Kidney beans, black	1 cup	245	15	16.3
beans, refried beans, lima beans	1 cup	215	15	14.3
Navy beans	1 cup	290	16	18.1
Soybeans, boiled	1 cup	420	29	14.4
Soybeans, roasted	1 cup	625	60	10.4
Tofu, firm	100 g	75	6	12.5
Tofu, soft	100 g	55	4	13..8
Cheese:				
Cheddar cheese	1 oz.	145	7	20.7
Mozzarella cheese	1 oz.	140	7	20.0
Swiss cheese	1 oz.	170	8	21.3
Cottage cheese, 1% fat	1 cup	150	14	10.7
Cottage cheese, 2% fat	1 cup	340	31	11.0
Cream cheese	2 Tbsp	30	2	15.0
Cream, Milk, Yogurt:				
Half and half cream	1 cup	230	7	33.0
Heavy cream	1 cup	150	5	29.8
Sour cream	2 Tbsp	30	1	30.0
Buttermilk	1 cup	220	8	27.5
Non-fat milk	1 cup	250	8	31.3
1% milk	1 cup	235	8	29.4
2% milk	1 cup	230	8	29.0
Whole milk	1 cup	230	8	28.7
Low fat yogurt	1 cup	340	12	28.3
Regular yogurt	1 cup	215	8	26.9
Fish and seafood				
Blue crab	3 oz.	175	17	10.3
Dungeness crab	3 oz.	150	19	8.0
King crab	3 oz.	240	16	15
Halibut	3 oz.	215	23	9.3
Oysters	3 oz.	195	13	15
Salmon	3 oz.	270	21	12.8
Shrimp	3 oz.	115	18	6.4

(Continued)

Table 5
(Continued)

Food	*Amount*	*Phosphorus (mg)*	*Protein (g)*	*mg Phos/g Protein*
Meat, poultry, eggs:				
Beef liver	3 oz.	390	23	17.0
Top sirloin	3 oz.	200	25	8.0
Chicken breast	3 oz.	195	27	7.2
Chicken thigh	3 oz.	150	22	6.8
Egg	1 large	85	7	12.1
Ham	3 oz.	240	19	12.6
Lamb chop	3 oz.	190	22	8.6
Pork loin	3 oz.	145	22	6.6
Turkey	3 oz.	190	27	7.0
Veal loin	3 oz.	190	22	8.6
Nuts and nut butters:				
Almonds	1 oz.	140	6	23.3
Macadamia	1 oz.	55	2	27.5
Peanuts, roasted	1 oz.	150	8	18.8
Peanut butter	2 Tbsp	110	8	13.8
Walnuts	1 oz.	100	4	25.0
Fast foods:				
Bean/cheese burrito	2 small	180	15	12.0
Breakfast biscuit (egg, cheese, bacon)	1 serving	460	16	28.8
Cheeseburger	1 serving	310	28	11.0
Chicken sandwich	1 serving	405	29	14.0
Pepperoni pizza	1 slice	225	16	14.0
Other Foods:				
Beer	12 oz.	40	1	40
Milk chocolate	1 oz.	60	2	30
Semi sweet chocolate	1 oz	35	1	35
Coffee	1 cup	2	0	
Colas	12 oz.	45	0	
Lemon lime soda	12 oz.	0	0	
Lemonade	1 cup	5	0.5	10
Root beer	12 oz.	0	0	
Tea	1 cup	2	0	

oz., ounce; Tbsp, tablespoon. Phosphorus and Protein Content of Common Foods. Developed by Kathy Shiro Harvey and adapted from the National Kidney Foundation.

Table 6
Common Phosphate Additives in Foods

Phosphate additive: Common Name	*Uses*	*Products*
Dicalcium phosphate anhydrous	Mineral source; dough conditioner	Bakery mixes; yeast-raised bakery products; cereals; dry powder beverages; flour; food bars; infant food; milk-based beverages; multivitamin tablets; yogurts. Used in powder form as an abrasive in toothpaste
Dicalcium phosphate dihydrous	Mineral source; leavening agent	Bakery mixes; cereals; dry powder beverages; flour; food bars; infant food; milk-based beverages; multivitamin tablets, yogurt.
Dipotassium phosphate	Nutrient in yeast culturing; sequestrant; buffer	Yeast-containing products; non-dairy creamers; casein-based creamers; processed cheese, meat products, mineral supplements, starter cultures.
Disodium phosphate anhydrous	Sequestrant; emulsifier; buffering agent; absorbent; pH control agent; protein modifier; source of alkalinity, stabilizer is used to adjust pH of cereal and pasta products to maintain quality color in final product. Accelerates the cook time of pasta and quick cooking cereals	Breakfast cereal, cheese; condensed milk; cream; evaporated milk; flavored milk powders; gelatin; half & half; ice cream; imitation cheese; infant food; instant cheesecake; instant pudding; isotonic drinks; nonfat dry milk; pasta; processed cheese; starch; vitamin capsules; whipped topping.
Disodium phosphate dihydrous	Same as disodium phosphate Anhydrous	Same as Disodium Phosphate Anhydrous

(Continued)

Table 6
(Continued)

Phosphate additive: Common Name	*Uses*	*Products*
Monocalcium phosphate monohydrate	Acidulant for foods and beverages; leavening acid; nutrient; dietary supplement; yeast food dough conditioner. Calcium source for fortification or enrichment	Biscuits, cakes, donuts, muffins
Magnesium phosphate	Nutritional source of magnesium and phosphorous; H control agent; dietary supplement; flow aid	Magnesium source in infant formulas and diet beverages.
Monopotassium phosphate	pH control agent; buffering agent; acidulant; leavening agent; nutrient source	Bread, doughs, dry powder beverages, eggs, isotonic beverages, mineral supplements, starter cultures, yeast cultures
Phosphoric acid	Acidulant; pH control agent; buffering agent; flavor enhancer; sequestrant; stabilizer; thickener; synergist	Carbonated and noncarbonated beverages
Sodium hexam-etaphosphate	Sequestrant; neutral salt; deflocculant; curing agent; dough strengthener; emulsifier; firming agent; flavor enhancer; flavoring agent; humectant; nutrient supplement; processing aid; stabilizer and thickener; surface-active agent; synergist; texture and buffering agent	Meat, seafood, poultry, vegetable proteins, processed cheese; sour cream; yogurt; eggs; table syrups; whipped toppings, vegetables, whey
Tetrasodium pyrophosphate	Buffer; emulsifying agent; stabilizer; protein modifier; provides meltability: in processed cheese; quickens cooking time of cooked breakfast cereals; color agent	Processed cheese and cheese products, cooked breakfast cereals, imitation cheese, isotonic beverages

Table 6
(Continued)

Phosphate additive: Common Name	*Uses*	*Products*
Monopotassium phosphate anhydrous	Dry acidulant; buffering agent; emulsifier; leavening agent; protein modifier; Sequestrant; gelling agent; color enhancer, flavor enhancer (tartness)	Cola beverages, dry powder beverages, egg yolks, gelatin, instant cheesecake, instant pudding, isotonic beverages, liquid egg mixtures
SAPP	Emulsifier; formulation aid; humectant; leavening agent, pH control agent; acidulant; buffering agent; coagulant; dispersing agent; protein modifier; processing aid; sequestrant; stabilizer; thickener; synergist; texturizer	Icing and frostings, processed meat, cured meats, processed chicken products, hotdogs, bologna, non-dairy creamers, processed potatoes, albacore tuna, processed cheese, vegetables, seafood, imitation cheese
Tetrapotassium pyrophosphate	Buffering agent; pH control agent; alkalinity source; dispersing agent; protein modifier; coagulant; sequestrant; nutrient source; antioxidant; texturizer.	Processed cheese, milk powders
Pentasodium triphosphate	Sequestrant; emulsifier; reduces oxidation; moisture retention; pH control; buffering agent; coagulant; dispersing agent; curing agent; flavor enhancer; humectant; thickener and stabilizer; texturizer	Meat, poultry, seafood
Tripotassium phosphate	Alkalinitysource; buffering agent; emulsifier; nutrient; protein modifier and stabilizer	Starter cultures, process cheese, dairy products, bread and dough conditioners, mineral supplements, isotonic beverages, cereals

Table 7
Biologic Values of Selected Animal and Vegetable Foods

Foods	*Biologic value range*
Red beans, lentils	45%
Wheat flour, wheat gluten, bean (avg), baker's yeast	50–59%
Sesame seed, white rice, black beans, peas, kale, cooked oatmeal, butter beans, wheat (avg), lima beans, brewer's yeast, chick peas, coconut	60–69%
Sunflower seeds, cheddar cheese, sardines, brown rice, soybeans, potatoes, wheat germ, beef, veal, chicken, pork, shrimp, fish (avg), rye, buckwheat	70–79%
Mushrooms, casein, barley, cod, haddock, milk, lobster	80–89%
Egg	>90%

Developed by Joni Pagenkemper and adapted from FAO Biological Values, 1970. Adequate protein quality >60 adults, >70 children.

Table 8
Food Sources of Oxalate

High oxalate content (>0.9% in the food)	*Moderate oxalate content (0.2–0.9%)*
Beet greens (and the tuber)	Beans, dried
Chocolate, cocoa	Blackberries
Figs	Carrots
Lamb's quarters	Celery
Pepper, black	Coffee, instant
Poppy seeds	Currants
Purslane	Endive
Rhubarb	Gooseberries
Sorrel	Grapes, Concord
Spinach	Green pepper
Swiss chard	Lemon Peel
Tea (instant)	Okra
Beer, draft	Onions, green
	Oranges, orange peel
	Raspberries
	Strawberries
	Sweet potatoes
	Tomatoes
	Wheat bran
	Nuts

Source: See ref. *(3)*.

Table 9
Renal Micronutrient Needs

Vitamin	*Pre-dialysis CKD*	*Chronic HD/PD*
Vitamin C	60 mg/day	60 mg/day
Vitamin B1	1–5 mg./day	1–5 mg/day
Vitamin B2	1.2–1.7 mg/day	1.2–1.7 mg/day
Niacin	13–19 mg/day	13–19 mg/day
Vitamin B6	5 mg/day	10 mg/day
Vitamin B12	2 mcg/day	2 mcg/day
Folic acid	1 mg/day	1 mg/day
Vitamin B5	4–7 mg/day	4–7 mg/day
Biotin	30–100 mcg/day	30–100 mcg/day
Vitamin A	None	None
Vitamin E	Unknown	Unknown
Vitamin K	None	None

CKD, chronic kidney disease; HD, hemodialysis; PD, peritoneal dialysis.

Table 10
Renal Micronutrient Supplement Comparison Chart

Product	Vit C (mg)	B1 (mg)	B2 (mg)	Niacin	B6 (mg)	B12 (mg)	Folic Acid	B5 (mg)	Biotin (mcg)	Vit A (IU)	Vit E (IU)	Zn (mg)	Se (mcg)
Dialyvite®	100	1.5	1.7	20	10	6 mcg	1 mg	10	300	–	–	–	–
Dialyvite® w/ Zinc	100	1.5	1.7	20	10	6 mcg	1 mg	10	300	–	–	50	–
Dialyvite® 3000	100	1.5	1.7	20	10	1 mg	3 mg	10	300	–	30	15	70
Diatx®	60	1.5	1.5	20	50	1 mg	5 mg	10	300	–	–	–	–
Diatx® Zn	60	1.5	1.5	20	50	2 mg	5 mg	10	300	–	–	25	–
Nephrocap®	100	1.5	1.7	20	10	6 mcg	1 mg	5	150	–	–	–	–
Nephron FA®	40	1.5	1.7	20	10	6 mcg	1 mg	10	300	–	–	–	–
Nephrovite®	60	1.5	1.7	20	10	6 mcg	0.8 mg	10	300	–	–	–	–
Nephrovite® Rx	60	1.5	1.7	20	10	6 mcg	1 mg	10	300	–	–	–	–
NephPlex® Rx	60	1.5	1.7	20	10	6 mcg	1 mg	10	300	–	–	12.5	–
PS Nephro-Aid	60	1.5	1.5	20	20	1 mg	950 mcg	10	300	3000	100	–	–
RenaPlex®	60	1.5	1.7	20	10	6 mcg	0.8 mg	10	300	–	–	12.5	–
Renax®	50	3.0	2.0	20	16	12 mcg	2.5 mg	10	300	–	35	20	70
Renal Caps	100	1.5	1.7	20	10	6 mcg	1 mg	5 mg	150	–	–	–	–

(Continued)

Table 10
(Continued)

Product	*Vit C (mg)*	*B1 (mg)*	*B2 (mg)*	*Niacin*	*B6 (mg)*	*B12 (mg)*	*Folic Acid*	*B5 (mg)*	*Biotin (mcg)*	*Vit A (IU)*	*Vit E (IU)*	*Zn (mg)*	*Se (mcg)*
Renaphro® Softgel	100	1.5	1.7	20	10	6 mcg	1 mg	5 mg	150	–	–	–	–
Albee plus Vitamin C	500	15	10.2	50	5	–	–	10	–	–	–	–	–

Diatx® Zn also contains 1.5 copper gluconate.
Nephron FA® also contains 200 mg ferrous fumarate.
Dialyvite®, Dialyvite® with Zinc and Dialyvite® 3000 are registered trademarks of Hillestad Pharmaceuticals, Woodruff, WI.
Diatx® and Diatx®Zn are registered trademarks of PamLab, LLC, Covington, LA.
Nephrocaps® is a registered trademark of Fleming & Company Pharmaceeuticals, Fenton, MO.
NephronFA® and NephPlex-Rx are registered rademarks of Nepro-Tech Inc, Shawnee, KS.
Nephrovite® and Nephrovite®Rx are registered trademarks of Watson Pharmaceuticals, Morristown, NJ.
PS Nephro-Aid is a registered trademark of Physiciain Select Vitamins LLC, Houston, TX.
RenalPlex® is a registered trademark of NutriMed Labs, Irving, TX.
Renax® is a registered trademark of Everett Laboratories, Inc, West Orange, NJ.
RenalCaps® is a registered trademark of Cypress Pharmaceuticals, Inc, Madison, MS.
Renaphro® Softgels is a registered trademark of Rising Pharmaceuticals, Inc, Allendale, NJ.
Allbee Plus Vitamin C® is a registered trademark of Inverness Medical Innovations, Inc, Freehold, NJ.

4. ASSESSMENT TOOLS

As discussed in Chapter 4, there are typically four components to conducting a comprehensive nutrition assessment: (i) Anthropometrics, (ii) Biochemical Indices, (iii) Clinical Symptomatology, and (iv) Dietary Intake Data. The KDOQI Practice Guidelines for Nutrition outline very specifically some of the methods for anthropometry, such as height, frame size, body circumferences, and so forth, which should be obtained. The reader is encouraged to refer to this resource at http://www.kidney.org/professionals/KDOQI/guidelines_updates/nut_appx07a.html.

As reported in KDOQI Guideline 9, the Subjective Global Assessment (SGA) is a valid and clinically useful measure of protein-energy nutritional status in maintenance dialysis patients. In 1994, SGA was presented at the annual Meeting of the American Society of Nephrology. Since that time, it has been used in growing numbers of dialysis clinics as another tool for assessment of the CKD patient's nutritional status. It is simple, subjective, and hands on *(4,5)*. SGA is a score-based assessment based on a medical history and a physical assessment. Medical history includes progression of weight loss, eating habits, gastrointestinal (GI) symptoms, physiological functioning, and simple analysis of metabolic stress. The physiological assessment includes loss of subcutaneous fat and muscle mass. Originally, patients were evaluated as (A) well nourished, (B) mild to moderately malnourished, or (C) severely malnourished. This ABC rating has been changed to a 7-point scale *(6,7)*.

- 6 or 7 = mildly nutritional risk to well-nourished.
- 3,4, or 5 = mild to moderately malnourished.
- 1 or 2 = severely malnourished.

The physical exam includes a visual assessment of muscle mass and subcutaneous tissue. The medical history reviews weight change during the previous 6 months. Given below is Sample SGA Form (used with permission from Kalantar-Zadeh, 2007)

The following information has been taken from http://www.nephrology.rei.edu/sga_instr.pdf.

SUBJECTIVE GLOBAL ASSESSMENT RATING FORM

Patient Name: **ID #:** **Date:**

HISTORY

WEIGHT/WEIGHT CHANGE: ***(Included in KDOQI SGA)*** 1. **Baseline Wt:** ________(Dry weight from 6 months ago) **Current Wt:** ________(Dry weight today) **Actual Wt loss/past 6 mo:** ________ **% loss:**_______(actual loss from baseline or last SGA) 2. **Weight change over past two weeks:** _______No change ______Increase ______Decrease	**Rate 1-7**
DIETARY INTAKE No Change_________(Adequate) No Change_________(Inadequate) 1. Change: Sub optimal Intake: _____ Protein ______ Kcal ______ Duration_____________ Full Liquid: __________ Hypocaloric Liquid ________ Starvation ___________	
GASTROINTESTINAL SYMPTOMS ***(Included in KDOQI SGA-anorexia or causes of anorexia)*** **Symptom:** **Frequency:*** **Duration:**+ _________ None _________ ________ _________ Anorexia _________ ________ _________ Nausea _________ ________ _________ Vomiting _________ ________ _________ Diarrhea _________ ________ *Never, daily, 2-3 times/wk, 1-2 times/wk +> 2 weeks, < 2 weeks	

FUNCTIONAL CAPACITY **Description** **Duration:** ___________ No Dysfunction ________________ ___________ Change in function ________________ ________ Difficulty with ambulation ________________ ________ Difficulty with activity (Patient specific "normal") ________________ ________ Light activity ________________ ________ Bed/chair ridden with little or no activity ________________ ________ Improvement in function ________________	b
DISEASE STATE/COMORBIDITIES AS RELATED TO NUTRITIONAL NEEDS Primary Diagnosis________________________Comorbidities___________________________ Normal requirements ____Increased requirements____ Decreased requirements _____ Acute Metabolic Stress: _____None _____Low _____Moderate _____High	
PHYSICAL EXAM	
________ Loss of subcutaneous fat (Below eye, triceps, biceps, chest) ***(Included in KDOQI SGA)*** ____Some areas _____All areas ________ Muscle wasting (Temple, clavicle, scapula, ribs, quadriceps, calf, knee, interosseous ***(Included in KDOQI SGA)*** ____Some areas _____All areas ________ Edema (Related to undernutrition/use to evaluate weight change)	
OVERALL SGA RATING	
Very mild risk to well-nourished=6 or 7 most categories or significant, continued improvement. **Mild-moderate** = 3, 4, or 5 ratings. No clear sign of normal status or severe malnutrition. **Severely Malnourished** = 1 or 2 ratings in most categories/significant physical signs of malnutrition.	

Fig. 3. Subjective global assessment rating form.

INTRODUCTION

Because nutritional assessment is difficult, a new technique called Subjective Global Assessment (SGA) was developed. Its ratings have been found to be highly predictive of outcome (1,2,3). The procedure is easy to learn and simple to implement. SGA requires no additional laboratory testing or capital outlay. In addition, SGA has been found to correlate strongly with other subjective and objective measures of nutrition.

Although originally used to categorize surgical patients, this nutritional classification system has been shown to be a reliable nutritional assessment tool for dialysis patients (4,5). Fenton (6) found that survival of patients classified as malnourished by SGA was significantly lower than patients classified as nourished, although he did not perform analyses to establish nutritional status as an independent risk factor.

SGA classifies the patient as:

A. Well-nourished

B. Mildly malnourished or suspected of malnutrition

C. Severely malnourished

Clinicians place the patient into one of these categories based upon their subjective rating of the patient in two broad areas: 1. Medical History, 2. Physical Examination. In general, 60% of the clinician's rating of the patient is based on the results of the medical history, and 40% on the physical examination (see SGA evaluation form in figure 3).

Medical History Section

The first SGA component, the medical history, involves asking questions and evaluating the patient's answers about the following four parameters:

- *Weight change*
- *Dietary intake*
- *Gastrointestinal symptoms*
- *Functional impairment*

The patient is rated as either nourished, mildly, moderately malnourished, or severely malnourished for each of the four parameters.

Physical Examination Section

Physical evidence of malnutrition is rated differently. There are four categories to select from: normal nutrition, mild malnutrition, moderate malnutrition, or severe malnutrition. Physical signs to examine include:

- *Loss of subcutaneous fat*
- *Muscle wasting*
- *Edema*
- *Ascites (in hemodialysis patients only)*

There are several body locations to examine for each parameter.

SGA SCORING GUIDELINES

The clinician rates each medical history and physical examination parameter as either an A, B, or C on the SGA Scoring Sheet. On the basis of all of these parameters' ratings, the clinical observer assigns an overall SGA classification which corresponds to his or her subjective opinion of the patient's nutritional status. SGA is not a numerical scoring system. Therefore it is inappropriate just to add the number of A, B, and C ratings to arrive at the overall SGA classification. The clinician should examine the form to obtain a general feel for the patient's status. If there seem to be more checks on the right-hand side of the form (more B and C ratings), the patient is more likely to be malnourished. If the ratings seem to be on the left-hand side, the patient is likely to be nourished. The severely malnourished (C) rating is given whenever a patient has physical signs of malnutrition such as, severe loss of subcutaneous fat, severe muscle wasting, or edema, in the presence of a medical history suggestive of risk, such as continuing weight loss with a net loss of 10% or more, or a decline in dietary intake. GI symptoms and functional impairments usually exist in these patients. Severely malnourished patients will rank in the moderate to severe category in most sections of the SGA form. When weight loss is 5–10% with no subsequent gain, in conjunction with mild subcutaneous fat or muscle loss and a reduction in dietary intake, the patient is assigned the mildly/moderately malnourished (B) rating. These patients may or may not exhibit functional impairments or GI symptoms. The B rating is expected to be the most ambiguous of all the SGA classifications. These patients may have a ranking in all three categories. In general, if the severely malnourished (C), or well-nourished (A) rating is not clearly indicated, assign the patient to the moderately malnourished classification. If the patient has no physical signs of malnutrition, no significant weight loss, no dietary difficulties, no nutritionally related functional impairments, or no GI symptoms which might predispose to malnutrition, the patient should be assigned to the well-nourished (A) category. If the patient has recently gained weight, and other indicators, such as appetite, show improvement, the patient may be assigned the A rating, despite previous loss of fat and muscle which may still be physically apparent. On the other hand, obese patients can be moderately or severely malnourished based upon their poor medical history and signs of muscle loss. Even patients with a normal appearance could be classified as mildly or moderately malnourished because of a poor medical history.

Information Provided by: Baxter Healthcare Corporation, Renal Division, 1620 Waukegan Road,

McGaw Park, IL 60085, Tel: (847) 473-6030, Fax: (847) 473-6935, e-mail: villanr@mor.baxter.com

REFERENCES

1. Jeejeebhoy KN, Detsky AS, Baker JP: Assessment of nutritional status. J Parent Nutr 55:193S–1965, 1990.
2. Baker JP, Detsky AS, Wesson DE, Wolman SL, Stewart S, Whitewa J, Langer B, Jeejeebhoy KN: Nutritional Assessment: A comparison of clinical judgment and objective measurements. N Engl J Med 306, 16:969–972, 1982.
3. Detsky AS, McLaughlin JR, Baker JP, Johnston N, Whittaker S, Mendelson RA, Jeejeebhoy KN: What is subjective global assessment? J Parent Nutr 1:813–817, 1987.
4. Young GA, Kopple JD, Lindholm B, Vonesh EF, De Vecchi A, Scalamogna A, Castelnova C, Oreopoulos DG, Anderson GH, Bergstrom J, DiChiro J, Gentile D, Nissenson A, Sakhrani L, Brownjohn AM, Nolph KD, Prowant BF, Algrim CE, Martis L, Serkes KD: Nutritional assessment of continuous ambulatory peritoneal dialysis patients: An intentional study. Am J Kidney Dis 17:462–471, 1991.
5. Enia G, Sicuso C, Alati G, Zoccali C: Subjective global assessment nutrition in dialysis patients. J Am Soc Nephrol 1:323, 1991.
6. Fenton SSA, Johnston N, Delmore T, Detsky AS, Whitewall J, O'Sullhan R, Cattran DC, Richardson RMA, Jeejeebhoy KN: Nutrition assessment of continuous ambulatory peritoneal dialysis patients. Trans Am Soc Artif Organs 23:650–653, 1987.

5. INTERNET SITES

There are a plethora of websites that may prove useful for the busy practitioner. The following represents a select few.

American Association of Kidney Patients (AAKP)	http://www.aakp.org
American Diabetes Association	http://www.diabetes.org
American Kidney Fund	http://www.akfinc.org
Culinary Kidney Cooks	http://www.culinarykidneycooks. com

(Continued)

(Continued)

Life Options Rehabilitation Program	http://www.lifeoptions.org
National Institute of Diabetes and Digestive and Kidney Diseases	http://www.niddk.nih.gov
National Kidney Foundation	http://www.kidney.org
The Nephron Information Center	http://www.nephron.com
DaVita Healtcare	http://www.davita.com
Fresenius NA	http://www.fmcna.com
Renal Dietitians Dietetic Practice Group of the American Dietetic Association	http://www.renalnutrition.org
United States Renal Data System (USRDS)	http://www.usrds.org
Council of Renal Nutrition (CRN)	http://www.kidney.org/professionals/CRN/
Council of Nephrology Social Workers (CNSW)	http://www.kidney.org/professionals/CNSW/
Council of Nephrology Nurses & Technicians (CNNT)	http://www.kidney.org/professionals/CNNT/
Centers for Medicare and Medicaid Services (CMS)	http://www.cms.hhs.gov/
National Kidney Disease Education Program (NKDEP)	http://http://nkdep.nih.gov/index.htm

6. SUMMARY

This chapter represents a collection of resources that are required for clinical practice in nutrition and kidney disease. It does not contain an exhaustive list of tools, but should serve to assist the practitioner in locating critical pieces of information necessary for appropriate care delivery.

REFERENCES

1. American Dietetic Association. National Renal Diet Professional Guide, 2nd edition, American Dietetic Association, 2002.
2. National Kidney Foundation. Potassium and your CKD Diet. Available from http://www.kidney.org/atoz/atozItem.cfm?id=103.
3. United States Department of Agriculture, Human Nutrition Information Service, Agriculture handbook Number 8–11, Composition of Foods: Vegetables and vegetable Products." Revised August 1984.
4. Assessing the Nutritional Status of Dialysis Patients Using Subjective Global Assessment. Educational Services Baxter Healthcare Corp, 1993.
5. Detsky AS, McLaughlin JR, et al. What is subjective Global Assessment of Nutritional Status? JPEN J Parenter Enteral Nutr 11: 8–13, 1978.
6. Jeejeebhoy KN, Detsky AS, Baker JP. Assessment of nutritional status. JPEN J Parenter Enteral Nutr 14 (5S): 193S–196S, 1990.
7. Lawson JA, Lazarus R, Kelly JJ. Prevalence & prognostic significance of malnutrition in chronic renal insufficiency. J Ren Nutr 1: 16–22, 2000.

Answers to Case Study Questions

CHAPTER 6

1. 27 ml/min, stage 4 CKD.
2. Mean plasma glucose 223.
3. No specific contraindications.
4. Initial dose would be 60 mcg (10 units) requiring separate injection from insulin up to three times daily before meals or snacks which must contain at least 250 kcal and 30 g CHO. Current insulin dose should be reduced by up to 50% to decrease the risk of hypoglycemia.
5. Based on 0.6–0.8 g/kg, if using actual body weight (168 kg) = 100–135 g/day; if using KDOQI adjusted body weight of 147 kg = 88–118 g/day; recommend 100 g/day.

CHAPTER 8

1. BH has anemia, but also iron deficiency. EPO therapy should not be initiated until iron stores are replete.
2. BH has diabetes, hypertension, and is physically inactive, putting him at high risk for CVD. However, he also has CKD stage 3, elevated phosphorus, potassium, and PTH. Priorities are to preserve kidney function and prevent complications of CKD. This can be accomplished with maintenance of blood glucose and blood pressure within normal range, by adjusting diet to decrease sodium, phosphorus, and potassium intake, maintaining appropriate protein and calorie intake, and balanced carbohydrate intake. At this first clinic visit, the dietitian presents the goals of good blood sugar and blood pressure control. As HbA1C and blood sugars have been good, no immediate change in medication or carbohydrates is indicated. The dietitian discusses how and why a lower sodium diet will help BH and works with him and his wife to plan simple changes they can make to decrease his sodium intake, including less use of processed foods (which will help decrease phosphorus intake also) and more low potassium fresh fruits and vegetables. BH is instructed to keep a food record before his next appointment.

3. Although BH's Hgb has improved, it remains below goal. Iron stores are now acceptable and EPO therapy is warranted to increase and maintain Hgb between 11 and 12 g/dL.
4. Serum phosphorus remains above goal level (2.7–4.6 mg/dL). Intact PTH is also above goal range (35–70 pg/mL), and literature shows that treating hyperphosphatemia will lower PTH level.
5. The dietitian discusses protein and phosphorus in more detail and prescribes 65 g of protein/day (0.8 g/kg body weight). She works with BH and his wife to plan how he can change his food choices to limit protein and phosphorus intake. The dietitian reviews the role that phosphate binders (Tums) will play in controlling his phosphorus level. BH leaves with a plan to continue with his previous diet changes plus decrease his protein and phosphorus intake toward goal levels.
6. At BH's usual weight of 83 kg, he is not overweight (BMI, 25). Considering the risk of malnutrition in people with CKD, especially those on low protein diets, weight loss is not recommended. BH has a history of adequate blood sugar control and his blood pressure is responding to diet and medication interventions. The dietitian discusses the need to maintain appropriate weight and ways to increase calorie intake without adding additional protein, potassium, or phosphorus. The dietitian also discusses how physical exercise can help maintain muscle mass and cardiac function, and BH develops a plan of how to increase his activity level.

CHAPTER 9

1. CT has no edema, so using 35 kcal/kg of his current body weight puts estimated kcal needs at 2894 kcal. Based on 1.3 g/kg body weight, his protein needs are 107.5 g. Based on a 60% absorption rate for CAPD, CT receives approximately 357 kcal from his dialysate.
2. The recommended diet is 107 g protein, 3–4 g sodium (as he has no overt cardiac issues or problems with fluid retention), 3–4 g potassium, 1380–1460 mg phosphorus (using 17 mg/kg ideal weight of 86.3 kg). Fluid intake should be kept to 2000–2500 mL (urine output plus 1000 mL) and adjusted based on UF. Based on CT's initial intake of 1800–2000 kcal and 70–80 g protein and allowing for the 357 kcal absorbed from dialysate, he will have to add a minimum 500–700 kcal and 27–37 g of protein to his diet. Sodium and potassium intake is acceptable, but he will need to decrease his intake of dairy foods to bring the phosphorus content of the diet within guidelines.
3. The pre-natal vitamin should be changed to a renal B and C complex with folic acid. His corrected calcium equal to 9.5 and to avoid any

elevation in calcium his Tums should be discontinued and another calcium-free medication prescribed for his indigestion. He is currently on a small dose of phosphorus binders (Renagel), and with the increased protein in the diet and elevated phosphorus, an increase in binders will be needed. Renagel was increased to 1600 mg at each meal, and patient was also instructed to take 800 mg with snacks which he had not been doing.

4. At 3 months, CT presents with a low potassium, low phosphorus, declining albumin and BUN secondary to poor oral intake due to dysphagia. His binders should be decreased by one-third and potassium intake liberalized along with being given an oral potassium supplement until his labs indicate normal levels and oral intake has improved. It should be suggested that he use a liquid nutritional calorie and protein supplement until after his dilation as his oral intake is limited.
5. Due to edema, his dry weight should be used to calculate his protein needs at 1.2 g/kg body weight. His protein needs will now change to 99 g. Because of edema, his sodium intake should be decreased to 2–2.5 g. His potassium supplement should be stopped and dietary potassium limited to 2–2.5 g/day. His phosphorus is elevated so his binders should be increased to at least 1600 mg with each meal and 800 mg with snacks and a dietary phosphorus restriction restarted. Corrected calcium is high, so his dialysate Ca bath should be lowered. Fluid restriction should be limited to 1500 mL or less (urine output plus 500 mL) and interdialytic weight gains monitored. His albumin level should be monitored for changes with improved appetite and loss of edema.

CHAPTER 10

1. ABW= 97 kg; Protein (1.3 g/ kg ABW) = 126 g/day; Calories (30 kcal/kg ABW) = 2910 calories/day may need to modify for DM and weight loss.

2a. Adjust doses/timing of immunosuppressive drugs (i.e., Cellcept qid vs. bid).

b. To address long-term concerns, which may be related to DM gastroenteropathy, soluble fiber supplements, such as psyllium-based products, could be tried.

3a. Walking per ability with slow gradual increase, and chair exercises with light hand weights.

b. When his incision is completely healed, perhaps water aerobics or stationary cycling will help with his weight management program.

4a. Avoid taking phosphorus supplement at the same time as his calcium supplement.

b. Dietary phosphorus intake could be increased but some high phosphorus foods are also high in potassium. Patient could be scheduled for bone density studies.

5a. Educate on the timing and action of currently used types of insulin, carbohydrate in foods, and CHO distribution as needed. Diabetes management could be updated.

b. HgbA1c is not considered accurate as patient's use of immunosuppressive drugs and the effect on blood glucose levels has been dramatic over the past 6 weeks. Reassess HgbA1c 3 months after prednisone and Prograf dosages are at maintenance levels.

c. Weight control (an issue for him in the past) may be a significant issue regarding his DM management.

6. Evaluate Multivitamin mineral (MVI) supplement to address megadosing of any component or inclusion of an "immuno-enhancing" item such as ginseng.

7. Fluid retention can be affected by steroid use; edema may also be affected by dietary sodium.

8a. Elevated levels of Prograf.

b. Neutra-Phos, containing 14 mEq/ K+/tab switch to K-Phos Neutral containing 1 mEq K+/tab.

c. Balance increased need for phosphorus intake with potassium education.

d. Avoid salt substitutes containing potassium.

e. Multivitamin mineral (MVI) may also contain potassium.

9a. Restart statin medications after immunosuppressive medications are at maintenance levels.

b. Review of sources of dietary fat would and weight management.

c. Soluble fiber and plant sterols could be introduced.

CHAPTER 12

1. An intervening clinical illness such as pneumonia.
2. It is administered two to three times per week in an inpatient or outpatient setting.
3. By an increase in the serum albumin and lean body weight.
4. Yes. This can occur with an improve sense of well-being which can result from aggressive nutritional support.
5. Increased quality of life that can be associated with traveling and attending various social functions.

CHAPTER 13

1. This patient's CKD is certainly severe enough (stage 4) that his anemia could be due to the erythropoietin deficiency that often accompanies

CKD of this degree. However, other causes of anemia need to be excluded. He should have a test of occult blood in the stool, a CBC with red blood cell indices (MCV, MCH, and MCHC), reticulocyte count, folate and vitamin B12 levels, TSAT, and serum ferritin level. Unless there is a specific reason to suspect some other cause of anemia, no further testing would be needed if these tests are all normal.

2. The laboratory studies are consistent with iron deficiency. Although the test for occult blood in the stool is negative, he still needs a thorough evaluation with colonoscopy and probably upper gastrointestinal tract endoscopy. The iron deficiency can be treated with a 2- to 3-month course of oral iron, such as ferrous sulfate 325 mg three times daily. Alternatively, he could be treated initially with intravenous iron, using either iron sucrose 300–400 mg over three occasions or ferric gluconate in sulfate complex 250 mg on four occasions to provide 1000 mg of iron. If the iron indices remain low, subsequent doses can be given.
3. His iron deficiency has been corrected, but he remains anemic. At this point, therapy with an ESA should be initiated. There are a number of ways to treat him; epoetin alfa may be administered once weekly or every other week, or darbepoetin alfa may be administered once every 2–4 weeks. Serum Hb should be monitored at least monthly, and iron studies should also be monitored periodically, at least every 3 months. It would be reasonable to prescribe oral iron also. Because he is not on hemodialysis, the ESA should be given by subcutaneous injection.

CHAPTER 14

1. Hypercalcemia and hyperphosphatemia are likely due, in part, to release of these minerals from the bone as a result of parathyroid hyperfunction and refractory secondary hyperparathyroidism. They may also be aggravated by significant doses of vitamin D/analogs.
2. Therapy options include calcimimetics, PTX, and/or more frequent dialysis. Calcimimetic therapy has the potential to suppress PTH secretion and to decrease serum levels of both calcium and phosphorus, even in refractory secondary hyperparathyroidism. Based on current research, it would be appropriate to initiate calcimimetic therapy prior to considering PTX. Based on the patient's biochemical profile, KDOQI would consider it appropriate to consider PTX as his SHPT is long term, non-responsive to significant doses of vitamin D analogs, and accompanied by hypercalcemia. If the patient is not able to control his hyperparathyroidism with appropriately titrated doses of calcimimetics, a PTX should be considered after performance of a bone biopsy. While an increased frequency of dialysis can help normalize mineral balance, the effect on refractory hyperparathyroidism has not been determined.

CHAPTER 15

1a. What time was insulin administered and how much? Wife informs you that he had fasting glucose >400 so she increased his 8:30 am 70/30 insulin dose by 10 units.

b. When did patient eat and what did he eat? Wife states that he had breakfast at 9 am consisting of one egg, one slice toast, half-cup hot cereal, and half-cup apple juice. She indicated that he did not have snack or meal after treatment, as he fell asleep upon arriving home and she did not want to disturb him.

c. When was glucose checked at home? Wife did not check glucose prior to patient falling asleep or until he appeared sweaty.

d. How did she treat his hypoglycemia? She gave patient a chocolate bar to eat.

2. Infection could be contributing to hyperglycemia. Residual renal function has probably declined given increases in BUN, creatinine, and potassium. Lack of residual renal function in addition to being on HD can prolong insulin's duration contributing to mid to late afternoon hypoglycemia. This was compounded by wife having increased the AM 70/30 dose and patient not having eaten for >5 hours. The effect of the extended duration of the NPH portion was then compounded by patient not having eaten for >12 hours. Chocolate candy is not a quick acting carbohydrate for hypoglycemia treatment.

3. Although the patient has had diabetes for 25 years, the patient and wife were not knowledgeable in how to make adjustments for blood sugar fluctuations, especially as patient was relatively new to being on insulin. Interventions that could be advised include the following:
   a. Review of peaks and durations of insulin with wife, so she can understand impact of insulin dosing and any changes that are made.
   b. Review of meal spacing and need of snacks pre- and post-HD treatment.
   c. Review low potassium treatment for hypoglycemia at home, advise carrying of glucose tablets or gel when away from home.
   d. If high blood sugars continue, changes in insulin regimen and type of insulin may be indicated.
   e. Review symptoms of hypoglycemia and hyperglycemia with wife, as patient may not always be reliable in his perception of these due to neuropathy and progressing dementia.
   f. Increase frequency of glucose monitoring to better detect trends and intervene accordingly.

CHAPTER 16

1. Commitment to increased and more frequent dialysis time.
2. 35 kcal/kg pregravida IBW + 300/day in second and third trimesters.
3. 1.2 g/kg pregravida IBW + 10 g/day.
4. Folic acid.
5. To create a less uremic environment for the fetus and allow the patient more liberal dietary intakes to meet nutrient needs of pregnancy.
6. Intravenous epoetin alfa and iron given during dialysis.

CHAPTER 17

1. Breast milk if available and if not, Good Start® or PM 60/40® infant formula. With normalization of his serum potassium and phosphorus levels, he can continue on breast milk or be switched to a standard infant formula. He will continue to have high urine output, requiring a high fluid (free water) intake and sodium supplements. He will innately prefer to drink water over formula and his high fluid intake will decrease his appetite for solids. He may need energy supplementation of his feedings or tube feeding to meet both his calorie and fluid requirements.
2. Explain a low phosphorus diet to AB's caregivers. Teach them about the phosphorus content of table foods and drinks and provide guidelines for keeping calcium and phosphorus intake within the KDOQI Pediatric Bone Guidelines.
   Pediatric enteral formulas have high calcium and phosphorus contents to support bone growth. Choose an adult renal enteral product (e.g., Suplenc®) and adjust the volume and strength to meet his energy and protein requirements. Monitor his serum magnesium concentration. Remind parents to increase his free water intake to compensate for the lower volume of tube feeding needed with the 2 kcal/mL renal product.
3. Work with parents to ensure day care providers become knowledgeable about his low phosphorus diet and timing of phosphate binders. An adjustment in enteral formula should be considered in order to meet his higher protein requirements. This may require a switch to a higher-protein content renal formula (e.g., Nepro®) or a mixture of a lower-and higher-protein renal formula. Work with parents to adjust his overnight feeding into his new daily routine and, if necessary, reduce the overnight volume and include a day-time bolus that someone at the school or daycare setting is able to administer.
   Start a renal water-soluble vitamin supplement.

4. While you are assessing whether he can reach his energy and protein requirements orally, change his tube feeding back to a standard pediatric product and if possible decrease the calories provided by the G-tube to stimulate hunger. Educate AB's parents about the nutrition-related side effects of his immunosuppressant therapy. Be sure that they stop restricting his dietary phosphorus intake. Educate them on a low sodium, low potassium diet. You may choose to discuss food hygiene/food safety. When AB is able to take approximately 75% of his calorie requirements orally, hold his G-tube feeds for at least 1 month to be sure he can do this continually before removing his G-tube. Monitor his weight and serum glucose, cholesterol and triglyceride levels. As he begins to gain weight, discuss portion sizes, highlight foods and drinks that are low in simple sugars and contain small amounts of "heart-friendly" fats and stress the importance of daily physical activity to support weight gain that is a healthy combination of muscle and fat mass.

CHAPTER 18

1. CV risk factors are HTN uncontrolled, A Fib, CAD, uncontrolled hyperlipidemia, uncontrolled DM and obesity with BMI 32.3.
2. Maria's eGFR is 47 ml/min/1.73 m^2 as client's race is considered black, the island of Cape Verde is off coast of Africa. Despite "slight" increase in Cr, patient has stage 3 CKD.
3. Psychosocial issues include fixed income with attempts at stretching medications, occasions of limited mobility, which impacts dietary intake, as well as risk of isolation and falls in persons who lives alone.
4. CKD issues of concern include hyperkalemia, anemia, metabolic acidosis and left ventricular hypertrophy that can worsen with uncontrolled cholesterol, HTN, A Fib and DM. Additional information which would be beneficial include serum folate and vitamin B12 level as well as iron-binding studies to assist with determining if iron deficiency is the cause of anemia, bone health status data including parathyroid hormone, calcium, phosphorus, vitamin D level and urinary protein level to determine evidence/level of proteinuria and subsequent protein loses.
5. Based on data, patient has secondary hyperparathyroidism, which may be due to a combination of vitamin D deficiency and hyperphosphatemia with several high phosphorus sources as staples in her diet. Patient also has proteinuria, which may have been heightened by uncontrolled DM, HTN and hyperlipidemia.

Index

D

E

About the Editors

About the Editors

Dr. Byham-Gray is an Associate Professor in the Department of Nutritional Sciences, School of Health Related Professions at the University of Medicine and Dentistry of New Jersey in Stratford, NJ. Prior to teaching, Dr. Byham-Gray practiced in the field of clinical nutrition, with specialty practice in nutrition support, kidney disease, and home care for over 20 years. She has held numerous elected and appointed positions at the national, state, and local levels of the National Kidney Foundation (NKF), the American Society of Parenteral and Enteral Nutrition (ASPEN), and the American Dietetic Association (ADA). Dr. Byham-Gray served on the *Clinical Standards Committee* for ASPEN because of her expertise in kidney disease, outcomes research, and evidence-based practice guideline development. She also serves on the editorial board for the *Journal of Renal Nutrition*. Presently, Dr. Byham-Gray is a Chronic Kidney Disease workgroup member for the ADA Evidence Analysis Library.

Dr. Byham-Gray has several peer-reviewed articles and over 30 professional presentations related to kidney disease, dietetics practice, and clinical decision-making as well as management. She has also authored one self-study publication entitled *Medical Nutrition Therapy in Renal Disease*, 3rd Edition with Wolf Rinke Associates, and she has co-edited the ADA publication, *A Clinical Guide to Nutrition Care for Kidney Disease*. Dr. Byham-Gray has received numerous awards, including the *Outstanding Service Award* by the ADA-Renal Dietitians dietetic practice group.

Dr. Burrowes is an Associate Professor in the Department of Nutrition, School of Health Professions and Nursing at the C.W. Post Campus of Long Island University in Brookville, NY. Prior to obtaining her doctorate in nutrition and dietetics from New York University, Dr. Burrowes was the research coordinator for the NIH-funded Hemodialysis (HEMO) Study and she practiced as a renal dietitian for 10 years prior to her involvement in the HEMO Study. Dr. Burrowes has held several leadership positions in the national and local chapters of the National Kidney Foundation (NKF) Council on Renal Nutrition (CRN). She was a member of the NKF Kidney Disease Outcomes Quality Initiative (NKF-KDOQI) Nutrition Work Group and a member of the NKF-KDOQI Advisory Board.

Dr. Burrowes has published numerous peer-reviewed articles and she has been an invited speaker at over 50 professional conferences on the topic of nutrition and kidney disease. She is currently a Contributing Editor for the Clinical Column in *Nutrition Today*, Editor of the Nutrition Section for

Advances in Chronic Kidney Disease, and a member of the editorial board for the *Journal of Renal Nutrition*. Dr. Burrowes has been the recipient of the Recognized Renal Dietitian Award and the Joel D. Kopple Award from the NKF-CRN.

Dr. Chertow is a Professor of Medicine and Chief, Division of Nephrology at Stanford University School of Medicine. Dr. Chertow's research has focused on epidemiology, health services research, decision sciences and clinical trials in acute and chronic kidney disease, with special interests in nutrition and mineral metabolism.

Dr. Chertow currently leads or participates in several clinical trials and population-based cohort studies sponsored by the National Institutes of Health-National Institute of Diabetes and Digestive and Kidney Diseases. He is Associate Editor for the *Journal of Renal Nutrition*, Co-Editor of Brenner & Rector's The Kidney, and on the editorial boards of *Journal of the American Society of Nephrology and Seminars in Nephrology*. Dr. Chertow was vice-chair of the NKF-K/DOQI Nutrition Work Group in 1998–2000, and a member of the K/DOQI Bone and Mineral Metabolism Work Group from 2000–2002. He is currently on the Scientific Advisory Board of the National Kidney Foundation.

Dr. Chertow was elected to the American Society of Clinical Investigation, and designated Fellow of the American College of Physicians, American Heart Association and American Society of Nephrology. He received the President's Award from the National Kidney Foundation and the National Torchbearer Award from the American Kidney Fund.

About the Series Editor

Dr. Adrianne Bendich is Clinical Director of Calcium Research at GlaxoSmithKline Consumer Healthcare, where she is responsible for leading the innovation and medical programs in support of several leading consumer brands including TUMS and Os-Cal. Dr. Bendich has primary responsibility for the coordination of GSK's support for the Women's Health Initiative (WHI) intervention study. Prior to joining GlaxoSmithKline, Dr. Bendich was at Roche Vitamins Inc., and was involved with the groundbreaking clinical studies proving that folic acid-containing multivitamins significantly reduce major classes of birth defects. Dr. Bendich has co-authored more than 100 major clinical research studies in the area of preventive nutrition. Dr. Bendich is recognized as a leading authority on antioxidants, nutrition and bone health, immunity, and pregnancy outcomes, vitamin safety, and the cost-effectiveness of vitamin/mineral supplementation.

In addition to serving as Series Editor for Humana Press and initiating the development of the 20 currently published books in the Nutrition and Health™ series, Dr. Bendich is the editor of 11 books, including *Preventive Nutrition: The Comprehensive Guide for Health Professionals*. She also serves as Associate Editor for *Nutrition: The International Journal of Applied and Basic Nutritional Sciences*, and Dr. Bendich is on the Editorial Board of the *Journal of Women's Health and Gender-Based Medicine*, as well as a past member of the Board of Directors of the American College of Nutrition. Dr. Bendich also serves on the Program Advisory Committee for HelenKeller International.

Dr. Bendich was the recipient of the Roche Research Award, was a Tribute to Women and Industry Awardee, and a recipient of the Burroughs Wellcome Visiting Professorship in Basic Medical Sciences, 2000–2001. Dr. Bendich holds academic appointments as Adjunct Professor in the Department of Preventive Medicine and Community Health at UMDNJ, Institute of Nutrition, Columbia University P&S, and Adjunct Research Professor, Rutgers University, Newark Campus. She is listed in *Who's Who in American Women*.

Printed in the United States of America